CSSPO International Conference 2018: Towards Inclusive and Sustainable Agriculture

Harmonizing Environmental, Social and Economic Dimensions: Is it Possible?

E3S Web of Conferences Volume 52 (2018)

Sarawak, Malaysia
9 – 11 July 2018

Editors:

D. Chalil
S.F. Sidique
R. Barus
A.H.I. Abdul Hadi
M. Khaliqi
M.I. Fatoni

ISBN: 978-1-5108-7008-6

Printed from e-media with permission by:

Curran Associates, Inc.
57 Morehouse Lane
Red Hook, NY 12571

Some format issues inherent in the e-media version may also appear in this print version.

Printed by Curran Associates, Inc. (2018)

For additional information, please contact EDP Sciences – Web of Conferences at the address below.

EDP Sciences – Web of Conferences
17, Avenue du Hoggar
Parc d'Activité de Courtabœuf
BP 112
F-91944 Les Ulis Cedex A
France

Phone: +33 (0) 1 69 18 75 75
Fax: +33 (0) 1 69 28 84 91

contact-edps@webofconferences.org

Additional copies of this publication are available from:

Curran Associates, Inc.
57 Morehouse Lane
Red Hook, NY 12571 USA
Phone: 845-758-0400
Fax: 845-758-2633
Email: curran@proceedings.com
Web: www.proceedings.com

TABLE OF CONTENTS

CSSPO INTERNATIONAL CONFERENCE 2018

Towards Inclusive & Sustainable Agriculture
Harmonizing Environmental, Social and Economic Dimensions: Is It Possible?
Kuching, Sarawak, 9-11 July, 2018

Editors:
Diana Chalil
Shaufique Fahmi Sidique
Riantri Barus
Ahmad Hanis Izani Abdul Hadi
Muhammad Khaliqi
R.B. Moh. Ibrahim Fatoni

Secretariat
Main Building of Agricultural Faculty, Universitas Sumatera Utara
Jl. A. Sofyan No. 3 Medan, North Sumatera, Indonesia
Contact Person : +6281361025887
Email : riantribarus@gmail.com
Website : www.csspo.or.id

FOREWORD

CSSPO Coordinator

Diana Chalil, Ph.D

This proceeding is a compilation of remarkable submitted paper to the 2018 CSSPO International Conference held in Kuching, Sarawak, 9-11 July 2018. The conference is conducted as an effort to contribute on the debatable academic discussion about inclusive and sustainable agriculture. In particular, the subtheme raises the topic on "Harmonizing the environmental, social and economic dimensions: Is it possible?". There are at least two different concepts based on the inclusive understanding. First comes from the Value Added Chain concept, which focuses on the increase in income and other economic aspects. Second stems from the empowering concept that is related to proportional access and participation. All keynote speakers focus in international and domestic market palm oil cases, which discusses (1) "Harmonizing Economic, Social, and Environment Sustainability: It's a Must, It's Possible, and It's Feasible" (Dr. Bayu Krisnamurthi, andgreen.fund Indonesia), (2) "Certifying Sustainable Palm Oil: The European Paradox" (Prof, Alain Rival, CIRAD, France), (3) "Economic and Social Dimensions of Import Demand of Edible Oil in Pakistan" (Prof. Muhammad Ashfaq, University of Agriculture Faisalabad Pakistan), and (4) "Agricultural Development and Environment: On Parallel Track or Collision Corse" (Prof. Mad. Nasir Shamsudin, UPM Malaysia). Abstracts of the keynote speakers' presentations are provided, while the full power point presentations can be accessed through www.csspo.or.id. Submitted papers are not limited to palm oil commodity, covering comprehensive aspects from environmental science, sustainable development, economics, energy, agronomy, political science, political science and social science. All of the findings are expected to contribute discussions on the inclusive and sustainable agriculture, both in theoretical an empirical aspects.

Vice Rector of Universitas Sumatera Utara

Prof. Dr. Ir. Bustami Syam, MSME

As the Representative of CSSPO Honorary Board, we would like to congratulate The Consortium Studies of Smallholder Palm Oil for conducting a conference on Inclusive and Sustainable Agriculture. A number of sustainability issues along the global supply chain, especially in the palm oil industry are broadly debated around the world. Bringing four exceptional keynote speakers from Indonesia, Malaysia, France and Pakistan, provides a good forum for sharing of information, knowledge, and ideas in enriching solutions for achieving an inclusive and sustainable palm oil industry.

The presentations of almost 100 participants' empirical researches in this conference will provide excellent complementary showcases and an excellent chance for positive networking. As a Vice Rector for Information, Planning and Development, Universitas Sumatera Utara, we believe that collaborations will play strategic role in the university development. Besides consortium studies, we are also committed in developing joint research, visiting Professors/Lecturers programs, and Student and Lecturer Exchange Programs with both local and international universities.

We believe that the outcomes of this conference will be a valuable reference for researchers and policy makers for implementing further excellent collaborative programs in the future.

Acting Consul General of The Republic Indonesia in Kuching

Muhammad Abdullah

It is a great honour for me to address the first international conference on Inclusive and Sustainable Agriculture which is held in Kuching, Sarawak, Malaysia. In many countries, oil palm cultivation has been proven playing an important role in achieving sustainable development targets Agenda 2030. Especially in Indonesia, oil palm cultivation contributes to the livelihoods of more than 17 million Indonesians, which are mostly smallholders. The real positive impacts of oil palm cultivation such as ; the level of education of the population increased, and the improvement of rural infrastructure that led to the rise of human development index (HDI). It also disclosed the pro-poor, pro-growth and pro-environment policies of the Indonesian oil industry in the framework of Indonesia Sustainable Palm Oil (ISPO).

This conference brings four exceptional keynote speakers from France, Pakistan, Malaysia and Indonesia, hence I believe this conference will provides a good forum for sharing information, knowledge and ideas in enriching solutions for achieving an inclusive and sustainable palm oil industry. Moreover, with almost 100 participants, this conference will be a good platform to exchange and sharing empirical relevant studies and an excellent chance for positive networking.

In this opportunity, I wish to congratulate The Consortium Studies of Smallholder Palm Oil for conducting a conference on inclusive and sustainable agriculture. It is my fervent hope that the conclusion of this conference can be a valuable reference for researchers, policymakers and all parties concerned in the development and improvement of a sustainable agriculture.

ABOUT CSSPO

Consortium Studies on Smallholder Palm Oil (CSSPO) is established in 2015 with DIKTI (Indonesian Higher Education) Fund No 5231/E2.4/LN/2015. This consortium is a collaboration of universities from main palm oil producing countries namely Indonesia, Malaysia and Thailand, with MoU on 7 June 2013 between USU and UPM, MoU on 27 March 2015 between USU and PSU, and MoU No 7155/UN5.1.R/KPM/2015 among USU, UNJA and UNIMAL. Universitas Sumatera Utara, Universitas Jambi and Universitas Malikussaleh are universities from Indonesia, Universiti Putra Malaysia is from Malaysia and Prince of Songkla University is from Thailand. Initially, the consortium was named as Consortium Studies on Palm Oil (CSPO). All members have long research experience and interest in palm oil agribusiness studies, which identified significant issues related to smallholders in the studies. With their increasing share, oil palm smallholders appear to have significant role in the palm oil industry. However, lack of technical, managerial and financial capability, in general smallholders are likely have weak position in the global value chain. Therefore, the vision is to become a leading consortium of academic institutions to conduct research and provide solutions to improve smallholders' position in the global palm oil value chain. The vision is break-downed into 3 missions namely (1) To build a global palm oil research network, (2) To produce high quality empirical research output and (3) To provide solutions for oil palm smallholder issues. The missions are detailed in a number of programs, covering researches, trainings for smallholders and producing policy briefs. In 27-28 May 2016 CSSPO conducted a seminar "Mencari Model Pemberdayaan dan Peremajaan Perkebunan untuk Sawit Indonesia yang Berkelanjutan" in Polonia Hotel, Medan, which received very positive response from 40 institutions include Universities, state and private companies, government offices, and research center. In the meeting, Mr Derom Bangun, as the Head of Indonesian Palm Oil Board (DMSI) explained that CSPO is well known as Certified Sustainable Palm Oil. Dr Bayu Krisnamurthi as the Head of Indonesian Palm Oil Funding (BPDPKS) suggested to add "Smallholder" in the consortium name, and to change CSPO to CSSPO. Since 12 June 2016, The consortium's name become CSSPO. In 2016 CSSPO were included as one of providers for conducting trainings for smallholders in Labuhan Batu, Labuhanbatu Utara and Labuhanbatu Selatan, North Sumatra. Twelve (12) experts from universities, state and private companies, and successful smallholders were invited to share their knowledge and experience in replanting preparation. CSSPO is also actively involved in a number of government smallholder policy formulations such as the Fresh Fruit Bunches pricing for smallholders, and Forum Kelapa Sawit Berkelanjutan Indonesia (FOKSBI). CSSPO is also involved in socializing the progress of sustainable palm oil, including to Pakistan Agriculture University, Peshawar and National Agriculture Research Center and Pakistan Agriculture Research Council, Pakistan in 2016, and Tasmanian University students in 2016, 2017 and 2018. Currently, CSSPO is also discussing the potential study of harvesting labor with Golden Agri Resource (GAR-SMART) to academically evaluate labor issues that have been broadly debated, especially among European countries. In January 2017, all Vice Rectors and Director of each university agreed to formally sign the USU organizational structure under Letter of Agreement .The structure consists of Joint Management Board, which members are Drs. Mahyuddin K.M. Nasution, M.I.T (USU), Ph.D, Prof. Dr. Ir. Zulkifli Alamsyah, MSc (UNJA), Jullimursyida, Ph.D. (UNIMAL), Prof. Dr-Ing. Ir. Renuganth Varatharajoo (UPM), and Asst. Prof. Dr. Pattara Aiyarak (PSU), CSSPO Coordinator Diana Chalil, and CSSPO members of each university namely Diana Chalil and Riantri Barus (USU), Zulkifli Alamsyah and Dompak MT Napitupulu (UNJA), Mawardati and M Authar (UNIMAL), Shaufique F Sidique and Yeong Sheng TEY (UPM), Sutonya F. Thongrak and Sirirat Kiatpathomchai (PSU). Each year the CSSPO conducts an annual meeting to evaluate previous programs and to prepare upcoming ones. In 2016, CSSPO piqued the interest of a CIRAD (Agricultural Research for Development, French) researcher, Dr Marcel Djama who joined the 1st annual meeting in Batam. This significantly improved CSSPO networking and studies, as CIRAD is a main and important research center in France and Europe. In 2017, Dr. Diana Chalil, Dr. Shaufique F. Sidique, and Dr. Marcel Djama were invited as Plenary Speaker in International Conference and Congress of The Indonesia Society of Agricultural Economics (ICC-ISAE) in Bali. In 2018, supporting by Sarawak Bureau Convention, CSSPO and CIRAD conducted a joint conference in Kuching, Sarawak with Dr. Bayu Krisnamurthi (andgreen.fund - Indonesia), Prof. Alain Rival (CIRAD - French), Pof. Muhammad Ashaq (University of Agriculture, Faisalabad – Pakistan) and Prof. Datuk Dr Mad Nasir Shamsudin (Universiti Putra Malaysia – Malaysia) as keynote speakers. All of the CSSPO activities are published through www.csspo.or.id

KEYNOTE SPEAKERS

Dr. Bayu Krisnamurthi

Andgreen-fund, Indonesia

Dr. Bayu Krisnamurthi is a Senior Lecturer at Department of Agribusiness, Faculty of Economic and Management, Institut Pertanian Bogor and Executive Chair of Bina Swadaya, Community Self-Empowered Foundation. He is also a Board of Adviser Andgreen-Fund (Netherland) from 2017, Senior Expert Panel Indonesia Chamber of Commerce from 2018, Board of Adviser PT East Weast Seed Indonesia from 2015, Invesment Committee, Millennium Challange Account (MCA) Indonesia from 2015, Founder, Chairman Board of Trustee Danamon Peduli Foundation from 2015, Founder, Initiator and Coordinator of Inclusive and Sustainable Agribusiness (ISA) Initiatives.

He was President Director / Chief Executive Indonesia Estate Crop Fund (IECF) for Palmoil, Ministry of Finance Republic of Indonesia (2015 – 2017), Vice Minister of Trade Republic of Indonesia (2011 – 2014), Vice Minister of Agriculture Republic of Indonesia (2009 – 2011), Deputy Minister for Agriculture and Marrine, Coordinating Minstry for Economic Affair Republic of Indonesia (2005 – 2009).

He has Republic of Indonesia Honorary Award “Bintang Mahaputra Utama”, Republic of France Honorary Award “Orde du Merite Agricole” , and Indonesia Presidential Honorary Award Tanda Penghargaan “Satra Lencana Karya Satya”.

He Received his BSc in Agribusiness (1987), MSc Agricultural Economics (1991) and PhD on Agricultural Economucs (1998) from Institut Pertanian Bogor.

Abstract

Harmonizing economic, social, and environment sustainability: it's a must, it's possible, and it's feasible

Dr. Bayu Krisnamurthi

With Sustainable Development Goals becoming global platform since 2015 for (almost) all development agenda in more than 180 member countries of UN, "sustainability" embark to a new meaning and a new dimension. Sustainability becoming center point of economic, social, and environmental activities. Learning from the establishment of AndGreen-Fund (&GreenFund), the collaborative effort within Partnership on Indonesia Sustainable Agriculture (PISAgro), and the operationalization of Indonesia Estate Crop Fund (IECF) for Palmoil; the key factors for harmonizing the three pillar of sustainability (economic, social and environment) are 1) determinative vision and principal that sustainability is a must for all economic and business activities, 2) implementing technology, management, and operational system that make multi-objective (economic, social, and environmental) possible; and 3) incentives system through business and market work in such that harmonizing economic, social, and environment sustainability in profitable, in the short term and long term.

Keywords: harmonizing, sustainability, vision, business and market orientation

Prof. Alain Rival

Cirad, French

Professor Alain RIVAL is the Resident Director for Southeast Asia Island Countries and the Focal Point for Oil Palm Research at Cirad, the French Centre for International Cooperation in Agricultural Research for Development.

Prof Alain Rival is an agronomist by training and a plant molecular physiologist; he received his PhD in Plant Biotechnology from the University of Montpellier, France in 1988 and he spent the first years of his career (1985-1995) running collaborative plant biotechnology programmers overseas in Ivory Coast, Indonesia, Malaysia and Costa Rica.

Alain Rival qualified for the Habilitation of Research Director in Life Sciences in 1997 from the University of Paris and he is a Qualified Professor in Biology and Physiology since 2002. Prof Rival spent two years as a Marie Curie International Fellow at CSIRO Plant Industry in Canberra, Australia, within the framework of a European Commission-funded project in plant epigenetics called EPIDEV (2004-2007). The project was aimed at understanding the molecular regulation of genes governing flower structure in higher plants.

Alain Rival is now based in Jakarta, Indonesia since September 2014 and he coordinates Cirad collaborative research and development programs and training partnership in the Region. Invited Professor at University Putra Malaysia, Kula Lumpur in 2015-16, Alain Rival is the author and co-author of more than 80 research articles and book chapters. Alain co-authored the book "La Palme des Controverses – *Palm of Controversies*" with Patrice Levang in 2013 (Quae Publisher, Versailles) and he was the scientific editor of the research book on "Achieving sustainable cultivation of oil palm" recently published by Burleigh Dodds Science.

Abstract

Certifying sustainable palm oil: the european paradox

Prof. Alain Rival

Since its origin, the RSPO certification system for the sustainable production of palm oil is rooted in Europe. Indeed, as a business to business initiative, the RSPO roundtable gathered at its early beginning several Europe-based NGOs and multinational firms like Unilever. At the same time, Europe is also a market for palm oil and - even more - the almost unique market for Certified Sustainable Palm Oil (CSPO). Indeed even if Europe is not a key player in the global palm oil market - which is led by India and China - it has a key role to play in the public acceptance of RSPO as a norm and in the updating of RSPO Principles and Criteria (P&C) in order to fit with changes in expectations from stakeholders and progress in science and knowledge. The European paradox in oil palm lies in the fact that Europe and its institutions must play a dual role:

i) pushing the concept of sustainability throughout the palm oil commodity chains and its actors, and

ii) dealing with European farmers who are producers in a very competitive vegetable oils global market.

Public opinions in Europe are more in favor of a "No palm oil" strategy - as reflected by the 2030 ban proposed by the European Parliament this year - rather than giving Europe a key role in the evolution of RSPO P&C towards more inclusivity and better alignment to national mandatory sustainability standards such as MSPO and ISPO.

The present paper explores the status and trends of CSPO market in Europe and the role and response of EU markets (importers, traders, processors and end consumers). It proposes future avenues for collaborative research and development aimed at linking Europe –through Research Institutions, NGOs and the transformation industry- with key stakeholders in producing countries, including oil palm smallholders.

Prof. Muhammad Ashfaq

University of Agriculture, Pakistan

Prof. Muhammad Ashfaq is a Professor at Institute of Agricultural and Resource Economics, University of Agriculture, Faisalabad. His primary areas of interest, ekspertise and experiance is Agricultural Policies, Impact Assassment of Interventions, and Irrigated Agriculture.

He began his career as a lecturer from 1988 in Department of Agriculture Economic, University of Agriculture, Faisalabad. He served as Assitant Coordinator of Departement of Business Management Science (Evening College), University of Agriculture, Faisalalabad (2000 – 2001), Agricultural Economist, International Water Management Institute (2001 – 2003), Director Institute of Agricultural and Resources Economics, University of Agriculture, Faisalabad (2012 – 2015), Director External Linkages, University of Agriculture, Faisalabad (2015 – 2016), Chairman, Agriculture Policy, USPCA-AFS (2016 – 2017), and Chairman Sports Board, University of Agriculture, Faisalabad from 2016 to the present.

He has Higher Eduction Commission of Pakistan (HEC) recognized supervisor for more than 100 M.Sc, M. Phil student and 8 Ph.D studies; International linkage award from HEC for Australia in 2004; Supervisor for ICDD, Kassel, University, Germany; Section Editor, Pakistan J. Of Agricultural Sciences; Member editorial Board, Journal of Soil and Environment (2013); TWAS Associateship award 2011; Pakistan Program for Collaborative Research (PPCR)" award to visit Reading University UK by HEC 2011; Australian Alumni Excellence Award 2013 for Innovation and Research – Pakistan; Resource Person in 'Decision Support System for Agro – Technology Transfer Training Program, held in Konya-Turkey, 2015, Organized by Bahri Dagdas International Research Institute and Ministry of Food, Agriculture and Livestock, Turkey.

Abstract

Economic and social dimensions of import demand of edible oil in Pakistan

Prof. Muhammad Ashfaq

Pakistan is the 6th most populous country of the world and agriculture is one of the dominant sectors. It contributes 18.9 percent to GDP, employs 42 of labor force and provide raw material to agro-based industries. Being a consumer based economy, to meet the growing demands for foods, Pakistan heavily depends on imports. Edible oil industry has grown with an impressive rate in Pakistan over the last three decades. Despite an increase in the domestic production, huge amount of foreign exchange is still spent on its import. Edible oil is the key ingredient of household food consumption. Pakistan has consistently been the third largest importer of edible oil. Domestic production of edible oil is unable to meet the needs of even half of the population. Hence to fulfil the requirement of 210 Million population, Pakistan imports 73 percent of its demand. Edible oil imports mainly composed of palm oil and soya bean oil. In terms of palm oil trade Indonesia and Malaysia are the biggest trading partners of Pakistan. Out of 73 percent of total edible oil imports, 92% of the imports composed of palm oil. This paper was aimed to determine the socio-economic dimensions of edible oil demand in Pakistan. Both primary as well as secondary data were used for arriving at on certain conclusions. This study employed Johansen Cointegration model (JCM) and Error Correction Model (ECM) approaches. The study concludes that in the short run, real GDP, domestic oilseeds production and exchange rate and in the long run, population has the main impact on palm oil import. The study suggested that for increasing the production of edible oil some policy measures should be taken to decrease the level of import bill by increasing local production, adoption of new processing technology, introducing the new varieties of oilseeds crops, stabilization of exchange rate, import substitution, population and inflation control.

Keywords: Edible oil, import demand, Pakistan

Prof. Datuk DR Mad Nasir Shamsudin

Universiti Putra Malaysia, Malaysia

Prof. Datuk DR Mad Nasir Shamsudin is a Professor in agricultural and resource economics. He received his Bachelor of Science from Louisiana State University, and PhD in Agricultural Economics and minor in statistics from Mississippi State University, USA. Prof Nasir has been teaching primarily in International Agricultural Trade, Development Economics and Policy, Environmental Economics, and Managerial Economics. His research areas include International Agricultural Trade Policy, Commodity Market Analysis, and Agro-environmental Economics. To date, he has authored and co-authored more than 250 publications in book, book chapters, journal articles, and conference proceedings, and presented more than 160 papers in both local and international seminars in his area of expertise.

In academic administration, he was appointed as Head of the Agricultural Economics Department (August 1996 – December 1998), Head of the Agribusiness and Information Systems Department (January 1999 – December 2003), Deputy Dean, School of Graduate Studies (March – August 2004), Dean, Faculty of Environmental Studies (September 2004 – August 2009), Dean, Faculty of Agriculture (September 2009 – August 2012), and Deputy Vice Chancellor (Academic & International) (December 2013 – November 2016).

Abstract

Agricultural development and environment: on parallel tracks or a collision course?

Prof. Datuk DR Mad Nasir Shamsudin

Sustainable agricultural development continues to be emphasized to ensure that the well-being of the present generation is not met at the expense of future generations. Economic, social and environmental aspects are increasingly integrated into the development process. Thus environmental consideration is now integrated into the agricultural sector policy in order to ensure a sustainable economic and social development. The primary interest in sustainable agriculture is to develop farming systems that promote equally farming profits, agro-ecosystem, and local communities. Unsustainable practices, as it is argued, often focus solely on farm profits, at the expense of ecosystems, farming communities, and externalities.

Agricultural activities and environmental interests, however, sometimes collide, as governments, industries, environmentalists and consumers push their competing agendas. There is no doubt that agricultural development has direct impact on natural resources and the environment. However, the physical and economic links between agricultural development and the environment are complex. There is no conclusive evidence to show that agricultural development in itself, harms the environment. Rather, agricultural development often magnifies the environmental affects, particularly on natural resources.

The relevant questions from the above scenario are: (i) since both agricultural development and environmental quality are the basic elements for sustainable development, are they on parallel tracks or a collision course? (ii) what are the linkages between the environmental and agricultural policies?; and (iii) what approach is feasible to have a balance between agricultural development, hence economic growth, and environmental quality?

Environmental protection is a basic element of sustainable agricultural development. The environment supplies natural resources for agricultural production activities and is shaped by these activities. Agricultural practices, however, can cause negative externalities on soil, water, air quality and biodiversity.

Environmental effects can also occur or be exacerbated by agricultural policies, which raise or lower input and output prices above or below market levels, which might have resulted in the absence of such policies. Examples include the agricultural support programs, such as import tariffs and quotas. These policies influence production patterns, farming practices and input uses, mainly by changing the relative costs and returns of using resources in agriculture, or by imposing direct restrictions on output and input uses. Although the objective of the policy is to optimize the returns from farm incomes and to ensure a stable food supply at reasonable prices, the policy unintentionally affected the environmental quality.

Environmental policies, on the other hand, influence the compositions of agricultural production. As the cost structure of a regulated industry or activity rises relative to that of other less-regulated industries or activities, productive resources will shift toward the less-regulated sectors. Resources may be encouraged to move out of agriculture, should environmental controls in agriculture become more restrictive than in other industries. These output composition effects also influence the relative mix of agricultural output and trade. For example, production and export of less fertilizer-intensive crops might be encouraged should a fertilizer tax be implemented. The composition effect may also influence the relationship between primary production and processing of agricultural products. Often, policies in the form of regulations (such as standards, bans, and restrictions on input uses) and incentive-based mechanisms (such as taxes, subsidies, and marketable permits) are implemented as corrective measures. While these policies may meet their environmental goals, they may also affect production, trade, investment, technological change, and consumption. Such effects may be particularly important to agricultural producers and food processors.

The preceding discussions highlight conflicts between the two sets of policy, agricultural development and the environment. So, if they are conflicting, could we have sustainable agricultural development? A major conflict arises for the diametrically opposed signals that agricultural and environmental policies send to farmers. Although environmental regulations often raise production costs and reduce competitiveness in the short term, it is found that the long term effects are less certain since agricultural producers have to adjust and innovate. Studies indicate that the regulations altered input values and imposed costs on producers, thereby inducing a change in input use and the subsequent choice of alternative technologies. Thus, changes in relative factor prices stimulate innovative activities. Research institutions will innovate to remedy the constraint imposed by the policy-induced factor scarcity. Hence environmental regulation can act as a signalling mechanism that stimulates research into environment-conserving technologies. In the irrigation and groundwater example, the regulations limiting water use increased its scarcity value and stimulated research into more efficient "drip-irrigation" technologies. Environment-saving technological change allows more goods to be produced with less damage to the environment.

Although there seems to be conflicts between agricultural development and the environmental protection, there is a convergence between the two. Higher investment in agricultural research to generate a range of adaptable technologies, particularly for small farmers, is thus a compelling public policy priority, along with the development of supporting institutions. Market solutions could be a means of giving proper signals to farmers in making choices about sustainable resource management. Border prices could be used to remove biases against sustainable farming practices. Decentralisation and community participation in resource conservation would be crucial in encouraging and fostering sustainable agriculture. This also calls for industrial ecology approach that integrates the agricultural sector development with environmental ecology. The key is to modify behaviour so that agricultural growth and environmental protection can become reinforcing rather than competing goals. Although the needed changes will take time to orchestrate, several initiatives being developed in nations around the world are moving public and private efforts in this direction.

ORGANIZING COMMITTEES

Chairman : Diana Chalil, Ph.D
Co-chair : Assoc. Prof. Dr. Shaufique Fahmi Sidique
Secretary : Riantri Barus
Member : Suryani Darham
Muhammad Khaliqi
R.B. Moh. Ibrahim Fatoni

Program Committee

1. Assoc. Prof. Sutonya Thongrak (Prince of Songkhla University, Thailand)
2. Dr. Sirirat Kiatpathomchai (Prince of Songkhla University, Thailand)
3. Dr. Mawardati (Universitas Malikussaleh, Indonesia)
4. M. Authar ND (Universitas Malikussaleh, Indonesia)

Honorary Board

1. Drs. Mahyuddin K.M. Nasution, M.I.T, Ph.D
(Vice Rector for Research, Community Service, and Cooperation, Universitas Sumatera Utara)
2. Prof. Dr. Ir. Zulkifli Alamsyah, MSc
(Vice Rector for Plan, Relationship, Communication and Information, Universitas Jambi)
3. Jullimursyida, Ph.D
(Vice Rector for Plan, Relationship, Communication and Information, Universitas Malikussaleh)
4. Prof. Dato' Dr.-Ing. Ir. Renuganth Varatharajoo
(Deputy Vice Chancellor for Industry and Community Relations, University Putra Malaysia)
5. Asst. Prof. Dr. Pattara Aiyarak
(Vice President for Global Relations and Information, Prince of Songkhla University)

International Board

1. Prof. Larry N Digal (University of the Philippines Mindanao, Philippines)
2. Prof. Christopher Franco (Flinders University, Australia)
3. Prof. Alain Rival (Cirad, France)
4. Prof. Fatimah Arshad (UPM, Malaysia)

Scientific Committee

1. Dr. Janice Ser Huay Lee (Nanyang Technology University, Singapore)
2. Dr. Mirza Hasanuzzaman (Sher–e–Bangla Agricultural University, Bangladesh)
3. Dr. Marcel Djama (Département Environnements et Sociétés, CIRAD, France)
4. Prof. Dompak N. Napitupulu (Universitas Jambi, Indonesia)
5. Dr. Sanath Kumaran (MPOCC, Malaysia) - TBC
6. Prof. Madya Dr. Normaz Wana Ismail
7. Prof. Dr. Ahmad Shuib (IKDPM, Malaysia)
8. Assoc. Prof. Dr. Sridar Ramachandran (IKDPM, Malaysia)
9. Dr. Ahmad Hanis Izani Abdul Hadi (IKDPM, Malaysia)
10. Dr. Rosmawani Che Hashim (University Malaya, Malaysia)
11. Dr. Nazli Madzir (Universiti Utara Malaysia, Malaysia)
12. Dr. Riswanti Sigalingging (Universitas Sumatera Utara, Indonesia)
13. Dr. Irda Safni (Universitas Sumatera Utara, Indonesia)
14. Dr. Yaya Hasanah (Universitas Sumatera Utara, Indonesia)
15. Dr. Ma'ruf Tafsin (Universitas Sumatera Utara, Indonesia)
16. Prof. Dr. Ningrum Natasya Sirait, SH, MLI (Universitas Sumatera Utara, Indonesia)
17. Dr. Edy Ikhsan, SH., M.Hum (Universitas Sumatera Utara, Indonesia)
18. Munawar Khalil, SPi, MSc (Universitas Malikussaleh, Indonesia)
19. Profil Dr. Ir. Ketut Sukiyono, MEc (Universitas Bengkulu, Malaysia)

PROGRAMS

Date	Time	Activities	Venue
09 July 2018	06.00 pm - 11.00 pm	Welcoming dinner	Imperial Hotel
10 July 2018	08.00 am - 08.30 am	Registration	Imperial Hotel
	08.30 am - 09.00 am	Opening Ceremony	
		Diana Chalil, PhD (CSSPO Coordinator)	
		Prof. Dr. Ir. Bustami Syam, MSME (Honorary Board – Vice Rector for Information, Planning and Development USU)	
		Muhammad Abdullah (Acting Konjen RI di Kuching)	
	09.00 am - 09.10 am	Welcoming Traditional Dance	
		Panel Session	
	09.10 am - 09.35 am	Dr. Bayu Krisnamurthi (andgreen.fund, Indonesia)	
	09.35 am - 10.00 am	Prof. Alain Rival (CIRAD, France)	
	10.00 am - 10.25 am	Prof. Dr. Muhammad Ashfaq (University of Agriculture, Faisalabad - Pakistan)	
	10.25 am - 10.50 am	Prof. Datuk Dr. Mad Nasir Shamsudin (Universiti Putra Malaysia, Malaysia)	
	10.50 am - 12.00 pm	Discussion	
	12.00 pm - 12.15 pm	Souvenirs awarding for keynote speakers	
		Photo session	
	12.15 pm - 01.00 pm	Lunch Break	
	01.00 pm - 06.00 pm	Parallel Session	Imperial Hotel
11 July 2018	08.00 am - 04.00 pm	Field Trip to Bako National Park	Bako National Park
		Closing Ceremony	
		Depart to Hotel	

Statement of Peer review

In submitting conference proceedings to *Web of Conferences*, the editors of the proceedings certify to the Publisher that

1. They adhere to its **Policy on Publishing Integrity** in order to safeguard good scientific practice in publishing.
2. All articles have been subjected to peer review administered by the proceedings editors.
3. Reviews have been conducted by expert referees, who have been requested to provide unbiased and constructive comments aimed, whenever possible, at improving the work.
4. Proceedings editors have taken all reasonable steps to ensure the quality of the materials they publish and their decision to accept or reject a paper for publication has been based only on the merits of the work and the relevance to the journal.

Title, date and place of the conference:

CSSPO International Conference 2018
Towards Inclusive & Sustainable Agriculture
Harmonizing Environmental, Social and Economic Dimensions: Is It Possible?

July 9 - 11, 2018, Sarawak, Malaysia

Proceedings editor(s) and editor's signature:

Diana Chalil

Shaufique F. Sidique

Riantri Barus

Ahmad Hanis Izani Abdul Hadi

Muhammad Khaliqi

Mohammad Ibrahim Fatoni

Date: 8 August 2018

Law enforcement against trade in counterfeit goods using well-known marks (a study in traditional markets in Medan)

Aflah[*]

Doctoral Program in Law, Graduate School, Faculty of Law, Universitas Sumatera Utara, Medan, Indonesia.

Abstract. The illegal use of well-known marks for counterfeit goods is qualified as a criminal act. The act may cause damages to the owner of a well-known mark. The protection of a registered mark and a well-known mark through Law Number 15 Year 2001 on Trade Marks should be able to address issues related to mark infringement. The number of cases of illegal use of well-known marks is high because, in a business, using a well-known mark is more profitable than using own mark. Other factor which influences trade in counterfeit goods using well-known marks is the lack of public legal awareness as consumers, producers and traders. An empirical research method was applied in the research, using primary and secondary data. Maximizing trade in local goods using registered marks is one of effective ways to minimize trade in counterfeit goods using well-known marks in traditional markets in Medan. There are two possible solutions to this problem. The first solution is socializing Law on Marks to local producers, consumers and traders. The second solution is implementing a supporting program to help local producers register their rights to a mark.

1 Introduction

The function of mark is as a sign which is used on goods produced by a person or by several persons jointly or a legal entity to distinguish the goods from other goods of the same kind. It also can be used as promotion media. A product can be easily promoted by mentioning its mark and quality warranty. It can be a goodwill, emblem, quality standard and selling value in different types of market. It also can give satisfaction, pride and good reputation to consumers [1].

As one of intellectual properties, mark plays a vital role in good and service trades and investment. Mark having a good brand image can meet consumers' need to self-identity and is a warranty of the quality of goods and services in the free competition. Therefore, it becomes an economic asset for its owner – an individual or a company – which can give them great benefit as long as they take the good business and management aspects into account [2].

[*] Corresponding author : aflah@usu.ac.id

Act Number 15 Year 2001 on Marks requires producers and holders of right to a mark to register their mark to ensure that they will get legal protection for goods or services they sell. It is vital to protect the rights to a mark. It is clear that mark infringements in good and service trades have disadvantaged various parties such as owners of the right to a mark, consumers and traders selling goods with well-known marks [3].

Various parties illegally and intentionally use well-known marks for their own benefits by misusing the good reputation of the well-known marks because many consumers want to use cheap or even very cheap goods using well-known marks. There are some factors, such as lack of public legal awareness, low purchasing power and low self-esteem of producers to use their own marks. This situation is then used by those who want to take advantages by producing and selling counterfeit goods using well-known marks.

Long economic crisis also becomes a factor that pushes many producers to combine original goods and counterfeit goods [4]. From their physical appearance, they are exactly same. Many people are interested in buying counterfeit goods because their price is cheaper than that is of the original ones. Furthermore, some counterfeit goods have super quality. According to consumers, the quality of counterfeit goods with super quality is almost same with that of the original goods. This situation, of course, benefits producers of counterfeit goods because they do not need to spend much money to build the image of their products and advertise them. Only by imitating the mark of an original product, they can produce and sell up-to-date product.

The use of well-known marks gives great economic benefit to producers of counterfeit goods because although communities have low purchasing power, they still want to look stylish and trendy. The counterfeit goodscan be easily found in big cities in Indonesia, including in Medan. They are sold freely on road side and traditional markets.

Sadly, it seems that the situation is not dealt with properly. Neither government nor law enforcers take measures to prevent and ban the production and trade of counterfeit goods. This shows that Law on Marks still cannot be implemented effectively.

2 Problem

Trade in counterfeit goods using well-known marks is rampant in Indonesia. Therefore, it is important to provide legal protection to owners of a well-known mark as an effort to enforce Law on Marks in the country.

This research aims to answer the following questions, What are factors influencing trade in counterfeit goods using well-known marks in traditional markets in Medan, What can be done to prevent trade in counterfeit goods using well-known marks in traditional markets in Medan, and How is law enforcement against trade in counterfeit goods using well-known marks.

3 Research methods

This research applied an empirical research method, relying on primary data from the field. The research respondents were traders selling counterfeit goods using well-known marks, especially counterfeit bags and shoes. The respondents were selected by using a purposive sampling technique [5]. Data in the research was collected by using a questionnaire which was distributed to traders and consumers. The questionnaire instruments were measured by using a 5-point likert scale [6].

This research was conducted in a number of traditional markets in Medan such as Petisah, Ramai, Cathay Shoe Central, Brayan and Central Market. There were total of 15 traders and 30 buyers who became the research respondents from each traditional market.

Based on the findings of our preliminary survey, there were five traditional markets which met criteria set by the researcher. The criteria were that they are run under the supervision of Local Company Market of Medan City (*Perusahaan Daerah Pasar Kota Medan)* and and sell counterfeit bags and shoes using well-known marks.

Criteria of respondents are traders who sell shoes and handbags using well-known brands and have been trading for more than 5 years. Buyers who subscribe to buy and use artificial shoes and handbags with a well-known brand are asked directly at the study site [6]. The survey was conducted by distributing a closed questionnaire to the respondents.

4 Discussions

4.1 Factors influencing trade in counterfeit goods using well-known marks

There were some reasons why traders sold counterfeit goods using well-known marks. First, they did not understand Law Number 15 Year 2001 on Marks. So far, they had never been involved in the socialization of the law. Thus, they hoped government would allow them to sell counterfeit goods using well-known marks. They did not know that therewas a legal sanction that could be imposed on those who intentionally produced and sold counterfeit goods using registered well-known marks.

Table 1. Traders' understanding of Law Number 15 Year 2001.

Q : Do you understand Law on Marks?

		Frequency	Percent	ValidPercent	Cumulative Percent
Valid	Yes	23	30.7	30.7	30.7
	No	52	69.3	69.3	100
	Total	75	100	100	

Second, they did not understand that selling counterfeit goods using well-known marks violated the applicable laws or their conduct was illegal. Due to the lack of understanding of Law on Marks which was applicable nationally in Indonesia, they did not know that selling counterfeit goods using well-known marks was a criminal act.

Based on the the research findings, 57.3% of traders who were the research respondents answered that they did not know that selling counterfeit goods using well-known marks was illegal and could be imposed with a legal sanction.

Table 2. Traders' understanding of law violations related to the use of well-known marks for counterfeit goods.

Q : Do you understand that selling counterfeit goods is illegal?

		Frequency	Percent	Valid Percent	Cumulative Percent
Valid	Yes	32	42.7	42.7	42.7
	No	43	57.3	57.3	100
	Total	75	100	100	

Third, traders sold counterfeit goods using well-known marks because selling such goods was more beneficial and easier. By selling counterfeit goods, traders gotmore profits. It was also easier for them to sell counterfeit goods using well-known marks because they were in high demand. Many traders in traditional markets sold counterfeit bags and shoes using well-known marks because many buyers were interested in such goods. Traders made more profits. Since the goods were in high demand, they could always renew their stock with up-to-date models.

Table 3. The advantages of selling counterfeit goods.

Q : Do you get advantages from selling counterfeit goods?

		Frequency	Percent	Valid Percent	Cumulative Percent
Valid	Yes. They are easier to be sold	29	38.7	38.7	38.7
	Yes. There are more customers	15	20	20	58.7
	Yes. We make more profits	27	36	36	94.7
	No	4	5.3	5.3	100
	Total	75	100	100	

Fourth, law enforcers did not take proper legal measure to deal with the issue. Actually, they could play a vital role in the prevention of trade in counterfeit goods using well-known marks. The absence of legal measure by law enforcers would perpetuate the illegal practice. In this context, the law enforcers were actually not only policemen as investigators but also prosecutors and judges. However, traders who were the research respondents assumed that the law enforcers were only policemen.

Table 4. The lack of legal action.

Q : Have law enforcers ever taken any action?

		Frequency	Percent	Valid Percent	Cumulative Percent
Valid	Yes	17	22,7	22,7	22,7
	Never	58	77,3	77,3	100,0
	Total	75	100,0	100,0	

4.2 Factors influencing buyers to buy counterfeit goods using well-known marks

There were also some reasons why consumers bought counterfeit goods using well-known marks. First, they did not understandthe applicable laws, especially Law Number 15 Year 2001. This could be seen from research findings presented in table below.

Table 5. Consumers' understanding of Law Number 15 Year 2001.

Q : Do you understand Law on Marks?

		Frequency	Percent	Valid Percent	Cumulative Percent
Valid	Yes	17	11.3	11.3	11.3
	No	133	88.7	88.7	100
	Total	150	100	100	

Second, consumers bought counterfeit goods using well-known marks because they could be easily found in traditional markets. Meanwhile, if they wanted to buy original goods, they had to go to a super market or a super mall which is quite far from their house. In addition, branded goods could also be bought from distributors, special dealers or agencies with a membership system.

Table 6. Places where counterfeit goods are sold.

Q : Where can you buy counterfeit goods?

		Frequency	Percent	Valid Percent	Cumulative Percent
Valid	Malls	32	21.3	21.3	21.3
	Road side shops	18	12	12	33.3
	Online	39	26	26	59.3
	Traditionalmarkets	58	38.7	38.7	98

	Others	3	2	2	100
	Total	150	100	100	

Third, consumers bought counterfeit goods using well-known marks because they were affordable and cheaper than original goods. Due to their low income, consumers prefered to buy counterfeit goods because they were affordable and cheaper than branded ones.

Since counterfeit goods could be easily found in traditional markets, consumers were more interested in buying them. Even, they intentionally bought counterfeit bags and shoes whose models and marks were same with those of the original ones for collection so that they could use them in various events. In addition, they thought that it was better to buy counterfeit bags and shoes because they could be used in their daily activities. The research respondents also said that they always bought counterfeit bags and shoes with the latest design and model because they were cheap. By doing that, they could always catch up with the latest trend in the fast changing world.

Table 7. Consumers' reasons for buying counterfeit goods.

Q : Why do you buy counterfeit goods?

		Frequency	Percent	Valid Percent	Cumulative Percent
Valid	To be used in daily activities	42	28,0	28,0	28,0
	To follow the trend	23	15,3	15,3	43,3
	Their price is cheaper than those of the original ones	85	56,7	56,7	100,0
	Total	150	100,0	100,0	

Fourth, consumers bought counterfeit bags and shoes because they wanted to avoid possible big loss. They said that they bought counterfeit goods because they could easily sell them again if they disliked them. By doing that, they could easily get their money back.

Table 8. The advantages of buying counterfeit goods for consumers.

Q : Do you get advantages from buying counterfeit goods?

		Frequency	Percent	Valid Percent	Cumulative Percent
Valid	Yes	127	84.7	84.7	84.7
	No	23	15.3	15.3	100
	Total	150	100	100	

4.3 Efforts to minimize trade in counterfeit goods using well-known marks in traditional markets in Medan

It is important to improve traders' and consumers' understanding of Law on Marks through socialization and legal extension. It is hoped that by improving their understanding of the law, their legal awareness will also raise and law enforcement is more effective.

It is also important to improve the self-esteem of local producers so that they want to use their own mark on their products. It is important to improve the quality of their products so that local consumers will be interested in buying them. It is also urgent to help local producers register their marks to ensure that they will get legal protection for the right to a mark used on their products. All of these activities can be implemented as an effort to prevent and minimize trade in counterfeit goods using well-known marks. Mark registration is a key requirement to get legal protection.

Chart below describes how local producers can be assisted to register their marks as an effort to increase the production of and trade in local products using their own marks in traditional markets in Medan, minimize trade in counterfeit goods using well-known marks and increase trade in local products using a registered mark.

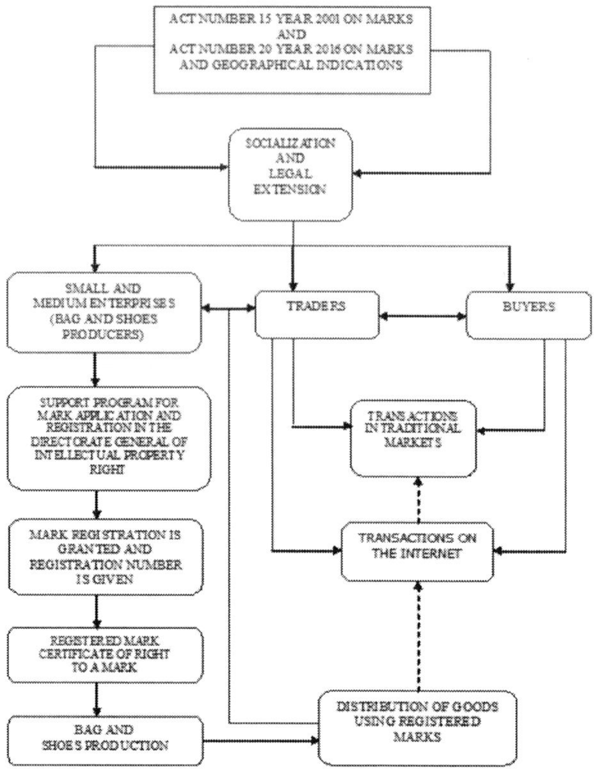

Fig. 1. Chart for increasing local products using a registered mark

Chart above can be further described as follows:

1. The first is socialization and legal extension on Law Number 15 Year 2001 on Marks and Law Number 20 Year 2016 on Marks and Geographical Indications.
2. The second is supporting local producers (small and medium enterprises producing bags and shoes) to apply for registration of a mark. It is important to support producers to increase the quality of their products in order to get consumer trust in using local products, especially shoes and bags. The support must continue until the mark registration is granted by and they get registration number from the Directorate General of Intellectual Property Rights, local products use registered mark and producers get the right to a mark.
3. Producers directly distribute the local products to traders. Then, traders sell the products in traditional markets or on the internet. Producers also can directly sell their products in traditional markets or on the internet without an intermediary (trader).

4.4 Law enforcement against the use of well-known marks on counterfeit goods

A constitutive system is currently used in mark registration system in Indonesia. This system has some strengths in term of legal certainty. According to the system, mark owner is the one who has registered it. This is known as presumption of ownership [7]. So, mark registration gives the one who has registered it a right to the registered mark. Thus, other parties have to respect it. According to the system, mark registration is a key requirement to get legal protection of the rights to a mark.

According to constitutive registration system, law enforcement against registered mark infringement in Indonesia can be done by filing a civil or a criminal lawsuit or through an alternative dispute resolution outside the court [8]. This is also applicable to law enforcement against trade in counterfeit goods using well-known marks [9].

As a comparison, we can see legal protection of the right to a mark in Benelux which also applies civil law like Indonesia. The protection of mark in Benelux is regulated by Benelux Law on Marks. Under the law, a mark can only be protected if it has been registered in the Benelux Office for Intellectual Property (BPOP) in The Hague. If there is an addition to name, logo or emblem and with certain requirements, shape, color, aroma or package of a product can also be protected as a mark [10].

Under Benelux Law on Marks, marks in Benelux are protected in Benelux countries. Since mark registration in Benelux is considered as a unit, an application applies in all territories [10]. In principle, mark registration system in both Indonesia and Benelux uses a constitutive system. The difference of mark registration in Indonesia and Benelux only lies in its procedures and phases.

Law enforcement against well-known mark infringement in Medan is done through dispute settlement in the Commercial Court. However, the research found that there was no well-known mark infringement case handled by Medan Commercial Court from 2016 to 2017. So far, the court only handled cases of registered mark. Most of well-known mark infringement cases were handled by Central Jakarta Commercial Court. Although Medan is one of big cities in Indonesia, well-known mark infringement cases were so far handled in Jakarta. In addition, since mark infringement is categorized as an ordinary offense, police can directly investigate a well-known mark infringer by using the criminal law and do not need to wait complaint or report from communities or parties concerned. Mark infringers use different motives in mark infringement cases such as imitating and counterfeiting well-known marks because theywant to get great advantages from the good reputation of a well-known mark. Criminal sanctions for the infringements of a well-known mark and a registered mark are regulated under Law Number 15 Year 2001 on Marks [11]. Different sanctions such as imprisonment, fine anddeprivation of the rights to a mark may be imposed on any party which is proven to have violated the law.

5 Conclusions

5.1 Summary

There are some reasons why traders sell counterfeit goods using well-known marks. The research found that traders sold counterfeit goods using well-known marks because they did not understand Law Number 15 Year 2001 on Marks. They also did not know that selling counterfeit goods using well-known marks was a crime. In addition, counterfeit goods were in high demand so that they could give them high profit.

There were also some reasons why consumers bought counterfeit goods using well-known marks. They said that counterfeit goods using well-known could be found easily. In addition, they were cheap and affordable. They also bought counterfeit goods for collection.

It is important to improve traders' and consumers' understanding of Law on Marks through socialization and legal extension. It is hoped that by improving their understanding of the law, their legal awareness will also raise and law enforcement is more effective.

It is also important to improve the self-esteem of local producers and the quality of their products so that local consumers will be interested in buying them. It is also urgent to help local producers register their marks to ensure that their rights to a mark used on their products can be protected legally. All of these activities can be implemented as an effort to prevent and minimize trade in counterfeit goods using well-known marks in traditional markets in Medan.

5.2 Recommendations

Local producer supposed to have a registered trademark of their products to get the legal protection. This legal protection is required to improve the reputation of the brand. It's time for the local government of Medan (*Pemerintah Kota Medan)* to help and provide the technical guidance for local producer in improving the products quality which impacts to the consumer trust enhancement to buy local product

Acknowledgements

The research was funded by the Directorate of Research and Community Service of the Directorate General for Research and Development of the Ministry of Research, Technology and Higher Education in accordance with the Funding Research and Community Service Agreement for the Fiscal Year 2018.

References

1. A. Herviandi, E. Susilowati, R. Njatrijani. Diponegoro Law Journal **6**, 1 (2017)
2. A. Sutedi. *Hak Atas Kekayaan Intelektual* (Sinar Grafika, Jakarta, 2009)
3. J. Mamahit, Jurnal Lex Privatum, **1**, 3 (2013).
4. Blogspot *Hukum Perlindungan Konsumen Perlindungan Merek* [online]. Retrieved from https://duniathoto.blogspot.co.id/2010/12/hukum-perlindungan-konsumen.html
5. R.H. Soemitro. *Legal Research Methodology and Jurimetrics*, Fifth Edition (Ghalia Indonesia, Jakarta, 1994)
6. B. Prasetyo, L.M. Jannah, *Metode Penelitian Kuantitatif* (PT. Raja Grafindo Persada, Jakarta, 2006)
7. Article 3 of Act Number 15 Year 2001 on Marks stipulates that the Right to a Mark is the exclusive right granted by the State to the owner of a Mark which is registered in the General Register of Marks for a certain period of time, to himself use said Mark or to grant permission to another party to use it.
8. See Article 76, Article 84 and Article 89 of Act Number 15 Year 2001
9. A. Sulistiyono. *The Existence and Resolution of Intellectual Property Right* Disputes (LPP and UPT Penerbitan dan Pencetakan UNS Press, Surakarta, p. 49, 2008).
10. I.B. Maulana, E. Flohil. *A Short Comparison of Mark Protection in the Netherlands and in Indonesia* (Alumni, Bandung, 2018)
11. See Article 100 of Law Number 15 Year 2001 on Marks.

Legal efforts to protect home workers in North Sumatra

Agusmidah *, *Suria* Ningsih, and *Erna* Herlinda

Faculty of Law, Universitas Sumatera Utara, Medan, Indonesia

Abstract Home workers without legal protection isan international phenomenon. North Sumatra Province encounters a similar problem because this province is a strategic area for investment. Its climate, land, natural resources, and the number of the labor force are several reasons which may attract investment in North Sumatra. This problem relates to the legal basis to establishlocal regulationconcerning protection to home workers. This is because the central government has not regulated this issue in the Act No. 13 of 2003 on Manpower. Therefore, this empirical juridical study will explain the legal efforts which have been done by the community to encourage the formulation of local regulation which can protect home workers. The analysis was conducted qualitatively by describing such efforts, including forming the communities of home workers so that they can be provided with training, advocation, accompaniment, and involvement in each stage of drafting the local regulation. In addition, political approaches were also performed continuously, such as voicing opinions to the Regional Representative Council (*DPRD*), Department of Manpower, and academics. Although these efforts have not received a positive outcome for the enactment of the local regulation, this article will give an idea that the legal issues of home workers particularly in North Sumatra still must be strived.

1 Introduction

ILO Convention No.177 have decribe home workers are the workers who are working from their homes or outside from the workplace of the company or corporation. A regulation to protect home workers is necessary in order to resolve unfit conditions at work. Wages and social insurance are two most basic things that should be protected. In average, home workers have low wages. In addition, they generally do not have social insurance, health insurance, occupational injuries insurance, or even an old age insurance. Other bad conditions of the home workers are including unsupervised working hours, an involvement of children in the work, and others

Home workers have a role in microeconomics because they increase productivity and economic growth for their employers, either individuals or companies (corporations). However, the assumption that home workers areconsidered as a lower-class job which does

* Correspondent Author: midahagus@gmail.com

not need to be regulated in the legislation is not in accordance with human rights and does not also support the purpose of being a state that protects all citizens (i.e. the Preamble of the 1945 Constitution of the Republic of Indonesia).

In Indonesia, the right to work is one of the basic rights of citizens (Article 27 paragraph (2) of the 1945 Constitution): "Every citizen has the right to work and to live in human dignity". It is further convinced as a human right (Article 28D paragraph 2 of the 1945 Constitution):"Everyone has the right to work and receive fair and reasonable remuneration and treatment in a working relationship." These reasons are quite convincing for Indonesia to formulatea regulation to protect all citizens regardless of their work type so that all citizens can get a job and live properly from the job, including home workers.

It is known that home workers have not been covered under the protection of labor laws as they are categorized into non-formal or informal economic sectors. In contrast, workers in the formal sector, such as workers in industrial sectors, are protected by the labor laws. It implies that home workers arenot considered employed by the "entrepreneurs" because the relationship between the home workers and the employers is generallybased on trust or habit.

The practice of hiring home workers in North Sumatra has quite significant numbers. The example of industries who hire home workers are in the manufacture of baby walkers, rubber slippers, fish rods, plastic mats, bedcovers, pillowcases, rolled incense, paper folders, and others. Although the home workers have been working for a long period of time, they have not received any protection as workers.

Therefore, this paper tried to describe the efforts which have been done to protect home workers, especially in North Sumatra Province. The purpose was to includehome workersin the category of workers who deserve thesupervision and protection of the statebecause their work also has a contribution to the community.

2 Problem

The underlying problem of this paper is the lack of protection for home workers, therefore needs to be learned the efforts that can be done to build a decent work for home workers.

3 Research Methods

The research location was North Sumatra Province. The data were obtained through literature study including primary legal materials,which consist of legislation and secondary legal materials such as books, scientific journals, previous research reports, and local regulation draft on the protection of home workers. In addition, primary?complementarydata, which supported the secondary data was also obtained from a Focus Group Discussion (FGD). The FGD was conducted at the Faculty of Law in the University of Sumatera Utara by presenting elements of the government (Regional Department of Manpower) as the excecutive and supervision in any labor regulation, the members of the Regional Representatives Council (abbreviated as DPRD) as the legislative and executive in North Sumatera (every province), home workers unions as the base for home workers, the non-governmental organization named "*Bitra Indonesia*" that organize and inisiate the protecting district regulation in North Sumatera, and academics from the Law who expert in State Law and Economics Department who expert in Manageent. Interviews were also conducted for the home workers to investigate their daily working conditions.

4 Discussions

4.1 Previous Research

A research by Bitra Indonesia found that home workers initially considered this system working relationship as advantageous for them because the work was regarded as an effort to earn extra money during their free time, and they did not have to leave the house, the children, and other domestic affairs. In addition, the work did not require a certain formal education as formal workers in the company.[1]

Siahaan[2] in his study found that home workers in North Sumatra have been there for more than fifty years, but they have never been included into the population data by the Central Bureau of Statistics (BPS). Since they did not know their work status, the homeworkers filled in their work status in the population data as unemployed (follow their husbands). The mapping conducted by Siahaan recruited homeworkers as research samples from 8 (eight) regencies/cities in North Sumatra, such as Asahan, Binjai, Deli Serdang, Medan, PematangSiantar, Serdang Bedagai, TanjungBalai, and Tebing Tinggi.

The research results of Bitra Indonesia[3]revealed the condition of home workers who havevery low wages, have no social insurance, and have no work health and safety insurance. The following Table 1 shows the data of wages of several home workers in North Sumatra.

Table 1 Wages and working hours of home workers per day (2017)

Home workers	Wages (IDR/day)	Working hours (hours/day)
Grill fish wire weaver	10,000	8-9
Prayer paperspacker	10,500	8-9
Air mineral pipettes/straws packaging	7,000	8-9
Rubber sandal straps cutter	3,000	9
*Baby walker*seat tailor	16,000	8-9
Belau packaging	17,000	10-12
Incense packer	20,000	8-9
Shrimps peeler	9,000	8-9
Pillowcase tailor	17,000	8-9

The results of the same research also found that the biggest problem experienced by the home workers in doing their job was the low pay or wages, followed by the problems shown in Table 2.

Table 2 Problems faced by home workers

Problems	Percentages (%)
Too low wages	40.76
Inconsistent work order	27.17
Unable to negotiate	9.13
Poor infrastructure access	6.24
Lack of skill development, education, and training	4.90
Expensive cost of materials and equipment	3.56
High transportation costs	1.78
Others	6.46

The study results of Haspels and Matsuura with ILO-Mampu[4]obtained a fact that home workers can be characterized according to whether they work for individual entrepreneurs or companies, often through intermediaries or agents, raw materials obtained from employers, wage paid per unit, and have no access to the market (not selling finished goods). However, the home workers, in fact, bear much production costs including their workplaces, equipment, electrical energy, and transportation.

Poverty and informal work system are two things, which are related to each other. Furthermore, the ILO[5]states that informal works occupy a large position in developing countries although informal works are correlated with the lack of protection, low wages, no insurance in old age or pensions, and no guarantee for work continuity.

Castel and Fortes[6]state that workers engage in the informal economy tend to be in a low-wage situation. Even when judging by the history of European countries, most occupations in the informal economy sector are occupied by undocumented immigrant populations.

Recently, the efforts to provide protection for workers/laborers through the state regulation are assumed to distort the market and cause a chaos in the market. [7] This opinion, however, is refuted by several previous studies' results. The ILO, for instance, finds a fact that there is no relevant link between regulation in employment through legislation that protects workers' rights and the rate of employment growth. Conversely, the important key seems to be how to create a conducive labor climate so that entrepreneurs or investors feel comfortable opening their business in Indonesia[8].

Another view, which breaks the paradox of free market and the rights of workers/laborers is stated by Uwiyono [9]. He evaluates that the policy of low wages and the weak implementation of legislation including restrictions on the fundamental rights of the workers in a free market economy becomes the factors that would hamper the free market. This is because the policy becomes an indication of social dumping which can lead to a disruption to the implementation of fair trade in the free market. As a result, this market economy model encourages autonomy innegotiating collectivelybetween the workers andthe entrepreneurs.Utz [10] in his writing states that the free-market economy emphasizes the economic balance which lies in the balance of supply and demand whereas social equilibrium lies in the balance of the power between the workers/laborers and entrepreneurs or entrepreneurs' organization. Therefore, workers need to be unionized so that the wage rates and other terms of employment can be negotiated with the entrepreneurs.

4.2 Work protection has not touched the non-formal sector workers

Act No. 13 of 2003 concerning Manpower (abbreviated as UUK) regulates the protection of work for the formal workingrelationship. A formal working relationship is characterized by an employment agreement between the employer and the employee. According to the Article 1 number 15 of UUK, a working relationship contains elements of work, orders, and wages.

Imam Soepomo[11] explained the scope of labor protection in three ways, namely: (a) economic protection, defined as the protection of workers in the form of sufficient income including when the workers work out of their will; (b) social protection, defined as the protection of workers in the form of occupational health insurance, freedom of association, and the protection of the right to form an organization; and (c) technical protection, defined as the protection of workers in the form of security and safety.

A weak bargaining position of workers/laborers causes the state must have a role in providing supervision and legal protection. In addition, Sinzheimer[12] asserts that the legal employment agreement is merely a 'voluntary adherence' of the workers/laborers to the conditions set unilaterally by the employers. This view, therefore, strengthens that need for the authorities (state) to minimize inequality through legislation.

4.3 Legal efforts to protect the home workers

Legal efforts in order to protect the home workers can be done by describing conditions that are often encountered and may be the causes of economic exploitation of the home workers.Following are the details:

1. The home workers provide or incur costs for working equipment and facilities, such as electricity, work equipment, storage (even at their own houses), the cost of maintaining the work equipment, and others;
2. The business owners think that it is unnecessary to do the obligations of some entrepreneurs as regulated in the labor laws and legislation because there is no employment agreement.
3. The delivery of work is usually through intermediaries, often through several levels which become the intermediaries between the employers and the workers.

The protection of home workers according to the ILO standards is by establishing the basic principles and rights at work, namely:[13]

(a) the freedom of association and the effective recognition of the right to collective bargaining;
(b) the abolition of all forms of forced or compulsory labor;
(c) the effective abolition of child labor; and
(d) the elimination of discrimination in terms of employment and position.

Decent work is central to poverty alleviation efforts and is a means of achieving equitable, inclusive, and sustainable development. A decent work also involves providing opportunities for productive employment and equitable incomes, providing security in the workplace, and providing social protection for workers and their families [14].

4.4 Efforts which have been done to protect home workers in North Sumatra

Efforts to protect home workers in North Sumatra have been done by both local governments and non-governmental organizations. The provincial government of North Sumatra undertakes efforts to protect home workers by proposing the formulation of local regulation (*Perda*). The authority to establish this regulation is based on the Act No. 23 of 2014 on Local Government, and it is also supported by the companion society of home workers, such as Bitra Indonesia and Alliance of Home Workers.

There were obstacles encountered during the process of drafting the local labor regulation in the DPRDof North Sumatra Province including the legislators' doubts concerning two reasons:

a. The legal basis because there is no law that provides instructions for the provision of the local regulation to protect home workers in the Manpower Act.
b. Regarding the investment climate, regulation can hinder investment.

The Act No. 13 of 2003 on Manpower is known to be limited in the protection of workers in the formal sector (i.e. work in the employment relationship), and it does not contain any protection for informal workers. Work relationships as legal relationshipsare important facts which provide several rights and obligations for all parties. However, in the case of home workers, one party works for another party through an intermediary or agent, so there is no working relationship between the worker and the employer[15].

Efforts were also made by convincing the DPRD in a hearing discussion (abbreviated as RDP) that the government has the right to formulate aregulation to protect its people without exception. In the process, however, the existing draft regulation was agreed not only for home workersbut also for all workers;thus, the name of the local regulation was changed from the protection of the home workers to the draft of the local labor regulation.

In addition to local governments, the protection of home workers was also done by Bitra Indonesia cooperated with *Bappenas* and the ILO in a program named *Mampu*.Bitra Indonesia made efforts by empowering the home workers who eventually form a home workers union. The benefits of this program were as an effort to improve the capacity of home workers as part of the workforce, and to provide an assistance, training about organizational knowledge, and decent work advocacy. This program has been conducted since 2014, and home workers unions has been formed in January 2015. Since 2014, at least there are 22 sectors that have been identified by Bitra Indonesia, but there are still four regions in North Sumatra, such as Deli Serdang Regency, Asahan Regency, Medan City, and Binjai. In these four regions, there are 1465 home workers participating in the home workers union.

5 Conclusions

Efforts made by the government and community in North Sumatra to provide protection for home workers used legal and non-legal approaches. The effort in the legal approach was making the local regulation draft, which will be a priority to be discussed in 2018. On the other hand, the non-legal approach was done by non-governmental organizations through several activities, such as: empower and train the home workers, gather them in workers union organizations, raise their awareness about the position of home workers, and increase their self-confidence as workers.

Acknowledgements

The authors would like to thank the Ministry of Research and Technology in the Directorate General of Higher Education who has provided a grant for an Applied Product Research scheme DRPM 2017under the contract No. 56/UN5.2.3.1/PPM/KP-DRPM/2017. We would also thank the Research Institute and the Rectorate of the University of Sumatera Utara who have given much support so that the DRPM 2017 grant can be obtained.

References

1. Bitra Indonesia, *Kerangka Acuan Kegiatan,* Bedah UUK No. 13 Tahun 2003 "Status Hukum Pekerja Rumahan", Medan, Desember, (2015).
2. E. Siahaan, *Bahan Paparan*, Melihat Pekerja Rumahan dari Sudut Pandang Ekonomi, disampaikan dalam FGD di FH USU pada 18 Juli 2017, (2017).
3. Bitra Indonesia, *Bahan Paparan*, Pekerja Rumahan: Potret Buram dalam Mata Rantai Industri, disampaikan dalam FGD di FH USU pada 18 Juli 2017, (2017).
4. N. Haspels dan A. Matsuura, *Pekerja Rumahan di Indonesia Hasil dari Penelitian Pemetaan Pekerja Rumahan di Sumatera Utara, Jawa Barat, Jawa Tengah, Yogyakarta, Jawa Timur dan Banten*, (ILO- MAMPU, Jakarta, 2015).
5. [ILO], http://www.ilo.org/global/topics/employment-promotion/informal-economy/lang--en/index.htm (retrieved at 2nd September 2017)
6. M. Castells, A. Portes, *World Underneath: The Origins, Dynamics, and Effects of The Informal Economy*,in Alejandro Portes (ed), The Informal Economy: Studies in Advanced and Less Developed Country, (John Hopkins University Press, Baltimore, MD, 1989).

7. S. Deakin, *The Contribution of Labor Law to Economic Development And Growth*, Working Paper No. 478, (Centre for Business Research University of Cambridge, 2016)
8. ILO**,** *Seri Rekomendasi Kebijakan: Kerja Layak dan Penanggulangan Kemiskinan di Indonesia, Dimensi Ketenagakerjaan dalam Kebijakan Makro dan Sektoral*, Jakarta, (2004).
9. A. Uwiyono,*Hak Mogok di Indonesia*, (PPs-FH UI, Jakarta, 2001).
10. A. F. Utz, Is The Right to Strike a Human Right?, *Jurnal Washington University Law*, **65**, (1987).
11. I. Soepomo, *Pengantar Hukum Perburuhan*, (Djambatan, Jakarta, 1995).
12. L. Wedderburn, Collective Bargaining or Legal Enactment: The 1999 Act and Union Recognition, *Industrial Law Journal*, **29**, (1) (2000).
13. ILO, *Naskah Konvensi Tentang Pekerjaan yang Layak bagi Pekerja Rumah Tangga*, Jenewa, (2011).
14. ILO, *Profil Pekerjaan Layak di Indonesia*, Jakarta, (2012).
15. Agusmidah, Proceeding Seminar Ilmiah Dies USU, Hak Ekonomi Pekerja Perempuan: Pekerja Rumahan dalam Jangkauan UU Ketenagakerjaan, Medan, (2016).

Motor vehicle lease agreement with fiduciary warranty

Agustining[*] and *Ningrum Natasya* Sirait

Doctoral Program in Law, Graduate School, Universitas Sumatera Utara, Medan, Indonesia

Abstract. Purchasing motor vehicles with instalment (credit) is currently becoming favourite choice by many people including people who live in plantation and agriculture area. There are several reasons, such as easy pre-requirements, low down payment and saving time. The financing institutions that generally deal with motor vehicle credit are financing companies and banks. Although, they are different in their operational legal basis, in practice, financing company and bank fall under the category of fiduciary warranty agreements. This study focus on the potential fraud on law in financing of capital goods. This study used the normative and empirical legal research methods. In the end of the study, it found that it used loan financing with fiduciary guarantee not a lease agreement. As a consequence, the use of fiduciary warranty by financing company in a lease does not accord with the law and considered to be illegal. Last but not least the study found that Ministry of Finance and the Financial Services Authority (*Otoritas Jasa Keuangan*/OJK) as supervisory agencies financing business must to ensure law enforcement, certainty and effective oversight. OJK should provides sanctions for financing companies which do not comply in accordance with the legal provisions.

1 Introduction

The increasing public need for motor vehicles either as a private transportation or as a main vehicle for working has driven a high public interest to own a motor vehicle. It appears in all areas in Indonesia including plantation and agriculture areas. Owning a motor vehicle has become a primary need for most people. There are some reasons why nowadys, owning motor vehicle became a primary need. First, using motor vehicle both 2 wheels or 4 wheels are more practicee rather than using rental vehicle for example for a farmer, he/she can bring directly the harvest back to her/his home. Secondly the public transport conditions are neither sufficient nor comfortable due to its condition and the expensive price. As an impact, it encourages people to have their own vehicle.[1]

Vehicles are motors driven by a technical equipment to move and mostly used for land transportation.[2] This study specifically discussed motor vehicle which use 4-wheels and 2-wheels motor vehicles for individuals as good renter. To be able to own a motor vehicle, people are provided with many ways on payment options. If they do not have cash to buy,

[*]Corresponding author: ning_mkn@yahoo.com

purchasing on credit becomes the most preferred choice.

The option of purchasing on credit can be realized with the financing help of banks or leasing companies. Both banks and leasing companies play the same role in helping people to own a motor vehicle through monthly instalment. However, they differ in the legal operation. Bank is a business entity that collects funds from the community in the form of savings and distributes them in order to improve the standard of living of many people. Commercial bank can provide services in the payment transactions.[3] On the other hand, leasing is a financing activity in the form of capital goods provision either under an optional lease (*finance lease)* or a non-optional lease (*operating lease)* to be used by the *Lessee* for a specified period of time on a periodical basis.[4]

Although they both have similar role, they differ in their legal basis in terms of regulation and procedures for vehicle acquisition. In banking practices, the bank's function is to provide bailout funds in cash. The bank as the creditor and the customer as the debtor are bound by a principal agreement in the form of a credit agreement. It is then followed by an additional agreement (*accesoir)* in the form of a fiduciary warranty. This situation is potential to make fraof of law happened.

Fiduciary is only an additional agreement (*accesoir*) which serves as a tool for the bank to execute if the debtor violates the agreed credit agreement. The bank does not hold the guaranteed goods physically. However, it is protected by the law under the fiduciary warranty registered with the Ministry of Justice and Human Rights of the Republic of Indonesia, in the form of a fiduciary warranty certificate.[5] The holders of fiduciary warranty are protected because the certificate has a written phrase "For the sake of justice under the Almighty God" that has similar executorial power as the decision of a court which bears legal force. If the debtor breaches the agreement, the fiduciary holder shall have the right to sell the object of fiduciary warranty on his/her own power.[6]

The concept of fiduciary warranty is the transfer of ownership on the basis of trust of the material right.[7] Material right here means the right to an object (a motor vehicle) which can be owned and transferred. (Insert references in endnote). The characteristic or nature of the transferable material right is absolute. It follows the object wherever and/or with whoever it's located. Furthermore, it has the preferential right.[8]

According to the prevailing regulations in Indonesia, the process or procedure of obtaining a motor vehicle through a leasing company involves 4 (four) parties namely the *Lessor* or leasing company, the *Supplier* or provider of goods, the *Lessee* and goods renter, and the insurance company.[9]

The leasing procedure can be described as follows: (1) the *Lessee* determines the required motor vehicle, then the *Lesse* selects and determines the *Supplier*. (2) the *Lessee* completes a leasing application and meets the requirements to be subsequently submitted to the *Lessor*. (3) The *Lessor* examines the financial condition and ability of the *Lessee,* if approved then the *Lessor* requires the *Lessee* to sign a leasing agreement and at the same time the *Lessee* will sign an insurance agreement with an insurance company designated by the *Lessor* for a good renter. (4) The contract and the goods renter purchasing payment shall be signed by the *Lessor* and *Supplier*. (5) The *Supplier* hands the goods renter over to the *Lessee*. (6) the *Lessee* pays the leasing fee periodically according to the agreed leasing agreement.[10] The procedure can be drawn in Fig 1.

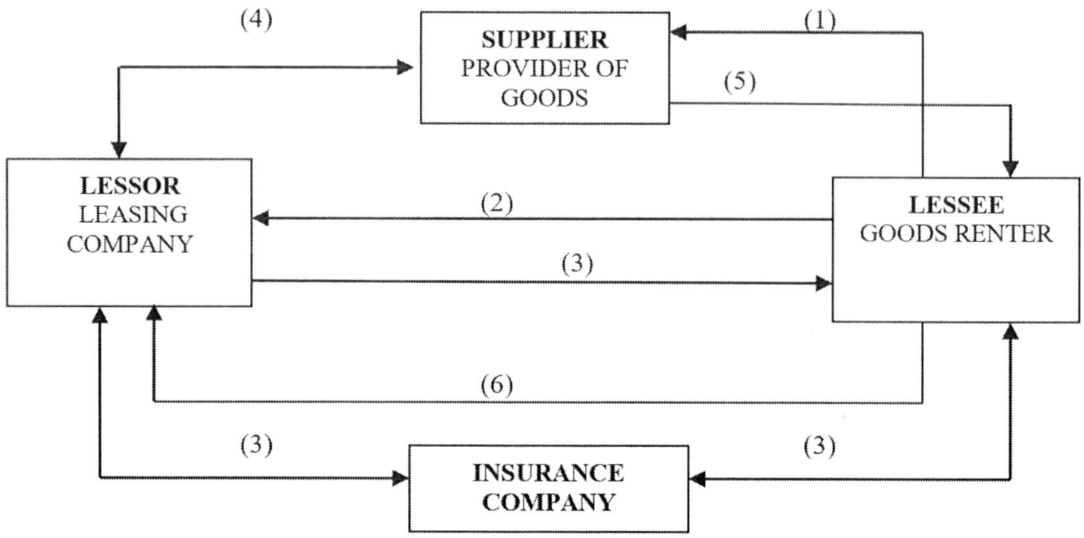

Fig. 1. Leasing Procedure

The nature of a *Leasing* contract is a rental activity. According to the procedure, it is clear that the buyer of good renter is the *Lessor* and during the course of the leasing agreement, the ownership of the vehicle goes to the *Lessor*.

At the signing of a leasing agreement, the *Lessee* as a motor vehicle renter may agree that he/she will purchase the leased motor vehicle at the end of the leasing term. As such, the last leasing payment is considered as a sale session between the *Lessor* and the *Lessee*. At that moment, a purchasing agreement between the *Lessor* and the *Lessee* should be proceed with the name transfer or registration of ownership right from the *Lessor* to the *Lessee*.

In fact, a mutual financing agreement or contract is made between the *Lessor* and the *Lessee,* with the *Lessee*'s obligation to make the *down payment* plus the cost of vehicle taxes and other expenses which follows the same financing bank procedure as. The registration of the motor vehicle has already been made on behalf of the *Lessee*, with a mutual financing agreement or contract made and a notary-signed fiduciary warranty subsequently registered at the Ministry of Justice and Human Rights of the Republic of Indonesia. Therefore, if the *Lessee* including the goods renter suffers a financial blockage in the instalments (*default*), the *Lessor* directly confiscates the good renter based on the executorial power of the fiduciary certificate as commonly found in banking practices.

With respect to the credit purchasing system, the Government regulates uniform amount of down payment in the Procedure for Motor Vehicle Loan as stated in the External Circular Letter of the Bank of Indonesia Number 14/10/DPNP dated 15 March 2012 on the Implementation of Risk Management in Bank Providing House Ownership Loan and Motor Vehicle Loan. The down payment, here in after referred to as Down payment (DP), is a pre-payment or advance cash to buy a motor vehicle on credit. This fund originates from the debtor (*self-financing*) and is set at least 25% (twenty five percent) for the purchasing of two-wheel vehicle and 30% (thirty percent) for the purchasing of four-wheel vehicle for non-productive purposes.[11] This rule applies equally to both leasing companies and banks.

Based on the facts above, this research conducted to respond to the following questions: a) Why leasing companies are able to use fiduciary warranty for motor vehicle in their leasing practice? b) What is the legal consequence of the fiduciary warranty agreement made by the motor vehicle leasing companies?

2 Research method

The study conducted based on an empirical legal approach. The empirical legal approach is the science of law which perceives law as a fact which can be construed or observed and is value-free.[12] The definition of *value-free* here is that the study of law should not be dependent or influenced by the researcher's personal assessment.[13] The empirical legal research aims at measuring how a law works in asociety.[14] Bahder Johan Nasution's in his book *Metode Penelitian Ilmu Hukum* (Research Methods in Law), the main character or feature of empirical legal research[15] is its emphasis on observation. This relates to the objective and empirical nature of the science itself. It is to include knowledge of empirical law which seeks to observe the prevailing legal facts in the society, in which it requires knowledge to be observed and openly demonstrated. The starting point of the observation lies in the facts or social facts that exist and live in the society as the living culture of the people.[16]

In addition to empirical study, this study also used a normative legal study, that is research on legal principles, legal sources, and scientific and theoretical laws and regulations that can analyze and discussed the issue.[17] The strength of the normative legal study lies in the sequential steps which are easily explored by other legal scientists.[18]

With the empirical and normative research methods, the extent to which a law works in the society can be studied. This study revealed or dealt with the fact that is can be observed with the rationale. The concepts or research terms must be firmly applied to the world of research. As such, it is expected to provide an answer to the fill in the gap that occurs amidst the existing regulations.

3 Discussion

3.1 The use of fiduciary warranty agreements on leased motor vehicles.

Fiduciary is the ownership transfer of an object on the basis of trust in a condition of the goods renter, whose ownership right is transferred, remains in the possession of the object owner.[19] It is clear that fiduciary (*Fidusia Eigendom Over dracht*) is a transfer of ownership based on trust.[20] The party gives full confidence to the other party to transfer their ownership right where the objects are considered as collateral.

Another definition of fiduciary (*Fiduciaire Eigendoms Overdracht*/FEO) is the transfer of ownership right on the basis of confidence in the debtor's goods used as a collateral to the creditor whereas the physical goods still remain in the control of the debtor.[21 From the definition, fiduciary can be distinguished from fiduciary warranty. Fiduciary is a process of transferring ownership rights and fiduciary warranty is a collateral given in the form of a fiduciary. A fiduciary warranty is *accessoir* (additional to) the principal agreement, that is a financing or credit agreement.[22]

The government [23] regulated that the nature of material right as a fiduciary object is absolute. The material right has a defendable possession against anyone; the material right follows the object (*droit de suite*) no matter who possess the good; the material right has *droit de preference*, the holder of the material security is entitled to repayment earlier than other creditors. In practice, purchasing a motor vehicle can be done in two ways:

a. Under bank's financing scheme

The debtor submits a request for funding to a particular bank. After being evaluated and meeting the requirements, the bank together with the debtor purchases the motor vehicle in cash (funded by the bank) at the dealer. At that time, the registration of vehicle is made on behalf of the debtor. Because the finance is supported by the bank, the debtor and the bank have signed a principal agreement that is a debt agreement or credit agreement. Further to guarantee the payment of debt in an orderly manner, the bank also makes an additional agreement in the form of a fiduciary agreement. Furthermore, the vehicle remains controlled by the debtor, while the bank as the creditor only keeps the credit agreement, fiduciary warranty agreement, Book of Motor Vehicle Owner (BPKB) and Certificate of Fiduciary Warranty.

In order to ensure the legal certainty in this fiduciary warranty, the law applies the principle of fiduciary warranty registration. Registering fiduciary warranty legally protects the interests of the guarantor, the recipient and the third party. The grant of material right to the creditor as a fiduciary recipient means the grant of a fiduciary guarantee certificate. The grant of *parate* execution right and separatist creditor status indicates the intent of the legislator to grant a strong position to the creditor.[24] If the debtor defaults (*wanprestatie*), the bank can confiscate the collateral for execution because the bank has the executorial power based on the fiduciary warranty certificate.

The creditor's authority to confiscate the collateral from the debtor in default cases cannot be done by force. For the safe, orderly, smooth and accountable execution of fiduciary warranty and for the safety and security of the fiduciary recipient, fiduciary guarantor, and/or the public from acts which may cause loss of property and/or personal safety, the creditor may submit an application for the execution safeguard in writing by the fiduciary guarantor or his/her lawyer to the Regional Police Head or District Police Head at the place of execution.[25]

b. Under a leasing company financing scheme

However, the scheme is different if a good renter obtained from a financing company. The *Lessor* as the owner of the goods rents the equipment for a certain period of time. So, it requires a leasing agreement. The *Lessee* is required to pay a rent periodically for a certain period of time. While the leasing company, as the owner of the goods, has the option to submit (sell) or buy the leased goods at the end of the leasing term.

Leasing is a financing activity in the form of capital goods provision through either an optional lease (*Finance Lease)* or a non-optional lease (*Operating Lease*) to be used by the *Lessee* for a specified period of time based on instalments.[26] Whereas rent-to-buy (*hire purchase)* is an act of buying and selling goods in which the seller sell goods by taking into account any payments made by the buyer through repayment of the goods price which has been mutually agreed upon and which is bound in an agreement. The ownership of the goods can only be transferred from the seller to the buyer after the price is paid in full.[27] Throughout the hire purchase period, the ownership right of the goods has not being transfer or still in the possession of the seller. Once the object is paid off, its registration is transferred to the buyer's name.

Fiduciary warranty in a lease is not necessary. In a lease, the creditor has full rights because the goods are formally owned by the *Lessor* although they are materially controlled and used by the *Lessee*. If fiduciary warranty is used, the leased goods must be registered on behalf of the renter so that the ownership of the goods can be guaranteed by fiduciary. However, as stated aboved that fiduciary warranty as the additional agreement in a *leasing* contract is not regulated by the Indonesian law which adheres to the civil law system. If there is an uncertain sale of goods, the right of ownership will not be granted until the goods comply with the contract.[28]

Hire purchase is an agreement not specifically regulated in the Civil Code. However, since Book III of the Civil Code embraces an open system, any parties may enter into agreements which are not specifically regulated in the Civil Code.[29] *Hire purchase* is a mixed agreement in which the terms of purchase and leasing agreement apply. The legal consequences of the two agreements are totally different. In a purchase agreement, the ownership of goods are immediately transferred upon the payment, while in a leasing agreement, there is no transfer of ownership.

3.2 The legal consequences of fiduciary agreements made by motor vehicle leasing companies.

The financing agreement is fall under an unnamed agreement. This means that all agreements made by the parties in accordance with the principle of freedom of contract, which has not been regulated under the Civil Code, are deemed valid as long as they are not contrary to the law, public order and morality.[30] Article 1319 of the Civil Code states, every *nominat* or *innominat* agreement is subject to the general provisions of the agreement law. The hire-purchase agreement as an *innominat* agreement is also subject to general terms of an agreement such as the requirements for a valid agreement and *default* cases.

According to Article 1 paragraph 5 of Presidential Regulation of Indonesia No. 9 of 2009, Leasing is a financing activity in the form of providing goods either by leasing with option rights (Finance Lease) or leasing without option rights (Operating Lease), to be used by the Lessee for a specified period of time based on instalment payments. Based on voluntary payments which accompanied by the right to choose for the company, either to purchase the capital goods concerned or to extend the leasing term based on the agreed remaining value.[31]

Hire purchase is an act of buying and selling goods in which the seller performs the sale of goods by taking into account any payments made by the buyer through repayment of the goods price which has been mutually agreed upon and which is bound in an agreement, and the ownership of the goods can only be transferred from the seller to the buyer after the total price is paid in full.[32]

Financing in the form of capital goods should be made lease agreement but in fact used loan financing with fiduciary guarantee Thus, if a fiduciary agreement is applied to a leasing guarantee, it will be determine as illegal. It is illegal since in a lease the ownership right has not been transferred from the seller to the buyer before the price is paid in full by the buyer. While the fiduciary warranty is used to guarantee the ownership right of goods in which the goods remain controlled by the owner. Because of its legal defects, fiduciary warranty in a lease should have no legal effect. If the *Lessee* defaults, the *Lessor* cannot execute on the basis of a fiduciary warranty certificate.

As a result of the *leasing* fiduciary, many people are disadvantaged by the debt collector who has unilaterally confiscated a motor vehicle from a defaulted debtor. The action is without taking into account the debtor's right to the payments previously made. The government seems to know nothing about the arbitrary practice of this lease and not yet respond to growing concern over this practice. In addition, the Financial Services Authority (OJK) as the supervisory agency in businesses, in this case supervision of banking services, must intervene..

4 Conclusion

4.1 Summary

Fiduciary warranty in a Lease is not necessary because in a leasing finance the creditor still has full rights. The goods formally belong to the company although they are materially controlled and used by the renter. If fiduciary warranty is used, the leased goods must be registered on behalf of the *lessee* so that the ownership of the goods may be guaranteed by a fiduciary.

Under the Indonesian law which adheres to the civil law system, fiduciary warranty in a lease is not regulated. In practice, if a lease of a motor vehicle uses a fiduciary warranty, the agreement will be deemed unlawful, because in a leasing agreement, the ownership right has not been transferred from the seller to the buyer unless the price is paid in full by the buyer.

4.2 Recommendation

The Indonesian Financial Services Authority (OJK) as the supervisory agency in businesses, in this case supervision of banking services, must intervene or against leasing practices that are not comply with. It is expected that OJK would provide sanctions on the violations. It is expected that the agency can provide legal certainty for the consumer as the *lessee* and the balance of rights between the vehicle renter (*lessee*) and the *leasing* company as the owner of the vehicle. Therefore there will be no arbitrary practice of leasing in handling a case of *defaulted lessee*.

References

1. I.W. Putra, *Minat masyarakat Indonesia akan transportasi umum*, web.unair.ac.id/.../f_71227_Minat_Masyarakat_Indonesia_Akan_Transportasi_Umu... retrieved on 30 April 2018.
2. https://id.wikipedia.org/wiki/Kendaraan_bermotor, retrieved on13 March 2018.
3. The Decree of the Minister of Finance of the Republic of IndonesiaNumber 1169/Kmk.01/1991 dated on 27 November 1991 on Leasing Activities Article 1 (a).
4. Law Number 7 of 1992on Banking Business Article1 (1) and (2) and President of the Republic of Indonesia Regulation Number 9 of 2009 on Financing Institution Article 1 (5).
5. Law Number 42 of 1999 on Fiduciary Warranty, Article 14.
6. Law Number 42 of 1999 on Fiduciary Warranty, Article 15.
7. I.D. Purnamasari, *Kiat-kiat cerdas, mudah dan bijak memahami masalah hukum jaminan perbankan*, p. 83, (Kaifa, Bandung, 2014).
8. The Letter of the Ministry of Law and Human Rights of the Republic of Indonesia Number: C.HT-1.10-74 dated on 27 September 2006 on Guidelines for Fiduciary Registration.
9. Z. Asyhadie, *Hukum bisnis, prinsip dan pelaksanaannya di Indonesia*, p. 107, (Rajawali Pers, Jakarta, 2017).
10. Z. Asyhadie, *Hukum bisnis, prinsip dan pelaksanaannya di Indonesia,* p. 108, (Rajawali Pers, Jakarta, 2017).
11. Circular Letter of the Bank of Indonesia Number 14/10/DPNP dated 15 March 2012, on Implementation of Risk Management in Bank Providing House Ownership Loan and Motor Vehicle Loan, point IV on Down Payment of Motor Vehicle Credit System.
12. B.J. Nasution, *Metode penelitian ilmu hukum,* p.81, (Mandar Maju, Bandung, 2008).

13. B.J. Nasution, *Metode penelitian ilmu hukum,* pp.81-82, (Mandar Maju, Bandung, 2008).
14. B.J. Nasution, *Metode penelitian ilmu hukum,* p. 123, (Mandar Maju, Bandung, 2008).
15. B.J. Nasution, *Metode penelitian ilmu hukum,* pp. 124-125, (Mandar Maju, Bandung, 2008).
16. B.J. Nasution, *Metode penelitian ilmu hukum,* p.124, (Mandar Maju, Bandung, 2008).
17. S. Soekanto and S. Mamudji, *Penelitian hukum normatif suatu tinjauan singkat,* p. 13, (Raja Grafindo Persada, Jakarta, 1995).
18. J. Ibrahim, *Teori & metodologi penelitian hukum normatif,* p. 278, 3rd Edition, (Bayumedia Publishing, Malang, 2007) (cited from P.M. Hajon).
19. Law Number 42 of 1999 on Fiduciary Warranty, Article 1.
20. Law Number 42 of 1999 on Fiduciary Warranty, Article 1.
21. H. Hadisaputro, *Pokok-pokok hukum perikatan dan hukum jaminan*, p.50, (Liberti, Jogjakarta, 1984).
22. I.D. Purnamasari, *Kiat-kiat cerdas, mudah dan bijak memahami masalah hukum jaminan perbankan*, p.88 (Kaifa, Bandung, 2014).
23. Ministry of Law and Human Rights of the Republic of Indonesia, The letter Number C.HT.-1.10-24 dated on 27 September 2006, in I.D. Purnamasari, *Hukum jaminan perbankan,* pp.83-84, (PTMirzan Pustaka, Bandung, 2014).
24. J. Satrio, *Hukum jaminan, hak jaminan kebendaan*, p. 180, (Citra Aditya Bakti, Bandung, 2007).
25. Regulation of the Police Department of the Republic of Indonesia Number 8 of 2011 on Fiduciary Warranty Execution Safeguarding, Article 2 jo Article 7 (a).
26. President Regulation Number 9 of 2009 on Leasing Companies ("Perpres 9/2009"), Article 1 (5).
27. http://www.hukumonline.com/klinik/detail/cl5033/perbedaan-leasing-dan-sewa-beli, retrieved on 14 March 2018.
28. P. de Cruz, *Perbandingan sistem hukum, common law, civil law and socialist law,*p. 553, (Nusa Media, Bandung, 2016).
29. Suharnoko, *Hukum perjanjian: Teori dan analisa kasus*, pp.64-64, (Kencana, Jakarta, 2004)
30. T. Kamelo, Filosofi dan aspek hukum fidusia dan pembiayaan (Masalah hukum, perjanjian, wanprestasi dan eksekusi), in a Law Seminar of PT Mandiri Utama Finance, Jogjakarta, 25 January 2018.
31. M. Djumhana, *Hukum perbankan di Indonesia*, p. 214, (PT Citra Aditya Bakti, Bandung, 1996).
32. Article 1 Letter a Ministry Decree 34/1980.

Biochar application for the improving soil fertility and paddy production on two systems of planting

Antonio Sipayung*, *Tengku* Sabrina, and *Revandy* Damanik

Faculty of Agriculture, Universitas Sumatera Utara, Indonesia

Abstract. A product of pyrolysis of residual biomass called biochar has several benefits into soil. Bio-charcoal formed by burning organic waste which will produce activated carbon and organic carbon and will be added to the soil. The purpose of this study was to evaluate biochar ability to improve soil fertility and increase the production of paddy crops with two planting systems. Legowo planting system is intermittent cropping between two or more rows of rice plants and one empty row. The research was conducted in Tanjung Garbus's Village, Deli Serdang District with an elevation of ± 50 Meters above sea level. This research used a Factorial Split Plot Design with two factors that was treatment of planting systems as main plot (S), consist of 2 (two) stages: Jajar Legowo 2:1 and Jajar Legowo 4:1 and the sub-plot was biochar application (B) consisting of 3 levels namely, without biochar application, 25.5 ton/ha, and 50.5 ton/ha. The results showed that planting system significantly affected shoot wet weight and shoot dry weight and biochar application had a significant effect on shoot wet weight and no significant effect on shoot dry weight. The interaction showed no significant effect on all parameters.

1 Introduction

Paddy is one of the main food crops and as a staple food for almost all of Indonesia's population and highly paid attention to productivity. However, the problems faced today are:(1) the problem of the increasingly narrow land area,(2) the decreasing rice field quality,(3) the difficulty of opening new land, and (4) the effect of climate change [1].

Paddy is a major food commodity that has a strategic role. Paddy field crop is a food crop that after going through various processes will produce rice. Rice is a staple food that is vital for everyone. That is why efforts to meet the needs of rice continue to be done the through various program, one of them is by intensification. Intensification of paddy with the intake of chemical fertilizers in large quantities and in the long term, as well as the use of organic materials in wetland rice production system has resulted in disturbed soil nutrient balance resulting in decreased quality of land resources [2].

The location of rice planting is in the village ofTanjung Garbus Kampung, Ujung Pertanian sub-village, Deli Serdang district. Based on the result of soil analysis at Socfindo Laboratory (2016), it is found that soil pH was 5.3 and the C-Organic soil content was 1.73% where the ideal soil condition for C-Organic is 3%. This means that soil fertility conditions on the research location was not in good conditions and needs to be done application of biochar to increase the productivity. These were evidenced by the yield of rice crops in two seasons of planting only 5 tons of grain with the use of inorganic fertilizer Urea fertilizer 150 Kg, 150 kg NPK, SP-36 150 kg, ZA 150 kg and yield of 4.8 tons of grain by using fertilizer inorganic amount of 100 Kg urea fertilizer, SP-36 200 Kg, ZA 150 Kg, KCl 50 Kg

The use of biochar is a new thing for the agricultural world particularly for paddy by utilizing biological charcoal formed from burning which will produce activated carbon, containing minerals such as calcium (Ca) or magnesium (Mg) and organic carbon and will be added to the soil. According to the research from the International Rice Research Institute (IRRI), biochar feeding on upland rice production in northern Laos increases topsoil hydraulic conductivity or surface layers', and improves the response of plants to fertilizer with nitrogen content [3].

Biochar can affect changes in the carbon cycle and CO_2 emissions through changes in characteristics of soil and microbial. In fact, biochar itself can be decomposed through microbes to some extent on the ground [4]. Using meta-analysis [5]. found that incorporation of biochar to soil increased crop productivity by an estimated 10%, however, yields from soil layers, land-processing practices plant experiments, and biochar feedstocks. The level of fertility of paddy fields in Indonesia was decreasing about 65% of the 5 million hectares of irrigated wetland have less than 2% organic matter while in normal condition the fertile land usually contains the cultivation of land, irrigation system and the return of organic material. In the paddy field, the most important source of organic material is the residual crop that has been harvested where the decomposition process occurs and organic material reshuffle. The rest of the harvest in the form of rice husk is the best source of organic material for soil fertility after decomposition process. [6].

Legowo row planting system is a cropping system that takes care of the crops', legowo row planting system is intermittent cropping between two or more rows of paddy field and one empty row. The advantage of the legowo row planting system is to make all plants or more plants become peripheral plants. Plants will get more sunlight and better air circulation, more nutrients, and easier maintenance of plants [7].

In principle the system of planting row legowo is to increase the population by adjusting the spacing. This planting system also manipulates the layout of plants, so that the clumps of crops are largely into peripheral plants. Rice plants located on the edge will get more sunlight, resulting in higher grain with better quality. In the legowo 2: 1 way of planting, every two rows of plants interspersed with an empty line with a width of twice the range of the row, but the spacing in the row is narrowed to half the original spacing [8].

Legowo 4: 1 planting system type 1 is a legowo cropping pattern with the whole line getting the insertion plant. This pattern is suitable to apply to less fertile land conditions. With this pattern, the plant population reached 256,000 clumps/ha with a population increase of 60% compared to the tile pattern (25x25) cm [9]. Legowo 4: 1 planting system type 2 is a cropping pattern by simply providing additional plant inserts on both rows of edge crops. The population of plants is 170,667 clumps/ha with the percentage increase of only 6.67% compared to the tile pattern (25x25) cm. This pattern is suitable to be applied to locations with high soil fertility. Even the absorption of nutrient by more plants. However, because the plant is more robust so as to minimize the risk of lumber during growth [9].

2 Research Method

This research was conducted in Tanjung Garbus Village, Sub-district of Pagar Merbau, Deli Serdang Regency. This research was conducted from December 2017 to March 2018. Paddy variety used were Inpari 33. The Seed was germinated, and after 21 days, they were planted in the experimental field. All treatment application used NPK fertilizer 45 kg N, 60 kg P2O5 and 50 kg K2O per hectare.

This research used Split Plot Design with two factors and four replicated (*n=4)*. Namely, the main plot was planting system consists of two treatments (S1): Legowo 2: 1, (S2): Legowo 4: 1 (S2). Subfactor: Biochar application consisting of 3 levels: without biochar application, (B1): 25.5 ton/ha Biochar, (B2): 50.5 ton/ha Biochar. Parameters were shoot wet weight and shoot dry weight.

Legowo 2: 1 row planting system has two lines of cropping with 1 lane removed. Distance between rows of plants 25 cm and distance between plants 12.5 cm and distance between lines is 50 cm. while the system of planting row legowo 4: 1 has a sense of four paths of cultivation with a path that is removed. Distance between rows of plants 25 cm and distance between plants 25 cm and distance between lines is 50 cm.

Stages of rice cultivation 1. Land preparation, 2. Application of biochar to paddy fields 3. Selection of varieties and nurseries, 4. Planting (two systems of planting legowo 2: 1, and jajar legowo 4: 1) 5. Watering, 6. Fertilization, 7. Weed control, 8. Pests and diseases control, 9. Harvest.

All data were reported as means and standard deviation (SD) of the means. Analysis of variance was used to determine the statistical significance of the biochar treatment effects on planting system. Post hoc comparisons were made using least significant difference (LSD) test, with the criterion for statistical significance being set at $P<0.05$. Pearson correlation was applied to analyze the correlations between the biochar and planting systems at the 5% and 1% levels of significance.

3 Result and Discussion

3.1 Shoot Wet Weight

Tabel 1 The comparison between main plot (two plant systems) to subplot (biochar application B0/Control).

No	Plant Systems	Average	S1 Legowo 2:1	S2 Legowo 4:1
			4.95	7.15
1	S1 Legowo 2:1	4.95a	0	2.2*
2	S2 Legowo 4:1	7.15b	2.2*	0
LSD = 1.50				

Explanation: The numbers followed by the same letter on the same row or column is not different significant according to Least Significant Difference Test at 5% level

Tabel 2 The comparison between main plot (two plant systems) to subplot (biochar application B1/22.5 ton/ha).

No	Plant Systems	Average	S1 Legowo 2:1	S2 Legowo 4:1
			4.95	7.15
1	S1 Legowo 2:1	5.5a	0	0.85^{ns}
2	S2 Legowo 4:1	6.4a	0.85^{ns}	0
LSD = 1.50				

Explanation: The numbers followed by the same letter on the same row or column is not different significant according to Least Significant Difference Test at 5% level

Tabel 3 The comparison between main plot (two plant systems) to subplot (biochar application B2/50.5 ton/ha).

No	Plant Systems	Average	S1 Legowo 2:1	S2 Legowo 4:1
			4.95	7.15
1	S1 Legowo 2:1	5.4a	0	0.85^{ns}
2	S2 Legowo 4:1	6.25b	0.85^{ns}	0
LSD = 1.50				

Explanation: The numbers followed by the same letter on the same row or column is not different significant according to Least Significant Difference Test at 5% level

The result of F-test showed that the treatment of planting system and biochar application had a significant effect on wet harvest weight, but it did not show any real effect on the interaction between cropping system treatment and biochar application for research parameter of shoot wet weight.

The results showed significant differences in planting system on shoot wet weight for biochar application B0 on the table 1 (2.2) and showed no significant effect for biochar application on planting system in B1 (table 2) (0.85) and B2 (table 3) (0.85)

3.2 Shoot Dry Weight

Tabel 4 The comparison between main plot (two plant systems) to subplot (biochar application)

No	Plant Systems	Average	S1 Legowo 2:1	S2 Legowo 4:1
			4.48	6.45
1	S1 Legowo 2:1	4.48a	0	1.97*
2	S2 Legowo 4:1	6.45b	1.97*	0
LSD = 1.13				

Explanation: The numbers followed by the same letter on the same row or column is not different significant according to Least Significant Difference Test at 5% level

Tabel 5 The comparison between main plot (two plant systems) to subplot (biochar application B1/25.5 ton/ha).

No	Plant Systems	Average	S1 Legowo 2:1	S2 Legowo 4:1
			4.48	6.45
1	S1 Legowo 2:1	4.65a	0	0.6^{ns}
2	S2 Legowo 4:1	5.25a	0.6^{ns}	0
LSD = 1.13				

Explanation: The numbers followed by the same letter on the same row or column is not different significant according to Least Significant Difference Test at 5% level

Tabel 6 The comparison between main plot (two plant systems) to subplot (biochar application

No.	Plant Systems	Average	S1 Legowo 2:1	S2 Legowo 4:1
			4.9	5.65
1	S1 Legowo 2:1	4.9a	0	0.75^{ns}
2	S2 Legowo 4:1	5.65a	0.75^{ns}	0
LSD = 1.13				

Explanation: The numbers followed by the same letter on the same row or column is not different significant according to Least Significant Difference Test at 5% level.

The result of F-test showed that planting system had the significant effect on shoot dry weight research parameter while biocar application had no significant effect on shoot dry weight, and interaction between planting system treatment and biochar application did not show the real effect for shoot dry weight parameter.

The results showed that planting system treatment to shoot dry weight significantly affected biochar application B0 on the table 4 (1.97) and showed no significant effect for biochar application on planting system in B1 (table 5) (0.6) and (table 6) B2 (0.75).

4 Conclusion

The results showed that the treatment of biochar increased the shoot wet weight but had no significant effect on shoot dry weight. The treatment of planting systems had the significant effect on shoot wet weight and shoot dry weight.

References

1. Mawardiana, Sufardi, and E. Husen, The Effect of biochar residue and NPK fertilizer to nitrogen dynamics, soil chemical character and paddy yield (Oryza sativa L.) on third planting season. *Jurnal Manajemen Sumberdaya Lahan* **2** (3). pp. 255-260, (2013)
2. J. Pramono, *Study on the use of organic material on paddy field*. J. A **6** (1). pp 11-14, (2004)
3. Rostaliana, P., Prawito, P., dan Turmudi, E, *Utilization of biochar for land quality improvement with the indicators of maize and gogo paddy on land and fire fuel system*. J. N. E. M. R **1**, No 3, (2012)
4. Czimczik CI, Masiello CA, *Controls on black carbon storage in soils*. G.B.C **21** : GB3005, (2007)
5. Jeffery S, Verheijen FGA, van der Velde M, Bastos AC, *A quantitative review of the effects of biochar application to soils on crop productivity using meta-analysis*. J. A. E. E **144**. pp175–187, (2011)
6. Oktalaseva, W., Hermansah., dan Nurwanita, E. P, *The characteristics of soil fertility and nutrient potential from organic materials waste paddy harvest at some locations in west sumatera*. Agriculture Faculty, Andalas University, Padang, (2013)
7. Mujisihono, R and T. Santosa, *Direct seed crop technology and Legowo Planting System*. Paper was presented on Engineering Production System of Rice Commodities and Palawija Seminar. Agriculture Institution, Province D.I Yogyakarta, (2001)
8. Ikhwani, Pratiwi GR, Paturrohman E, and Makarim AK. 2013, *Increasing Rice Productivity through the Implementation of Jajar Legowo Planting System*. J.P.F.S **8,** No 2, (2013)
9. Research Centre for Agricultural Technology Jambi, *Legowo planting system for paddy*, (2013)

Environmental policy commitment of North Sumatera provincial government in climate change mitigation efforts (case study of mangrove protection)

Antonio Marro Sipayung[1*] dan *Henri* Sitorus[2]

[1]Researcher, WALHI (Indonesia Forum for Environment) Province North Sumatera, Indonesia
[2]Researcher at Postgraduate of Sociology Program, Universitas Sumatera Utara, Medan 20155, Indonesia

Abstract. The purpose of the study is to evaluate the policy commitment of North Sumatera provincial government in mitigating climate change (case study of mangrove protection). The Regional Action Plan for Greenhouse Gas (RAD-GRK) has been adopted in 2012 by the Province of North Sumatra. The action plan identifies action needed to rehabilitate mangroveswithin or outside of forest conservation areas. The action plan aimed for reforestation of 50,000 ha mangroves areas. However, data showed that by 2017 the North Sumatra government has only been able to rehabilitate 4,765 Ha of mangrove forest. On the other hand, mangroves conversion to non forest land continues including conversion to oil palm plantations in Langkat District. Based on the qualitative data of interview with policy makers at the Provincial of North Sumatera and document analysis particularly from the government agencies, the research discovers that the policy commitment of the local government for mangroves protection is weak.

1 Introduction

The increasing concentration of greenhouse gases (CO_2, CH_4, CFC, HFC, N_2O), particularly the increase of CO_2 concentration, in the atmosphere cause global warming which trigger global climate change. This phenomenon has severely affected and hamper the sustainability of human life and other living organism, including the shifting of seasons and changes in the pattern/distribution of rain that trigger floods and landslides during the rainy season and dry season drought, seawater that potentially drowns small islands and floods, and storms/waves that often devastate life-sustaining infrastructure in coastal areas [1].

The climate change has been in the policy agenda from local to global scale. It has been realized that many aspects of life are disrupted including by the climate change. One of the

* *Corresponding author* Phone: +62812 6346 4307, Email: sipayungantonio@gmail.com

simplest examples is an increase in the surface temperature of the earth which causes many people to start feeling hot or hot. This is one of the simplest examples of climate change.

Global warming is the occurrence of increasing average temperatures of the atmosphere, the ocean and the earth's land. Historically, the planet Earth has warmed (and also cooled) over and over for 4.65 billion years. At this time, the earth faces rapid warming, which scientists consider to be attributable to human activity. This activity is the burning of fossil fuels, such as coal, petroleum, and natural gas, which release carbon dioxide and other gases known as greenhouse gases into the atmosphere. As the atmosphere gets richer in greenhouse gases, it becomes an insulator that holds more heat from the reflected sun from the earth [2].

The excessive release of emissions from the activity of organism metabolism or the result of decomposition into the free air is a form of contamination for the atmosphere which is basically negative to environmental conditions that will ultimately affect the weather or climate on a large scale. From the accumulation of small changes will lead to major changes. Carbon emissions in nature can have diverse forms in the form of CO_2 (carbon dioxide) from the respiration of the organism and CO (carbon monoxide) from fossil or petroleum burning. Other activities such as burning, smoking, discharge of CH_4 (methane) natural gas, greenhouse gases such as HFC (hydrofluorocarbon) or PFC (perfluorocarbon) and others also contribute to increased carbon gas in nature [3].

Indonesia as a country on the path of the world's crossroads and on the equator has felt the vast impact associated with extreme climate change. The rising sea surface causes many small sinking islands and the occurrence of abrasion/floods that impact on residential areas.

The Government of Province North Sumatra is one of 33 provinces in Indonesia. Policies in climate change mitigation efforts are contained in the Green House Gas Regional Action Plan (RAD-GRK) with Pergub No 36 of 2012 for the period of 2010-2020. RAD-GRK is derived from Perpres No. 61 of 2011 related to the National Action Plan for Greenhouse Gases (RAN-GRK). This is evidence of the Government of Indonesia's serious efforts to address climate change issues.

The Government of Province North Sumatra is one of 33 provinces in Indonesia. Policies in climate change mitigation efforts are contained in the Green House Gas-Regional Action Plan (RAD-GRK) with Pergub (Governor Rules No 36 year 2012 for the period of year 2010-2020. RAD-GRK is derived from Perpres (Presidencial Rules) No. 61 year 2011 related to the National Action Plan for Greenhouse Gases (RAN-GRK). This is evidence of the Government of Indonesia's serious efforts to address climate change issues.

The Government of Indonesia is committed to reducing emissions by 29% under business as usual by 2030, or 41% with international assistance. "To reach a Paris agreement, all parties must contribute more, in mitigation and adaptation actions, especially developed countries, that is, by mobilizing funding of USD 100 billion USD by 2020" President Jokowi's speech at the UNFCC Conference Paris 2015.

With an area of 473,607 km^2,Sumatra is known as having extensive biodiversity. Because of its rich resources, UNESCO declared the Sumatran rainforest a world heritage site in 2004. In this island, between 9,000 and 10,000 species of plants have been identified, including 17 species of mangroves, which contribute to the rich diversity of fauna [4]. According to FAO (2007), mangroves are coastal forests found in sheltered estuaries and along riverbanks and lagoons in the tropics and subtropics, which have very important ecological and economic functions.[5] note that mangroves are "shrubs or trees that live adjacent to the intertidal zone and have adapted to a regime of widely varying salinities, and to periodic and sometimes prolonged inundation".[6] Elaborates this function by detailing the benefits of mangroves in their branches, root trunks, and bark, which all have particular functions.

Mangrove forests are coastal forests that are always or regularly inundated by sea water and are affected by tides [7]. Mangrove is a general term used to describe tropical beach community varieties dominated by distinctive tree species and shrubs, which have the ability to grow in salt waters [8]. The mangrove ecosystem is dominated by plants of Rhizophora, Avicennia, Bruguiera, and Sonneratia In addition, the mangrove ecosystem also found plant types Ceriops, Xylocarpus, Acrostichum, Lumnitzera, Aegiceras, Scyphyphora, and Nypa [9].

The role of mangroves is critical to sustaining production in coastal fisheries, because they act as important nursery areas for fisheries species (Burke et al., 2001). In mitigating disaster, mangroves are able to reduce storm surge, hence potentially reducing the number of casualties. Another ecological contribution of mangroves is the ability to prevent erosion and abrasion. In this role, mangroves also stabilize soils and protect against erosion, as well as reducing the strength of winds from the sea [10].

As an ecosystem in the transition region between the sea and the land, mangrove will be the first type of ecosystem affected by the impact that will occur due to global climate change. As expressed by some experts, among them Field (1995) [11], most mangrove in various parts of the world will experience the condition of air temperature increase, hydrological regime change, sea level rise, and the increase and frequency and frequency of tropical storm disaster.

Mangrove is a plant community or an individual plant species that forms the community in tidal areas. Mangrove forests are a type of forest that is naturally affected by tidal seawater, flooded during high tides and free from puddles during low tides. The mangrove ecosystem is a system consisting of biotic and abiotic environments interacting within a mangrove habitat. Mangrove has a very high physiological adaptability, resistant to environment with high water temperature, wide area wide salinity fluctuation and anaerobic soil. One important factor in such physiological adaptation is the system of broadcasting in its roots [12].

2 Research problem

The research on "Policy Commitment of North Sumatera Provincial Government in Climate Change Mitigation Measures" is intended to know about: (1) How is the commitment of climate change mitigation policy in Provice North Sumatera? (2) How is the protection policy of mangrove forest in Province North Sumatera?

3 Literature Review

Climate is high closely related to weather changes [13], climate change and global warming can reduce agricultural production between 5-20 percent. Climate change is a condition marked by changing world climate patterns that result in an uncertain weather phenomenon. Climate change occurs due to changes in climate variables, particularly air temperature and rainfall that occur continuously over a long period of time between 50 to 100 years [14]. Climate change is also affected by unstable weather conditions as examples of unpredictable rainfall, frequent storms, extreme temperatures, drastic wind direction, and so on [15].

Mangrove has four specific functions that can affect the quality of coastal waters, namely: 1. its ability to supply nutrients for the surrounding waters. In a study conducted by ecoton carrying more than 7 tons/ha/year of litter (dried leaves) produced by mangrove ecosystems on the coast of Surabaya. This result is equivalent to the productivity of mangrove ecosystems generally scattered from the tropics to sub-tropical litter. Mangrove plays an important role in the process this is because the litter contains 40% of water-soluble compounds converted into bacterial biomass less than 8 hours after falling into mangrove waters. This makes the mangrove area frequented by a variety of animals to get the nutrients. 90% of the number of fish caught within 10 km of the coast contains mangrovefragments in the gut. 2. Mangroves as aquatic bird habitat. As a fertile, nutrient-rich ecosystem, the area is bustling with birds, even in the western (October-December) seasons of more than 5,000-20,000 bird populations that make up the region. 3. The existence of mangrove plays an important role in the life cycle of some economically valuable biota such as crabs, shrimps, milkfish and other marine fish, because at the time of laying and breeding his children most of the biota cycles in coastal areas of mangrove, just after they will mature back to the high seas. This can be demonstrated by the high population of zooplankton (an important chain in the food web) its existence can connect between producer I and consumer I) most of these organisms will grow into fish, shrimp, crab and shellfish. 4. In addition, several types of mangrove trees such as Raw Tree (Rhizophora mucronata) and Fire-fire Tree (Avicennia marina) have the ability to accumulate (absorb and store in leaf organs, will and stems) of heavy metals contaminants, resulting in the presence of mangroves in brackish waters it plays a role to filter and reduce the level of heavy metal pollution in marine waters [16].

Therefore, an increase in atmospheric CO_2 concentration is expected to stimulate crop growth under conditions of limited availability of water, carbon or nitrogen, but plant growth will not increase if the salinity level is too high for smooth water-taking by plants or when the availability of some nutrients is limited in availability

In the mangrove ecosystem there are several types of biological resources that can be utilized for human welfare. The economic benefits that can be obtained from mangrove forests are wood for building materials, firewood, charcoal materials, and pulp materials. Other products are honey, can also be processed into organic fertilizers, foodstuffs, medicines, beverages, home appliances, textile raw materials, and leather [17].

The provincial government of North Sumatra has a strategy to reduce greenhouse gas emissions for 2010-2020 period in 6 (six) sectors, namely agriculture sector (with 15 activities of mitigation action plan), forestry and peatland sector (with 6 core activities and 3 supporters), the energy sector (6 core activities and 8 supporting activities), the transportation sector (17 core activities and 5 supporters), the industrial sector (8 activities and 5 supporters), and the waste treatment sector (6 core activities and 5 supporters). Particularly in the forestry and peatland sectors, the North Sumatra provincial government plans to rehabilitate 50,000 ha of mangrove forest within the forest area and beyond the mangrove forest area [18].

4 Research Method

This research used qualitative descriptive method. The subject of research is related to the Policy Commitment of North Sumatera Provincial Government in mitigating climate change. There are two types of data used in this research, namely primary and secondary data, with the following details: (1) Primary Data: policy makers at the level of the Province of North Sumatra namely the Office of Environment, Forestry Service and Regional Development Planning Agency, (2) Secondary Data: reports, archives, documents, or notes, especially from government agencies that have to do with the focus of research studies. Technique of data retrieval by through observation, depth interview, and study of secondary material.

5 Result and Discussion

The commitment of North Sumatra government to climate change mitigation efforts is far from good. The policy found in the field has not been in accordance with what is planned in the local action plan particularly in the forestry and land (mangrove forest). According to Ahmad Syafei Hutahuruk as Head of Forest and Land Rehabilitation Division of North Sumatra Province Forestry Office said that, the commitment to realize the regional gas-gas action plan is still not fully implemented. Prior to the change of regulation related to authority at the provincial level, there was a Mangrove Agency, specializing in mangrove forest. However, until 2015 there is still no significant activity done. Mangrove according to him has a function for the protection of the land from the threat of abrasion, play a role in reducing the greenhouse effect. Mangrove generates a lot of oxygen due to the density of mangrove forests is greater so that the absorption of carbon becomes larger. Reduced areas of mangrove forest also affect climate change. Ahmad Syafei added the forestry service needs funding support for the implementation of greenhouse gas (RAD-GRK) action plan.

In the regional action plan in the field of forestry and peatlands it is mentioned that the North Sumatra government is committed to rehabilitate mangrove forests covering 50,000 Ha in 2010-2020 period. However, from data in the forestry service per year 2015, mangrove rehabilitation activities that have been done covering 4,765 Ha (table 1). If estimated per year, should the North Sumatra provincial government to rehabilitate mangrove forest per year of 5,000 ha. The condition of mangrove forests in northern Sumatra is critical and needs to be rehabilitated. Data on the condition of mangrove forest in north sumatera per year 2015 indicates that there is an area of 59,584.90 Ha of mangrove forest in extremely critical condition and 98,301.21 Ha in critical condition (table 2).

Tabel 1 Mangrove Rehabilitation year 2011-2015(Ha)

No	District/City	Planting per year/Ha					Total
		2011	2012	2013	2014	2015	
1	Deli Serdang	58	90	610	-	20	778
2	Langkat	500	1,754	1,436	-	-	3,690
3	Karo	-	-	-	-	-	-
4	Dairi	-	-	-	-	-	-
5	Simalungun	-	-	-	-	-	-
6	Asahan	11	32	14			57
7	South Tapanuli	-	-	-	-	-	-
8	North Tapanuli	-	-	-	-	-	-
9	Central Tapanuli	-	157	-	-	-	-
10	LabuhanBatu	-	-	-	-	-	-
11	Mandailing Natal	-	-	-	-	-	-
12	TobaSamosir	-	-	-	-	-	-
13	HumbangHasundutan	-	-	-	-	-	-
14	Pakpak Bharat	-	-	-	-	-	-
15	Samosir	-	-	-	-	-	-
16	SerdanBedagei	-	-	-	-	-	
17	Batubara	42	-	-	-	-	-
18	North Padang Lawas	-	-	-	-	-	-

No	District/City	Planting per year/Ha					Total
		2011	2012	2013	2014	2015	
19	Padang Lawas	-	-	-	-	-	-
20	South LabuhanBatu	-	-	-	-	-	-
21	North LabuhanBatu	41	-	-	-	-	-
22	Nias	-	-	-	-	-	-
23	North Nias	-	-	-	-	-	-
24	West Nias	-	-	-	-	-	-
25	South Nias	-	-	-	-	-	-
26	Sibolga	-	-	-	-	-	-
27	TanjungBalai	-	-	-	-	-	-
28	P. Siantar	-	-	-	-	-	-
29	Tebing Tinggi						-
30	Medan	-	-	-	-	-	-
31	Binjai	-	-	-	-	-	-
32	Padang Sidempuan	-	-	-	-	-	-
33	GunungSitoli	-	-	-	-	-	-
	Total	652	2,033	2,060		20	4,765

Source: Forestry Institution, (2018)

Tabel 2 Mangrove Forest condition at Province North Sumatera per year 2015(Ha)

No	District/City	Width and Mangrove Forest Condition			
		Extremely Critical	Critcal	Good	Total (Ha)
1	Deli Serdang	6,300.91	8,170.84	3,326.83	17,798.58
2	Langkat	13,526.90	23,564.93	13,559.11	50,650.94
3	Karo	-	-	-	-
4	Dairi	-	-	-	-
5	Simalungun	-	-	-	-
6	Asahan	940.17	7,506.74	2,624.64	11,071.55
7	South Tapanuli	186.97	479.39	29.64	696
8	North Tapanuli	-	-	-	-
9	Central Tapanuli	3,889.61	2,664.91	376.71	6,932.23
10	LabuhanBatu	7,181.19	8,383.39	4,099.15	19,663.73
11	Mandailing Natal	620.84	2,261.94	455.49	3,338.27
12	TobaSamosir	-	-	-	-
13	HumbangHasundutan	-	-	-	-
14	Pakpak Bharat	-	-	-	-
15	Samosir	-	-	-	-
16	Sergei	7,962.99	4.524,05	508,22	12.995,26
17	Baubara	6,553.64	12.561,10	517,29	19.632
18	Paluta	-	-	-	-

No	District/City	Width and Mangrove Forest Condition			
		Extremely Critical	Critcal	Good	Total (Ha)
19	Padang Lawas	-	-	-	-
20	South LabuhanBatu	-	-	-	-
21	North LabuhanBatu	11,834.45	10,129.05	2,817.40	24,780.90
22	Nias	-	-	-	-
23	North Nias	-	-	-	-
24	West Nias	-	-	-	-
25	South Nias	512.54	16,383.11	372.76	17,268.41
26	Sibolga	-	-	-	-
27	TanjungBalai	74.59	2.22	-	76.91
28	P. Siantar	-	-	-	-
29	Tebing Tinggi	-	-	-	-
30	Medan	-	-	-	-
31	Binjai	-	-	-	-
32	Padang Sidempuan	-	-	-	-
33	GunungSitoli	-	-	-	-
	Total	59,584.90	98,301.21	29,435.96	187,322.07

Source: Foresrty Institution, (2018)

Faisal Abrani Siregar as head of the data processing and information control, unit of climate control said that the focus on climate change control units is on the waste sector in climate change mitigation and adaptation efforts. Mitigation activities in the form of garbage bank establishment and encourage the utilization of methane gas (waste of oil palm). Faisal added related to the condition of mangrove forests, mangrove land should be controlled. Mangrove has a function as a carbon catcher and abrasion holder then mangrove forests should be preserved.

Andriadi as staff of the economic planning, maritime and natural resources of the Regional Planning and Development Agency (Bappeda) said that the total reduction of total GHG emissions has been reported as 12.50 M ton CO_2eq from 536 mitigation actions that have been implemented from various institutions. Total approved mitigation actions are 280 actions from 536 actions. For the forest and land sector, there are as many as 198 mitigation action and emission reduction activities recorded as 11.91 CO_2 eq. Priority of land on critical land for development there is an area of 279.864 Ha.

Findings in the field showed that the condition of mangrove forest has decreased due to the land conversion into oil palm plantation and fish farming Cases in district of Langkat (bubun villages and kwala serapuh villages) indicate that mangrove forests are massively converted into oil palm plantations of approximately 260 ha and illegally done. Mangrove forests are dammed and then dried making it easier to be cleared into oil palm plantations. Similarly, mangrove land is converted to shrimp ponds.

Damage to mangrove forests is also caused by the mining of marine sand. Marine sand mining activities have been ongoing since 2008. According to data from the Provincial Investment and Licensing Office of North Sumatra Province as of December 2016, in the district of Deli Serdang (subdistrict pantai labu) there are 8 marine mining permits granted permission to mine sea sand of 38 Million m3 with the need for airport expansion and port reclamation. Approximately there were 2,000 Ha mangrove forest had been destroyed by the marine sand mining activities. The condition of the mangrove forest has been destroyed and the distance of the settlement from the shoreline is only 50 metres. Communities living on the coast are threatened by drowning due to abrasion and flooding. This was the effect of the destruction of mangrove forests in their area.

6 Conclusion

The Government of Province North Sumatra does not yet have a good commitment in mitigating climate change (the case of mangrove forest). The large number of cases of mangrove land conversion into illegal oil palm plantations and the presence of sea sand mining activities can accelerate the rate of damage to existing mangrove forests. Preventive action and serious concern in rehabilitation of mangrove forest area need to be taken.

References

1. C. Kusmana, Mangrove response to global climate change: Biological and ecological aspects of mangroves. Department of Silviculture. Faculty of Forestry IPB. Bogor, (2010)
2. V. Darsono, *Introduction to Environmental Science. Revised edition*, (Yogyakarta, 1993)
3. H. Purnobasuki, *The Role of Mangrove In Climate Change Mitigation*. (Department of Biology, Airlangga University Surabaya. Bulletin PSL Airlangga University 18: 9-10, 2006)
4. H. Purnobasuki, *Peranan Mangrove dalam upaya mitigasi perubahan iklim. Departemen Biologi.* (Buletin PSL. Universitas Airlangga Vol 18: 9-10, 2006)
5. T. Whitten, S.J. Damanik, J. Anwar, N. Hisyam, *The Ecology of Sumatra.The Ecology of Indonesia Series,* (Oxford University Press, 2000)
6. L. Burke, Y. Kura, K. Kassem, C. Revenga, M. Spalding, and D. McAllister, *Coastal ecosystems*, (World Resources Institute, Washington DC ,2001)
7. C.L. Angell, *Review of Critical Habitats: Mangroves and Coral Reefs. Bay of Bengal Large Marine Ecosystem (BOBLME) Theme Report* (FAO BOBLME Programme. Regional Planning and Development, 2004)
8. W.A.M. Mae, A General Account of the Fauna and Flora of Mangrove Swamp and Forest in the Indo-West Pacific Region, *Adv Mar. Bio*, **6**: 73 – 270. (1968)
9. J.W. Nybakken, *Marine Biology, An Ecological Approach*, (Gramedia, Jakarta, 1988)
10. Supriharyono, *Conservation of Biodiversity Ecosystems in Coastal Areas and Tropical Seas. Learning Library,* (East Celeban, Yogyakarta, 2009)
11. FAO, State of the world's forests. Rome: Food and Agricultural Organisation of the United Nations, (2005)
12. C.D. Field, Impact of expected climate change on mangroves, *Hydrobiologia* **295**, 75-81, (1995)
13. S. Subekti, *Global Warming and mangrove Management Efforts*, (Faculty of Engineering, University of Pandanaran, Semarang, 2018)
14. Suberjo, *Adaptation of Agriculture in Global Warming*. (2009)
15. The Indonesian Ministry of Environment. Global Climate Changes, http://climatechange.menlh.go.id, (2004)
16. Ratnaningayu, *From timor to krui: how farmers and fishermen face the impacts of climate change, Climate action*, (Pelangi Indonesia, Jakarta, 2009)

17. D.G. Bengen, *Coastal and Marine Ecosystems and Resources and Integrated and Sustainable Management.* (Center for Coastal and Marine Resource Studies IPB, Bandung, 2003).
18. Bappeda/ Regional Planning and Development Agency, Policy and Effort to Achieve Low Carbon Society in North Sumatera Province, (2016).

Rights to restitution for child victim of criminal act under government regulation number 43 year 2017

Azmiati Zuliah* and *Madiasa* Ablisar

Doctoral Program in Law, Universitas Sumatera Utara, Medan, Sumatera Utara, Indonesia

Abstract. The criminal act against children is a gross violation of human rights. The criminal act committed by the offenderis intolerable because it can affect their survival as well as growth and development. Therefore, the victim deserves legal protection. Restitution is one of forms of protection and fulfillment of the rights of the child to compensate any damages. So far, child victim of criminal act suffers not only material loss (countable) but also immaterial loss (uncountable) such as shame, loss of self-esteem, depression and/or traumatic anxiety.It is the fact that so far the rights of child victims to restitution are very often violated. Therefore, the child victim and his or her family feel that they do not given justice. It is hoped Government Regulation Number 43 Year 2017 will be able to give legal certainty to the victim to ensure that he or she can enjoy his or her rights to restitution for any loss he or she suffers as stipulated in Law Number 35 Year 2014 on the Revision of Law Number 23 Year 2002 on Child Protection. This research uses theory of justice as grand theory, law enforcement as middle range theory and theory of victimology as applied theory.

1 Introduction

Law Number 35 Year 2014 on the Revision of Law Number 23 Year 2002 on Child Protection gives special protection to child victim of criminal act. The special protection is given to children in conflict with the law, especially child victim of economic and/or sexual exploitation, pornography, kidnapping, sale, and/or trafficking, physical and/or psychological abuse, and sexual crimes, for harm or loss he or she suffers in the form of a compensation from the offender or his or her family, especially if the offender is a child as a result of the committed criminal act. [1]

Every child victim is entitled to file to court his or her right to claim for restitution imposed on the offender.[2] Article 71 D of Law Number 35 Year 2014 on the Revision of Law Number 23 Year 2002 on Child Protection stipulates that (1) Every child victim as referred to in Article 59 paragraph (2) point b, point d, point f, point h, point i, and point j is entitled to file to court his or her right to claim for restitution imposed onthe offender; (2)

* Corresponding author: emi_zuliah@yahoo.com

further provisions on the implementation of restitution as referred to in paragraph (1) shall be regulated by a government regulation.

In the elucidations of article above, "restitution" shall mean the payment of compensation imposed on the offender based on court ruling having permanent legal force for the material and/or immaterial damages suffered by victims or their beneficiaries. [3]. Especially for children in conflict with the law, the one who is entitled to receive restitution is the child victim.

Unfortunately, up to now there are still various problems in the payment of restitution at police, judiciary and court levels. From victimology point of view, restitution aims to minimize victim's sufferings by claiming the offender be held responsible for physical or morallosses or death suffered by the victim.

This principle affirms that victim recovery must covered various aspects raised by the criminal act. The victim and his or her family must receive just and proper compensation from the guilt or a third party that is held responsible for the criminal act. The compensation will include the return ofproperty or payment of any damage or loss suffered by the victim, payment of costs arising as a result of death, provision of services and rights to recovery. [4]

2 Formulation of the Problem

The problem of this research are: how is restitution regulated in Government Regulation Number 43 Year 2017 on payment of restitution to child victim of criminal act and also what are the roles of law enforcers in the implementation of restitution and weaknesses of the implementation of restitution.

3 Research methods

This research is a juridical normative research which is qualitative in nature. The juridical normative research aims to analyze concepts of laws and regulations which are closely related to the topic of the research.

The primary data in the research was collected through field research. Meanwhile, the secondary data was collected through documentary research. The documentary research aims to find various concepts, theories, principles, doctrines and various documents which are related to the topic of the research.

4 Discussions

4.1 Regulations on Restitution in Government Regulation Number 43 Year 2017

On 17 October 2017, the Government of Indonesia made a regulation on restitution for child victim of criminal act through Government Regulation Number 43 Year 2017. The making of this Government Regulation is a mandate of Article 71 D paragraph (2) of Law Number 35 Year 2014 on the Revision of Law Number 23 Year 2002 on Child Protection which stipulates that every child victim as referred to in Article 59 paragraph (2) point b, point d, point f, point h, point i, and point j is entitled to file to court his or her right to claim for restitution imposed on the offender. [5]

The victim and his or her family must receive just and proper compensation from the guilt or a third party that is held responsible for the criminal act. The compensation will include the return of property or payment of any damage or loss suffered by the victim, payment of costs arising as a result of death, provision of services and rights to recovery.[6]

In principle, rule of law which regulates restitution in Government Regulation Number 43 Year 2017 defines restitution as the payment of compensation imposed on the offender based on court ruling having permanent legal force for the material and/or immaterial damages suffered by victims or their beneficiaries.

Under Government Regulation Number 43 Year 2017, claim for restitution can be submitted by the victim's parents or guardian; or victim's beneficiary; or someone given an authority by victim's parents, guardian, or beneficiary. Claim for restitution can also be submitted by an institution in case the victim is the offender of the criminal act. Claim for restitution can be submitted to investigator during investigation phase or to public prosecutor during prosecution phase. It also can be submitted through the Witness and Victim Protection Agency (*Lembaga Perlindungan Saksi dan Korban* – LPSK) in accordance with provisions contained in the applicable laws and regulations.

Government Regulation Number 43 Year 2017 also regulates the authority of investigator and public prosecutor to give information on rights to restitution to the victim and regulates the technical matters of the implementation of restitution by judge. In addition, it gives authority to the Witness and Victim Protection Agency to assess loss suffered by the victim. [7]

4.2 The Roles of Law Enforcers in the Implementation of Restitution

Actually, there are a number of parties that play important roles in the implementation of restitution to ensure that the rights of victim of criminal act to restitution can be fulfilled. Submission of claim for restitution for victim of criminal act and filing of the criminal case must be done by law enforcers. So, they play vital roles in restitution. [8]

Below are law enforcers that play vital roles in restitution.

4.2.1 The Roles of Investigator

Claim for restitution can be submitted when the victim reports his or her case to local police and this delt with by investigator at the same time with the handling of his or her criminal case. So, the roles of the investigator start when the victim report the criminal case. The investigator must immediately inform the victim on how he or she can receive compensation from the offender. For example, the victim must collect evidences that can be used as a basis to receive restitution such as proof of payments for medical service, etc. These evidences must be enclosed to or included in his or her case reporting documents. [9]

4.2.2 The Roles of Public Prosecutor

Public prosecutor can play a role by informing the victim about his or her right to claim for restitution. Then, the public prosecutor inform about losses suffered by the victim and his or her prosecution. [10]

It is particularly important for the public prosecutor to inform the victim about his or her rights because the victim, who is usually a community member, is not familiar with legal proceeding and does not know his or her rights as a victim. So, sharing this information to the victim will be very helpful for him or her. In addition to having done his or her legal responsibility, the public prosecutor also has helped the victim, as a justice seeker, enjoy his or her rights to restitution.

In his or her prosecution, the public prosecutor has to submit a request for restitution based on evidence brought to the court and inform the amount of the requested restitution to the Witness and Victim Protection Agency after the request or the complete document is submitted.

4.2.3 The Roles of Judge

Judge plays a vital role in considerating the amount of restitution for victim's material and immaterial losses. It is his or her obligation to include it in the dictum/court decision, make just decision and impose severe punishment on the offender in accordance with the severence of his or her criminal act as well as give compensation for victim's material and immaterial damages and losses.

Therefore, it is hoped judge will make court decision by considering not only legal justice but also moral justice to get precise justice because the defendant/the offender also has a right to justice. It must be noted that communities have to get justice from judge's decision. This may take the form of sense of safety, peace of mind and relief because the judge's decision is acceptable and just.

Judge's decision is the end of a case handled by judge. Therefore, judge has to consider various aspects in decision making such as the importance of carefulness, avoidance to possible carelessness about formal and material aspects of the case and technical capability in decision making.

It is hoped that judge also will get moral satisfiction when his or her decision is used as a reference for similar or same cases by law experts and legal practitioners and when his or her decision is upheld and not dismissed by a higher court. [11]

4.3 Challenges in the Implementation of Government Regulation Number 43 Year 2017 on Restitution

There are still various legal issues after the issuance of Government Regulation Number 43 Year 2017 as mandated by Law Number 35 Year 2014 on the Revision of Law Number 23 Year 2002 on Child Protection.

This is because the Government Regulation does not explicitly explain sanction for offender who is unable to pay the restitution and how the prosecutor shall seize the assets of the convicted offender.

This provision should be amended to be adjusted to losses suffered by the victim and supervision of the offender's assets. This aims to prevent the convicted offender from not paying the restitution. Unlike the Government Regulation, Law on the Eradication of the Criminal Act of Trafficking in Persons contains provisions on asset seizure and imposement of a subsidiary criminal saction if the offender is unable to pay the restitution although, in fact, a victim of the criminal act of trafficking in person rarely suffers from financial loss.

Other obstacle in the implementation of Government Regulation Number 43 Year 2017 is that victim and his or her family do not know how to submit claim for restitution. This process actually should be facilitated by police, prosecutor and Witness and Victim Protection Agency.

Procedure to claim for restitution is complicated and victim is usually asked to trace or search for the offender's assets, both those which are still in the name of the offender and those which have been converted to other parties as an effort to make their ownership unclear or avoid them from possible legal measure by law enforcers.

It is the Witness and Victim Protection Agency that can determine whether or not a criminal act case deserves restitution. The problem is that the agency is located in Jakarta and it cannot reach all cases throughout Indonesia.

Other problem is the mechanism and procedure for submitting claim for and implementing restitution because the Witness and Victim Protection Agency is not a part of criminal justice system and it is not a law enforcement agency which has executorial authority. Additionally, it is difficult for the Witness and Victim Protection Agency to prove the material lose suffered by the victim because he or she does not have any proof. [12]

However, it is a fact that victim has spent his or her money for medical service, lost property or spent some money during the legal proceeding. Unfortunately, he or she does not have any written proof of these costs.

There are also some obstacles in the implementation of restitution by during law enforcement. At investigation phase, for example, police investigator very often does not include restitution in the police investigation report (*Berita Acara Pemeriksaan* – BAP) because not all police officers understand restitution well. It is also quite difficult to collect relevant evidence to enable the victim to submit a claim for restitution for any material or immaterial loss he or she has suffered. [13]

At prosecution level, prosecutor very often does not include restitution in his or her prosecution because there is no mechanism or technical procedure for submitting claim for restitution when the prosecutor submits his or her prosecution. Additionally, it is difficult to assess immaterial loss suffered by the victim. Laws in Indonesia do not clearly state or determine the amount of restitution which can be claimed to the victim through the court. In addition, it is difficult to assess or calculate immaterial loss suffered by the victim. [14]

At conviction level, it is difficult for the judge to decide whether or not the offender has to pay restitution to the victim because the offender has neither mobile asset or immobile asset. In addition, under the applicable laws, a subsidiary punishment can be imposed on the offender if he or she is unable to pay the restitution to the victim.

Therefore, we need to advocate law enforcers to ensure that they will facilitate victim to submit claim for restitution. Besides that, the Government Regulation gives a strong mandate to the Witness and Victim Protection Agency to help the victim assess any damage or loss he or she suffers. This mandate must be done by the agency maximally without condition.

5 Conclusions

5.1 Summary

Rights to restitution for child victim of criminal act are regulated in various laws and regulations in Indonesia. One of them is Law Number 35 Year 2014 on the Revision of Law Number 23 Year 2002 on Child Protection. In practice, the implementation of restitution is regulated through Government Regulation Number 43 Year 2017 on the Implementation of Restitution for Child Victim of Criminal Act.

Under the government regulation, the provision of restitution to child victim of criminal act aims to help the victim pay recovery costs and get justice although there are still various obstacles in its implementation.

The roles of law enforcers in restitution are greatly dependent on their understanding of the implementation of the applicable laws at investigation, prosecution and conviction levels. They also have to inform child victim about their rights as a victim of criminal act, especially his or her rights to restitution, to ensure that he or she can enjoy these rights.

There are still various obstacles in the provision of restitution to child victim of criminal act. These obstacles are caused by not only law enforcers' lack of understanding of the applicable laws and regulations, ambiguous legal sanction that can be imposed on the offender if he or she refuses or is unable to pay the restitution but also complicated procedure for submitting claim for restitution and difficulties to prove the material and immaterial losses suffered by the victim.

5.1 Recommendations

Government has to ensure that victim of criminal act can enjoy his or her rights by better formulating draft of the Indonesian Criminal Code.

Request for restitution must be filed at the time the victim reports his or her case to police and it is handled by investigator together with the case handling. The investigator informs the victim his or her rights to request for restitution and includes details of loss suffered by the victim in the investigation report. Then, the investigator submits the report to public prosecutor for further assessment or study. If the public prosecutor finds that the prosecutor does not include details of loss suffered by the victim, he or she will advise the investigator to revise the report. The public prosecutor has to consult the victim to ensure that he or she has proof of his or her loss which is prepared or legalized by relevant authority. After that, public prosecutor who is appointed to handle the case has to ask the victim to file his or her request for restitution which will be submitted to the public prosecutor together with the prosecution. This is done to fill gaps and address inclarity in the criminal law procedure in Indonesia.

References

1. Elucidations on Government Regulation Number 43 Year 2017 on the Implementation of Restitution for Child Victim of Criminal Act
2. Law Number 35 Year 2014 on the Revision of Law Number 23 Year 2002 on Child Protection, Article 71D paragraph 1
3. Elucidations on Government Regulation Number 43 Year 2017. Op.cit
4. Muliyawan, Artikel ,*paradigma baru hukum perlindungan anak, pasca perubahan undang-undang perlindungan anak,*(PN Palopo, 2015)
5. Elucidations on Government Regulation Number 43 Year 2017. Op.cit
6. Y.P Sibuea,Harris, *Persoalan restitusi terhadap anak korban tindak pidana,* majalah info singkat, Vol No 21/Puslit/November/2017.
7. Supriyadi Widodo Eddyono, et.al, "Inputs for Revision of Law Number 13 Year on Witness and Victim Protection" p.16(Koalisi Perlindungan Saksi dan Korban,Jakarta, 2006)
8. Article 9 of Government Regulation Number 43 Year 2017
9. Article 14 of Government Regulation Number 43 Year 2017
10. Sujoko, *The Implementation of Claim for Compensation in Article 98 of the Criminal Code for the Criminal Act of Rape in Semarang Territory,* p. 170(A Thesis of Magister Program of Faculty of Law of Diponegoro University, Semarang, 2008)
11. http://www.hukumonline.com/berita/baca/lt59eef5e356c54/ini-poin-poin-pp-pelaksanaan-restitusi-bagi-anak-korban-tindak-pidana
12. Muhadar, E. Abdullah, dan H. Thamrin, *Perlindungan saksi & korban dalam sistem peradilan pidana*, p.20 (Putra Media Nusantara, Surabaya 2009)
13. http://sp.beritasatu.com/home/lpsk-pidana-perdagangan-orang-harus-perhitungkan-restitusi/84254

14. Marlina, A. Zuliah, , *Hak Restitusi Terhadap Korban Tindak Pidana Perdagangan Orang,* p.89 (PT Refika Aditama, Bandung 2015)

The application of empty fruit bunch (EFB) fibers of oil palm for the development of light structure materials and products

Bustami Syam[1,*], *Maraghi* Muttaqin[1], Mawardi[2], *Fachrur* Rozi[2], Martgomi, and *James* Hutagalung

[1]Impact and Fracture Research Center, Department of Mechanical Engineering, Faculty of Engineering, Universitas Sumatera Utara
[2]Graduate Student, Mech. Engineering Study Program, Universitas Sumatera Utara
[3]Undergraduate Student, Department of Mechanical Engineering, Universitas Sumatera Utara

Abstract. We have been involved in developing the use of EFB fibers for several light structural products, such as speed bumps, parking stoppers, helmets, and horse shoes. This paper presents our current research work on parking bumper made of EFB fibers to be utilized for power generation. The products are designed and produced in full scale covering the whole width of vehicles passing over the bumpers. The ideas are how to reserve kinetic energy harvested from the vehicles to convert into electricity to be stored in a battery. To obtain a structural integrity-proof material, the speed bumps were subjected several tests, e.g. static and impact tests as well as live body test using the kinetic energy obtained from vehicles pass over the speed bumps. For this, a speed bump station parallel to street zebra-cross was prepared. It was obtained that A-5 concrete foam type is suitable for 3B class road. The structural response of speed bumps were checked using ANSYS 17.0.

1 Introduction

Palm oil industries processing crude palm oil (CPO) or kernel palm oil (PKO) produce a huge capacity of biomass in terms of empty fruit bunch (EFB) and tree trunk, and leaves. Utilization of the energy that comes from the by-products in the oil palm industry can be done by using biomass and biogas technology. EFB fibers, for instance, has been used for strengthening polymeric and concrete materials [1-10]. The EFB cannot directly be used. A series of processing of the EFB to reduce oil contents are necessary. Ready used fibers may reinforce the concrete or polymeric based materials. In our laboratory, some new materials reinforced with EFB fibers have been widely used for several light structures due to it light weight but they are still strong enough to sustain external loading.

In this paper, we extend our research on developing a large scale speed bump the concrete foam reinforced EFB is used as a speed bump to generate electricity. The speed bump is placed on to a mechanical system, as shown in Figure 1. Mechanically, the translatory motion

* Corresponding author: bstsyam@gmail.com

of speed bump is changed into circular motion at a fly wheel connected to electrical generation apparatus. With this connection, the speed bump works well when accepting the weight of the vehicle weight (speeds up to 10 km/h).

Fig. 1. Power generation apparatus using speed bump.

2 Materials and method

2.1 Material and speed bump fabrication

In our previous research we have developed classes of confoam materials [1, 10]. In this study, as per conclusion and suggestion in our previous research [10], we choose A5 type-confoam to produce a full scale-speed bump. The physical and mechanical properties of the material are shown in Table 1.

Table 1. Confoam mechanical properties (MP3EI research report).

Type spesimen	S_{uc}(MPa)	S_{ut}, (MPa)	□	E (MPa)
A5	2.1	0.18	0.2	10.1

The speed bumps were produced in the accordance with the geometry that meets the requirements set forth in Kemenhub (Ministry of Transportation decree), number 3, 1994, as shown in Figure 2. Several parts of the speed bump geometry, i.e., the upper surface configuration and the base speed bump, had been revised to meet a structure suitable for the passing vehicle load. In this paper, we propose geometry and dimensions as shown in Figures 3. The main dimensions of the speed bump are as follows: (a) type 1 (length, L = 1900 mm, witdh, w = 400mm, thickness, t = 52mm); (b) type 2 (length, L = 1900 mm, width, w = 400mm, thickness, t = 40mm). The dimensions are difference from those of we have published in our previous research work [10].

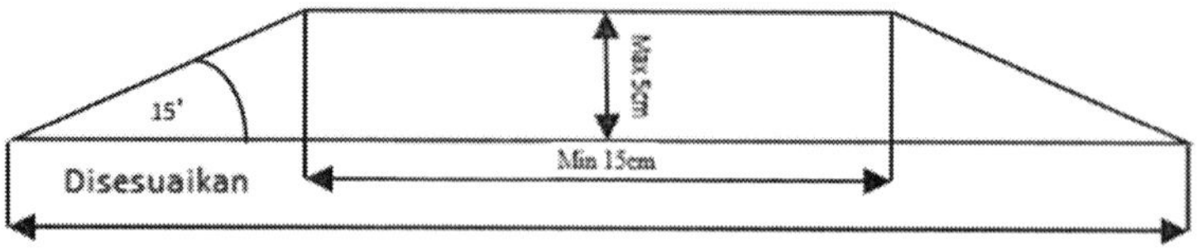

Fig. 2. Kemenhub 2 year 1994 (Ministry of Transportation standard size).

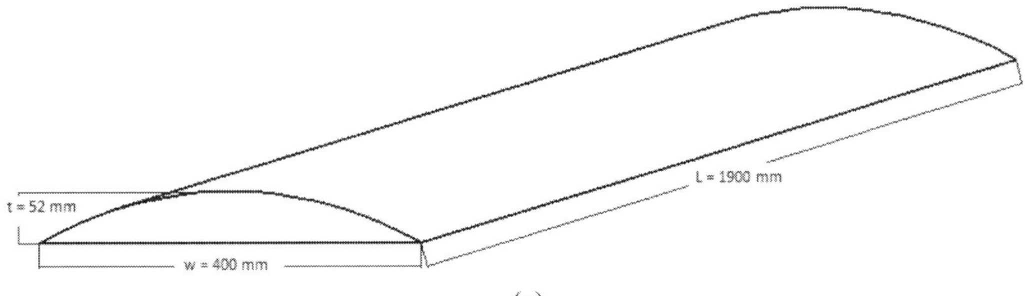

(a)

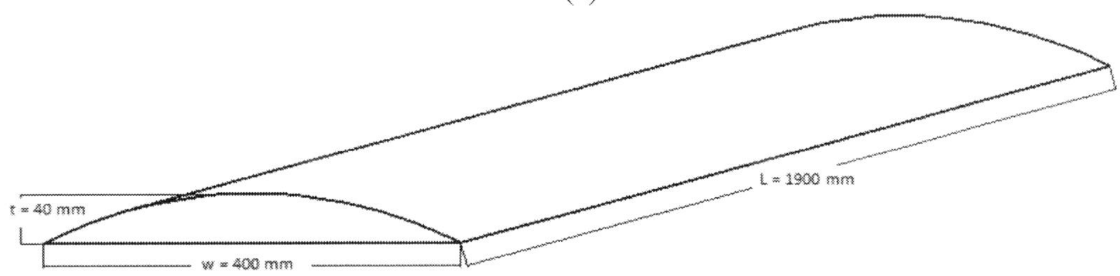

Fig. 3. Geometry and dimensions of the speed bump.

2.2 Method

As shown in Figures 1 and 2, the speed bump is placed on a steel bed attached to the top of the mechanical system of power generation. The speed bump accepts the load from the wheels of cars that pass through it. Thus, the speed bumps are designed to be able to receive static and dynamic loads in such a way as to meet technical and manufacturing requirements. For this, we selected newly developed materials at our strong enough to withstand both static and impact loads (not reported in this paper). A work station was prepared to place the mechanical system as well as the speed bumps. The static load simulated a station is obtained from vehicles pass over it in range of 5-10 km/hour.

(a)

https://doi.org/10.1051/e3sconf/20185200007

(b) (c)

Fig. 4. Experimental setting of speed bump power generation vehicle entering the speed bump (b) vehicle on top the speed bump.

3 Numerical simulation

3.1 Geometry and load model

The load, applied to the speed bump contact area (Figure 5) representing the vehicle-tyre track, is obtained from a quarter of the total vehicle weight allowed for class III road. Numerical calculations are focused on stress distributions at x, y, and priciple points using commercial FEM software, with 3-D elements. For both models we also calculated von-Mises stress. Let's examine the stress contours of the models one by one.

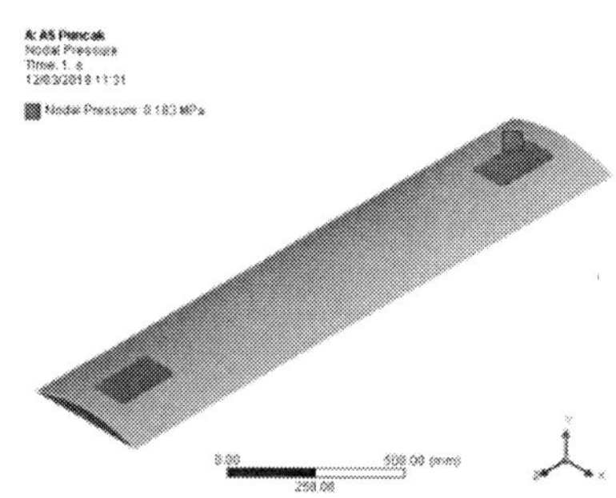

Fig. 5. Loading model at contact area.

3.2 Contours of stress

We focus the discussions on the stress contours on x, y, z, and principle directions both for type 1 and 2 of the speed bump models. The x and z directions may show the initiation of speed bump failures. All calculations as shown in Figure 6 are presented in Table 2. It is interesting to observe that the model reduces the stresses in all directions. The stress contour in the area remote from the contact surface show us that it is uniformly distributed in which for all cases the tensile stress is lower than that of tensile strength of the speed bump material. However, for a series of tests conducted in our work station with higher velocity of vehicles the failure may also occur (Figure 8). We may say that the impact tensile strength of the concrete foam should also be known (our current research).

Let us check normal stress responses in z direction on the entry base and the upper surface of the speed bump type 1 and 2. It is obviously indicated that the z direction stress is lower at the entry base and then slightly increase when the tyre approaching the top surface of the speed bump. However, it is interesting to note that the height of the speed bumps are not significantly increase the equivalent stresses.

From Figure 9. We observe that the von Mises stress is quite low compared to the static speed of type 1 bump material. The results conclude that the case of the speed bump type is susceptible to load resistance. Thus, they will have the potential to be used to accelerate power generation.

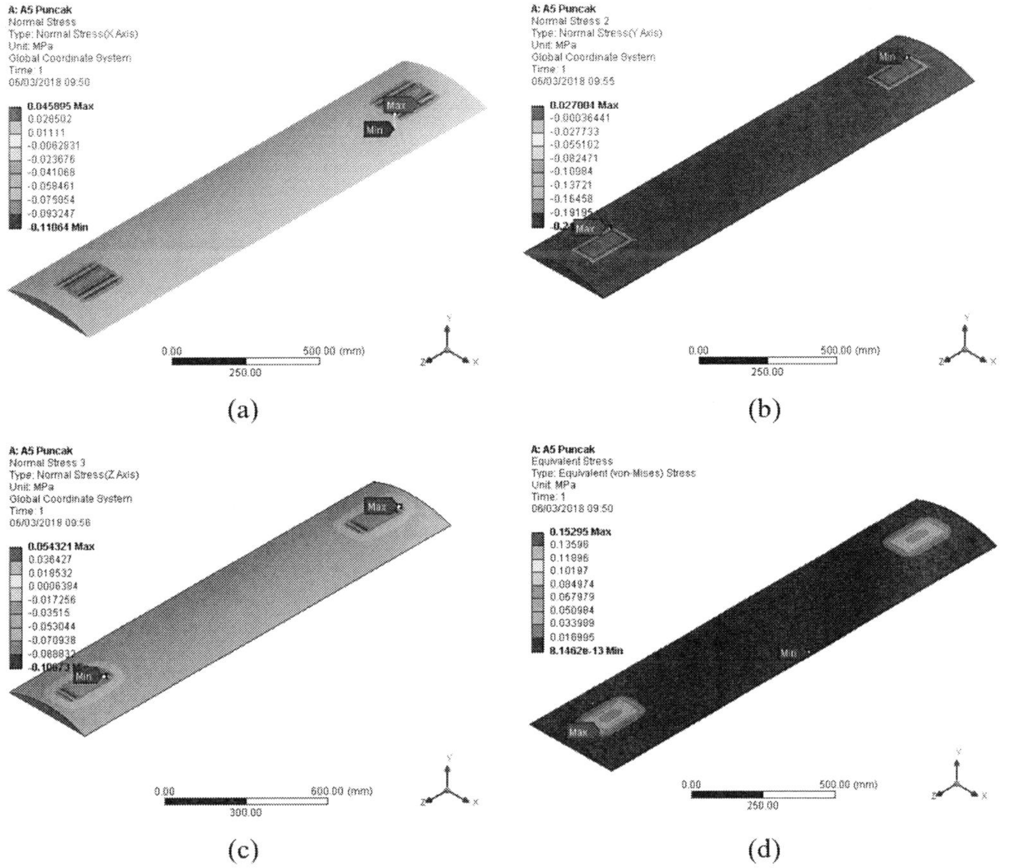

Fig. 6. Stress contour due to loading at top surface for type 1 model.

(a) Normal stress in x direction (b) Normal stress in y direction

(c) Normal stress in z direction (d) Equivalent stress

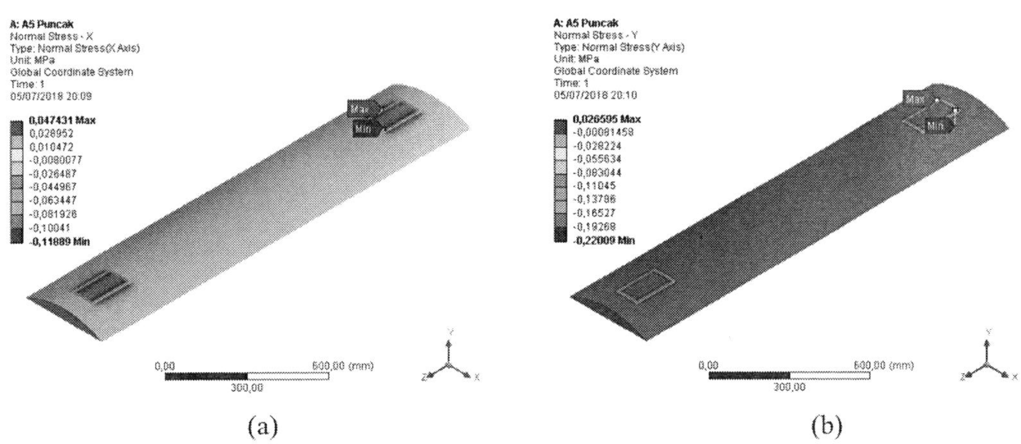

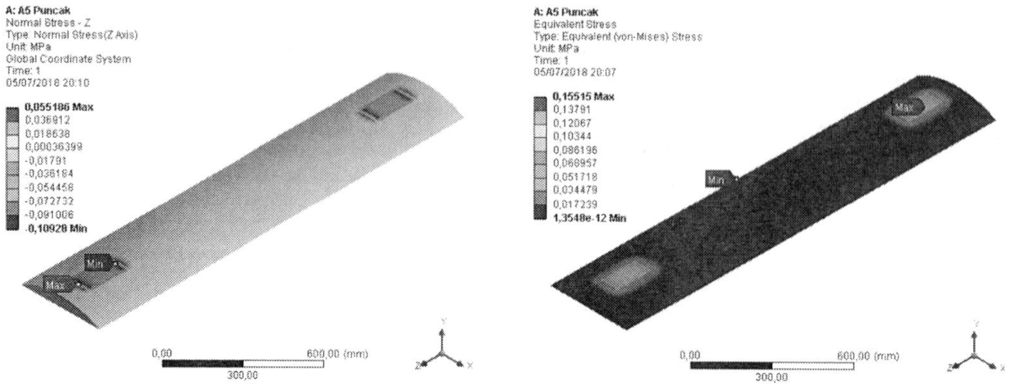

(c) (d)

Fig. 7. Stress contour due to loading at top surface for type.

(a) Normal stress in x direction (b) Normal stress in y direction

(c) Normal stress in z direction (d) Equivalent stress

Table 2. Stress calculation at certain locations on speed bump (h=40mm).

Speed bump material/Contact area	Equi. Stress (MPa)	Stress in x axis (MPa)	Stress in y axis (MPa)	Stress in z axis (MPa)
A5/Entry base*	0.164	0.033147	0.021	0.0357
A5/Top surface	0.152	0.045	0.027	0.054321

Table 3. Stress calculation at certain locations on speed bump (h=52mm).

Speed bump material/Contact area	Equi. Stress (MPa)	Stress in x axis (MPa)	Stress in y axis (MPa)	Stress in z axis (MPa)
A5/Entry base*	0.1748	0.0334	0.0205	0.0373
A5/Top surface	0.1551	0.0474	0.0265	0.0551

*Note: computer simulation not shown

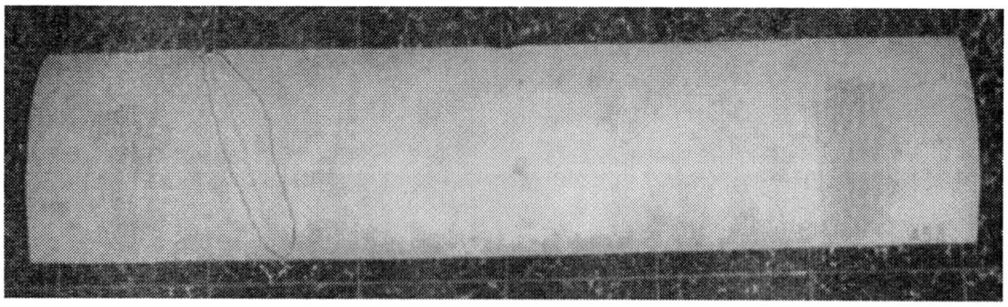

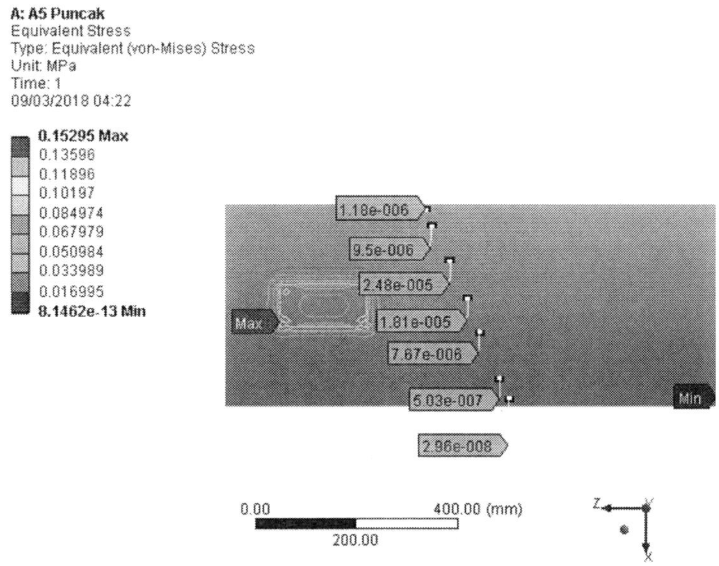

Fig. 8. Specimen failures and the corresponding stresses.

4 Conclusions

This paper discussed the analysis of speed bump made of concrete foam composite which is used to generate electrical power. The concrete foam is reinforced with oil palm EFB fibers. Two models were proposed, they are: (a) type 1 (length, L = 1900 mm, witdh, w=400mm, thickness, t=52mm); (b) type 2 (length, L = 1900 mm, witdh, w=400mm, thickness, t=40mm). To observe the structure integrity of the speed bumps the models were analyzed using a FEM-based numerical softwares, ANSYS17.0. It was obtained that the z direction stress is lower at the entry base and then slightly increases when the tyre approaching the top surface of the speed bump. However, it is interesting also to note that the height of the speed bumps are not significantly increases the equivalent stresses.

From Figure 8, we observe that the von Mises stress is quite low compared to the static strength of the speed bump material. The results conclude that concrete foam composite may also be potential for use as power generation speed bumps.

Acknowledgements

The authors acknowledge the research fund obtained from USU TALENTA scheme 2018.

References

1. B. Syam, B. Wirjosentono, S. Rizal. Pengembangan bahan komposit busa polimer (polymeric foam composite) diperkuat serat tandan kosong kelapa sawit (TKKS) untuk pembuatan berbagai produk industri transportasi dan olah raga. *Final Report PENPRINAS MP3EI*, DIKTI, (2013)
2. A. Irwan, B. Syam, W.S. Basuki. Analisa respon mekanik genteng dari bahan confoam diperkuat serat tkks akibat beban impak hujan es. *Prosiding Seminar DIES USU*, (2016)

3. B. Syam, A. Sebayang, I. Gunawan, M. Muttaqin, H. Darmadi, W.S. Basuki, M. Sabri, S. Abda. Mechanical response of open channel cover made of concrete foam due to external loadings. Journal of Physics: Conference Series **801** 012097, (2017)
4. B. Syam, M. Muttaqin, D. Hastrino, A. Sebayang, W.S. Basuki, M. Sabri, S. Abda. Analysis of power generating speed bumps made of concrete foam composite. IOP Conference Series: Materials Science and Engineering **180,** 1, (2017)
5. B. Syam, A. Sebayang, S. Sebayang, M. Muttaqin, H. Darmadi, W.S. Basuki, M. Sabri and S. Abda. Design and production of stopper made of concrete foam composite used for open channel conduit cover and parking bumper. IOP Conference Series: Earth and Environmental Science **126** 012053, (2018)
6. M Sabri, J. Lauzuardy, B. Syam. Design mechanic generator under speed bumper to support electricity recourse for urban traffic light. IOP Conference Series: Earth and Environmental Science **126** 012014, (2018)
7. D. Hastrino, I. Pasi, M. Muttaqin, B. Syam, W.S. Basuki, M. Sabri, S. Abda. Analisa struktur speed bump paduan bahan concrete foam dan polymeric foam dengan variasi rongga menggunakan sofware Ansys 14.5. Prosiding Seminar DIES USU, (2016)
8. S. Pranoto. Desain dan pembuatan kerucut lalu lintas dan bahan polymeric foam diperkuat serat tandan kosong kelapa sawit. Universitas Sumatera Utara, (2011)
9. Y. Muhammad. Desain dan pembuatan helmet sepeda bahan komposit poymeric foam diperkuat serat TKKS. Universitas Sumatera Utara, (2011)
10. Ahmadyani S. Performasi respon mekanik bola golf polymeric foam yang diperkuat serat tandan kosong kelapa, (2012)

Position of employees in the state-owned enterprises in achieving a welfare state

Christian Orchard Perangin angin* and *Bismar* Nasution

Doctoral Program in Law, Faculty of Law, Universitas Sumatera Utara, Medan, Indonesia

Abstract. Indonesia is a welfare state where the people are guaranteed to get a decent standard of living. The establishment of State-Owned Enterprises (SOEs) is part of the responsibility of the State in the prosperity of its people. The SOEs management cannot be separated from the aspect of labor. Law No. 13 of 2013 on Manpower (UUK) opens an opportunity for entrepreneurs, including state-owned enterprises, to practice the outsourcing recruitment. This research is conducted based on normative and supported by bibliographic data. The theories of Justice and Welfare State were used to analyze the problems. *Outsourcing* not providing certainty of employment as well as violating the workers' normative rights. In conclusion, the practice of the *outsourcing* recruitment at SOEs that violates the workers' normative rights will have a profound effect. The absence of certainty guarantee during work will only turn workers as an object of commodity. This activity will hinder people from achieving the justice and prosperous state which has become the goal of the state. The fulfillment of SOEs workers' rights in achieving the welfare state will only be attained if the company can treat workers as partners.

1 Introduction

The fourth Paragraph of The Preamble to the 1945 Constitution of the Unitary State of the Republic of Indonesia (UUD 1945) declares that the purpose of Indonesia as a state is to promote the common welfare and educate the life of the nation based on the principle of social justice for all Indonesian people. The formulation is then manifested into the Constitution serving as guidelines for the national life and state administration.

Article 27 Paragraph 2 of the 1945 Constitution states that every citizen deserves the right to work and get a decent standard of living for humanity. Then Article 28A stipulates that everyone has the right to live and maintain his life and living. Furthermore, other rights are also regulated in Articles 28B, 28C, 28H, 31, 33 and 34. Article 34 Paragraph 1 asserts that the poor and neglected children are kept by the state. After the fourth amendment, Article 34 Paragraph 2 clarifies that the state task in the field of social welfare is expanded with additional responsibility for developing social security system and empowering the poor, as well as providing health services and public service facilities for the people.

The above descriptionshave clearly defined Indonesia as a *welfare state* where the people are guaranteed to get a decent standard of living. Welfare state is defined as the

*Corresponding author: orchard27christian@gmail.com

concept of government in which the state plays a key role in safeguarding and promoting the economic and social welfare of its citizens.[1]

The word *welfare* contains at least three subclassifications: (1) *Social welfare*, which refers to the collective acceptance of welfare; (2) *Economic welfare*, which refers to the security guarantee through formal markets or economies; and (3) *State welfare*, which refers to the guarantee of social welfare services through agents of the state. The welfare state is briefly defined as a country in which state governments are held accountable for ensuring minimum welfare standards for every citizen.[2]

In a welfare state, the power (in terms of economic and political policies) is aimed at: 1) ensuring that every citizen and his family earn the minimum income according to the standards of eligibility, 2) providing social services for every problem that citizens experience (due to illness, old age, or unemployment), as well as other conditions such as economic crisis, 3) ensuring that every citizen gets his rights regardless of differences in status, economy class, and other differences.[3]

The establishment of State Owned Enterprises (SOEs) is part of the state's responsibility for the prosperity of its people. Article 33 Paragraph 2 of the 1945 Constitution mandates that the production sectors that are important for the State and affect the livelihood of the public are controlled by the State. SOEs are public companies that contribute to the development of the state economy/income, pioneer business activities and support government policies in the field of economy and development.[4] In addition, SOEs also serve as a tool to foster profits. To do such mission, SOEs adopt several forms such as *Persero, Perjan* and *Perum*. Thus, the functions and roles of these SOEs are essential to maintaining the economic stability of the country and can affect government policies including the state political affairs.[5]

In the beginning, SOEs were the nationalized former foreign companies (Dutch companies) which were then designated as State enterprises. Then, Law No. 1 of 1969 divided the State Enterprises into three types: State Agencies (*Perjan*), Public Service Companies (*Perum*) and State Limited Liability Companies (*Persero*). This division was formed according to the task, function and mission of the businesses at that time.

Thus, the first task of the state in establishing business entities is to meet all the needs of the people when those sectors cannot be performed by the private companies. Then such tasks are translated into a form of "pioneering" businesses by the State which turns SOEs into agents of development.

As corporate organizations, the aspect of workers can not be separated from the management of SOEs. Workers act as the front guards of a company in realizing its ideals. Manpower is a factor supporting the economy of a country. To promote the economy of a country, a quality relationship between employers and workers is required and becoming mandatory.

Law Number 13 of 2003 on Manpower provides norms in the form of opportunities and convenience for companies to perform efficiency, both in management and finance. The companies may delegate some of the internal work to other companies through a service contract and worker supply agreement. The implementation is regulated in the Minister of Manpower and Transmigration Regulation No. 27 of 2014 on Amendment to the Regulation of the Minister of Manpower and Transmigration No. 19 of 2012 on Terms of Submission of Some of the Internal Work to Other Companies (*Permenakertrans*). This condition then triggered the issues of *outsourcing* problem in Indonesia.

2 Research method

This research employs a normative[6] legal [7] approach by analyzing legal materials[8] and conducting literature studies[9] to be used as secondary data covering official documents, library books, rules and regulations, scientific papers, articles, and documents related to the research. All subsequent data are then analyzed using theories to answers the problems in this research.

3 Literature review

The idea of a welfare state has been introduced since around the 18th century. According to Bessant, Watts, Dalton and Smith, the basic idea of the welfare state dated back to the 18th century when Jeremy Bentham (1748-1832) promoted the notion that government has a responsibility to guarantee *the greatest happiness (or welfare) of the greatest number of their citizens*. Bentham used the term 'utility' to explain the concept of happiness or welfare. Based on the principle of utilitarianism that he developed, Bentham argued that something that can lead to extra happiness is good. Conversely, something that causes pain is bad. In his opinion, the actions of the government should always be directed to increase happiness toward as many people as possible. Bentham's notions of legal reform, role of constitution and social research for the development of social policieshas made him known as the "*father of welfare states*." [10]

The keyword in the welfare state is the issue of people's welfare guaranteed by the state. Jurgen Habermas argued that the welfare of all people is essential to the modern state. Furthermore, according to Habermas, the welfare of all the people is manifested in the following protection: *the risk of unemployment, accident, illness, old age and death of the breadwinner must be covered largely through welfare provisions of the state.*[11]

The basic concept of a welfare state can be seen from the state's efforts to manage all existing resources in order to achieve one of the state's goals, that is to improve the welfare of its people. These ideals are then translated in a policy that has previously been consulted to the public. As such, it can be seen whether or not a country has really realized the welfare of its citizens. The poverty and public health are some of the problems that the government must respond to in the formulation of a welfare policy.

The economic development undertaken by a welfare state must correlate with the welfare and prosperity of the people. This principle is the main task that must be realized in the welfare state. According to Habermas, there are two things that are directly related to economic development efforts: *First,* the realization of a welfare state cannot be separated from economic development efforts. As stated earlier, an economic development should make people more prosperous not the other way around. *Second*, the goal of the realization of a welfare state is not only for equality, but also for efficiency in the economic process. Ideally, the reason for equality or equity should not conflict with the purpose of efficiency in the economy. These two things become part of the welfare goals. In designing a welfare state, there are two important questions that must be addressed: *Firstly*, what is the purpose of a policy?, and *secondly*, by what method can the goal be achieved? These two questions can at least explain whether or not a country's policy seeks to realize the welfare of the people.[12]

The essence of a welfare state can be described as being the influence of human desires which expect to assure security, peace and prosperity in order not to fall into misery. The reason can be described as a driving force as well as a goal for humans to constantly seek various ways to achieve prosperity in life. So, when the wishes have been guaranteed in the constitution of a country, then they must be guaranteed by the state and the state must realize the wishes. In this context, the state exists in the stages of being a welfare state.

The Unitary State of the Republic of Indonesia also embraces the welfare state ideology. This was confirmed by the Pioneers of Independence and the Founders of the Unitary State of the Republic of Indonesia who declared that the democratic state to be established was the "*Welfare State*" (*walvaarstaat*) not the "*the State of the Night-Watch*" (*nachtwachterstaat*). In determining the concept of a welfare state for Indonesia, Moh. Hatta used the term "*Governing State*."[13] In a similar view, Jimly Asshiddiqie said that "... Indonesia is a welfare state that idealizes the nature of state interventionism in the dynamics of the social economy, solely for the welfare of the people".[14]

4 Result and discussion

Workers or employees play an important position in the production process of a company. Every company needs to make an employment plan according to the needs of the company which should match to the needs of the employees. This has become the goal of formulating the Law on Manpower.

With regard to the employment regulation between employers and workers, the Law on Manpower ranks at the top position. Under this Law, the employment relationship occurs due to an employment agreement between employers and workers. Article 1313 of the Civil Code states that an agreement is an act by which one or more persons commit themselves to one or more persons. The definition of agreement, as written down in Article 1313 of the Civil Code, only mentions the clause *one party or more binding itself to the other* but does not clearly specify for what purpose an agreement is made. Wiwoho argued that an agreement can also be interpreted as the relationship between a person who acts as a worker/laborer with someone who acts as an employer.[15]

The definition of the employment agreement is firstly mentioned in the provisions of Article 1601a of the Civil Code which reads: *An Employment Agreement is an agreement whereby one party (the worker) binds himself to be under the command of the other party (the employer), for a certain period of time, performs work and receives wages.* The phrase "*under the command of the other*" means that there is a relationship between workers and their employer, the relationship between subordinates and their superior where employers give orders to workers to do a certain work. There is a governing authority in an employment agreement which distinguishes it from other agreements.According to R. Imam Soepomo, the employment agreement is an agreement in which the first party, the laborer, binds himself to work by receiving wages from the other party, the employer, who binds himself to hire the laborer by paying the wages.[16]

The nature of a worker-employer relationship is basically contradictory, where on the one hand it seeks *colaborative relations*, but on the other hand there are potential trends which can lead to *adversarial relations*. When signing an *individual contract*, the worker and the employer are legally equal before the law in accordance with the principle of *equality before the law*, but sociologically the employer occupies a higher socio-economic position compared to the worker. Even in the context of a company, there are some opinions assuming that labor is only an external factor. The worker has a similar position as the supplier or buyer who serves as a supporter of the company's continuity, not as an internal part of the company which is inseparable or as a constructive element that builds the company.[17]

In the management of SOEs, *outsourced* labor is an issue that needs attention. This statement is based on research and surveys conducted by management experts over the last twelve years in more than 1,200 companies. They expect to gain significant business benefits from *outsourcing* recruitment practice.[18]

Outsourcing employment in Indonesia is conducted only for supporting work or activities or commonly called *non-core business* activities. It is not for the main work or activities or commonly referred to as the *core business*. Article 64 of the Law on Manpower states that a company may subcontract some of its work to another company under a written contract for the provision of service/labor.

However, employers make use of this system not for the efficiency of work management or to seek out the experts in their field. They make use of this employment system solely for cost-cutting reason. In fact, violation to the Law is still going on. Referring to the Regulation (*Permenakertrans*), there are only five types of work that can be outsourced and they exclude the main activities. However, a number of state-owned companies noticeably outsourced sometypes of work outside the provisions.[19]

Another problem is the uncertainty experienced by the outsourced workers because of an unclear working relationship. There is no legal relationship between outsourced workers and their employing company (user). There are no rights and obligations that are directly related between the two, for example, orders come from outsourcing companies to outsourced workers.

Based on Article 56 of the Manpower Law, there are 2 (two) types of employment agreements, namely the Certain Time Working Agreement (*PKWT*) and Non-Specific Time Working Agreement (*PKWTT*). PKWT is an employment agreement between the worker/laborer and the employer to establish a working relationship within a certain period of time or for a particular job or for a certain temporary job. PKWTT is an employment agreement between the worker/laborer and the employer to establish a permanent employment relationship.[20] PKWTT can be made literally or orally without requiring an approval from the relevant government employment agencies. If the PKWTT is made verbally, then the clauses apply between them (between employers and workers) are clauses as set out in the Manpower Law.

The PKWT is a conditional agreement that is (among other things) required to be written and made in the Indonesian language, under the sanction that if it is not made in writing or not made in the Indonesian language, it is deemed as PKWTT (Article 57 Paragraph (2) of Manpower Law). PKWTT may require a maximum period of 3 (three) month probation. During the probation period, the employer is required to pay the wage which must not be lower than the prevailing minimum wage.

Work relationships are created when there is an agreement between employers and workers through a contract. This requirement also applies when subcontracting companies work for other parties. The types of work that can be subcontracted must meet the following requirements: a) done separately from the main activities; b) carried out by direct or indirect orders from the employer; c. serves a supporting activity of the company as a whole; and d) does not inhibit the production process directly.[21]

The types of work which are not directly related to the production process include: [22] a) *cleaning service*; b) food providing for workers/laborers (*catering*); c) security service; d) supporting services in mining and petroleum; and e) providing transportation for workers/laborers.

In practice, the applied *outsourcing* system does not accord with the laws and regulations. *Outsourcing* tends to be applied in jobs which are directly related to the production process. In addition, there are also neglected workers' normative rights. Laborers who work for jobs directly related to the production process should get the same rights as obtained by the permanent employees.

Based on the Law on Manpower and the Regulation of the Minister of Manpower, the rights of permanent and outsourced employees can comparatively be described as follows:[23]

Tabel 1. Comparison of rights between permanent employees and outsourced employees

Worker's rights	Permanent employee	Outsourced employee
Basic Wage	• Minimum Wage • Scheduled working hours	• Minimum Wage • Scheduled working hours
Premium Presence	Yes	No
Work Insurance, Health	Yes	Yes

Insurance, and Death Insurance		
Payment for food and transportation	Yes	No
Right to Rest	Yes	Yes
Holiday Time	Yes	Yes
Work Termination	Yes	Yes
Worker's Social Security	Yes	Yes

Indonesian laws and regulations do not restrict the ability of an outsourced employee to transition to permanent status. A restriction may, however, be imposed by the labor-outsourcing contract between the employercontractor and the labor provider; such a restriction is permitted by the freedom of contract principle in Article 1338 of the Indonesian Civil Code. Absent such a contractual restriction, an employee of the labor provider may concurrently work as a permanent employee of the primary employer.

Based on the Report of the Working Committee of Commission IX of the House of Representatives regarding Outsourcing Recruitment, it is stated that there are few violations of legal norms by companies include the absence of status change guarantee for contract workers to become permanent workers (PKWTT) pursuant to Article 59 Paragraph (2) of Law No.13 of 2003; workers' wage is paid lower than the regulated minimum wage; the absence of health/safety insurances; unclear relationship between the duration of employment and the status and welfare of the workers; the normative rights they deserve being ignored by the company in terms of *THR* (Special Holiday Allowance) payment; wage adjustment that must be taken into account from the accumulation of work tenure; social security or other rights regulated in the legislation (Article 29 of *Permenakertrans RI* No.19 of 2012). Workers' wage is paid lower than the minimum wage (Article 90 Paragraph (1) of Law No.13 of 2003).[24]

The working rules carried out by state-owned enterprises through their directors often violate the provisions of Article 59 of the Manpower Law whereby working agreements for a certain period of time may only be made for certain jobs whosetype, nature or activity will be completed within a specified time.In reality, the jobs performed by the worker through the contracting agreement or the provision of a worker's services were continuous and carried out for the same jobs,neither temporary nor seasonal, or related to new products, new activities or additional products which werestill in a trial process.[25]

5 Conclusion

The government as the administrator of the state must be able to create legislation in favor of the people. [26] Any government policy in the form of laws and regulations should be able to create opportunities for the welfare of the people. Thus, the task of the state is to ensure that every regulation should be able to realize the goals of a social welfare. This way, people can clearly feel that the existence of law is instrumental in creating their welfare. This should also become a basic milestone for a law-based country to realize the concept of welfare state in its governance.

The state-based status has turned SOEs into companies managed under the regulation of Limited Liability Companies, but within certain limits, its management is still regulated in the regulation of SOEs which gave birth to SOEs. However, labor-related arrangements and their relation to policy making within the company have not been explicitly regulated in both laws.

When applying the outsourcing system under the Manpower Law, violations are frequently found. The main types of violations that SOEs often commit activities including the continuous use of the outsourcing system which is inconsistent with the legislation and ignores the workers' normative rights. The violationsof the outsourcing system against the

Manpower Law can be categorized into 3 (three) kinds. *First*, outsourcing is used for core business activities. *Second*, people have worked for a relatively long period, an average of over 3 (three) years. In fact, some workers have already workedfor more than 9 (nine) years. And *third*, the employment contract is repeated for similar type of job and workplace.[27]

Another violation is the absence of protecting the workers' normative rights by the vendors (outsourcing companies). Violation to BUMN labour normative rights covering: severance pay right (reported at PT DI, OS worker at PT PLN Bali, PT Jamsostek, PT Pertamina, others), termination process pay right, occupational allowance right (reported at PT ASKES and ASDP), pension fund benefit right (reported at Perum Peruri, Perum Damri, PTPN II Sumut-Langkat), overtime and annual leave pay right (reported at PLN, PT Petrokimia), payment under minimum wage (reported at PT Kimia Farma, PT Petrokimia, Perum Damri Kantor Surabaya, others), occupational health and safety right, and social security right (reported at PT PLN, PT Petrokimia Gresik), and other rights.[27]

The Law on Limited Liability Company, as a reference for SOEs in its management, has not placed workers as a strategic partner in the management of the state businesses although the SOEs Law has provided a space for the management of SOEs to refer to the Manpower Law for matters relating to employees as state-owned enterprise workers.

SOEs as a legal company are not only seen to have an organ as stipulated in the Law on Limited Liability Company. In carrying out their activities, workers have a leading role in implementing the company policies. In the development history of Limited Liabilty SOEs in Indonesia, manpowerhas received less attention.As such, it is necessary to change the paradigm that has set workers as controlled variables (factors of production) to be one of the stakeholders or strategic partners in the management of Limited Liability SOEs.

A quality relationship will create a conducive situation and synergy which will eventually improve the company's performance. Comfortable and secure working atmosphere are the benchmark in creating conducive company.Thus, it will not be difficult for a company to achieve its goal, if the relationship between workers and their employer runs harmoniously. If the harmonious relationship continues to work, the goals mandated by the Constitution to create a welfare state in Indonesia will not be difficult to achieve.

References

1. Welfare State, *Encyclopedia Britannica*, http://www.britannica.com/print/topic/639266, retrieved on 27 April 2018.
2. P. Christopher,*Welfare state: The new political economy of welfare,* p. 9 (Pennsylvania State University Press, Pennsylvania, 2007).
3. A. Briggs, Welf.St. His.Pers, E.J.S, **Vol. 2 (2)**, pp. 221-258, (1961).
4. K.S. Aziza, I Semester of 2017 of BUMN Revenue at 936 trillion IDR, https://ekonomi.kompas.com/read/2017/08/29/202938326/semester-i-2017-pendapatan-bumn-rp-936-triliun., retrieved on 27 April 2018
5. B. Rothstein, *Just institution matter: The moral and political logic of the universal welfare state.* p.8 (Cambridge University Press, Cambridge UKs, 1998)
6. J. Ibrahim, *Teori dan metodologi penelitian hokum normatif,* p. 295 (Bayumedia Publishing, Malang, 2006).
7. R.H. Soemitro, *Metodologi penelitian hukum dan jurimetri*, pp. 13-14 (Ghalia Indonesia, Jakarta, 1988).
8. S. Soekanto, and S.Mamudji, *Penelitian hukum normatif, suatu tinjauan singkat*, p.12 (Raja Grafindo Persada, Jakarta, 2011).
9. Ibid, p. 13

10. J. Bessant, R.Watts, T.Dalton, and P.Smith,*Talking policy: How social policy in made, crows nest: allen and unwin*, (2006 in E. Suharto, *Peta dan dinamika welfare state di beberapa negara: pelajaran apa yang bisa dipetik untuk membangun indonesia?*, paper presented at a Seminar entitled "Mengkaji ulang relevansi welfare state dan terobosan melalui desentralisasiotonomi di indonesia", (IREYogyakarta and Perkumpulan Prakarsa,Jakarta, Yogyakarta 25 July 2006)
11. G. Poggi, *The development of the modern state "sosiological introduction"*, p.126 (Standford University Press, California, 1992).
12. B. Nicholas.*The economics of the welfare state*. (Stanford University Press, California, 1998).
13. M. Yamin, *Naskah persiapan UUD 1945: Risalah sidang BPUPKI/PPKI*, p.299, (Sek.Neg.RI, Jakarta, 1959); compare with M.Kusuma, *Lahirnya UUD 1945*, p.355 (Badan Penerbit FH UI, Jakarta, 2004)
14. J. Asshiddiqie, *Gagasan konstitusi sosial: Institusionalisasi dan konstistusionalisasi kehidupan sosial masyarakat madani*, p.122 (Pustaka LP3ES, Jakarta, 2015).
15. W. Soedjono, *Hukum perjanjian kerja*, p.9 (Rineka Cipta, Jakarta, 1991).
16. I. Soepomo, *Hukum perburuhan bagian pertama hubungan-kerja*, p.75 (Bhayangkara, Jakarta, 1968).
17. H.P. Rajagukguk, *Peran serta pekerja dalam pengelolaan perusahaan (co-determination)*, A talk on a discussion panel: "75 Tahun Dr. H.P. Rajagukguk, S.H"p.3 (Fakultas Hukum UKI, Program Kajian Wanita dan Gender Pascasarjana UI, dan Fakultas Hukum UI, Jakarta, 2000)
18. R.E. Indrajit, R. Djokopranoto, *Proses bisnis outsourcing*, pp.4-5 (2003).
19. N.C. Burkholder, *Outsourcing (the defitive view, applications, and implications)*. p.6 (J. Wiley and Sons, Inc, Hoboken, New Jersey, 2006)
20. Article 1 point 2 A Decree of Man Power and Transmigration Minister of RI No.KEP. 100/MEN/VI/2004 on the Implementation of Temporary Appointment Work Agreement (Kepmenakertrans 100/2004)
21. Article 65 Paragraph 2 of Manpower Law
22. Article 17 Paragraph (3) Regulation of the Minister of Man Power.
23. R. Bales, S.B. Audyanto, S. Guven, M. Jockel, C.B. Juniarto, J. Junlu, & D. Langmead, A Comp. An. of Lab. Outsourc, AJICL **Vol. 31, No. 3**, pp.580-622 (2014)
24. Laporan Panitia Kerja Outsourcing BUMN Komisi IX DPR RI Masa Sidang 2013-2014, DPR RI, (2013).
25. Meeting Report between BUMN Labour and Komisi IX DPR RI (House of the Representative)attended by Ministry of Labour, Ministry of Finance, Ministry of BUMN, Representatives of the Boards of Directors of PLN, Pertamina, PGN, Krakatau Steel, Indofarma, JasaMarga, and Garuda Indonesia, on 7 February 2018, (https://www.bantuanhukum.or.id/web/geber-bumn-menanti-dpr-tuntaskan-kasus-outsourcing-di-bumn/retrieved on, 1 May 2018).
26. Y. Suhardin, Perananhukumdalammewujudkankesejahteraanmasyarakat, JHPJ, **Vol. 25 No. 3**, p.273 (July 2007)
27. K.S. Cahyono, *Perbudakan modern di BUMN*, p.4 (PT Leutika Nouvalitera, Yogyakarta, 2015)

The responsibility of the regional heads against the management of regional finances

Dani Sintara* and *Faisal Akbar* Nasution

Doctoral Program in Law, Graduate School, Universitas Sumatera Utara, Medan, Indonesia

Abstract. In administering governance, Regional Heads are given an authority to manage the local finances independently. This authority is regulated in Law Number 9 of 2015 and Government Regulation Number 3 of 2007. Accountability of Regional Heads is a form of a democratic government. In a democratic government, the government has limited power and is not justified to act arbitrarily against its citizens. In reality, the implementation of the accountability is often the result of a political compromise. Political compromises occur due to the behavior of the political elite and the weakness of the existing legislation. This weakness relates to the mechanism of the accountability against the management of regional finances which opens an opportunity between the Regional Heads and the Regional House of Representatives (DPRD) to conduct a negative relationship. The political compromises in the accountability of the Regional Heads have had a negative impact on the local governance. Whereas, the purpose of implementing a regional autonomy is to improve the quality of justice, democracy and prosperity. Therefore, before the DPRD rejects the accountability of the Regional Heads, it must first be tested in the Supreme Court.

1 Introduction

The State of Indonesia is a unitary state in the form of a republic. It is stipulated in Article 1 Paragraph (1) of the 1945 Constitution. As a unitary state, all power/authorities and responsibilities of governance in order to realize the prosperity and survival of the nation is under the control of a centralized power in the central government.[1]

In fact, the implementation of power centralization provides efficiency in decision making.[2] Yet, the founders of the state are aware of the natural conditions living in the diverse Indonesian society. So the type of government in Indonesia is based on the power-sharing system between the central government and community groups in the region[3]. This will, in the end, create the Regional Government based on the decentralization system as regulated in Article 18 of the 1945 Constitution.

Technically, decentralization gives rise to arguments that can support the smooth running of local government, namely: the people of a region are obliged to advance the region, are more closely related and more familiar with the needs of the region.[4]

* Corresponding author: danisintara@yahoo.co.id

In implementing Regional Governance, the Regional Head will be given an authority to organize things including the management of regional finances independently because regional autonomy generated through a devolution of power will have a positive effect when given financial management autonomy.[5]

However, although the regions are given the authority in financial management, the implementation must be accompanied by accountability. The responsibility should certainly goes to the Regional Head, given that the accountability of the Regional Head in the financial management is a manifestation of a democratic government. Without such a means, the meaning of democracy is lost because the government has turned into a ruler who does not need to give accountability.

The accountability of the Regional Head against the management of regional finances is intended to determine the level of success and failure that occurs after the regional planning is made and implemented. That is why responsibility is required as a form of control on the implementation of the Local Government. The goal is to get feedback for improvement, if there are errors or aberrations, before they become worse and more difficult to fix.

In reality, the accountability of the Regional Head against the financial management has given rise to an assumption: the accountability of the Regional Head in the management of regional finances is a political compromise. The process is full of interesting nuances of political interest among existing political institutions in the region.

Before the enactment of Law Number 9 of 2015 on Regional Government, the regulation governing the Regional Government is regulated in Law Number 22 of 1999. Both Laws bring different consequences in the implementation of the Regional Head's accountability to the financial management area.

At the time of the enactment of Law Number 22 of 1999, the Regional Head is supervised by Regional House of Representatives (DPRD). It proved that the Regional Head is responsible to the DPRD. The Regional Head is elected and determined by the DPRD and authorized by the President. In carrying out the duties and obligations as the Regional Head of the Governor shall be responsible to the Provincial DPRD, and in its capacity as the government representation, Governor is under supervision and responsible to the President. In carrying out its duties and authority as Regional Head, Regent/Major is responsible to the Regency/Municipal DPRD. After the enactment of Law Number 9 of 2015, the Regional Head is not accountable to the DPRD, because the Regional Head is directly elected by the voters in that area in a democratic manner. The relationship between the Regional Head and the DPRD is an equal as partner and working relationship. Equal position means that among that both local government institutions have the same position and parallel or not mutually supervisory.[6]

Furthermore, in relevancy to the responsibility of the Regional Head on the management of regional finances, in Law Number 22 of 1999, the Regional Head is obliged to submit the Accountability Report (LPj) on the management of regional finances to the DPRD. If the DPRD assesses that the Head of Regional Head is not in accordance with the Regional Strategic Plan (Renstra) then the DPRD can reject it. The DPRD can put forward the dismissal of the Regional Head in case of LPj is twice rejected. The juridical consequences is placed the Head of Region under supervision of the DPRD, when it should be between the Head of Regional and DPRD are equal partners in the implementation of Regional Government.

After the enactment of Law Number 9 of 2015, the Regional Head no longer gives LPj to DPRD. However, it is only limited to provide the information submitted through Accountability Report (LKPj). With respect to LKPj of the Regional Head, in case the DPRD judges it is not in accordance with RKPD, the DPRD cannot refuse it, but only gives a recommendation to the Head of Region for improvement.

However, in the implementation of Law Number 9 of 2015, even though the DPRD can no longer refuse LKPj of the Regional Head, but with the rights attached to the DPRD in carrying out its supervisory functions, namely the right of interpellation, the right of inquiry and the petition rights allow the DPRD to investigate LKPj of the Regional Head. Even suggested the dismissal of the Regional Head. In the implementation of the Regional Head politically hosted by the power of DPRD.

The political compromise will have a devastating effect on the implementation of Regional Government. Whereas the purpose of regional autonomy implementation is to improve the quality of justice, democracy and prosperity for all elements of a diverse nation within the Unitary State of the Republic of Indonesia.

This study was intended to see the accountability of the Regional Head against the management of regional finances.

2 Problem

The problem of this study was: How does the accountability of the Regional Head against the management of regional finances work?

3 Research methods

The type of research conducted is standard legal research. This study is only aimed at normative regulations. This study is closely related to the library because the standard legal research requires data that is secondary to the library.

The nature of the research is descriptive analytical. This means that this study illustrates how a legal provision in the context of legal theories in the presentation describes the various issues related to the accountability of the regional head on the management of regional finances.

The data was secondary data consisting of (a) primary legal materials in the form of legislation, (b) secondary law materials in the form of writings, both books and articles containing comment or analysis related to the subject matter, and (c) tertiary legal materials in the form of dictionaries. The data was collected using literature reviews and document analysis for further analyzed using a qualitative method based on a logic deductive thinking.

4 Discussions

4.1 Liability mechanism of the management of regional finances

In the Law on State Finances, there is an affirmation on financial management saying that the power of state financial management is the power of the Government. The power of state financial management from the President is partly handed over to the Governor/Regent/Mayor as head of the Local Government to manage local finances. This provision certainly has implications for the regional financial management arrangements where the Governor/Regent/Mayor is responsible for the management of regional finances.[7]

The accountability of the Regional Head against the management of local finances might come in the form of responsibility against the use of Regional Revenue and Expenditure Budgets (APBD). This responsibility is included in the Liability Report (LKPJ) and presented to the Regional House of Representatives (DPRD) to be discussed and heard by each political faction. Then, a joint discussion with the Regional Work Units (SKPD) is conducted before a conclusion is made.

If the accountability mechanism of the local financial management is further elaborated, then it all starts with the Heads of SKPD as the budget users organizing the accounting of financial transactions, assets, and debts, which are under their responsibility.

The accounting report clarifies the recording/administering financial transactions in SKPD working areas in order to prepare financial statements. The financial statements consist of budget realization reports, balance sheets, and notes to the financial statements submitted to the Regional Head through the Regional Financial Management Officers (PPKD). Heads of SKPD as users of budget/users of goods give a statement that the management of regional finances under their responsibility has been conducted based on the internal control system which is adequate and practiced in accordance with the provisions of the legislation.[8]

The financial statements will be attached with the performance overview report and the Regional Financial Institution/Company report. The performance overview report is compiled from the summary of the accountability report of the Regional Heads. The financial report of the Regional Government above is prepared based on SKPD's financial report and submitted to the Regional Head as the implementation of the accountability of the management of regional finances. Then the Regional Head submits the Drafts of Local Regulations in the form of the implementation accountability of APBD to DPRD through financial reports which have been examined by the Financial Audit Board (BPK).[9]

The Regional Head's LKPJ is presented in a plenary session of DPRD and then discussed internally by DPRD. Based on the discussion, DPRD determines a decision no later than 30 (thirty) days after the LKPJ is received. The decision of DPRD will be submitted to the Regional Head in a special plenary meeting as a recommendation to the Regional Head for improvement in the implementation of Regional Government in the future. If the LKPJ has not been addressed by DPRD within 30 (thirty) days after the report is accepted, it is considered that there is no recommendation for improvement needs to be performed.[10]

It can be said that in the management of accountable and transparent local finances, the Regional Government must submit accountability in the form of financial statements. The financial statements are prepared in accordance with government accounting standards. Before the financial statements are reported to the public through DPRD, they need to be audited by BPK.

4.2 Responsibility of the management of regional finances as a result of political compromises.

Power tends to corrupt and absolute power corrupts absolutely. The statement that Lord Acton, a British scholar who lived between 1838 and 1902, said was originally a mere hypothesis. The hypothesis that eventually became undeniable destiny in a long history of power.[11]

Furthermore, what is the relationship between the sentence expressed by Lord Acton and the responsibility of the Regional Head against the management of regional finances?

Law Number 9 of 2015 on Second Amendment to Law Number 23 of 2014 on Local Government hereinafter abbreviated as UU PEMDA and Government Regulation Number 3 of 2007 state that DPRD only gives recommendation to LKPJ submitted by the Regional Head. In other words, DPRD can no longer reject the accountability of the Regional Head against the implementation of the regional financial management.

In fact, although it is impossible for DPRD to reject LKPJ, with the rights attached to the DPRD, there is a great possibility for DPRD to take action if LPKJ is considered to violate the benchmark of the Regional Development Plan (RKPD) conducted under the supervisory function of DPRD. This happens in connection with the operationalization of the theory of checks and balances in order to create a direct supervision of one branch of government against other branches.[12]

The supervisory function of DPRD on the implementation of the management of local finances can be done through the rights of DPRD: *First*, the right of interpellation, that is the right of DPRD to request a clarification to the Regional Head about an important and strategic Local Government policy that has a wide impact on the life of the community, region and state which is assumed to violate the laws and regulations.

Second, the inquiry right, that is the right of DPRD to investigate a particular policy of the Regional Head that is important and strategic and has a wide impact on the life of the community, region and state which is assumed to violate the laws and regulations. *Third*, the petition right, that is the right of DPRD to express an opinion toward the policy of the Regional Head or about extraordinary events occurring in the region accompanied by a recommendation for completion or as a follow-up to the implementation of the rights of interpellation and inquiry.[13]

The rights attached to DPRD can result in the impeachment of the Regional Heads. However, the supervision carried out by DPRD against the Regional Government is not balanced with the supervision carried out on the institution of DPRD itself. Until now there is no specific regulation governing the control of DPRD.

Consequently, in carrying out his/her accountability against the management of regional finances, the Regional Head often becomes hostage to the power and/or political interests existing in DPRD. If the Regional Head comes from the minority Political Party or was an independent candidate, then in the process of accountability, he/she will get much pressure from the DPRD. Whereas, if the Regional Head comes from the majority party in DPRD, it will be difficult for DPRD to supervise the management of the regional finances.

As such, the Regional Head's accountability against the management of regional finances will result in a political compromise between DPRD and the Regional Head. In the end, this will bring about a bad impact on the implementation of Regional Government as the purpose of regional autonomy implementation, namely to improve the welfare of the community in the region, will not be fully implemented.

The above opinion is consistent with what J. Kaloh put forward: The legislative and executive relationship in a region can lead to a negative one-way relationship when regional executives and legislators collaborate in governance and collectively conceal the collaboration from the public.[14]

Therefore, there must be an external control of DPRD which will be a medium to accommodate, channel and fight for the complaints of the community toward the actions taken by DPRD members and its institution. In order to implement the check and balances system, in the future the Local Government should also be given an authority to conduct supervision on DPRD.

The rights that attached to the DPRD, namely the right of interpellation, the right of inquiry and the right of petition is in order to oversee the implementation of Regional Government. It is intended that the administration of the government is no despotism occurs.

However, to maintain the balance in the implementation of the Regional Government, especially in accountability of the Regional Head to the management of regional finance, so that the Regional Head politically hosted by the power of DPRD, there is ideal mechanism stages shall be done. *Firstly*, after the Head of Region submits LKPj to the DPRD, the DPRD shall discuss whether the LKPJ is in accordance with the RKPD or not. In case it is not yet appropriate, the DPRD shall give a recommendation to the Regional Head for improvement. In conducting the discussion, the DPRD must provide space for the society to participate in issuing opinions as a manifestation of civil and political rights owned by the society. *Secondly*, if the recommendation given by DPRD has been improved by the Regional Head and in accordance with the RKPD, then the LKPj is considered to be completed. *Third*, if the recommendation given by DPRD has not been or has been improved by the Regional Head but still not in accordance with RKPD, the DPRD then uses its supervisory function through the rights attached to it, namely: interpellation rights, the right of inquiry and petition rights. *Fourth*, the results of the DPRD supervision function were brought to the Supreme Court for testing, wheter the accountability of the Regional Head to the management of regional finances in accordance with RKPD or not. *Fifth*, the Supreme Court as an institution that has the authority as the final decision against the potential responsibility of the Regional Head whether has been in accordance with the RKPD or not.

4.3 Comparative law system approach to accountability of the management of local finances.

From the legal system comparison view, especially the *civil law* and *common law* systems related to the Legal Institutions, there is no such a thing as legislative tradition in *common law* that seeks to amend or fix the law through legislators. This is totally different from what happens in *civil law*. Therefore, a significant historical fact is that the *common law* has been developed within and by the Court to give a legal status to the decision made by the Judge. While the *civil law* is formulated, compiled and improved and then codified and given the legal power by the Legislators. Therefore, within the *common law* jurisdiction, in relation to the legal institutions, the typical things are the beliefs, the principles of error, and judgment. While in *civil law* jurisdiction, there is a strong tendency towards formalism and rules that protect moral integrity.[15]

In conclusion, the *civil law* system promotes legal positivism. Legal positivism has several meanings. Most of its supporters adhere to 2 (two) basic propositions: *First*, the definition of law should not depend on the question of moral validity. *Second*, the law should only be identified according to concrete formal provisions such as legislation. H.L.A. Hart basically states that law is a human command and there is no need for a relationship between the law and morality.[16]

Therefore, the political compromises between the Regional Head and DPRD in relation to the management accountability of regional finances cannot be separated from formal procedural errors that exist in the legislation. The legislation established by the legislature provides an opportunity for political compromises, i.e. the accountability of the Regional Head against the management of regional finances cannot be rejected by DPRD. However, with the existing authority in DPRD (supervisory function), DPRD can give pressure to the Regional Head so that he/she will inevitably have to collude with DPRD by ignoring the moral responsibility.

In the *civil law* system adopted by the Indonesia (prioritizing legal positivism), there is no obligation to be morally responsible. The most important thing is that everything should be done in accordance with the laws and regulations even if they are contrary to the sense of justice in the society. Because one characteristic of legal positivism is: there is no need for a relationship between the law and morality.

Should it happen in a country with the *common law* system, the Regional Head should not worry about getting a pressure from the DPRD because the measure does not refer to the formal procedure contained in the legislation, but it is the verdict of the Judge that decides whether or not the responsibility of the Regional Head against the management of regional finances has met the ideals of the regional autonomy.

5 Conclusions

5.1 Summary

In principle, the accountability of the Regional Head to the management of regional finance is a manifestation of a democratic government. Because without accountability then the meaning of democracy disappears because the government has turned into a ruler who does not need to give accountability. This is deemed necessary because the identity of constitutional democracy is a government that is limited its power from despotism. However, in its application the accountability mechanism of the Regional Head's management of regional finances is currently based on the political interests of the existing political institutions in the region. As a result, the purpose of the implementation of regional autonomy is not achieved, one of which is to provide a better prosperity for the existing communities in the region. The political compromises amongst the political institutions in the region are caused by the disproportion of power between the DPRD and the Regional Head.

5.2 Suggestions

It is recommended that the mechanism of accountability of the Head of Region towards the management of regional finances is implemented by giving space for the society. The provision of space for the society is to convey his opinion in order to conduct a joint discussion of LKPj. In case the LKPj is inconsistent with the RKPD, then the DPRD recommends the Regional Head for improvement. If the Head of Region has not made any improvements, or has made improvements but still not in accordance with RKPD, the DPRD shall use its supervisory function through interpellation rights, inquiry rights and petition rights. Then the result of the DPRD oversight function is tested in the Supreme Court as the authorized institution as the final decision whether the LKPj has been in accordance with RKPD or not. Besides, the Regional Head must also have the authority to supervise the performance of DPRD in order to check and balances system in the implementation of the Local Government is running well.

References

1. T. Triwulan Tutik, *Pokok-pokok hukum tata negara*, (Prestasi Pustaka, Jakarta, 2006).
2. F. Akbar Nasution, *Dimensi hukum dalam pemerintahan daerah*, (Pustaka Bangsa Press, Medan, 2003).
3. K. Marijan, *Sistem politik Indonesia, konsolidasi demokrasi pasca orde baru*, (Kencana Prenada Media Group, Jakarta, 2010).
4. F. Akbar Nasution, *Pemerintahan daerah dan sumber-sumber asli pendapatan daerah*, (PT Sofmedia, Medan, 2009).
5. C.E. Kaunang, A.T. Naukoko, A.T. Londa, Analisis kinerja pengelolaan keuangan daerah di era otonomi daerah: Studi pada kota Manado (Tahun 2010-2014), *Jurnal Berkala Ilmiah Efisiensi,* **16**, 02 (2016).
6. B. N. Marbun, DPRD dan Otonomi Daerah Setelah Amandemen UUD 1945 dan UU Otonomi Daerah 2004, (Pustaka Sinar Harapan, Jakarta, 2005).
7. A. Syafrudin, *Kapita selekta hakikat otonomi dan desentralisasi dalam pembangunan daerah*, (Citra Media, Yogyakarta, 2006).

8. A. Yani, *Hubungan keuangan antara pemerintah pusat dan daerah di Indonesia*, (PT Raja Grafndo Persada, Jakarta, 2008).
9. A. Yani, *Hubungan keuangan antara pemerintah pusat dan daerah di Indonesia*, (PT Raja Grafndo Persada, Jakarta, 2008).
10. S. Sunaryo, *Hukum pemerintahan daerah di Indonesia*, (Sinar Grafika, Jakarta, 2008).
11. D. Thaib, *Teori dan hukum konstitusi*, (PT Raja Grafindo Persada, Jakarta, 2003).
12. M. Fuady, *Teori negara hukum modern*, (PT Refika Aditama, Jakarta, 2009).
13. H. A W. Widjaja, *Penyelenggaraan otonomi di Indonesia*, (PT Raja Grafindo Persada, Jakarta, 2008).
14. J. Kaloh, *Mencari bentuk otonomi daerah*, (Penerbit Rineka Cipta, Jakarta, 2002).
15. P. de Cruz, *Perbandingan sistem hukum common law, civil law and socialist law*, (Nusa Media dan Diadit Media, Bandung dan Jakarta, 2016).
16. W. Menski, *Perbandingan hukum dalam konteks global sistem Eropa, Asia dan Afrika*, (Nusa Media, Bandung, (2016).

Effectiveness of blower and HVLS fan in feedlot to performans of brahman, belmont red and charolais cattle

David Isaac Yasin, *Ma'ruf* Tafsin*, and *Simon P* Ginting

Agricultural Faculty, Universitas Sumatera Utara

Abstract. The purpose of this study is to determine the impact of heat stress on Brahman Cross cattle performance which is maintained by feedlot with three treatments (non using fan, blower fan and HVLS fan) using split plot design, Giving the fan is treated as the main plot factor, while the cattle is treated as a plot factor factor (sub plot factor). Knowing the financial value of each treatment: feed consumption, daily weight gain, feed conversion, feed cost, cost per gain of Brahman, Belmont red and Charolais is fattened in feedlot. The result showed that using HVLS fan is most effective to cattle performance than blower fan and non using fan for belmont red and charolais cattle.

1 Introduction

Indonesia is one of the countries with very high meat needs, the average annual requirement in Indonesia reaches 3.5 million head or equivalent to 680,000 tons per year which can be fulfilled by local supply about 54% of total requirement, so that the government grants import permits in the effort to fulfill the national meat requirement in the form of meat and beef, almost 100% of imported cattle are from Australia. The age of cattle imported to Indonesia is 1.5 years to 3 years with an average weight of about 250 kg to 340 kg. Cattle imported to Indonesia are delivered by ship with a comfortable design for cattles, a long journey from Australia to Indonesia between 5-10 days of travel.

Imported cattle imports from abroad are entirely private, and intensive feedlot feed for three to four months. Cattle imported into Indonesia is a brahman cross (BX) cattle species originally from Australia, there are many types of Brahman cross cattle from Australia crossed with the nation of cows with superior genetics, especially European beef cattle (Bos Taurus). The type of cattle that is commonly sent to Indonesia from Australia, 75% is brahman cattle genetic that has been crossed with European and American cattles, while 25% are European cattle that are kept in cold areas (southern Australia) and almost all cows are delivered to Indonesia is a brahman crosses cattle. Constraints faced by feedlotter usually is if they get cattle from the south (cold areas) that temperature between (6 ° C to 18 ° C) not from the north or west of a temperature that is similar to the temperature of the tropics (21 ° C to 32 ° C) with an average humidity of 60%. These obstacles include respiratory disease due to overheated cattles (average temperature 32 ° C) and high

* Correspondent Author: martafsin@yahoo.com

humidity reaches 80% on average, this greatly affects the productivity of cattle that we will keep. Another case if we sent cattles from northern Australia or western Australia cattles are usually already resistant to the temperature in Indonesia, because there has been a cross with brahman cattles that have proven resistant to hot temperatures.

There are four elements of microclimate that can directly affect the productivity of livestock, namely: temperature, air humidity, radiation and wind speed, while the other two elements of evaporation and precipitation affect indirect livestock productivity. Temperature and air humidity are two climate factors that affect cattle production, because it can cause changes in heat balance in the body of cattle, water balance, energy balance and balance of livestock behavior [1]. Life and production in livestock require optimum environmental temperature *Robert* E.[2]. Comfortable temperature thermonetral zone for European cattle ranges from 13 - 18 ° C [3]; 4 - 25 ° C [4], 5 - 25 ° C. Daily air and humidity in Indonesia are generally high, ranging from 24 - 34 ° C and 60 - 90% humidity. This will greatly affect the level of cattle productivity. At these temperatures and moisture, the evaporation process of the cattle's body will be inhibited so as to experience heat stress. The effects of cattle due to heat stress are decreased appetite, increased consumption of drinking, decreased metabolism and increased catabolism, increased heat release through evaporation, decreased blood hormone concentration, increased body temperature, respiration and heart rate [5] and changes behavior [6], increased intensity of shelter. Environmental modification is necessary if there are genetically dominant cattles of European cattle, such as; Angus, Charolais and Belmont red, although the cattle are all completely crossed with Brahman cattles. Modified environments necessary to optimize the growth of the cattle. One of the success factors of cattle feedlot is optimization of sheed management, including sheed design and maintenance support facility. In the feedlot sheed management one of the optimization management of the sheed is by the addition of a fan, can be a blower or a hanging large fan or often called HVLS fan (High Volume Low Speed). The purpose of the fan usage is to lower the room temperature and speed up the release rate of ammonia in the sheed, thus making the cattle feel comfortable especially for the cattle that come from cold regions (south Australia). So far there has not been much research literature on cattle maintenance in feedlot using fan, so it is necessary to do research to know the effectiveness of different fan usage

2 Materials and Methods

The cattle used in this study were 15 Brahman steers ,15 Charbray / Charbrah steerers and 15 Belmont Red steerers with an initial body weight of 310 ± 10 kg were used in this study. Gang way connected with cattle crush, individual scale digital cattle brand True Test, a computer unit, the sit scales. This research was conducted for ± 4 months starting from November 2017 until February 2018 at Fattening Feedlot PT. Juang Jaya Abdi Alam, Deli Serdang District, Noth Sumatera.

The design used in this study is the Design Split plot Design. This design is used for experiments intended to investigate major influences and interactions with unequal degrees of accuracy. Giving the fan is treated as the main plot factor, while the cattle is treated as a plot factor factor (sub plot factor). Place of experiment divided into block, number of replication. The division of the block is in accordance with the principle of local control. Each block is divided into main plot (PU). The amount of PU in each block is the same as the type of treatment. The placement of treatment into PU is done randomly, and coded according to the treatment given.

3 Results and Discussion

Table. 1 Influence of fan and cattle breed on performance during maintenance period

Cattle breed	Type of fan			Average ± SD
	Control	Blower	HVLS	
1. Consumption (Kg/head/day)				
Brahman	13,66	13,86	13,68	13.73^{a} ± 0,11
Belmont Red	13,26	13,40	13,78	13.48^{b} ± 0,27
Charolais	13,78	13,82	13,82	13.81^{a} ± 0,02
Average ± SD	113.57^{a} ± 0,27	13.69^{a} ±0,25	13.76^{a} ± 0,07	
2. ADG (Kg/head/day)				
Brahman	1,85	1,81	1,80	1.82^{a} ± 0,03
Belmont Red	1,67	1,86	2,04	1.86^{a} ± 0,19
Charolais	1,71	1,87	2,04	1.87^{a} ± 0,17
Average ± SD	1.74^{c} ± 0,095	1.85^{b}± 0,032	1.96^{a}± 0,139	
3. FCR				
Brahman	7,42	7,71	7,62	7.58^{a} ± 0,15
Belmont Red	7,98	7,24	6,77	7.33^{a} ± 0,61
Charolais	8,11	7,39	6,77	7.43^{a} ± 0,67
Average ± SD	7.84^{a}± 0,37	7.45^{ab}± 0,24	7.05^{b}± 0,49	
4. Feed cost (IDR/Head/day)				
Brahman	31.418	31.878	31.464	31.586.67^{a} ± 253,35
Belmont Red	30.498	30.820	31.694	31.004.00^{b} ± 618,87
Charolais	31.694	31.786	31.786	31.755.33^{a} ± 53,12
Average ± SD	31.203^{a}± 626	31494^{a}± 586	31648^{a} ± 165	
5. Cost of gain (IDR/kg PBB)				
Brahman	17.076	17.725	17.517	17.438.95^{a} ± 331,49
Belmont Red	18.361	16.656	15.567	16.861.24^{a} ± 1.407,92
Charolais	18.656	17.007	15.567	17.076.67^{a} ± 1.545,91
Average ± SD	18030^{a} ± 840	17129^{ab} ± 544	16216^{b}±1125	

Description: unequal letters in rows show significant differences in ($P < 0.05$), unequal letters in the columns show significant differences ($P < 0.05$)

3.1 Feed Consumption

Table 1 shows that fan usage has no significant effect ($p < 0.05$) on feed consumption with average value of 13.57 kg / head / day (pen with non fan), 13.69 kg / head / day (cage with blower fan) and 13.76 kg / head / day (enclosure with HVLS fan). It can also be seen that each cow has the same average value for Brahman 13,73 kg / head / day, Belmont Red 13,48 kg / head / day and Charolais 13,81 kg / head / day. Cattle reaction to temperature changes seen from the respiratory response and heart rate is a mechanism of the cow's body to reduce or release heat received from outside the body of the cattle. Increased heart rate is a response from the cattle body to spread the heat received into the cooler organs. Breathing is the response of the cattle body to remove or replace heat with the surrounding air. If these two responses do not succeed in reducing the extra heat from outside the body of the cattle, the temperature of the livestock organs will increase so that the cattle experience heat stress (*Anderson* .B. Pleasant, 1983).

The ongoing heat stress on cattle will have an impact on increased consumption of drinking water, decreased production, increased urine volume, and decreased feed intake.

3.2 ADG (Average Daily Gain)

Table 1 shows that the use of enclosures with HVLS fans is an effective fan use of ADG growth in each breed of cattle with an average of 1.96 kg/heads/day, blower fan average 1,85 kg/heads/day and non-fan pen has an average of 1.74 kg/heads/day for the cattle breed. Can also be seen in each cattle breed has the same average with the average value of each for Brahman 1.82 kg/heads/day, Belmont Red 1.86 kg/heads/day and Charolais 1.87 kg/heads/day.

3.3 FCR (Feed Conversion Ratio)

Table 1 shows that the use of non-fan enclosure is the most ineffective fan usage of the increase of FCR in each cattle breed with an average of 7.84, blower fan at 7,45 and the use of the pen with the most effective is HVLS fan with average of 7.05 for the cattle. In this table can also be seen in each cattle breed has the same average with the average value of each for Brahman 7.58, Belmont Red 7.33 and Charolais 7.43.

3.4 Feed Cost

Table 1 shows that fan usage for each cage has the same effect for each cattle breed on feed cost with 31,203,33 IDR/heads/day (control/non-fan) 31,494,67 IDR/heads/day (blower fan) 31,648 IDR/heads/day (HVLS fan). In this table can also be seen in each cattle breed has the same average with the average value for each Brahman 31,586.67 IDR/heads/day, Belmont Red 31.004 IDR/heads/day and Charolais 31.755,33 IDR/heads/day.

3.5 Cost per Gain (Cost of Increase Weight Gain)

Table 1 shows that the use of non-fan pen is the most ineffective usage of each cattle breed with 18.030.76 IDR/heads/day ,blower at 17.129 IDR/heads/day and the most effective use of pen with HVLS fan has an average of 16.216.85 IDR/heads/day for the cattle. In this table can also be seen in each cattle breed has the same average with the average value of each for Brahman 17.438.95, IDR/heads/dayBelmont Red 16.861,24 IDR/heads/day and Charolais 17.076.67 IDR/heads/day.

4 Conclusion

There are some significant differences performance using Blower and HVLS fan with non using fan from belmont red and charolais, but there is no different with brahman with this treatment, there are big differenet in ADG, Cost of Gain and Feed Convertion Ratio (FCR). Using HVLS fan most effective performance and could make high margin than using blower fan and non using fan. So feedlotter has to use this method to optimize cattle performance, if they are import cattle with mostly european breed such as Belmont Red and Charolais, but you can also not using this method if you are import cattle with mostly brahman breed, because using this method not effective.

References

1. B. Pleasants dan R.A. Barton, Comparison of the carcass characteristics of steers of different breeds and pre-weaning environments slaughtered at 30 months of age. *New Zealand Journal of Agricultural Research* 40: 57- 68. (1997)
2. Aus-Meat, *Aus-Meat Languange. 2nd Edition*. (Aus-Meat, Sydney, 1987)
3. B. Brahmantiyo, Karkas, sifat fisik dan kimia daging tiga bangsa (Brahman, Angus dan Murray Grey). [Tesis]. Program Pasca sarjana. Institut Pertanian Bogor, Bogor (1996).
4. Blakkely dan Bade, Ilmu Peternakan 3rd Edition. Terjemahan: B. Srigandono dan Soedarsono, (Gajah Mada University Press, Yogyakarta, 1994)
5. B. Sarwono dan H. B. Arianto, *Penggemukan Sapi Potong Secara Cepat* (Penebar Swadaya, Jakarta, 2003)
6. B.S. Siregar, *Penggemukan Sapi*, (Penebar Swadaya, Jakarta, 2006).

Smallholders' willingness to pay for dura marking oil palm seeds

Diana Chalil[1], *Mohammad* Basyuni[2,*], *Riantri* Barus[1], and *Lollie* AP Putri[3]

[1]Department of Agribusiness, Faculty of Agriculture, Universitas Sumatera Utara, Medan, North Sumatera 20155, Indonesia
[2]Department of Forestry, Faculty of Forestry, Universitas Sumatera Utara, Jl. Tri Dharma Ujung No. 1 Medan, North Sumatera 20155, Indonesia
[3]Department of Agrotechnology, Faculty of Agriculture, Universitas Sumatera Utara, Medan, North Sumatera 20155, Indonesia

Abstract. In 2018, around 4 million ha of oil palm smallholdings need to be replanted. Indonesia could utilize this moment to improve smallholdings' productivity by ensuring the usage of good quality seeds. For the industry, Dura is considered as bad seeds as its thick shells can damage the machine and the oil content is low. Therefore, most mills do not want to buy it, hence the selling price is low. In fact, most smallholdings' use uncertified seedlings that include many contaminated Dura. Unfortunately, smallholders can only detect this 3 years after the planting and spending a considerable amount of production costs. Actually, to avoid the unnecessary loss a Dura marking technique can be applied to the oil palm seeds. However, most smallholders are unlikely consider Dura usage as an important issue due to the higher weight and higher sales revenue. This study is conducted to analyse smallholders' willingness to pay (WTP) for seed marking service. Data is collected from 60 independent smallholders in Labuhan Batu Utara, which mostly use the uncertified seedlings. The willingness to pay is measured with Conjectural Variations Approach, revealing 71% of the respondents are willing to pay the Dura seed marking service, although still with a minimum level of WTP. Pearson Chi Square Correlation Test shows that the WTP for Dura Marking Service is only significantly correlated with the smallholders' knowledge and perception, but not with their land size and experience. Based on the results, a number of policy implication are included.

1 Background

Oil palm growers in Indonesia can be differentiated into 3 groups, namely state companies, private companies and smallholders. Smallholdings are the most recently developed group, just started in 1979 with 3,125 ha, but exponentially grew to 4.7 million ha in 2017 or around 38.64% of the total Indonesian oil palm areas [1]. Initially smallholdings were developed within government program as plasma of big companies. Successfully increasing the schemed smallholders' income, many independent smallholders then started developing

* Corresponding author: m.basyuni@usu.ac.id

their oil palm plantations. However, without big companies' supervision, most of the independent smallholdings did not apply the recommended cultivation techniques, including the usage of good quality seeds. In fact, the usage of counterfeit seed tends to be particular constraint for oil palm smallholders all over the world [2]. There are 3 types of oil palm seeds, namely Dura, Pisifera and Tenera. Tenera is considered as the high quality seeds, thus determined as the certified seed, which has higher productivity and oil content. It is the cross type from Dura and Pisifera. In 2012 previous study found in North Sumatra Province, 80.28% of independent smallholders used the uncertified and illegitimate seedlings [3]. In 2015 similar condition still appeared, in which 60% of smallholdings in Sumatra used uncertified and illegitimate seedlings. Among them 90% are independent smallholders [4]. In addition, [5] found that the smaller the smallholding area, the less Tenera trees are found. Dividing 243 smallholdings' size into small, medium and large group, each has 12%, 24% and 49% Tenera of the total seeds, respectively. As a result, independent smallholdings' productivity appeared to be the lowest comparing to the schemed smallholdings, state and private plantations, which are 2.92, 3.52 and 3.52 ton CPO/ha/year, respectively.

On average Dura productivity and oil content are 10-12 ton FFB/ha/year and 12-18%, respectively (Lubis 2004 in [4] [6]). Although with good treatment, the productivity of the uncertified seedlings could be improved, but their maximum productivity are only 60-70% of those the certified optimal productivity. With a significant smallholdings' share in the Indonesia oil palm plantations, such a condition makes the average of Indonesia's oil palm plantation productivity only 3.19 ton CPO/ha/year, while the potential is 8-9 ton CPO/ha/year [4]. The low productivity increases the average cost, which at the end could reduce the Indonesian palm oil competitiveness. However, as the average of oil palm productivity 5 times higher than other vegetable oil productivities, currently palm oil prices are still the lowest and recorded as the highest traded vegetable oil in the world.

The main factor for smallholders not using the certified seeds is the big price difference between the certified and non-certified. In fact, seeds' costs in only 3% of the total production costs, while the impact will take place for the whole 25 years of the production period. [7] predicted that the usage of high quality seeds and improving the agricultural practice could increase oil palm smallholding yield around 26%. Another reason is the lack of knowledge among smallholders, which presumed that all Tenera trees will always produce Tenera FFB [8][9]. In fact, the crossed of these seeds will produce around 25% Dura 50% Tenera, and 25% Pisifera [6][10][11]. Unfortunately, smallholders can only detect Dura from the FFB, which first harvest is 3 until 4 years after the planting. If smallholders do not have to wait until the early fruit to marking the good seeds, the unnecessary loss could be avoided.

On average, the world palm oil production is around 3 ton CPO/ha/year, while the maximum could reach 12 ton CPO/ha year. Theoretically the potential could even reach 18.5 ton CPO/ha/year [7]. This means, closing all production gaps could increase the world production without further increasing land clearing. However, most smallholders do not concern this as an important issue and will to expend some extra costs. This study is conducted to analyze smallholders' willingness to pay (WTP) for seed marking service. The results are expected to give some information for policy makers in improving oil palm smallholdings' performance.

2 Method

This is study is conducted in Kabupaten Labuhanbatu Utara, North Sumatra Province, which is one of the oil palm plantation centers. Primary data is collected from 60 independent smallholder samples, which are selected with purposive sampling method.

There are 2 types of Willingness to Pay (WTP) measured, namely WTP for using legitimate seed and WTP for Dura marking. WTP is measured with the Conjectural Variation (CV) approach where samples were directly asked about their WTP for using the seed marking service and for buying certified/legitimate seeds. WTP were revealed using two types of questionnaire, namely the open-ended and the bidding game approach of maximum amount that samples' would be willing to pay. The amount is divided into 6 classes, which are 0 for those who do not want the service, WTP < IDR10,000 (USD0.72), IDR10,000 (USD0.72) < WTP < IDR20,000 (USD1,45), IDR20,000 (USD1,45) < WTP < IDR30,000 (USD2.17), IDR30,000 (USD2.17) < WTP < IDR40,000 (IDR2.89), and WTP > IDR40,000 (IDR2.89). Then the 4 possible correlated factors with the Dura marking WTP, namely land size, experience, knowledge and perception were tested using the Pearson Chi-Square cross tabulations. Land size is measured in hectare. Experience refers to samples' year of working at or managing oil palm plantation. Knowledge is measured with scores on smallholders' awareness about seed and seedling types, characteristics and benefit of not using Dura. Perception refers to samples' view about the benefit on using Dura marking service.

3 Results and discussion

The survey shows that 89% of the respondents can differentiate Dura and Tenera, which all know from the shell size of their FFB. This means that respondents revealed it after the first harvest, which is 3 to 4 years after planting. These respondents know that Dura is heavier, while Tenera has higher oil content. However, only 5% agree that Tenera is beneficial for both smallholders and mills, while the rest prefer Dura because all of them sell their FFB to traders that do not grade the FFB. The Pearson Chi Square value of 3.418 shows that no significant difference between the average productivity of smallholdings using the certified and uncertified seeds. On average both receive yield in between 1 to 2 ton FFB/ha/month. Similarly, The Pearson Chi Square of 0.804 also shows no significant income difference between smallholders that use the certified and uncertified seeds. However 29% of the certified have enjoyed more than IDR2,000,000 (USD144.59) per month, while only 15% of the uncertified receive the similar income amount.

Among the samples, 75% suggest Dura is more profitable than Tenera. Therefore, most (77.97%) smallholders still use the uncertified/illegitimate seedlings, mainly because the price is significantly lower than the certified, which are IDR 15,000 (IDR1.08) and IDR 40,000 (IDR2.89), respectively. The study shows that all respondents are willing to pay the certified seedlings, but only 29% with WTP IDR 25,000 (USD1.81) or more for the average price difference between the certified and non-certified seedlings. This is in accordance with the fact that 77.97% of the samples use uncertified seedlings, as on average the price difference between the uncertified and certified seedlings is about IDR25,000 (USD1.81). More than 80% of the smallholders got the seedlings from other smallholders and illegitimate traders. Only 16.67% of them bought the seedlings from palm oil companies, input production shops and nurseries. Such a condition indicates that smallholders and illegitimate breeders still have important roles in supplying seeds and seedlings for smallholders. Therefore, Indonesian Oil Palm Research (IOPRI) includes them in their mentoring and coaching program, so that smallholders could obtain good quality seed and seedlings [6].

In contrast with the WTP for the certified seedlings, 29% of the samples do not willing to pay for the Dura marking service. In general, smallholders WTP for certified seeds are greater than their WTP for Dura marking service. The complete comparison of the WTP for getting certified seedlings and WTO for having marking service can be seen in Figure 1.

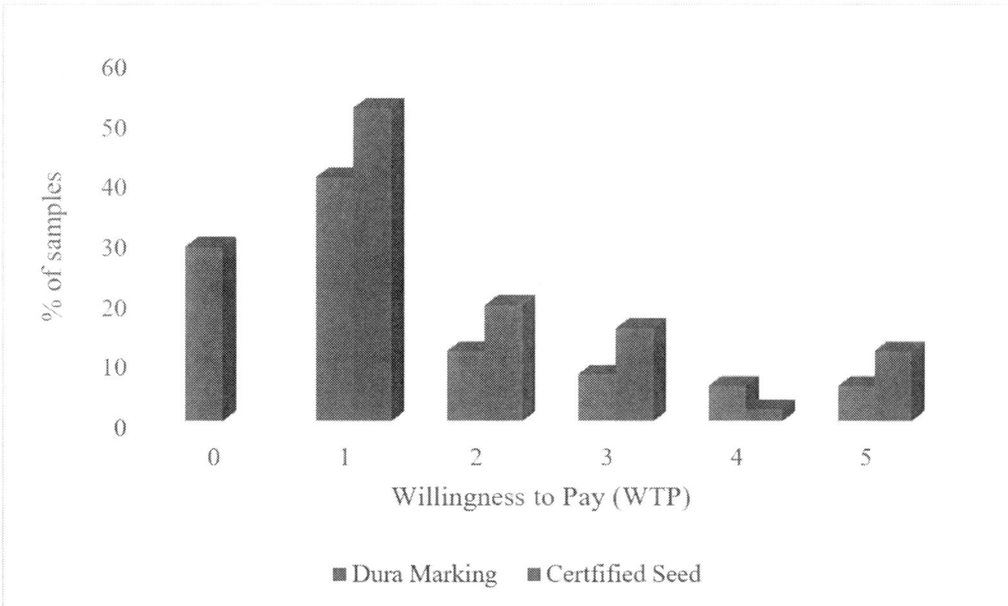

Fig. 1. Sample distribution based on 6 classes of WTP for certified seeds and dura marking service.

Figure 1 shows that for Dura marking service, there are still 29% of the sample have 0 score, indicating their unwillingness to pay for the service. This might indicates the smallholders' view about Dura, which is not the best seeds but still beneficial for its heavier weight. In other words, these smallholders do not in purpose pay for the Dura but they would be happy to enjoy the benefit if coincidentally get the Dura. Therefore, they do not want any extra expenditure to get the marking service. However, 71% of the respondents are already willing to pay the Dura marking service, although most (40%) with minimum amount (maximum IDR10,000 or USD0.72), and only 3% will to pay the maximum amount (IDR40,000 or USD2.89).

Table 1. Compare means test results between WTP for dura marking service and possible correlating factors.

Description	**Range**	**Average**	**Pearson Chi-Square**	
			Value	**Significance**
Land Size	1.00 – 5.50	2.22	47.074	0.206
Experience	1.00 – 56.00	16.00	125.092	0.245
Knowledge	0.00 – 4.00	1.79	31.770	0.046*
Perception	0.00 – 0.10	0.73	17.752	0.003**

Table 1 shows that land size and experience are not correlated with the smallholders' WTP for Dura marking service. With the range of 1.00 ha to 5.50 ha and 92.31% are less than 4 ha, most of them sell their FFB to village traders that did not differentiate the Dura and Tenera FFB. Therefore, land size is not significantly correlated with the WTP for marking service. Similarly, without involving sufficient learning process related to Dura seed, smallholders' experience is also not correlated to their WTP. In contrast, knowledge and perception are significantly correlated with the smallholders' WTP for the marking service. The smallholders' knowledge are mostly obtained from parents and friends (around 70%), 25% autodidact, and only 3.45% get it from palm oil companies through trainings or experience as the companies' workers. None got the information from extension workers. In such condition, information about Dura might be unclear and incomplete, thus might partly explain the zero and low score of their WTP for the marking service. Smallholders rarely receive complete knowledge about good seeds and seedlings, and Good Agricultural Practice. Some smallholders have known some indications of the bad seedlings, but only few of them understand various types of the bad seedlings. Most obtain comprehensive seedling knowledge through field trips to pre-nurseries and nurseries. Most smallholders still assume that certified seedlings are expensive, until they know the seedling expenditure share to the total production costs. They also know that Tenera has higher productivity, but most assume that Dura productivity could be improved with good fertilization. None understood that even with complete recommended fertilization, uncertified seedlings could only reach 70% of the certified seedlings productivity. Instead their willingness to pay, 70% of the samples are willing to bring the marking test material if the marking service is provided in the same district, otherwise only 30% are willing to do so. IOPRI provided mobile seedlings shop to improve smallholders' access on good quality seedlings. They periodically visit 8 smallholding centers in Sumatra every 3-6 months, and has distribute 1,692,650 seeds from 2011-2015. In fact, after the visit, some smallholders even directly come to IOPRI office in Medan to buy certified seedlings [6].

4 Policy implications

Improving the usage of good quality seeds and seedlings is important among smallholders. Currently, most smallholders still use illegitimate seedlings with Dura dominance, thus having low FFB productivity with low oil content. Most smallholders know the Tenera quality, but still many prefer Dura due to its heavier weight. In open market Dura and Tenera FFB selling price is not differentiated. In government FFB selling price, oil content has been concluded as one of measurement component. However, this price only applies to schemed smallholders. In fact, independent smallholders share is increasing. Therefore, similar selling price components also need to be applied in open market, which can be interfered through the Plantation Directorate General. On the other hand, similar program of the certified seeds and seedlings could be modified for the Dura marking service. This is not only limited for smallholders as consumers but also small scale oil palm nurseries, which mostly supply seeds for smallholders.

Acknowledgements

This work was supported in part by a Penelitian Strategis Nasional Institusi (PSN Institusi 2018) from the Directorate for Research and Community Service, Ministry of Research, Technology and Higher Education, Republic of Indonesia.

References

1. Tree Crop Estate Statistics of Indonesia 2015-2017: Palm Oil, Secretariate of Directorate General of Estate Crops, Directorate General of Estate Crops, Ministry of Agriculture
2. L.S. Woittieza, M.T. Van Wijk, M. Slingerlanda, M. van Noordwijk, K.E. Giller. Yield gaps in oil palm: A quantitative review of contributing factors. *European Journal of Agronomy* **83**, (2017)
3. D. Chalil, Z. Zein Analisis implemetasi kesepakatan RSPO pada perkebunan sawit rakyat. Research Report. National Strategic Fund (Hibah Strategis Nasional),

Direktorat Jenderal Pendidikan Tinggi, Kementerian Pendidikan dan Kebudayaan. (2012)

4. M.A. Agustira, R. Amalia, R. Nurkhoiry. Program sawit untuk rakyat (Prowitra) sebagai upaya peningkatan produktivitas, pemberdayaan, keberlanjutan, dan kesejahteraan pekebun kelapa sawit rakyat. Prosiding Seminar Nasional "Perlindungan dan Pemberdayaan Pertanian dalam Rangka Pencapaian Kemandirian Pangan Nasional dan Peningkatan Kesejahteraan Petani". Pusat Sosial Ekonomi dan Kebijakan Pertanian, Kementerian Pertanian. (2015)
5. I. Jelsma, G.C. Schoneveld. Mewujudkan petani kecil sawit mandiri yang lebih produktif dan berkelanjutan di indonesia pandangan dari pengembangan tipologi petani kecil, Pusat Penelitian Kehutanan Internasional (CIFOR), Universitas Utrecht. Working Paper 217, (2016)
6. K. Setiawan. Pemuliaan kelapa sawit untuk produksi benih unggul: tanaman pendek, kompak, dan minyak tak jenuh tinggi (Plantaxia, 2017)
7. T. Soliman, F.K.S. Lim, J.S.H. Lee, L.R. Carrasco. Closing oil palm yield gaps among Indonesian smallholders through industry schemes, pruning, weeding and improved seeds. *Royal Society Open Science Aug 3*, 8 (2016)
8. E. Akpo, P.V. Vissoh, R.C. Tossou, T. Crane, D.K. Kossou, P. Richards. T.-J. Stomph, P.C. Struik. A participatory diagnostic study of the oil palm (Elaeis guineensis) seed system in Benin. *Wageningen Journal of Life Sciences* **60–63**, (2012)
9. Aprizal. Analisis daya saing usahatani kelapa sawit Kabupaten Mukomuko. Tesis, Program Studi Magister Agribisnis, Fakultas Pertanian, Universitas Bengkulu. (2013)
10. M. Basyuni, N. Amri, L.A.P. Putri, I. Syahputra, D. Arifiyanto. Characteristics of Fresh fruit bunch yield and the physicochemical qualities of palm oil during storage in North Sumatra, Indonesia. *Indonesian Journal of Chemistry* **17**, 2 (2017)
11. M. Basyuni, R. Wati, I. Deni, A.R. Tia, B. Slamet, E.S. Siregar, I. Syahputra. Cluster analysis of polyisoprenoid in oil palm (*Elaeis guineensis*) leaves in different land-uses to find the possible cause of yield gap from planting materials. *Biodiversitas* **19**, 4 (2018)

Fiscal legitimacy of environmental taxation: challenges for green growth policy

Edi Slamet Irianto[1], *Haula* Rosdiana[2,*], *Lucas Filberto* Sardjono[2], and *Maria R. U. D.* Tambunan[2]

[1]Graduate Program of tax Policy and Administrative Science, Faculty of Administrative Science, Universitas Indonesia, Depok, Indonesia
[2]Faculty of Administrative Science, Universitas Indonesia, Depok, Indonesia

Abstract. In Indonesia the fiscal legitimacy of environmental taxation has not received serious attention from tax experts. This is proved by the limited number of studies investigating it. There are at least three interesting perspectives. First, the Indonesian perspective as a country, this research is very necessary and the results are awaited in order to get answers to the failure of Indonesia in application of environmental taxes. Second, policy maker perspective, the result of this research is needed to build policy based on evidence and Third, scientific perspective to introduce contemporary tax paradigm that should be well understood by government and society. It is in this context that the tax essence adequately can be well known in the structure, the posture of state levies and its impact on the productivity and competitiveness of society. This research aims to explain the implications if the government implements new environmental taxes (e.g. environmental taxes) and challenges of fiscal legitimacy from environmental taxes within the framework of green growth policy. By using constructivism paradigm, this research is done through documentation and literature study, and technical data analysis in the form of meta-analysis. The research results show that there are some pseudo environmental taxes and charges applied in Indonesia. That is, if the government impose a new tax type, then the government will add compliance costs and distort entrepreneur cash flow. However, if the government can manage revenue from environmental taxes and charges through good spending quality, it will create tax justification that will rise the tax trust. Therefore, spending quality should be done transparently, accountably and consistently, to ensure that environmental taxes that do not cause high economic costs, do not reduce competitiveness, and ensure the stability of state revenues. Thus, the restructuring, convergence and harmonization of environmental taxes and charges must be conducted in order to achieve a connection between tax payments and inclusive and sustainable economic growth can be created.

* Corresponding author: h.rosdiana@ui.ac.id

1 Introduction

Taxes are often regarded as effective social, economic and political engineering, so that governments often use it as the instruments in achieving government programs. Although its effectiveness is still often questionable, yet belief that taxes can overcome the problem of negative externalities led to theories like pigouvian tax [1] still considered relevant today. Environmental taxes had been increasingly used to influence the behaviour of economic actors whether producers or consumers [16], can encourage more environmentally responsible behaviour [17] and make the economic actors using resources and energy more efficiently [19]. However, taxation would be a boomerang for the government if the government is too excessive without being able to build a fiscal legitimacy for the various taxes that imposed [2]. In this context, it will be explained why environmental taxes are opposed and fail to apply in Indonesia.

Currently, there are five theories that discuss environmental tax, namely (1) polluter-pays principle; (2) precautionary principle; (3) least-cost abatement; (4) double dividend theory; and (5) microeconomic approaches [1]. Environmental tax is rooted in the tax function as a regulatory instrument to overcome market failures due to negative externalities [18], i.e. "situations where economic activity results in social costs which are not paid for the producer or the consumer who causes them."[3]. In practice, often, producers do not fully bear the costs incurred due to hazardous factory waste, which is an excess of a goods production process. In some cases, factory waste has an impact on the destruction of the environment [4] and then the communities who bear such costs or side effects. If that happens, then the government must also be responsible to overcome it. Smoke from factory as an excess of iron production process, can pollute the air. If there is no government intervention, massive iron production will aggravate air pollution and may cause respiratory illness. The market does not deal with such a complex problem and the market has no authority to limit its adverse effects and to punish every person or institution that causes it. Therefore, the state that must function as a Regulator, among others by requiring entrepreneurs to make an analysis of environmental impacts as a form of environmental behavior [5], create a waste disposal site and/or by tax collection.

In dealing with negative externalities, economists generally advocate tax collection instruments because they are considered more efficient to reduce pollution than if the government only makes regulations about pollution. Tax collection will affect the price thus affecting the shifting supply curve (see Figure 1).

However, tax theory to overcome negative externalities should not stand alone and isolated with other tax theories, so that the policy formulation is not partial, not comprehensive and not holistic so that the purpose of tax collection is not achieved. It is in this context that a fiscal legitimacy perspective is needed so that taxation can strengthen the relations between state and people in achieving the goals of the nation and the state.

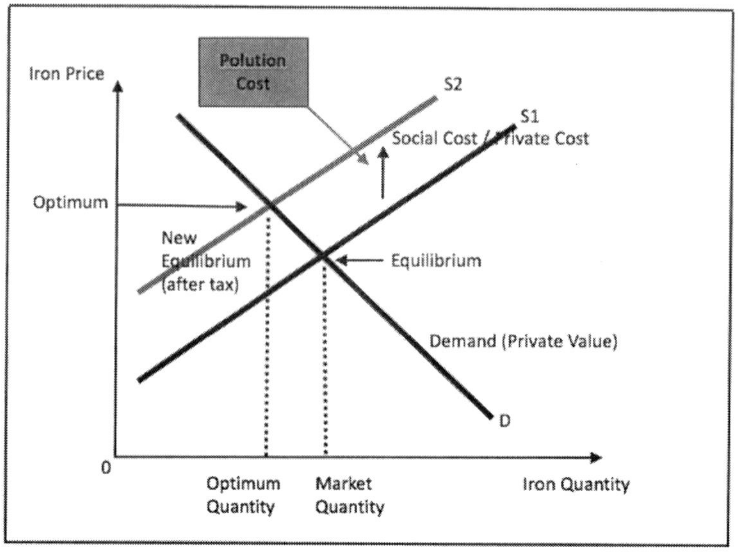

Fig. 1. Iron market after government intervention (through tax collection).

Source: Mankiw, idea modified by Rosdiana and Irianto, 2015

2 Research Methodology

This research is conducted by using constructivism paradigm. Through this paradigm the researcher understands, interprets, and analyzes social phenomena from the collected data. The meaning of social reality is built and then constructed with a cognitive scheme through a structured concept. The research method occupies qualitative approach by using inductive principle. Data collection is done through documentation and literature studies. The data collected comes from various regulation that related to pseudo environmental taxes & charges in Indonesia. Data analysis technique is meta-analyzes, with retrospective observational studies to provide a holistic overview of a phenomenon under review. Meta-analysis was raised to address research issues in various social sciences. Meta-analysis can make visible findings contradictory and difficult to accumulate, becoming more integrative and systematic [6].

3 Result and Discussion

Although environmental taxes have failed to apply in Indonesia, but essentially, there are several types of taxes that are pseudo environmental taxes, such as Street Lighting Tax (PPJ), Motor Vehicle Fuel Tax (PBBKB), Motor Vehicle Tax (PKB), Surface Water Tax (PAP), and Underground Water Tax (PABT). The tax rate, the tax base, and the tax revenue utilization allocation are shown in Table 1.

Table 1 describes the various taxes that can be categorized as environmental taxes that have been applied in Indonesia. In general, it can be seen that earmarking has not been applied clearly and firmly. Therefore, it is understandable why people have not yet felt strong fiscal legitimacy over these tax levies.

Table 1. Environmental taxes in Indonesia.

Variety of Taxes	Tax Rate	Tax Base	Spending Allocation
Motor Vehicle Tax (PKB)	1. The first ownership is at least 1% and the is highest 2%; 2. The second and subsequent ownership is set progressively at the minimum 2% and at the maximum 10%.	1. Selling Value of Motor Vehicle; and 2. Weight that reflects the relative degree of road damage and / or environmental pollution caused by the use of motor vehicles.	10%
Motor Vehicle Fuel Tax (PBBKB)	1. Private vehicle = the highest is 10%. 2. Public vehicles = the lowest is 50% lower than the cost of private vehicle's PBBKB.	• Vehicle Fuel Sale Value before Charged by Value Added Tax.	Not Clear
Underground Water Tax (PABT)	• The highest is 10%	• Surface Water Acquisition Value.	Not Clear
Street Lighting Tax (PPJ)	**1.** The highest is 10% for street lighting. **2.** The highest is 3% for the use of electricity from other sources by industry, petroleum and natural gas mining. **3.** The highest 1.5% for the use of self-generated street lighting.	• Selling Value of Electricity	Not Clear
Ground Water Tax	• The highest is 20%.	• Acquisition of Ground Water	Not Clear

Source: Law No.28/2009 concerning Local Tax and User Charge

Besides to the above taxes, there are several other types of state charges, essentially also pseudo environmental charges, such as cost for Environmental Impact Analysis (AMDAL), Life Environmental Management (UKL) and Environmental Monitoring Efforts (UPL) and also CSR as required by GR No. 27 of 2012 [8]. The cost of the various taxes and other state levies that can be categorized as pseudo environment tax and charges, by constructing the model by using the system dynamic analitical thinking. System dynamic analitical thinking commonly used to analyze hte implementation of public policy. The speciality of system dynamic is understanding the dynamics generated by systems composed of closed-loop relations, which has ability of representing economic, social, and environmental issues through simple simulation models. The system dynamics that shows the relation between pseudo environment tax & charges and its impact to the fiscal legitimacy, can be seen in Figure 2.

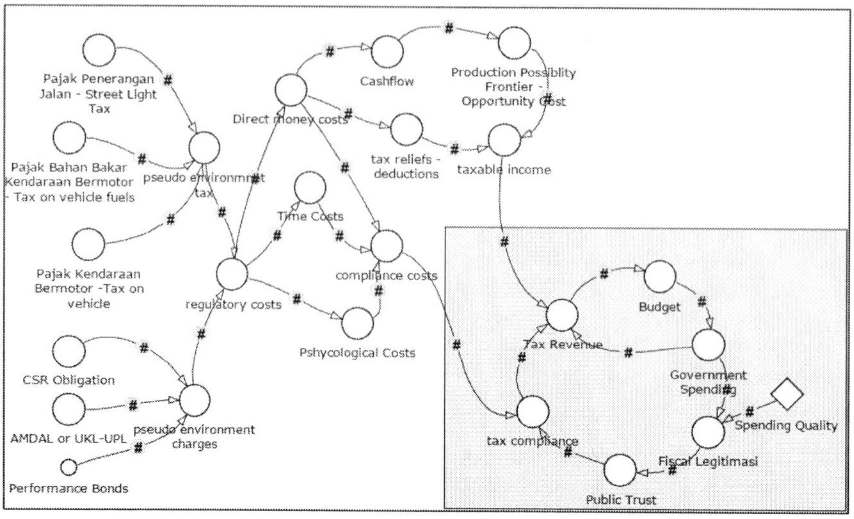

Fig. 2. Relationship between pseudo environment tax & charges and fiscal legitimacy
Source: data processing, 2018

From the figure above, it can be seen that various types of taxes and charges, in addition to disrupt the company's cash flow, also add direct money costs that must be issued by the company / taxpayer. Direct money costs are one factor that adds to compliance costs. The amount of compliance costs will affect to the tax compliance [9], because as the economic law states that "rational people think about profit / margin". Any addition of compliance costs will reduce the margin / profit to be received by the company / Taxpayer.

Implications if the government implements new types of taxes (e.g. environmental taxes) without making any simplification, harmonization and convergence of all pseudo environmental taxes and other levies which are pseudo environmental charges, can be described in Figure 3.

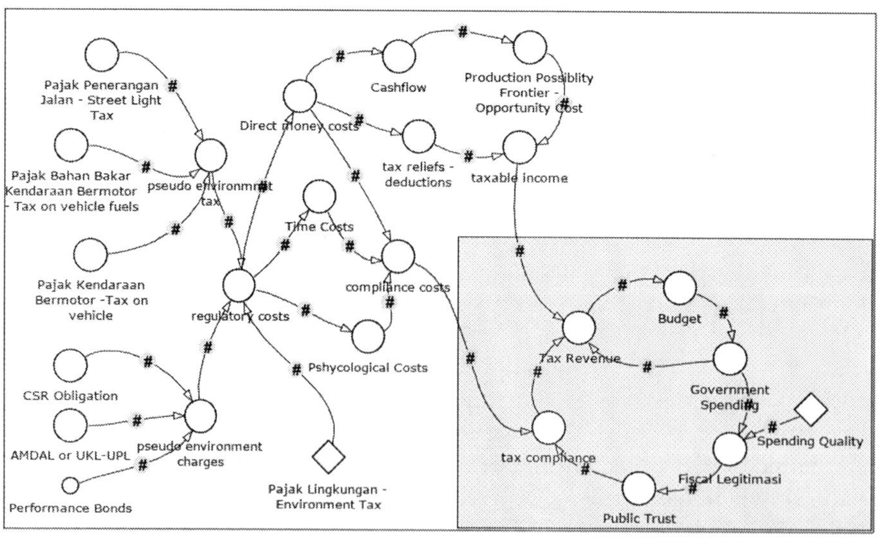

Figure 3. Implication of environment tax to cash flow and tax trust
Source: Rosdiana, 2018

Instead of imposing new taxes to increase tax revenues, the application of new taxes (e.g., environmental taxes) will actually lower tax legitimacy which will ultimately lower government tax revenue. This can happen if in the implementation there are 2 (two) paradoxical phenomena, first the amount of regulatory cost increases and taxpayers consider cash flow is more limited to do productivity since the tax obligation will add new burden for business that eventually will retain them to invest into other business due to new state levies obligation. Secondly, if the government fails to obtain tax trusts from the people due to low spending quality as the function of government realocation of money collected, the government fails to get tax legitimacy from environmental taxpayers.

Figures 2 and 3, in fact, explain the phenomenon of why there is a rejection of environmental taxes that actually exist in the Law of the Republic of Indonesia Number 32 Year 2009 [10] About the Protection and Management of the Environment (UU PPLH). In Paragraph 8 of the Environmental Economics Instrument, Article 42 provides that "in order to preserve the function of the environment, the Government and regional governments shall develop and apply environmentally friendly economic instruments [11] by using environmental economic instruments including: a. development planning and economic activities; b. environmental funding; and c. incentives and / or disincentives. Furthermore, in Article 43 it is stipulated that incentives and / or disincentives among others are applied in the form of: a. procurement of goods and services that are environmentally friendly; b. application of taxes, user charges, and environmental subsidies.

In the elucidation of Article 43, it is explained that what is meant by "environmental tax" is levies by the Government and regional government on any person utilizing natural resources, such as underground water collection tax, fuel oil tax, and swallow bird's nest tax. Elucidation of Article 43, then PBBKB and PABT are actually part of environmental taxes. Therefore, the justification for enacting a new tax called the Environmental Tax is weak. Moreover, as seen in Table 1, earmarking is not fully applied properly and correctly to the pseudo environmental tax/and or charges, so that the public only sees it as an additional new tax burden.

From the perspective of evidence-based tax policy, mathematically, the calculation of compliance costs is not only based on the money spent on tax compliance, but also the intangible costs such as time costs and psychological costs, so that the following mathematical equations can be made:

$$Compliance\ Costs = f\ direct\ money\ costs + time\ cost + psychological\ costs \qquad (1)$$

The greater the compliances costs the more reluctant people pay taxes, and they will seek to avoid taxes. From the government side, there are 2 (two) costs to be incurred are collection costs and policy cost. The combination of these three costs is called the cost of taxation. To obtain fiscal legitimacy, the government must ensure that cost of taxation is comparable to spending quality [12]. If spending quality is low, while cost of taxation is high, it is certain that the fiscal legitimacy will be low, which can be constructed with the following mathematical equations:

$$Fiscal\ Legitimacy = f\ compliance\ costs + collection\ costs + policy\ costs + spending\ quality \qquad (2)$$

On the other hand, the low level of fiscal legitimacy in the form of public reluctance to pay taxes (tax trusts) is also caused by spending quality government that has not been good and not as expected by the community. Good quality of public services, community satisfaction, and public trust, all three are related and closely related to each other. Guidance of the state to its citizens to abide by tax resulting in the consequences of increasing demand for public services [13]. That is why it becomes very important for the government not merely allocating various environmental taxes revenue for the preservation and maintenance of the environment, but proving that the use of budget allocations is realized with clear evidence. In addition, the government should also convince the public that tax revenue from environmental taxes that are not specific to the environment, will continue to be allocated to support the green growth policy, that is to rearch the growth that in the same time it has to preserve the environment.

Public money management is a domain where public trust [14] is at stake. Fiscal transparency and how the dialogue is built by the government with its citizens in determining spending decisions is the key in maintaining the trust of its citizens. According to Wagner's law, public expenditure and economic growth have a positive relationship [15]. In the present context, economic growth is not generally the goal, but leads to a more specific, namely green growth and green economic to create inclusive and sustainable development.

4 Conclusion

Indonesia has actually set Environmental Taxes and charges policy. This is evidenced by the existence of several types of taxes among others; PPJ, PBBKB, PKB, PAP and PABT. Therefore, Article 42 and Article 43 of the PPLH Law should not be interpreted by adding new types of tax levies in the form of environmental taxes. In addition to increasing compliance costs and distorting cash flow entrepreneurs, it will also impact on the lack of public trust in this government called distrust tax. If the government fails to manage revenue from environmental taxes and charges through good spending quality, then this new tax will cause high economic costs that reduces competitiveness, thereby reducing the source and economic activity that becomes the basis of tax revenue which leads to disruption of stability of state revenue. Therefore, instead of the government imposing new taxes, so that restructuring, convergence and harmonization of environmental taxes and charges must be made to achieve a connection between tax payments and inclusive and sustainable economic growth can be realized, in particular to achieve green growth policy in Indonesia. For the further work, it need to establish a model of expenditure management from the revenue mobilized for the purpose of environmental preservation.

References

1. A. Bakker. *Tax and the environment: a world of possibilities* (IBFD, Amsterdam, 2009)
2. M. Moore. How does taxation affect the quality of governance?. Working Paper 280, Institute of Development Studies. (2007)
3. R.A. Musgrave, P.B. Musgrave. Public finance in theory and practice (McGraw-Hill Book Company, New York, 1988)
4. H. Herdiansyah, T. Jokopitoyo, A. Munir. Environmental awareness to realizing green Islamic boarding school (eco-pesantren) in Indonesia. *IOP Conference Series: Earth and Environmental Science* **30** 012017, (2016)
5. H. Herdiansyah, A.S. Utama, Safruddin, H. Hidayat, A.G.Z. Irawan. Analysis of environmental friendly library based on the satisfaction and service quality: study at library "X". *IOP Conference Series: Earth and Environmental Science* **88** 012030, (2017)
6. H. Widhiastuti. Studi meta-analisis tentang hubungan antara strees kerja dengan prestasi kerja. *Jurnal Psikologi* **29**, 1 (2002)
7. Undang-undang Pajak Daerah dan Retribusi Daerah No. 28 Tahun 2009, Republik Indonesia.

8. Peraturan Pemerintah Republik Indonesia Nomor 27 tahun 2012 Tentang Izin Lingkungan.
9. OECD. Co-operative tax compliance: building better tax control frameworks, OECD Publishing, Paris. http://dx.doi.org/10.1787/9789264253384-en. (2016)
10. Undang-Undang Republik Indonesia Nomor 32 Tahun 2009 Tentang Perlindungan Dan Pengelolaan Lingkungan Hidup.
11. Dewan Energi nasional 2016. Ketahanan Energi Indonesia 2015. Jakarta: DEN.
12. B. Welham, E. Hedger, P. Krause. Linkages between public sector revenues and expenditure in developing countries. ODI: Shaping policy for development. (2015)
13. UNFPA. *State of world population 2011: people and possibilities in a world of 7 billion.* (United Nations Population Fund, New York, 2011)
14. P. Arizti *et al. Results, performance budgeting and trust in government* (The World Bank, Washington D.C., 2010)
15. M. Afzal, Q. Abbas. Wagner's law in Pakistan: another look. *Journal of Economics and International Finance* **2**, 1 (2010)
16. D. Streimikiene. Impact of environmental taxes on sustainable energy development in Baltic States, Czech Republic and Slovakia. Economie, XVIII, 4 (2015)
17. R. Diemer. Green taxation: a contribution to sustainability. Paper presented on The European Semester and Green Tax Reforms (environmental taxation and the removal of environmental harmful subsidies) – a Contribution to the wider fiscal consolidation process. (2013)
18. R. C. Wiliams III. Environmental taxation. Washington DC, Resources for the Future, Discussion Paper, RFF DP 16-24. (2016)
19. UNESCAP. Environmental tax reform in Asia and the Pacific. (2017)

On quest of environmental tax implementation in Indonesia

Edi Slamet Irianto[1], *Haula* Rosdiana[2,*], and *Maria RUD* Tambunan[2]

[1]Graduate Program of Tax Policy and Administrative Science, Faculty Administrative Sciences, Universitas Indonesia, Depok, Indonesia
[2]Faculty of Administrative Sciences, Universitas Indonesia, Depok, Indonesia

Abstract. The number of research evaluating the implementation of environmental tax using regulatory cost perspective is still limited. Whereas, this type of research is quite crucial in formulating tax policy to find the balance between regulatory function, revenue productivity and its impact on maintaining conducive business climate. Therefore, this research is aimed to elaborate types of environmental taxes applied in Indonesia, to map the problems arised of their implementation and their impact on regulatory compliance cost. This research occupied postpositivism paradigm with mixed method approach. The data were gathered through qualitative and quantitative technique. The result shows that eventhough Indonesia has not implement explicity a new environmental tax, there are a lot of quasi-environmental tax scatterly levied by local government and provincial government due to their authority given by the decentralization besides the corporate social responsibility performed by the business entities. Unfortunately, the reallocation budget framework on the tax/charged levied from the quasi-environmental tax has not been clear. Thus, it is essential to establish a comprehensive environmental tax with an integrated roadmap considering and accomodating the former quasi-environmental tax into a one package environmental tax policy with clear concept of earmarking with efficient collection cost.

1 Introduction

The issue of environmental tax re-arised in Indonesia in a parallel with the shifting paradigmn of economic development framework during Susilo Bambang Yudhoyono presidential era. During that era, to rearch the high quality of economic development, the government took a lead to optimize the intensity of inclusive economic growth and sustainable development by taking attention on environmental aspects. To reachthe goal, inclusive economic growth strategies were based on four fundamental principles namely(a) to optimize the high quality of economic growth (pro-growth), (b) to exentesify employment opportunity (pro-job), (c) to improve social welfare by implementing social security-related programs with specific low economic target (pro-poor) and (d) to improve the waste management and environment preservation (pro-environment) [1].

* Corresponding author: h.rosdiana@ui.ac.id

Factually, taxation has been occupied as an instrument to control and to prevent the environment degradation as stated in the Environmental Act No.4 year 1982. However, the type of disincentives/levies or incentives on environmental management were not clearly stated in the act [2]. The government's authority to levy environment-related tax implicitly solely was given through State Financial Act, Local Government Act and Local Tax and User Charge Act then the implementation authority of levying environmental-related taxes can be undertaken by Local Government [3].

Eventhough the legal and conceptual framework of levying environmental tax has existed, its implementation is arduous. From business perspective, the double payments on the tax (general tax as well as other taxes) is burdensome.. Similary, the former Minister of Industrial Affrair also rejected the initiation to levy a specific environmental tax by arguing that the enforce the law for activities causing environmental damage was more effective than impose environmental tax. In addition, the Minister also highlight that mobilize revenue throught environment tax will lead double tax burden [4].

It is important to note that in reality, there are number of quasi-environment taxand user charge levied, should be paid by the business. In addition, unfortunately, the imposition of such quasi-environment tax and/or user charge has not been place on a comprehensive framework collected budget allocation through earmark projects with strong legal basis [5]. Therefore, this article will comprehensively examine the why the environmental tax failed to be implemented in Indonesia. Further, this article also intended to discuss the implementation of existing quasi-environmental tax in Indonesia.

The nature function of levying taxes is as budgetary and regulatory function. With regard to its regulatory function, citing Zolt saying that the tax system can be used to encourage or discourage certain activities [6]. On public finance theory, Mankiew said that " negative externalities lead market to produce a larger quantity that is socially desirable, whereas positive externalities lead the market to produce a smaller quantity that is socially desirable [7]. To remedy the problem, the government can internalize the externality by taxing goods that have negative externalities and subsizing goods that have positive externalities.In addition, Seidman (2009, 22) posed that when negative externalities exist means that market could not optimally produces goods and services, for the supply of public goods or public services which could not produce by the market mechanism, there will be the potential of free rider. Seidman further stated that the feasible solution to minimize free rider is through imposing tax[8]. Similarly as Hinman stated that "the government should use taxes or subisies to correct for failure of self-interested buyers and sellers to account for the effect on their decision on others" [9].

Based on KPMG survey in 2013, large number of countries occupied tax mechanism to influence and to enforce the business behavior or people specifically to internalize the environment preservation and sustainable development aspect on their business plan [10]. In implementing environmental tax, The UN-ESCAP (2012) propose that the idea of taxing the negative externalities with the premises of revenue neutrality and "double dividend" is a powerful tool toward green growth. On the UN-ESCAP perspective, imposing environmental tax means there will be a shift of tax burden from conventional taxes levying from environment damaging activites into a new concept of environmental taxes. The imposition of environmental taxes is a appropriate mechanism to bring the market prince together with cost of environment externalities [11]. To desing the environmental tax, OECD provide the principle and guidance, the highligh of them are as follow [12].

a. The target of enviromental tax subject should be clear for example, the pollutant or polluting behavior with particular exception.
b. The scope of environmental damage catagories should not be larger than the scope of environmental tax base
c. The tax rate should be cover the externality in exchange of

d. The tax burden should be predictable as the base calculation for future environmental quality improvement
e. The environmental tax burden should be able to consolidate with other fiscal obligation
f. The impact realocation of revenue mobilized should be addressed through other policy instruments
g. The environmental tax imposed should not prevent the business to be competitive.
h. The government need to comprehensively socialize the levies in order to get acceptance from public
i. Environmental taxes should be combined with other instrument aiming to address the certain similar issues

In fact, the success of policy implementation including the initiation of enviromental tax implementation is affected by the resiliency of enviroment strategic both external state and/or internal state toward the achievable sustainable development objective. A State resilency is determined by natural resiliency aspect (trigatra) and socio-political resiliency (pancagatra). The natural resiliency aspect consisted of geographical aspect, natural resources availability and number of citizen. Whereas, the socio-political resiliency is constructed from ideology, politics, economy, social culture and national security [13]

Recalling the aim of this research to examine why the environmental tax failed to be implemented and to examine the existence of quasi-environmental tax in Indonesia, the perspective of natural resiliency aspect (trigatra) and socio-political resiliency (pancagatra) will be the base premises of research guidance. In order to conduct the research, it use post-positivist paradigm with mix method approach. The data was collected through qualitative and quantitative technic. The qualitative technice has been done through literature study, in-depth interview and focus group discussion, whereas the quantitative technice has been done through the construction of model depict the phonomenon by applying sistem dynamic as a research tool.

2 Method

This research occupied postpositivism paradigm with mixed method approach. The data were gathered through qualitative and quantitative technique. Data and information was collected through literature study and field research. A series of in depth interview has been conducted to the relevant informants; Directorate General of Taxes and Fiscal Policy Office Ministry of Finance, Ministry of Energy and Mineral Resources, Ministry of Environmental Affairs, Business Practititoners.

3 Results and discussion

In the initial plan of environmental tax, by injecting the concept of "polluter pays principle", the subject of environmental tax is the manufacture business entities that its omzet morethan IDR 300 million per year.Whereas, the object of environmental tax is the production activities that give an added value from raw material into new product and waste by using natural resources as main materian of supporting material. For the production volume below IDR 300 million, the environmental rate tax would be 0,5% from production cost [14].

Refering to the basic principle of tax policy, a good tax policy should be clearly stated the subject, object, exemption of object, tax base, tax rate and other related to the element of tax imposition. Examining the proposed environmental tax, for example, the entitled subject has not been clearly defined to what extend the scope of manufacturing should be, similarly the exemption of tax subject and criteria of subject and exempted subject. If the determination of tax subject solely rely of legal formal of industry, it will be easily faded on their legal corporate documents. In designing a tax policy if these elements has not been clear, there will be a potential shifting environmental tax burden into production cost.

It means the consumer indirectly has to pay the externalities perform by business entity as part of their production process which should not be paid by customer, means like another indirect tax character, it would bear regressive effect. In fact, when the initiated environmental tax went to public hearing, it was revelead that besides the direct money cost has to be paid for environmental tax it self, this type of tax is a trigger to generate another cost, for example environment impact analysis fees and other fees has to paid with related to environmental affairs[15]. Economically, there will be higher shifting burden into production cost then have to paid by the consumer lead to less competitive price level of products. Unfortunately, the incentive will be given if a business implement or comply with the initiated regulation has not been precisely stated.

For the practical aspect, there are range of options as the convergence of environmental tax. The OECD in 2011 offer several implementing policy preferable to reduce specific environmental problem such as minimum fuel efficiency rate for machinery or vehichle, the subsidy for efficient energy used. Of course, it will need specific comprehensive data including the business changing or even technoly replacement into more sophisticated ones to get fair calculation of burden. It cannot be blamed that cost of levying of the "fair" environmental tax is considerably high.

In Indonesia, the debate of environmetal tax relate to the issue; the imposition of enviromental tax would have been legitimized by the Act Bill of Environmental Tax seems abuse its regulatory function to preserve the natural resource and prevent environmental degration, instead it would be intended to optimize the budgatary function. In addition, for the developing and industrial countries, the high economic growth with intensive goods and service production is more priority. It means high priority economic growth unavoidable will produce another by product with high used of energy. For the infant industry and small and medium enterprises which most use less sophisticated technology, it probably will leash them to operate efficient business.

Eventhough Indonesia has not been imposed environmental tax, several quasi-environmental tax has been imposed as regulated in Local Tax and User Charge Act. The quasi-environmental tax consist of tax on the used or upper soil water, tax on vehichle used, tax on gasoline, user charge/fees for disturbance, charge for light waste recycle and other types of fees stipulated by particular local government or provincial government. Taking example in Jakarta province, up to 2013, the Governor with Provincial House of Representative has agree to relase regulation for imposing 12 types of local tax including quasi-environmental taxes. Annually, the contribution of tax on the used or upper soil water, tax on vehichle used, tax on gasoline, tax on road electricity on Jakarta potential revenue province is about 60% from the total Own-source Revenue [16].

However, the provincial government has not set the allocation management framework of money collected which specifically go to environmental affair. The provincial do not have clear purposes and expected outcomes of earmark concept eventhough the Local Tax and User Charge Act mentioned that minimum of 10% from tax revenue shall be allocated to environmental and natural preservation affair. In addition, up to 70% of total budget annually goes to civil servent salary expenditure item. Whether this pattern of spending trend is similar with other local government or provincial government, it needs to examined comprehensively [17].

Besides imposition of taxes, business entities also are obliged to comply with other environmental related obligation such as Corporate Social Responsibility (CSR), performance bonds, AMDAL (stands for Analisis Mengenai Dampak Lingkungan Hidup), obligatory fees to conduct environmental analysis, and UKL-UPL (stands for Upaya Pengelolaan Lingkugan Hidup (UKL) dan Upaya Pemantauan Lingkungan Hidup), obligatory fees to manage and to monitor environmetal externalities.. The obligation to perform CSR is entitledto entreprises which its main work relate to natural resources, energy used, environmetal degradation. This obligation has been legally mentioned in Corporate Act No.40 year 2007 [18]. Failure to comply with this obligation will lead the entities on verdict. On the other hand, whether the fees paid as CSR obligation balance to the cost of environment degradation is need to reexamined eventhough no special incentives given to the comply enterprises or disincentive. Another problem is that no specific treatment taken by the government if CSR project is not inline with the initial goal.

Important to note that, environmental tax, quasi-environmental tax and other state levies related to environmental problems in fact will bear regulatory compliance cost adding the burden of business. As OECD stated that, each regulation will arise its regulatory system. The term regulation means the varies of instrument package completed by requirements applicable to business or citizen. The regulation exists in form of formal and informal rules, subordinate rules issued by all level of government, non-governmental institution or self-regulatory agency which has authority given by the government.

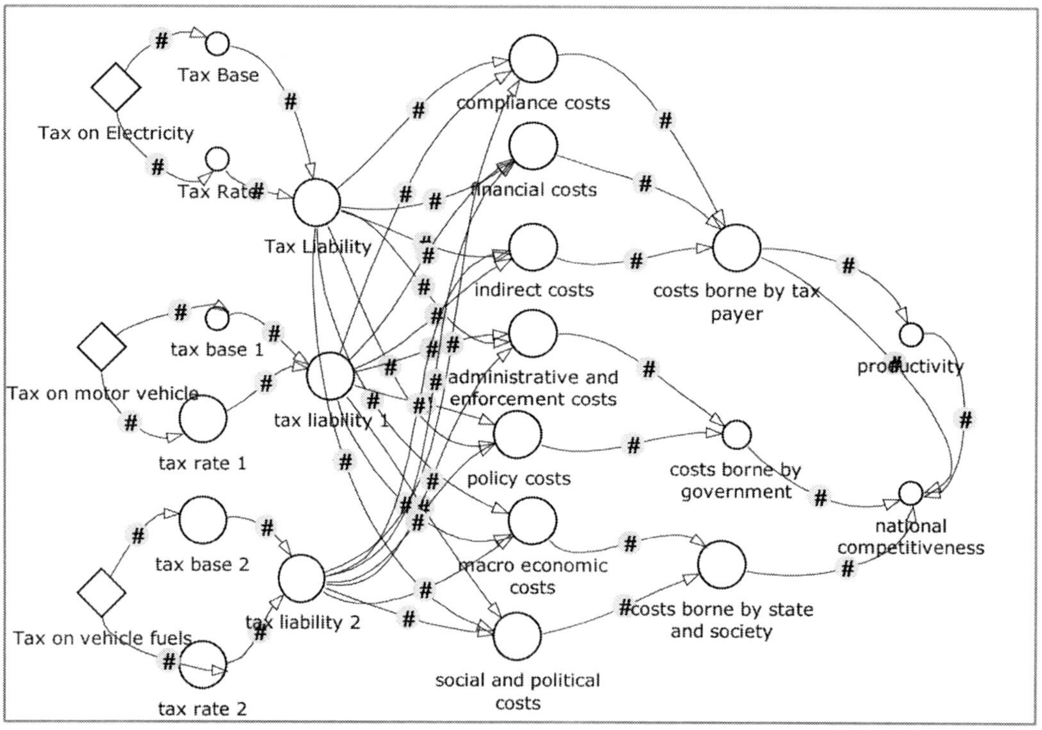

Fig. 1. Regulatory costs arised by implementing environmental taxes.

Figure 1. describes how environmental taxes arise regulatory costs that affect national business competitiveness. While formulating tax policy, the government would determine tax base and tax rate as a base to impose amount of tax obligation which in the same time it should not prevent the business competitivenes due to cost bear by government, by business as taxpayer and by state/society. Practically, tax formula (tax rate x tax base) on environmental taxes will never be simple. For example, Art. 54(1) Local Tax and User Change Act stipulated that the tax base for road electricity tax is the amount of money for the sale of electricity, then Art. (2) regulated that the amount of money for the supply of electricity as mentioned in Art. (1) [19]:

a. If the electricy supplied by entities differ from the entities to whom the payment will be made, the value of electricity service supplied is that the amount of bill plus the amount of money converted to the use of each kWh/variable as the total of the payment.

b. If the use of electricity is self-supply, the value of electricity is calculated based on total capacity, the rate of use, time frame of use, the price of unit used applicable for each region.

The general formula to calculate the tax should be levied based on composition of total cost is related to the composition of total cost and composition of environmetal cost [20]:

$$\text{Total Cost (IDR/kWh)} = f\,\text{Production costs (IDR/kWh)} + \text{Enviromental costs (IDR/kWh)}^* \quad (1)$$

$$\text{Enviromental costs (IDR/kWh)} = f\,\text{Pollutant emission factors (t/kWh)} + \text{Cost Equivalent (IDR/t)}^* \quad (2)$$

$$\text{Enviromental costs} = f\,\text{resources consumption costs} + \text{pollution control cost} \quad (3)$$

Based on these situation, the local government should draft the implementing regulation which is dynamically adjust to the change of tax base determinant factors such as composition of total cost and composition of environmental cost. This complicated formula describes the policy cost. In addition, the government also get burden to ensure that tax regulation has been implemented as its goal. Thus, the burden should not not solely the administrative cost, but also other cost such as collection cost, Distress Warrant or dispute settlement.This model also apply for other use of resources such as the use of gas supply [21].

Performance bonds is subjected to legal person who has forestry management permit as mentioned in Law No.35 year 1999 regarding Forestry [22]. The performance bonds is treated like a warranty of acceptable forestry management business. The bonds will be liquid once the business work meet the standardized quality of work. The performance bonds schemes seems effective enough to diminish the practice of unacceptable foresty related business and to slow down forests degradation. Then, AMDAL and UKL-UKP is a obligation of assessing whether the business comply to run their activity in the minimum level of the environmental externalities. If the business fail to comply with the minimum level externality standard, they have to pay the fines/penalty.

Based on the field study, eventhough Indonesia has applied several quasi-environmental tax, however how the money collected shall be realocated to fulfill the earmarked function, has not been clear. In addition, Indonesia has not has comprehensive legal framework to enforce if the fulfilment of quasi-environmental obligation could not meet its purpose. Similarly, as the OECD noted that the achievement of the goals are challenging enough even in the situation a particular government has comprehensive framework of collecting and allocating system of environmental tax. The OECD also noticed that it take a lot of effort to stricly enforce the regulation for the shake of fruitful result.

In imposing taxes, quasi-environmental tax, types of fines/penalities, it is very important to ensure that the amount of money paid has to be larger than the environmental degradation. It means the marginal environmental damage has to be smaller than marginal cost paid by the business to fund the preservation toward sustainable development. On the other part, if the government can convince the reliable and valid calculation formula of the amount paid commensurate with the damage, then the allocation framework comprehensive set with periodical outcomes of the money used, probably it will encourage the business or people to comply with the rule.

Considering numbers of quasi-environmental tax that probably is quaty variable on each local government/provincial government, the government need to construct a comprehensive environmental tax. The comprehensive environmental tax means that the tax tax policy, tax law and tax administration framework are in one package with clear framework of spending/earmarking system. The proposed comprehensive environmental tax will be less reluctant to comply by the business if its imposition do not add to high burden and less regulatory cost.

4 Conclusion

The government of Indonesia has been aware of the need to preserve the sustainable environment and prevent the environmental damage by offer the initiation of environmental tax. However, the initiation get against by business entities for the potential additional tax burden it make that in the same time the government has not has the clear framework of environmental tax and its earmark concept. On the other hand, the government also has imposed the quasi-environmental tax through local/provincial government, similarly with less budget allocation concept. Therefore, it need to redesign the environmental tax system to minimize its convergences into other increasing number of quasies with clear earmark concept. The environmental tax system will be less reluctant if it not add high burden will less regulatory cost.

Ackonwledgments

This article is a part of Environmental Tax Series research conducted by Cluster Research Political Tax, Welfare and National Resiliency, Faculty of Administrative Science University of Indonesia lead by. Haula Rosdiana. This article is the further discussion of the article entitled "Fiscal Legitimacy of Environmental Taxation: Challenges for Green Growth Policy". These series of research are funded by Professorship Research Fund provided by Faculty of Administrative Science University of Indonesia

References

1. Handbook of National Plan Development 2012-2013. Memperkuat Perekonomian Domestik. National Planning and Development Agency
2. D. Hasan, D.E. Puspitasari. Tinjauan terhadap rencana penerapan pajak lingkungan sebagai instrumen perlindungan lingkungan hidup di Indonesia. *Mimbar Hukum* **20**, 3 (2008)
3. K. Hardjasoemantri. *Hukum Tata Lingkungan*. (Gadjah Mada University Press, Yogyakarta, 2005)
4. C. Septiviani. Penetapan Nilai Pajak Lingkungan untuk Industri Tekstil. *Laporan Hasil Penelitian*. Institut Pertanian Bogor. (2009)
5. Indonesia Insititute of Science. Pajak untuk Peningkatan Perbaikan Lingkungan, downloaded from http://lipi.go.id/berita/pajak-untuk-peningkatan-perbaikan-lingkungan/4664, (2005)
6. Zolt. Tax incentives and tax base protection issue, papers on selected topics in protecting the tax base of developing countries. United Nations Department of Economics and Social Affairs. (2013)
7. G.N. Mankiew. *Principle of microeconomics* (Singapore: Harvard University Press, 2004)
8. L. Seidman, K. Lewis. Compensation contribution under international carbon treaty. Working Paper Seris No.03 University of Delaware. (2009)
9. J.S. Hinman. The green economics recovery: wind energy tax policy after financial crisis and the American recovery and reinvestment tax act of 2009. *Journal of Environmental Law & Litigation* **24**, 1 (2009)
10. KPMG. The KPMG Green Tax Index 2013, An Exploration of Green Tax Incentives and Penalities. (2003)

11. UN-ESCAP. Low carbon green growth roadmap for asia and the pacific, environmentl tax reform: major findings and policy implication from a multi-regional economic simulation analysis. (2012)
12. OECD. Environmental taxation a guide for policy makers. (2011)
13. E. Anom. Pemahaman ketahanan nasional dalam upaya memperkokoh keutuhan NKRI. *Forum Ilmiah Indonusa* **5**, 3 (2008)
14. E.M. Pratiwi, G. Setyawan. Tinjauan dalam rencana pemberlakuan green tax atau eco tax di Indonesia untuk menghadapi ASEAN Economic Community 2015. *Jurnal Riset Akuntansi dan Keuangan* **2**, 3 (2014)
15. Indonesia Chamber of Commerce. Kalangan dunia usaha menolak pajak lingkungan. (2015)
16. Suwardi D. and Ayu A.C. Formulasi kebijakaan pajak lingkungan (environmental taxation) sebagai instrumen perlindungan lingkungan hidup menurut prinsip regulerend. Depok: Universitas Indonesia. (2013)
17. Jakarta Open Data. Data rencana dan realisaasi penerimaan pajak 2015
18. Corporate act No.40 year 2007
19. Local Tax and user charge act No. 28 Tahun 2009, Republik Indonesia.
20. W. Han, D. Liu, Y. Fan, Y. Hu, Y. Xu, X. Liu. Methods research to calculate equivalent of environmental cost in power industry. *Journal of Power and Energy Engineering* **2**, 4 (2014)
21. Nuryadin, Tb R.R. Nitibaskara, H. Herdiansyah, R. Sari. The Model of Simulation In Regional Energy Security Framework: Policy Studies PT. Cirebon Area. *IOP Conference Series: Earth and Environmental Science **88** 012029*, (2017)
22. Law No.35 year 1999 regarding Forestry

Jakarta Bay reclamation policy: an analysis of political ecology

Elok Faiqotul Mutia[1] and *Donna* Asteria[2,*]

[1] School of Environmental Science, Universitas Indonesia, Salemba, Indonesia
[2] Communication Department, Faculty of Social and Political Science, Universitas Indonesia, Depok, Indonesia

Abstract. The Jakarta Bay reclamation policy is an issue that is currently a public debate. This debate is strengthened by environmental issues, the marginalization of certain groups, the regulation of authority and also corruption in the Jakarta Bay Reclamation project. Some central and provincial government institutions take different stances on the issue. The first objective of this research is to analyse environmental policy in reclamation of Jakarta Bay with the approach of political ecology. This research uses a qualitative approach. The data were obtained from in-depth interviews with stakeholders involved in the policy making of the Jakarta Bay reclamation environment and the literature study. This study concludes that the Jakarta Bay reclamation policy creates a conflict between the central and provincial governments. This is due to the overlapping regulation of Jakarta Bay reclamation authority as well as differences in attitude and interests among stakeholders. In the approach of political ecology, the Jakarta Bay reclamation policy has failed to accommodate the three interests, are social, economic and environmental interests. The conflicts that occur also cause an omission that not only harms the economy, but worsens the social and economic impact.

1 Introduction

1.1 Background

The Jakarta Bay reclamation policy was initiated from the General Spatial Plan of DKI Jakarta from 1985 to 2005 which stated that small-scale reclamation in Penjaringan, Pademangan, Ancol, and Pluit is required to meet the growing demand for the population of Jakarta. in 1994 issued Presidential Decree. 17 of 1994 which states Pantura Jakarta as a Mainstay Area. in 1995 issued Presidential Decree no. 52 of 1995 on the Implementation of North Jakarta Beach Reclamation, which was followed up by the Jakarta Provincial Regulation no. 6 of 1999 on Spatial Planning of DKI Jakarta. Reclamation objectives are increasing for international trade and services, housing for the upper middle class and tourist ports.

* Corresponding author: donna@ui.ac.id

The Jakarta Bay reclamation policy is now a policy that raises debates among stakeholders. This debate is strengthened by environmental issues, the marginalization of certain groups, the regulation of authority and also corruption in the Jakarta Bay Reclamation project. The reclamation of Jakarta Bay is considered to threaten the ecosystem in the coastal area of Jakarta Bay. Mangrove forest in Jakarta Bay, which have an important role, is one of the threaten by reclamation project [1]. Same as, the mangrove forest, the water quality will be threatened by Jakarta Bay reclamation. It will affect not only the environment, but also economic loss for fisherman [2]. In addition to the regulatory process for the implementation of the Jakarta Bay reclamation policy, there are corruption issues involving investors and members of the DPRD. This then raises the reaction from stakeholders who feel competent and interested in the project.

Minister of Marine Affairs and Fisheries, Susi Pudjiastuti, stated that the reclamation developers in the Jakarta Bay must complete the permits of the Ministry of Marine Affairs and Fisheries (KKP). Prior to the permit, the project must be stopped. Not only the KKP, Minister of Marine Affairs Coordinator for the period of 2015-2016, Rizal Ramli, also took a stand by cancelling the development of G Island. The decision was taken by considering the violations committed by developers such as the development of the island which is done on the PLN power lines as well as the act of shutting down and disrupting the access road of fishermen. In addition, the technical development of G island is considered to potentially damage the marine biota. However, his stance to the Jakarta Bay reclamation changed when the position of the minister was replaced by Luhut Panjaitan in July 2016. In his leadership, he decided that the Jakarta Bay reclamation would be continued. Ministry of Environment and Forestry (KLHK) also took a position in this issue. KLHK took their steps on the basis of their authority over the environment which is threatened by the Jakarta Bay reclamation.

DKI Jakarta Provincial Government took their stance based on Presidential Decree (Presidential Decree) Number 52 Year 1995 and Presidential Regulation (Perpres) Number 122 Year 2012 in the debate of Jakarta Bay reclamation. Both regulations give the authority to the provincial government for the implementation of the Jakarta Bay reclamation. DKI Jakarta Provincial Government affirmed that the Jakarta Bay reclamation could no longer be stopped because some islands such as C and D have been formed. The Jakarta Bay reclamation is considered not only a solution to the land limitations and flood management, but also to the fishermen's socio-economic interests. This is manifested into strategic environmental assessment (KLHS) a spatial plan for Jakarta north coast strategic area which has been validated by the Ministry of Environment and Forestry. The government institutions different stances indicate the different interests in the Jakarta Bay reclamation. To date, the Jakarta Bay reclamation still has no clear decision, be it to cancel or to continue.

1.2. Conflict

Conflict happen when two or more subject have difference interest with the some object [3][4]. It also can happen because subjects have difference access and inequalities to resources [5]. In Jakarta Bay Reclamation, this happened with the conflict over the rejection of the Jakarta Bay reclamation which considered that there were no interests of the coastal communities accommodated, especially the people who relied on the Jakarta Bay as the main source of income [6]. According to Mitchell et al. (2003), the roots of arising conflict include four aspects, namely: differences in knowledge and understanding, differences in values, differences in interests, as well as personal issues or historical background [7]. In this research, conflict can be arise because of management of resources. Resource conflicts occur often due to differences in perception or interpretation of the resource is manifested in their attitude [8]. Conflict over natural resources can occur at various levels, from local, regional, to global. Conflicts can also occur between these levels. Conflicts occurring primarily in local contexts can extend the traditional, regional and global levels due to their specific legal relevance [9][10]. In Indonesia, resources conflict that happened between central and provincial government because of several things, such as decentralization [11], overlapping of regulation [12], and conflict of interest in management of resources [13]. In the global level, resource conflicts caused by gaps of demand and supply [14], which makes resources value increased and attract many actors to be involved. Not only gaps demand and supply, inequality access to resources can also lead to conflict [15]. In this research, resource conflicts occur due to high value resources and different perceptions and interests of stakeholders at Jakarta Bay Reclamation.

1.3 Political ecology

The political ecology approach is an interdisciplinary approach to analysing environmental conflicts about the social relationships of actors, who are often dubbed "stakeholders", in conflict [16]. Martinez-Alier (2002) defines political ecology as the study of the distribution of ecological conflicts, such as conflicts over access, control, natural resources, livelihoods, and including environmental damage costs [17]. Methodologically, political ecology deals with the theory of social practice that understands society through the practice of resource users, or the actors, and the social relations that accompany the process of production in line with their existence in time and space, or certain spatiality, and in relationships social welfare [18]. Although the political ecology approach has been widely used to analyse human and environmental relations, there is a growing political ecology focus on what affects human action itself. Two influential theoretical models that have become the theoretical lenses in political ecology derive from poststructuralism and Marxism. Despite its unique theoretical foundation, the resulting conceptual approach becomes a highly influential lens in human learning, nature, and power.

Political ecology approach also began to be used as a way of looking at the problems that exist in Indonesia. Arifin (2012) in his research entitled "political ecology: environmentally friendly as justification" gives one conclusion that environmentally friendly discourse is used as a tool by certain groups to influence the behavioural patterns of society including behaviour in the consumption of certain goods [19]. Arifin's research focuses on political ecology as an analysis of discourse formation. Research conducted Tarigan (2016) using a political ecological approach to analyze agricultural transformation and water crisis that occurred in Bali. This study shows that the water crisis is causing conflict between stakeholders and marginalizing agriculture. This is due to the politics of tourism-oriented development that brings a widespread impact on the utilization of water resources [20]. Political ecology approach used by Tarigan is to understand the relationship between actors in the utilization of water resources. Teak (2013) uses a political ecological approach as a just paradigm of natural resource governance. According to Jati, natural resource governance should put people and nature on the same level. Community involvement is also necessary in the management of natural resources [21]. This study uses the same approach with the previously mentioned research approach. However, in this research the political ecology approach is used as a perspective to describe environmental policy in reclamation of Jakarta Bay.

2 Methodology research

The purpose of this study is to analyse the policy of Jakarta Bay reclamation with the approach of political ecology. This study will analyse the policy of each institution that related to Jakarta Bay Reclamation. It's include the statement of stakeholder in in-depth interview. Political ecology will used to approach to describe the relation of stakeholder ini Jakarta Bay Reclamation. This research uses a qualitative approach by conducting in-depth interviews to stakeholders. The stakeholder consists of the central government and the provincial governments with some criteria that show in Table 1. Selected ministries and agencies based on their involvement and authority in the reclamation of the bay of Jakarta. In addition, selected ministries and agencies are those involved in policy debates. Questions in in-depth interviews are open so that researchers can explore questions according to stakeholder responses. Exploration in literature studies has been done for adding, comparation and clarifying information.

Table 1. Criteria of Stakeholder.

No.	Code of Informant	Institution	Criteria
1	A001	Ministry of Environment and Forestry	Having knowledge of Jakarta Bay reclamation and being or having been involved in the Jakarta Bay reclamation policy process
2	A002	Ministry of Marine Affairs and Fisheries	
3	A003	Coordinating Ministry for Maritime Affairs	
4	B001	Regional Development Planning Agency DKI Jakarta	
5	B002	Environmental Agency DKI Jakarta	
6	B003	Office of Public Works - Human Settlements and Spatial Planning DKI Jakarta	

The result of the interview will write a word into a verbatim transcript which will then be analyzed using Nvivo. Transcripts will be categorized based on certain variables such as knowledge, interests, authority and conflict. The results of the analysis are validated by comparing discourse studies and policy analysis.

3 Result and discussion

The results of in-depth interviews and literature studies indicate that there are different views among stakeholders regarding Jakarta's reclamation policy. This difference in view is due to two things, differences in authority and different attitudes and interests.

3.1 Authority regulation

To date, there are 3 regulations that are used as the basis for the authority of Jakarta Bay reclamation. The regulation gives authority to 3 different stakeholders. First, the authority of the Provincial Government of DKI Jakarta. Article 4 of Presidential Decree 52 of 1995 on North Coast Reclamation Jakarta to date still remains the basis of the Provincial Government of DKI Jakarta that the authority and responsibility of the Pantura Reclamation reside in the Governor of DKI Jakarta. But there is a debate because in article 3 verse 1, the notion of reclamation is not a new island, the reclamation covers the waters of Jakarta which is measured from the north coastline of Jakarta perpendicularly towards the sea until the line connecting the outermost point shows the depth of 8 meters. It then reap the debate whether the current reclamation is in accordance with the Presidential Decree. However, the Presidential Regulation No. 54 of 2008 on Spatial Planning of Jakarta, Bogor, Depok, Tangerang, Bekasi, Puncak, Cianjur regulates reclamation should be done by forming the island, where there are lateral canals ± 200-300 meters between land with reclaimed island, depending on the zonation. In this presidential regulation, there is no article which states the change of authority.

Second, the authority of the Ministry of Marine Affairs and Fisheries. Presidential Regulation No. 122/2012 on Reclamation in Coastal Areas and Small Islands Article 16 paragraph 2, said the Minister grants location permits and permits for the implementation of reclamation in certain National Strategic Areas, reclamation activities across provinces, and reclamation activities at fishery ports managed by Government, while Presidential Regulation No. 54 of 2008 establishes Pantura as part of the National Strategic Area. So in this case the Minister also has the authority to grant permission. However, in article 2, paragraph 2a, there is an exclusion for reclamation in the Work Area (DLKr) and Regional Environment Interests (DLKp) of the main ports and harbor collectors as well as in the territorial waters of special terminals. With this article and the release of a letter from the Minister of Transportation, the Ministry of Marine Affairs and Fisheries withdrew from the authority of the Jakarta Bay reclamation.

Third, the authority of the Ministry of Transportation. Government Regulation of the Republic of Indonesia Number 5 Year 2010 On navigating derived from Minister of Transportation Regulation no. 136 of 2015 Second Amendment to Regulation of the Minister of Transportation No. PM 52 of 2011 on Dredging and Reclamation states that reclamation in the Working Zone (DLKr) and Regional Environment Interests (DLKp) of the main port and gathering port as well as in the territorial waters of special terminals shall obtain permission from the Minister of Transportation. The reclamation of Jakarta Bay enters DLKr and DLKp Sunda Kelapa and Tanjung Priuk.

This overlapping regulation of authority makes conflicts among stakeholders who feel that the reclamation of the Bay of Jakarta is the authority of their institution. The settlement of this matter of authority is then resolved by forming a joint committee consisting of several stakeholders, which is chaired by the Coordinating Ministry for the Ministry of Marine Affairs, and comprising the Ministry of Marine Affairs and Fisheries, Ministry of Environment and the Provincial Government of DKI Jakarta. In the Committee it is stipulated that the authority of the Jakarta Bay reclamation is in the Provincial Government of DKI Jakarta

3.2 Differences in attitudes and interests

The continuation of the reclamation of the Jakarta Bay is still unclear. Whereas on the other hand, there are 2 islands that have been formed even already have a number of buildings. The policy on the continuation of the reclamation of the Bay of Jakarta is based on the importance of what the stakeholders are interested in and wants to achieve. The Jakarta Bay reclamation policy was initially based on the need for land for settlements and the desire to create waterfront city. In 1999, the reclamation objectives increased for international trade and services, housing for the upper middle class and tourist ports. However, environmental problems that occurred in Jakarta made the government prioritize the handling of the environment. The condition of the waters of Jakarta Bay has now exceeded the quality standard and polluted. The results of monitoring conducted by the Environmental Agency, 11 sample points expressed exceeded the quality standard. This shows that the ecosystem in the waters of Jakarta Bay is being threatened, therefore, it needs a policy that is able to solve environmental problems in Jakarta Bay.

The Jakarta Bay reclamation policy is judged by some stakeholders to still not accommodate environmental and social interests and is still dominated by economic interests. The results of the field visits conducted by the Ministry of Environment, the Ministry of Marine Affairs and Fisheries, the Ministry of Coordinating Ministry of Marine Affairs and the Provincial Government of DKI Jakarta in May 2016. In the visit, found several violations, including permission violations and considered not to improve the quality of the environment, it will exceed carrying capacity and environmental capacity in the long term. In addition, the Ministry of Marine Affairs and Fisheries assessed there were violations in the technical aspects and neglect of the interests of the fishermen community. Although a temporary suspension or moratorium was suspended, the final decision of the joint committee decided that the reclamation of the Jakarta Bay was still in place. Although this decision has been conveyed through the Coordinating Ministry for the Ministry of Environment, Ministry of Environment and Forestry, stressed the need for regulation to oversee the policy of Jakarta Bay reclamation. This is necessary because until now the reclamation of Jakarta Bay is dominated by the investor's economic interest, so the position of the country is weak. It's stronger with the statement of two of informant that said the government need stronger rules and position as a controller of the Jakarta Bay Reclamation, and same as in the project of NCICD [22]. The same thing is also conveyed by the Ministry of Marine Affairs and Fisheries which is still on the same view and attitude, namely Jakarta Bay reclamation policy should be cancelled.

Table 2. Results of Jakarta Bay Estuary Water Monitoring.

No.	Sample Location	Fulfillment Quality Standards	Pollution Index
1	Muara Kamal	Exceeds	Heavy
2	Cengkareng Drain	Exceeds	Heavy
3	Muara Karang	Exceeds	Heavy
4	Muara Angke	Exceeds	Heavy
5	Gedung Pompa Pluit	Exceeds	Heavy
6	Ancol	Exceeds	Medium
7	Sunter	Exceeds	Light
8	Muara BKT	Exceeds	Heavy
9	Cilincing	Exceeds	Light
10	Marunda	Exceeds	Light
11	Muara Gembong	Exceeds	Heavy

Source: Environmental Agency, 2016

3.3 Jakarta Bay reclamation in political ecology approach

Political ecology is an environmental science-based approach, which emphasizes human interaction with nature. Through a political ecology approach, this research looks at how interactions among stakeholders in the Jakarta Bay reclamation policy. This interaction in the reclamation of the Jakarta Bay resulted in differences in authority and different attitudes and interests that resulted in a conflict. In Political ecology the imbalance of authority and interest between actors over resources makes the existence of conflict. It is also a source of conflict in the Jakarta Bay reclamation policy. Overlapping authority regulations and the interests of different stakeholders have led to unclear policies of the Jakarta Bay reclamation to date.

In addition, political ecology also emphasizes the influence of actors in decision making. In the policy of Jakarta Bay reclamation, the interests of actors who play a strong role in an institution both the central government and the provincial government influence the attitudes and policies taken by the institution. This leads to more complex conflicts and the more important elements of environmental policy, the principle of sustainability. Pollution that occurred in Jakarta Bay requires an action. However, the current Jakarta Bay reclamation policy does not prioritize the principle of sustainability. This protracted conflict ultimately not only makes financial losses such as total value of estimated losses arising from the activities of Jakarta Bay reclamation perceived by fishermen and cultivators is Rp. 207.153.292.610, - / year [23], but also adds to the deterioration of the environment in the Jakarta Bay due to the omission of reclaimed lands [24].

Political ecology in the paradigm of an environmental management policy should prioritize a justice-based policy. A public policy on the environment must accommodate three interests, namely social, economic and environmental [25]. It can be done by community participating on decision making, and open data publicly. So, public can knows and participate to control the policy. The reclamation of the Bay of Jakarta has failed to become a public policy that can accommodate social, economic and environmental interests because of the dominance of one interest over other interests. This policy will continue to be a conflict between government institutions until there is awareness to equate knowledge and balance authority for the common good, ie sustainable environmental management for both people and the environment itself.

4 Conclusion

This study concludes that the Jakarta Bay reclamation policy creates a conflict between the central and provincial governments. This is due to the overlapping regulation of Jakarta Bay reclamation authority as well as differences in attitude and interests among stakeholders. In the approach of political ecology, the Jakarta Bay reclamation policy has failed to accommodate the three interests, are social, economic and environmental interests. The conflicts that occur also cause an omission that not only harms the economy but worsens the social and economic impact.

Acknowledgements

This research is funded by the Grant of Indexed International Publication for Final Project of Students/Publikasi Terindeks Internasional Untuk Tugas Akhir Mahasiswa (PITTA) Universitas Indonesia 2018 with contract number 2579/UN.R3.1/HKP.05.00/2018

References

1. H.S. Alikodra. The impacts of land reclamation of Jakarta Bay on mangrove ecosystem. *Media Konservasi* **5**, 1 (1996)
2. R. Puspasari, S.T. Hartati, R.F. Anggawangsa. Analisis dampak reklamasi terhadap lingkungan dan perikanan di Teluk Jakarta. *Jurnal Kebijakan Perikanan Indonesia* **9**, 2 (2017)
3. S. Fisher. *Working with conflict: skills and strategies for action* (Zed books, 2000)
4. S.P. Robbins. "Conflict management" and "conflict resolution" are not synonymous Terms. *California Management Review* **21**, 2 (1978)
5. F. Stewart. *Horizontal inequalities and conflict: Understanding group violence in multiethnic societies* (Springer, 2016)
6. Arsyad, A. Coastal reclamation projects in Indonesia: the weaknesses which lead to severe socio-environmental impacts. Proceedings International Conference on Environmental Research and Technology. Universiti Sains Malaysia. (2008)
7. B. Mitchell, D.H. Rahmi, B. Setiawan. *Resource and environmental management* (Gadjah Mada University Press, 2003)
8. A.S. Cheng, L.E. Kruger, S.E. Daniels. "Place" as an integrating concept in natural resource politics: propositions for a social science research agenda. *Society & Natural Resources* **16**, 2 (2003)

9. L. Talaue-McManus, A.C. Yambao, S.G. Salmo III, P.M. Alino. *Cultivating Peace: Conflict and Collaboration in Natural Resources Management* (International Development Research Centre, Canada, 1999)
10. V. Weitzner, M.F. Borrás. *Cultivating Peace: Conflict and Collaboration in Natural Resources Management* (International Development Research Centre, Canada, 1999)
11. J.F. McCarthy. Changing to gray: decentralization and the emergence of volatile socio-legal configurations in Central Kalimantan, Indonesia. *World Development* **32**, 7 (2004)
12. D. Dirhamsyah. Indonesian legislative framework for coastal resources management: a critical review and recommendation. *Ocean & Coastal Management* **49**, 1-2 (2006)
13. A. Kustanti, B. Nugroho, C. Kusmana, D. Darusman, D. Nurrochmat, M. Krott, C. Schusser. Actor, interest and conflict in sustainable mangrove forest management—a case from Indonesia. *International Journal of Marine Science* **4**, 16 (2014)
14. M.F. Giordano, M.A. Giordano, A.T. Wolf. International resource conflict and mitigation. *Journal of Peace Research* **42**, 1 (2005)
15. T.F. Homer-Dixon. Environmental scarcities and violent conflict: evidence from cases. *International Security* 19, 1 (1994)
16. J.A. Douglas. What's political ecology got to do with tourism?. *Tourism Geographies* **16**, 1 (2014)
17. J. Martinez-Alier. *The Environmentalism of the poor: a study of ecological conflicts and valuation* (Edward Elgar Publishing, 2002)
18. C.A. Kull, X.A. de Sartre, M. Castro-Larrañaga. The political ecology of ecosystem services. *Geoforum* **61**, (2015)
19. Z. Arifin. Politik ekologi: ramah lingkungan sebagai pembenaran. *Jurnal Ilmu Sosial Mamangan* **1**, 1 (2012)
20. H. Tarigan. transformasi pertanian dan krisis air di bali dalam perspektif ekologi politik. *Forum Penelitian Agro Ekonomi* **34**, 2 (2016)
21. W.R. Jati. Manajemen tata kelola sumber daya alam berbasis paradigma ekologi politik. *POLITIKA: Jurnal Ilmu Politik* **3**, 2 (2012)
22. E. Colven. Understanding the allure of big infrastructure: Jakarta's great garuda sea wall project. Water Alternatives 10, 2 (2017)
23. A. Ramadhan, M. Firdaus, R.A. Wijaya, I. Muliawan. Estimasi kerugian nelayan dan pembudidaya ikan akibat reklamasi di Teluk Jakarta (Economic loss of fisher and fish farmer due to reclamation of Jakarta Bay). *Jurnal Sosial Ekonomi Kelautan dan Perikanan* **11**, 1 (2016)
24. R. Puspasari, S.T. Hartati, R.F. Anggawangsa. Analisis dampak reklamasi terhadap lingkungan dan perikanan di Teluk Jakarta (Analysis of reclamation impacts on environment and fisheries in Jakarta Bay). *Jurnal Kebijakan Perikanan Indonesia* **9**, 2 (2018)
25. A. Brotosusilo, D. Utari, A.A. Satria. Sustainability of water resources in the upstream watershed-based community engagement and multistakeholder Cooperation. *IOP Conference Series: Earth and Environmental Science* **30** 012018, (2016)

Community participation based on social capital to enhance tourist visit at Lake Toba Parapat North Sumatera

Erika Revida[1*] and *Sukarman* Purba[2]

[1]Public Administration Departement, Universitas Sumatera Utara, Medan, Indonesia
[2]Education Managenent, Universitas Negeri Medan, Medan, Indonesia

Abstract. The The purpose of this research was to analyze the participation of community based on social capital to enhance tourist visit at Lake Toba Parapat North Sumatera. The study utililized both qualitative and quantitative methods. 80 respondents (head of family) were interviewed at Lake Toba Parapat North Sumatera. Thte analysis technique of data used quantitative research used descriptive statistics, and multiple regression analysis, whereas qualitative research started from the data organization, reduction and interpretation of data and took conclusion on the results of research. The results of the research showed that the hypothesis of research that there was a positive and significant influence between social capital on the community participation to enhance tourist visits in Lake Toba Parapat North Sumatra was accepted. Participation of community based on social capital to enhance tourist visit was in moderat and it can be further improved by social activities including communication, information sharing and education on the importance of community participation based on social capital such as cohesiveness, altruism, networking, coorperation, unselfish feeling, and trust.

1 Background

Tourism development in various region including Lake Toba Parapat North Sumatera is an important and urgent sector, especially in the era of decentralization now on. Tourism is a journey undertaken by one or groups of people from one place to another in other to enjoy the beauty of nature with the aim of seeking pleasure. Tourism development is a variety of tourism activities undertaken by community, private and government supported by tourism facilities and services to enhance the number of tourist visits both domestic and foreign countries. Tourism not only enhance state/central revenue but also enhance community income and local revenue. Therefore, the participation of all elements include government, private and community around tourist site become a very important strategic issue to enhance the tourist visit. Therefore the participation of community based on social capital to enhance tourist visit will be a good concept in Lake Toba Parapat North Sumatera.

* Correspondent Author: erikarevida@yahoo.com

According to Law of Tourism number 10 of 2009 [1], the destination of tourism development are to promote economic growth, prosperity of the people, eradicate poverty, overcome unemployment, conserve nature, environment, and resources, promote culture, elevate the image of the nation, nurture the love of the country, strengthen the identity and unity of the nation, and strengthen friendship among nations. It means that the development of tourism conducted with the participation of community-based social capital is an absolute requirement at the Lake Toba Parapat North Sumatera.

But, even central and the local government had already undertaken a various tourism development programs to enhance tourist visit, but in fact that up to now at tourism sector especially Lake Toba Parapat North Sumatra did not have a significant results to increase tourist visit. Decreasing number of tourist visits both domestic and foreign countries in North Sumatera since the last 3 (three) years can seen from the following table.

Table 1 Number of Tourist Visit to Lake Toba North Sumatera 2014-2017

No	Tourist Visit (Year)	Total (people)
1	2014	270.837
2	2015	229.288
3	2016	233.643
4	2017	167.037
	Total	900.805

Source : Badan Pusat Statistik Sumatera Utara, 2017[2]

The development of the tourism sector to enhance tourist visit is not only a mere of government affair alone, but also required the participation of private and the community especially around of tourist site. This was the reason why this research conducted by taking the title "Participation of Community Based on Social Capital to Enhance Tourism Visit In Lake Toba Parapat North Sumatera".

2 Literature Review

Participation of community is a role of community in every stage of activities in development including the tourism development starting from the process of identifying the problems and potentials in the community, the selection and decision-making on alternative solutions to handle problems, implementation of the program and the effort to solve the problem, and also evaluating the programs[3].

While social capital define as a set of values and norms that are recognized and shared by the local community and be used as guidelines in behaving in community everyday and implemented by a good cooperation in the community [4] stated that the components of social capital were trust, cohesiveness, altruism, unselfish feeling, cooperation, and network. Thus, the participation of community based on social capital is the active participation of community that is driven by social capital such as trust, cohesiveness, altruism, unselfish feeling, cooperation, and networks.

In this research the meaning of participation of community based on social capital to enhance tourist visits is the community involvement by giving contribution of thoughts, energy, time, funds, and facilities based on trust, cohesiveness, altruism, unselfish feeling, cooperation, and networks to enhance tourist visit at Lake Toba Parapat North Sumatra.

3 Research Method

The study employed both quantitative and qualitative methods [5]. The respondents consist of 80 local community people who lived in Lake Toba Parapat North Sumatera. The research informant were bureaucratic apparatus in head of distric, head and village apparatus around Lake Toba Parapat North Sumatera.

Data were collected from literatures, questionnaire, observations and interviews. The data were then analyzed by using descriptive statistic, and simple regression analysis. Qualitative research methods were done from data organization, reduction, and interpretation of data and took conclusiona of the result [6].

4 Result and Discussion

4.1 Participation of Community Based on Social Capital to Enhance Tourist Visit at Lake Toba Parapat North Sumatera

The participation of community based on social capital to enhance tourist visit at Lake Toba Parapat North Sumatera was the active involvement of community by giving by giving contribution of thoughts, energy, time, funds, and facilities based on the trust, cohesiveness, altruism, unselfish feeling, mutual cooperation, and network among community, community with government, and the public with the private sector to enhance tourist visit at Lake Toba Parapat North Sumatera. The result of study id shown in table 2.

Table 2 Participation of community based on social capital to enhance tourist visit at Lake Toba Parapat North Sumatera, 2017.

No	Instrument	Responden Answers				
		SA	A	S	N	VN
1	The trust of government can increase your participation to enhance tourist visits in Lake Toba Parapat North Sumatera	10 12,50%	20 25%	30 37,50%	20 25%	-
2	The trust of community can increase your participation to enhance tourist visit in Lake Toba Parapat North Sumatera	7 8,75%	20 25%	29 36,25%	20 25%	4 5%
3	The trust of private sector can increase your participation to enhance tourist visits in Lake Toba Parapat North Sumatera?	8 10%	22 27,5%	32 40%	18 22,50%	-
4	Do you have any confidence that the government will enhance tourist visit in Lake Toba Parapat North Sumatra?	10 12,50%	19 23,75%	36 45%	15 18,75%	-
5	Do you have any confidence that the local people will enhance tourist visit in Lake Toba Parapat North Sumatera?	15 18,75%	20 25%	33 41,25%	12 15%	-
6	Do you have a belief that private sector will enhance tourist visit in Lake Toba Parapat North Sumatera?	6 7,50%	30 37,50%	27 33,75%	17 21,25%	-
7	Is the cohesiveness with the government can increase your participation to enhance tourist visit in Lake Toba Sumatera Utara?	10 12,50%	25 31,25%	29 36,25%	16 20%	-
8	Is the cohesiveness with the private	12	31	30	7	-

	sector can increase your participation to enhance tourist visits in Lake Toba North Sumatera?	15%	38,75%	37,50%	8,75%	
9	Does the cohesiveness with local people increase your participation to enhance tourist visits in Lake Toba North Sumatera?	7 8,75%	30 37,50%	28 35%	15 18,75%	-
10	The attention of tourists (altruism) can enhance tourist visits in Lake Toba Parapat North Sumatera	5 6,25%	35 43,75%	30 37,50%	10 12,50%	-
11	The attention of the facilities of tourism such as infrastructure and so on can enhance tourist visit in Lake Toba Parapat North Sumatera	11 13,75%	20 25%	29 50%	20 25%	-
12	Did you have any attention to tourists who come to Lake Toba Parapat North Sumatera?	10 12,5%	28 41,25%	33 41,25%	12 15%	-
13	Did you have any attention of tourist facilities/infrastructure in Lake Toba Parapat North Sumatera?	7 8,75%	20 25%	29 36,25%	20 25%	4 5%
14	Did you do a prioritize the interests of tourists rather than your own in Lake Toba Parapat North Sumatera?	8 10%	32 40%	22 27,5%	18 22,50%	-
15	Did you pay attention to the needs or desires of tourists to enhance tourist visit in Lake Toba Parapat North Sumatera?	10 12,50%	10 12,50%	50 75%	10 12,50%	-
16	Did you work without expecting a rewards to enhance tourist visits in Lake Toba Parapat North Sumatera?	12 15%	18 18,75%	32 40%	18 18,75%	-
17	Did you do prioritize mutual needs to enhance tourist visit in Lake Toba North Sumatera?	10 12,50%	20 25%	32 40%	18 22,50%	-
18	Did you do a prioritize to the need of government to enhance tourist visit in Lake Toba Parapat North Sumatera?	11 13,75%	27 33,75%	28 35%	14 35%	-
19	Did you notice the more urgent needs of tourist to enhance tourist visit in Lake Toba Parapat North Sumatera?	9 11,25%	16 20%	31 38,75%	22 27,50%	2 2,50%
20	Do you want to work together with community to enhance tourist visits in Lake Toba Parapat North Sumatera?	9 11,25%	30 37,50%	30 37,50%	11 13,75%	-
21	Did other communities cooperate to enhance tourist visits in Lake Toba Parapat North Sumatera?	8	20 %	38 47,50%	12 15%	-
22	Did you have network with government to enhance tourist visits in Lake Toba Parapat North Sumatera?	8 10%	12 15%	32 40%	25 31,25%	3 3,75%
23	Did you have networking with the privat sectors to enhance tourist visits in Lake Toba Parapat North Sumatera?	10 12,5%	30 37,5%	33 41,25%	7 8.75%	-
24	Did you have networking with other communities to enhance tourist visits in Lake Toba Parapat North Sumatera?	12 15%	28 41,25%	33 41,25%	10 12,5%	-

25	Did you urge other community around the tourist sites to enhance tourist visits in Lake Toba Parapat North Sumatera?	10 12,5%	19 23,75%	35 43,75%	11 13,75%	5 6,25%

Note :

SA : Strongly Agree
A : Agree
S : Sometimes
N : Never
VN : Very Never

4.2 The Trend Level of Participation of Community To Enhance Tourist Visit At Lake Toba Parapat North Sumatera

The trend level of community participation based on social capital to enhance tourist visit at Lake Toba Parapat North Sumatera could be seen in table 3.

Table 3 The trend level of participation of community based on social capital to enhance tourist visit at Lake Toba Parapat North Sumatera, 2017

Range	F. Absolute	F. Relative	Category
89 - 110	20	25.00%	Good
66 - 88	46	57.50%	Moderate
43 - 65	14	17.50%	Less
20 - 42	0	0.00%	Low
Total	80	100.00%	

As shown in table 3, it can be said that the trend level of participation of community to enhance tourist visit at Lake Toba Parapat North Sumatera was in moderate category and need to enhance by participation of community based on social capital.

Participation of community to enhance tourist visits can be done in various forms, as an example by developing the potential of local art and culture such as culinary, fashion, cultural performing arts, and so on. The show of potential of local art and culture will bring something uniqueness for tourists and also makes local community proud of them.

5 Conclusions and Recommendations

The hypothesis of research stated that there was a positive and significant influence between Social Capital on the Participation of Community to enhance tourist visits in Lake Toba Parapat North Sumatra was accepted. Participation of community based on social capital to enhance tourist visit was in moderat category. From the six indicators of Social Capital, the greatest influence towards Participation of Community to enhance tourist visits was cohesiveness, followed by Altruism, Networking, Coorperation, Unselfish Feeling, and the weekness was Trust

The hypothesis of research stated that there was a positive and significant influence between Social Capital on the Participation of Community to enhance tourist visits in Lake Toba Parapat North Sumatra was accepted, therefore social capital namely cohesiveness, altruism, networking, coorperation, unselfish feeling, and trust need to increase by socialization and role playing. Participation of community based on social capital to enhance tourist visit was in moderat category, therefor need to increase by involving community leader, religious leader and educational leader in all situations. From the six indicators of Social Capital, the greatest influence towards Participation of Community to enhance tourist visits was cohesiveness, followed by Altruism, Networking, Coorperation, Unselfish Feeling, and the weekness was Trust. It means that cohesiveness was an important thing to increase among society, leaders and privat sectors by increasing emotional closeness, forming groups that share common goals and have caring leaders.

Acknowledgements

Finnally, I thank to the honorable : 1. Ministry of Research and Technology and higher education of the Republic of Indonesia who has funded this research. 2. Rector of Universitas Sumatera Utara, Dean of Fakulty of Social and Political Science and Chairman of Public Administration Departement which has given permission to do this research. 3. Head of tourism department simalungun regency and his staff. 4. Respondents and informants of this study who had been willing to provide data needed in this study. Hopefully this article useful to enhance tourist visit at Lake Toba Parapat North Sumatera.

References

1. Undang-undang Nomor 10 tahun 2009 tentang Kepariwisataan. Jakarta : Sekretariat Negara.
2. Badan Pusat Statistik Sumatera Utara. 2017. Medan : Sekretariat Daerah
3. R.A. Isbandi, *Perencanaan Partisipatoris Berbasis Aset Komunitas : Dari Pemikiran Menuju Penerapan*, (FISIP UI Press, Jakarta, 2007)
4. F. Fukuyama, *Trust : Kebajikan Sosial dan Penciptaan Kemakmuran,* (Penerbit Qalam, Yogyakarta, 2007)
5. J.W. Creswell, *Research Design. Qualitative & Quantitative Approaches* (Sage Publication, Inc, New York, 2004)
6. L.J. Moleong, *Metodologi Penelitian Kualitatif. Edisi Revisi.* (Penerbit PT Remaja Rosdakarya, Bandung, 2006)

Palm oil sustainability partnership: implementation and connection with farmers income

Ernawati Hamid[1], Zakky Fathoni[2], and Mirawati Yanita[3]

[1]Agribusiness Department, Faculty of Agriculture, University of Jambi, Jalan Raya Jambi-Muara Bulian Km 15, Jambi Indonesia
[2]Agribusiness Department, Faculty of Agriculture, University of Jambi, Jalan Raya Jambi-Muara Bulian Km 15, Jambi Indonesia
[3]Agribusiness Department, Faculty of Agriculture, University of Jambi, Jalan Raya Jambi-Muara Bulian Km 15, Jambi Indonesia

Abstract. Partnership is a business strategy that performed by two or more parties in a certain period to obtain the benefits together with the principle of mutual need and mutual rearing. The study aims to observe and assess the implementation of the oil palm agribusiness partnerships and analyze the level of farmers' income. This research used a survey method. Descriptive analysis is used in data analysis, to provide an overview of the implementation of partnerships applied by oil palm plantation companies in Jambi Province. The results showed that agribusiness partnerships that implemented by the palm oil company basically has managed to create independent farmers who can canalize the aspirations of farmers, both in KKPA and PIR Trans pattern. Empirically, the maximum value and benefit aspects of process management partnerships as an indicator of the level of performance partnerships oil palm plantation companies in Jambi Province is quite high, has reached 82.5%. But these facts have not been fully supported by a partnership of cooperation actors, in the sense that the level of achievement of the implementation aspects of the value of partnership firm activities and performance of oil palm plantations has reached approximately 71%

1 Introduction

Agribusiness partnership is a form of cooperation between small and medium-sized businesses or large businesses along with business coaching and development by medium/large businesses with the principle of mutual need, mutually reinforcing and mutually beneficial. Agribusiness partnership aims to 1) increase the income of small businesses and communities, 2) increase the benefit for the acquisition of partnership actors, 3) improve equity and the empowerment of communities and small businesses, 4) improve rural economic growth and national territory, 5) expand job opportunities and 6) increase the resilience of the national economy [7].

Partnership of oil palm plantation companies and industries in Jambi Province has started with a design such as PIR, creditor and is now famous for its independent partnerships (partnerships that historically generation II / PRP), which form partnerships that have developed between the company (core) with farmers via cooperatives (plasma). When cooperative or farmer stronger, the corporate exploitation to the farmers are not much and vice versa if the farmer or cooperative are weak, the partnership will give more benefits to oil palm company that will be reflected from the letter of partnership agreement [4].

Asymmetrical and exploitative practices in plantation business partnership relationships, such as decision-making in business activity is determined more by the core companies or other parties (such as governments) that have greater strength, also not optimal on distribution of value-added benefits that should be enjoyed by the farmers make the imbalance of the system led to a partnership. Farmers only act as a complement to the partnership structure [14] .

Farmer participation, both individually and collectively are still very rare (especially in post-harvest activities) due to limitations in the ability of farmers to follow a high-tech process. Meanwhile, the development of the plantation business partnership is needed, because: 1) the demands of society (local) including redistributing business opportunities, asset production, business benefits to farmers, 2) the global challenges of conducting plantation business, which is "seized" that controls the downstream industry and seize the largest margin industrial production inputs burden of production costs of farmers and planters [5, 10, 14].

The challenges which are faced in the development of plantation system in the framework of the development of agribusiness among others are the limited supply of data and information technology, resources and markets, lack of support and initiatives from various related functions at all levels in the growing interest and participation of the community and the business world. Other challenges in agribusiness development are lack of support infrastructure such as roads, ports, means of communication and transportation in the area of development; tariff and non-tariff barriers, including the application of international trade standards are more rigorous; institutional planters who have established businesses, as well as the emergence of a new competitor countries [5, 10, 14].

In the processing and quality product aspects, the challenges are the absence of an integrated policy between upstream and downstream globally, unavailability of commodities development road map that agreed by all parties and being the reference for them. Product which has good quality still not received proportional incentives/ reasonable, strict demand of consumer for quality requirements, and the low interest of investors to develop the downstream industry since the absence of guarantee for certainty of sustainable business [6, 8, 9, 13, 15].

In the institutional aspect, the challenges are the community cultures that still individuals in managing their plantation, the demand to increase government revenue as VAT for primary products estates, levies, and charges. Other challenges in institutional aspects are absence of institutional commodity with authority / full authority in the development of commodity; and absence of commodity fund raising policies for commodities [4, 11, 12].

Method

The design used in this study is descriptive verification, the research aims to gain an overview of the characteristics of the study variables and conduct a careful examination of all variables/indicators of oil palm agribusiness partnerships. The data of this research consists of primary and secondary data. The source of primary data is taken from the farmers participating in the partnership, and the age of their oil palm trees is more than 5 years. Interviews, observations, and study literatures are used in this research in order to gather the data. Descriptive analysis is used to provide an overview of the implementation of the partnership implemented by oil palm plantation companies in Jambi Province [4].

Descriptive analysis method is used to give an overview of the implementation of partnership pattern implemented by oil palm company in Jambi Province. To find out the implementation of the partnership pattern can be seen from the conversion of plasma plantation, the determination of MOU, the division of plasma plantation, the installment/credit of plasma plantation, the institution, the pricing of FFB and the production of FFB. Descriptive data analysis is an ongoing, repetitive and continuous effort. Data analysis in this study took place along with the data collection process, which includes three lines, namely data reduction, data presentation and conclusion (Miles and Huberman, 1994). To calculate the income level of farmers, which is the third objective of this study, the different pattern of partnerships in the oil palm plantation company of Jambi Province can be mathematically used as follows: a. Revenue, operationally it is calculated by: TR = PX x Q where: PX = The price of fresh oil palm bunches received by plasma farmers at the revised corporate level (Rp / Kg). Q = Average production that farmers have sold to KUD based on January - December 2010 period. TR = Total revenue received by smallholders in the period January - December 2010. b. Income, operationally it is calculated by: I = TR – TC, where: TR = Total revenue received by smallholders in the period January - December 2010. TC = Total cost incurred to pay the costs to be paid by farmers to plasma cooperatives every month during the period of January - December 2010. I = Gross revenue earned by oil palm farmers from January - December 2010. Data analysis was done by describing the gross income obtained by farmers participating in PIR Trans partnership pattern in oil palm plantation company of PT. Agrowiyana and PT. Kirana Sekernan for a period of one year [4].

Results and discussion

A successful of partnership business strategy is largely determined by the compliance between the partners in running the business ethics. Detailed implementation partnership PIR-Trans (PT Agrowiyana) and creditor (PT Kirana Sekernan) with plasma farmers can be explained as follows: [4, 8, 9, 15]

Table 1. Aspects and indicators of implementation on KKPA and PIR trans patterns.

No.	Variable Implementation Partnership	Partnership	
		PIR-Trans	KKPA
1.	Conversion smallholding	Before the conversion was not charged loan payments. After the conversion gets 30% cost burden for installment credit.	Before the conversion is charged to credit payments begin on the first harvest. After the conversion gets 30% cost burden for installment credit.
2.	Determination MOU	Bank credit can be transferred from the account of the cooperative / plasma to account for distribution to the plasma core in the form of agricultural inputs are channeled through the company.	Bank credit can be transferred from the account of the cooperative / plasma to account for distribution to the plasma core in the form of agricultural inputs are channeled through the company.

3.	Distribution smallholding	Determining plots by means drawn from a stretch of up to individuals.	Distribution plots established by cooperatives in collaboration with the company.
4.	Installment/credit smallholding	Installment loans with 30% of the cutting garden products smallholders and farmers receive 70% of the gardens, lasted until the end of the loan installments.	Installment loans with 70% of the cutting garden products smallholders and farmers receive 30% of the garden, done before the conversion. After conversion, the farmer receives 70% and 30% of the product is cut garden installment payment credit.
5.	Institutional	PIR Trans has a cooperative 5 to 5 sub units, in cooperation with companies in the areas of: sales TBS, payment for farmers, interest-free loans, and credit deposit.	Patterns KKPA cooperate with one cooperative in the areas of: sales TBS, payment for farmers, lending money without interest, channeling subsidized and non-subsidized fertilizer, and credit deposit. Provide basic food and some snacks.
6.	FFB pricing	67% PIR Trans farmers are satisfied with the price set by Disbun (Regional Estate Service), 33% other farmers want the price as high FFB time before the global crisis hit.	All farmers KKPA pattern was quite satisfied with the price set by Disbun (Regional Estate Service)
7.	Production	Production smallholding very varied, due to maintenance activities of each farmer is different.	Production has roughly the same, due to the maintenance of the garden to harvest simultaneously in a single stretch.
8.	Payment Systems	Payments or the sale of TBS depend on each KUD, there are 2 weeks and there is 1 month.	Payments or the sale of FFB performed 1 month and salary taken by the head of each farmer groups with payment details and pieces if it has debt, credit, or other purchases through the cooperative.

The income of the farmers is the revenue from the oil palm production that is sold to palm oil factory. It will be processed and produces Crude Palm Oil (CPO) and palm kernel. One of the most decisive factors on farmers income is the production of oil palm.

Differences in income received by farmers PIR-Trans and KKPA patterns due to the difference in average production, where the average farmer production PIR-Trans is greater than the average production of KKPA. The difference is due to the production of the differences between the applications of patterns during maintenance until harvest. In this case a partnership with farmers PIR-Trans plantation company more profitable for farmers in terms of production [4].

Table 2. Farmers average revenue per month PIR-Trans and KKPA.

Variables	PIR-Trans	KKPA	Difference
Receipts (IDR)	5,529,893	5,392,992	136,901
Cost (Rp)	443,597	414,280	29,317
Revenue (IDR)	5,086,296	4,978,712	107,584

In its application, the company has always been help farmers to increase production. The company always supervise and provide necessary assistance to farmers to increase production or also to assist farmers in overcoming problems that can interfere with the production of Oil Palm. In addition KUD also plays an active role as an intermediary between the company and farmers. KUD also can help farmers so that farmers can apply for a loan to be able to expand its land bank so that production also increased [4, 8, 9].

In KKPA patterns, the company conducts surveillance on farmers while harvesting, while in a maintenance company did not provide special assistance for farmers to increase production of oil palm. KUD also plays an active role as an intermediary between farmers and companies, it is just the location of the garden with KUD far enough to make farmers cooperatives when it comes to sharing only. Conditions of KKPA farmer who worked in his garden together in one plot also affects farmers' production. In this case,farmers' creditor patterns always share equally the amount of production obtained in each expanse. So, the amount of production that farmers obtained is relatively similar [4, 8, 9].

Conclusion

Agribusiness partnerships implemented by the palm oil company has managed to create independent farmers who can channel the aspirations of farmers, both the pattern KKPA and PIR Trans. Farmers 'income who has become participants in partnership oil palm plantation companies in Jambi Province is quite high, as demonstrated by the average income of farmers PIR Trans and farmers' income patterns each creditor per month or $ 410.91 USD 5,086,296 and USD 4,978,712 or $ 402.22 (calculated based on the value of $ 1 = USD $ 12.378 in December 2014).

Reference

1. Badan Penelitian dan Pengembangan Pertanian. *Prospek dan Arah Pengembangan Agribisnis Kelapa Sawit.* Edisi Kedua. Departemen Pertanian. (2007)
2. Direktorat Jenderal Perkebunan. *Pedoman Umum Program Revitalisasi Perkebunan (Kelapa Sawit, Karet dan Kakao).* Departemen Pertanian. (2007)
3. E. Hamid. *Peranan PIR Khusus II Perkebunan Kelapa Sawit Sungai Bahar dalam Pengembangan Wilayah Kabupaten Daerah Tingkat II Batang Hari. Tesis Program Pascasarjana Universitas Padjadjaran Bandung.* (1994)
4. E. Hamid. Implementasi Kemitraan Agribisnis Kelapa Sawit dan Analisis Pendapatan Petani Peserta Di Provinsi Jambi. Penelitian Disertasi Doktor, Universitas Padjadjaran Bandung. (2012)
5. H. Haeruman. *Kemitraan dalam Pengembangan Ekonomi lokal: Bunga Rampai.* Jakarta: Yayasan Mitra Pembangunan Desa-Kota. (2001)
6. I. Pahan. *Panduan Lengkap Kelapa Sawit Manajemen Agribisnis dari Hulu hingga Hilir.* Penebar Swadaya, Jakarta. (2010)
7. M. J. Hafsah. *Kemitraan Usaha: Konsepsi dan Strategi.* Pustaka Sinar Harapan, Jakarta. (2000)
8. Peraturan Menteri Pertanian Nomor : 33/Permentan/OT.140/7/2006 tentang *Pengembangan Perkebunan Melalui Program Revitalisasi Perkebunan.* Ditetapkan di Jakarta pada tanggal 26 Juli 2006.
9. Peraturan Pemerintah Republik Indonesia Nomor 44 Tahun 1997 tentang *Kemitraan.* Ditetapkan di Jakarta pada tanggal 17 November 1997.

10. Rante Tondok, A. *Peranan Perkebunan Besar Sebagai Mitra Perkebunan Rakyat.* Majalah media Perkebunan, Nomor 15 April 1997. (1997)
11. Sihombing, R.S.M. *Kemitraan usaha dalam Pemberdayaan UKM (Usaha Kecil Menengah) (Studi tentang pelaksanaan kemitraan usaha dalam upaya memberdayakan UKM tenun ulos di Kabupaten Toba Samosir, Sumatera Utara).* Tesis Pascasarjana Administrasi Negara Fakultas Ilmu Sosial dan Politik Universitas Gadjah Mada Yogyakarta. (2004)
12. S. Mangoensoekarjo dan H. Semangun. *Manajemen Agrobisnis Kelapa Sawit* (Gadjah Mada University Press, 2005)
13. Sunarko. *Budidaya dan Pengelolaan Kebun Kelapa Sawit dengan Sistem Kemitraan.* Penerbit PT. AgroMedia Pustaka, Jakarta. (2009)
14. U. Fadjar. *Kemitraan Usaha Perkebunan: Perubahan Struktur yang Belum Lengkap.* Forum Penelitian Agro Ekonomi **24**, 1, (2006)
15. Undang-Undang Republik Indonesia Nomor 18 Tahun 2004 tentang *Perkebunan.* Ditetapkan di Jakarta pada tanggal 11 Agustus 2004.

The potential of geographical indications and Its legal protection

Faradila Yulistari Sitepu*

Doctoral Program in Law, Graduate School, Universitas Sumatera Utara, Medan, Indonesia

Abstract. The legal protection of Geographical Indication (GI) is given to Indonesian typical products to prevent them from being registered as overseas brands, as previously happened in the case of Toraja Arabica coffee registered by a Japanese company and the Gayo coffee by a Dutch company. With the foreign trademark registration, the products harvested from the land of Indonesia can no longer be recognized as belonging to Indonesia because they can hardly be claimed to have come from Indonesia. Therefore, it is the foreign party that will enjoy the benefits of the famous Indonesian products. Geographical indication certainly requires a legal protection. This is important so that the natural rights owned by the original holder can be guaranteed and a sense of security can be realized. A product yielded from the nature and human creativity that has distinctive features, cannot be found elsewhere, and has a reputation that can increase the competitive power of the original region should get more attention from the government, especially the local government. The legal protection of Geographical Indication does not only provide economic benefits to the right holder, but also to the geographical environment of the product origin.

1 Introduction

Indonesia has a diverse culture and biodiversity. The issue of geographical indication is set forth in the *Trade Related Aspects of Intellectual Property Rights* (TRIPs) agreement, which requires member states to draft regulations on geographical indication in order to provide a legal protection against fraudulent competitions or practices[1]. Geographical indications are governed by the *Trade Related Aspects of Intellectual Property Rights* (TRIPs) agreement, which requires member states to draft regulations on geographical indication in order to provide a legal protection against fraudulent competitions or practices[2].

By ratifying the establishment approval of the World Trade Organization (WTO), Indonesia is obliged to accept WTO agreements and its attachments, including the TRIPs. The existing protection in the framework of the TRIPs, through the ratification of the States toward the establishment of the WTO, is expected to provide a systematic and sustainable international protection scheme for States that are likely to sell the geographical indication products. Indonesia is obliged to adapt its national legal framework in such a way that it complies with the degree of protection of the six types of intellectual property rights as

* Correspondent Author: faradila.yulistari@usu.ac.id

contained in the TRIPs agreement, including protection of geographical indication into the Law No.15 of 2001 on Trademarks[3].

TRIPs defines Geographical Indication as a sign that identifies a territory of a member country, or a region or area within the territory as the origin of the goods, in which the reputation, quality, and characteristics of the goods concerned are determined by geographical factors. Thus, the origin of a particular product which is attached to the reputation, characteristics and quality of a product associated with a particular area should be legally protected.[4]

Geographical Indication is a branch of Intellectual Property Rights in addition to patents, copyrights, confidential/secret trade information and other types of intellectual property rights. Geographical indication is a sign used for goods having a certain geographical origin and also has the quality or reputation generated by the place of origin.

The information technology is developing so rapidly nowadays that it has made the world smaller. Information can easily and quickly spread to all corners of the world. Under these circumstances, intellectual property rights are becoming increasingly important. Since the intellectual property rights act as monopoly rights that can be used to protect investments and can be transferred. But the most important aspect, when associated with the protection of intellectual work, is the legal aspect.[5]

Discussing geographical indication cannot be separated from studying trademark since Geographical Indication, under Law No.15 of 2001, is considered as one type of a trademark[6]. There are several notions of a trade mark, one of which is defined by H.M.N Purwo Sutjipto, namely: "Trademark is a mark by which a particular item is personalized to distinguish it from other similar product".[7]

Ironically, the protection of geographical indication in Indonesia until now has not been sufficient to provide a legal protection for the potential geographical products. Internally, the number of geographical indication certificates is still relatively small compared to the potential of geographical indications.[8] At the same time, a number of violations of Indonesia's geographical indications have externally happened such as the case of Gayo Coffee in which the trademark was claimed by a Dutch trading company as the right holder. In fact, Gayo coffee is a typical product from Nanggroe Aceh Darussalam, Indonesia. The Dutch company (Holland Coffe B.V) claimed that the company is the owner of the coffee trademark and the trademark has been registered internationally under the name Gayo Mountain Coffee.[9]

The protection of geographical indication is given to Indonesian typical products to prevent them from being registered as overseas brands, as previously happened in the case of Toraja Arabica coffee registered by the Japanese company and the Gayo coffee by the Dutch company. In addition to coffee, there are other Indonesian products that need to be protected through geographical indications, such as nutmeg from the Banda Islands, Nilam from Aceh, and Bilih Fish from Lake Singkarak before they are claimed by other countries. With the foreign trademark registration, the products harvested from the land of Indonesia can no longer be recognized as belonging to Indonesia because they can hardly be claimed to have come from Indonesia and it is the foreign parties that will enjoy the benefits of the famous Indonesian products.[10]

The law should provide solutions to various problems arising in relation to the intellectual property rights. The law should be able to protect the intellectual works in order to develop the people creativity which ultimately leads to the goal of successful intellectual property protection.[11]

The geographical indication is very important because it serves as a sign indicating the origin of an item which, due to geographic environmental factors including natural factors, human factors, or a combination of these two factors, gives certain characteristics and qualities.

2 Problem

The problem in this study: how can Geographical Indications as part of the intellectual property rights be legally protected?

3 Research method

This research applied a normative legal approach. A normative legal method is a legal research that treats law as a norm system. The data was secondary data consisting of: a) primary legal materials in the form of legislation; b) secondary legal materials in the form of writings, both in the form of books and articles containing comments or analysis related to the subject matter; c) tertiary legal materials in the form of dictionaries. The data was collected using a literature review and document analysis. The data was analyzed using a qualitative method based on deductive reasoning.

4 Discussion

Geographical indication was initially unknown in Indonesia's legal terminology. This term emerged from the inclusion of the Trademark Law No. 15 of 2001 adopted from Article 22 to 24 of The Agreement on Trade Related Aspects of Intellectual Property Rights (TRIPs Agreement). The definition of geographical indication by law *is a sign indicating the origin of an item and/or a product which, due to geographical factors including natural factors, human factors or a combination of these two factors, gives reputation, quality, and certain characteristics to the product and/or item.* (Law No. 20 of 2016 Article 1 Paragraph (6)). Soon after Indonesia ratified the convention - which was the outcome of the 1994 World Trade Organization (WTO) agreement in Marrakech, all laws and regulations regarding Indonesian intellectual property rights, including trademarks, were formulated in accordance with the TRIPs Agreement. The clauses on geographical indication are placed in Articles of the Trademark Law No. 15 of 2001.

After including the clauses of geographical indication, Indonesia began to identify, record and list the domestic products that have a potential for a geographical indication. The universities start to perform various researches and scientific works related to geographical indication. Almost all provinces in Indonesia conducted studies on goods and handicraft products that have a potential for geographical indications and the research findings were then recorded and archived at various university libraries. However, only a small number of the research results was registered at the Directorate General of Intellectual Property of the Ministry of Law and Human Rights of the Republic of Indonesia to obtain a legal certainty or certification.

In response to the Article on geographical indications in the Trademark Law, the Government issued a Government Regulation (PP) No. 51 of 2007 which technically regulates matters pertaining to geographical indications. Indonesia is one of the countries that signed the agreement on the establishment of the World Trade Organization. Inside the WTO, there are provisions of the TRIPs (Trade Related Aspects of Intellectual Property Rights) which regulate Intellectual Property Rights and ratify the agreement through Law Number 7 of 1994 on Legalization of Agreement on WTO establishment, which was enacted on 2 November 1994. One form of the Intellectual Property Right protections in the TRIPs Agreement is the geographical indication as set forth in Articles 22 to 24.

In the Indonesian system of Intellectual Property Rights, geographical indications are part of the trademark law. Geographical indication is the concept of Intellectual Property Rights that indicates the origin of a product in relation to the quality, characteristics, and reputation of the product. Geographical indications show the relationship between goods, place of origin, quality and characteristics of the goods. The quality and characteristics of the product are related to the place where the product is originally generated. Another term used to protect products from a region is *Original Indication* (Designation of Origin). The difference between geographical indication and original indication is that the original indication only shows the origin of an item without displaying any characteristics or superiority of the item. In contrast, geographical indication shows not only the origin of a product, but also demonstrates the characteristics and advantages of the product in comparison to other similar items.

Geographical indication is a sign indicating the origin of an item and/or a product which, due to geographical environmental factors including natural factors, human factors or a combination of both factors, gives reputation, quality and certain characteristics to the products and/or goods.[12] Thus, the reputation of quality and specific characteristics are generated by the presence of the geographical factors.[13]

Based on the definition, there are two main points referring to geographical indication, namely a sign indicating the origin of an item and/or a product due to geographical environmental factors and the generated items and/or products have certain characteristics.

The sign, as referred to in the geographical indication, may come in the form of a label placed on the goods. It may also be the name of a region or area, word, image, letter or a combination of such elements. While the goods and/or products possessing certain characteristics can be in the form of agricultural products, processed products, handicrafts, or other processed products which have the characteristics of the region.[14]

Actually, geographical indication has its own characteristics that make it distinctive from other products. The characteristics of geographical indications include: first, Geographical indications can be requested by institutions representing communities living in the areas which produce natural products or natural resources; manufacturers of agricultural goods; makers of handicraft or industrial products; the merchant who sells the goods, the agency authorized to do so or the consumer group. Second, the exclusive right to geographical indication is not regulated and this logically means that there is no time limit for the right owner of the geographical indication.

The protection of intellectual property rights will be closely linked to aspects of science, technology and economics.[15] However, the protection of geographical indications is different from other intellectual property protection. The protection of intellectual property rights is generally individual, but the protection of geographical indications is collective in the sense that the protection is given to a product manufactured in a particular area and is requested by the community group or the provincial or district/city government in the area where the goods/products are generated.

Protecting geographical indication has a great economic value and benefits for the place of origin. The area can be popular both nationally and internationally. New geographical indication will get a protection only if it is already registered. Learning from European countries, geographical indication products can provide substantial benefits to the economy of the country. Wine sales in France rose about 230% of its sales to other countries. Cheese sales grew between 158-203%.[16] The sales of the local Florida oranges from the American State of Florida, which is widely known for its typical taste, was amounted to 9 million US Dollars as well as opening about 80,000 new jobs and expanding 230, 670 hectares of land.[17]

The benefits of the existing potential can be attained if countries (including Indonesia) can protect their typical products with the protection system of geographical indications. From this point on, the protection of international geographical indications is deemed necessary. With an international protection, some benefits can be gained as follows: first, geographical indication protection can be used as a product marketing strategy at domestic and international trade levels. Second, it adds value to the product and improves the producer's welfare. Third, it improves the reputation of geographical indication products at international trade level. Fourth, equality of treatment as a result of promotion from abroad, and fifth, geographical indication protection acts as a tool to avoid fraudulent competitions.

Geographical indication protection aims at protecting uniqueness from counterfeiting acts or unnecessary use while providing opportunities and protection for the community living in the region of the typical product to obtain maximum benefits from that particular product. In addition, the protection of geographical indication is also beneficial for the consumer as it provides product quality assurance.[18]

5 Conclusions

Geographical indication certainly requires a legal protection. With such protection, the natural rights owned by the right holder can be guaranteed and a sense of security can be realized. A product yielded from the nature and human creativity with certain distinctive characteristics, cannot be found elsewhere, and has a reputation that can increase the competitive advantage of a region, should get more attention from the government, especially the local government. The legal protection of geographical indication not only provides economic benefits to the right holder, but also to the geographical environment where the product is originally generated.

References

1. Adolf, *Dasar-dasar hukum kontrak internasional*, (Refika Aditama, Bandung, 2007).
2. U.H. Sanjaya, Problematik penerapan undang-undang hak kekayaan intelektual tentang indikasi geografik pada undang-undang merek, *Jurnal Panorama Hukum* **Vol.1 (2)**, (2016).
3. Sudarmanto, *kepentingan negara berkembang terhadap hak atas indikasi geografis, sumber daya genetika dan pengetahuan tradisional*, (Lembaga Pengkajian Hukum Internasional Fakultas Hukum Universitas Indonesia, Jakarta, 2005).
4. O.K. Saidin, *Aspek hukum hak kekayaan intelektual (intellectual property rights)*, (Raja Grafindo, Jakarta, 2015).
5. B.A. Riswandi and M. Syamsudin, *Hak kekayaan intelektual dan budidaya hukum*, (Raja Grafindo Persada, Jakarta, 2004).
6. T.S. Utomo, *Hak kekayaan intelektual (HKI) di era global sebuah kajian kontemporer*, (Graha Ilmu, Yogyakarta, 2010).
7. S. Margono, and A. Angkasa, *Komersialisasi aset intelektual aspek hukum bisnis*, (Grasindo, Jakarta, 2002).
8. I. Rahmatullah, Perlindungan indikasi geografis dalam hak kekayaan intelektual (HKI) melalui ratifikasi perjanjian Lisabon, *Jurnal Citra Hukum* **Vol.1 (2)**, (2014)
9. S. Mawardi, Establishment of geographical indication protection system in Indonesia, case in coffee, Worldwide Symposium on Geographical Indications jointly orginized by the World Intellectual Property Organization (WIPO) and the Patent Office of the Republic of Bulagaria, Sofia, June 10 – 12, (2009).
10. A. Sutedi, *Hak atas kekayaan intelektual*, (Sinar Grafika, Jakarta, 2009)

11. N. Anggraeni, *perlindungan terhadap indikasi geografis (produk yang disertai nama tempat) dalam kerangka hukum nasional dan hukum internasional*, Tesis Hukum, (2013).
12. A. Muhammad, *Hak ekonomi hak kekayaan intelektual*, (PT Citra Aditya Bakti, Bandung, 2001).
13. E. Purwaningsih, *Perkembangan hukum intellectual property rights*, (Ghalia Indonesia, Bogor, 2005).
14. M. Djumhana, Djubaedillah, *Hak milik intelektual sejarah, teori dan prakteknya di Indonesia*, (PT Citra Aditya Bakti, Bandung, 2003).
15. A.M. Ramli, *Hak atas kepemilikan intelektual teori dasar perlindungan rahasia dagang*, (CV. Mandar Maju, Bandung, 2000).
16. J.A. Clarke, The public policy objectives of geographical indications, Worldwide Symposium on Geographical Indications, Lima 22-24 June (2011).
17. K. Keck, Florida orange juice healthy, pure and simple, Worldwide Symposium on Geographical Indications, Lima, 22-24 June (2011).
18. A. Miru, *Hukum merek cara mudah mempelajari undang-undang merek*, (PT Raja Grafindo Persada, Jakarta, 2005).

Human dimensions of gain and loss of plant species diversity surrounding oil palm plantations

Farras Imantoko[1*], *Arzyana* Sunkar[2], and *Yanto* Santosa[3]

[1]Department of Forest Resources Conservation and Ecotourism, Faculty of Forestry, Bogor Agricultural University P.O Box 168 Bogor, Indonesia
[2]Protected Area Management Division, Department of Forest Resources Conservation and Ecotourism, Faculty of Forestry, Bogor Agricultural University P.O Box 168 Bogor, Indonesia
[3]Ecology and Wildlife Management Division, Department of Forest Resources Conservation and Ecotourism, Faculty of Forestry, Bogor Agricultural University P.O Box 168 Bogor, Indonesia

Abstract. Changes in plants diversity occurring the establishment of oil palm plantations, have been considered as the reasons for the changing pattern in community livelihoods. Hence, the values of plants diversity to the local community worth considering. This research was conducted to identify the local communities' perceptions related to gain and loss of plant species diversity due to the establishment of oil palm plantation. The study was carried out in the indigenous Dayak Villages of Tajok Kayong and Nanga Tayap, West Kalimantan Province of Indonesia in March 2018. Data were collected using field observation and interviews. Our results indicate that the establishment of oil palm plantation have resulted in the gain of 3 plant species and the loss of 8 plant species functional values (medicine and food). Nevertheless, as many as 75.68% of interviewed person stated that the presence of oil palm plantations increased plant diversity, in terms of flowering plants and other non-functional plants. These findings reveals that local community were not affected by the species loss since prior to the plantation's establishment, their living patterns were already transformed into semi modern and modern living styles in which the people have lower dependence on their surrounding plants.

1 Background

* Corresponding author: farrasimntk@gmail.com

The existence of biodiversity has a very important value for communities just like the Dayak people who rely on life pattern and life fulfilment system to the environment. Soni [1] asserts that the Dayak people consider their land or territory as a means for their survival and culture thus showing that lands and territories containing natural resources are very important. Rifai & Walujo [2], stated that ethnobotanical studies should be able to describe the cultural relations of the community, especially the perception and conception of the community in understanding natural plant resources in the environment in which they live. So this makes the presence of oil palm plantations necessary to be seen through community perspectives because this phenomenon often gets accusations that lead to the loss of biodiversity, especially in plants species.

The expansion of oil palm plantations in Indonesia is very significant indeed, according to Ditjenbun [3] the estimated area of oil palm plantations in Indonesia reached 12.3 million hectares in 2017. This area supports the status of Indonesia as the largest palm oil exporting country in the world (Budidarsono et al) [4]. West Kalimantan is one of the tropical rain forest area in Indonesia which harbours a high diversity of plants and it is estimated to contain various types of plants that have potential as medicine (Budiasih) [5].

The rapid development of oil palm has led to several allegations [6]. Various research results suggest that oil palm expansion is thought to be the main cause of deforestation and biodiversity decline in Indonesia [7, 8, 9, 10]. Koh and Wilcove [11] assert that the conversion of primary forests or secondary forests to oil palm plantations has the effect of reducing species diversity. However, on the other hand the results of research from Santosa et al. [12] stated that there is an increase in the diversity of bird species, butterflies, herpetofauna and the density of earthworms on oil palm plantations based on the history of land cover. Such contradictory findings are interesting because they show that the impact of the oil palm plantations need to be investigated more deeply.

So the phenomenon of acquisition and loss of species of plants diversity after the establishment of oil palm plantations needs to be investigated while studying the of social, economic and cultural aspects point of views based on community perception in the field.

2 Research Methodology

2.1 Time and Study sites

The research was conducted around NTYE Palm Oil Plantation, Tajok Kayong Village and Nanga Tayap Village, Nanga Tayap Sub-District, West Kalimantan Province. The duration of our research was 1 month, starting from March 12, 2018.

2.2 Instruments

The instruments used in this research consists of audio recorder, Microsoft Excel, Microsoft Word, and camera. We used an interview guide that contains a list of structured questions and five-scale Likerts statements.

2.3 Data Collection

Methods of data collection used to identify species diversity based on community perception were based on the interview. Interviews were conducted through open-ended interviews. Data were recorded under the the form of species composition from the knowledge of the community about the types of plants, both gain and lost or the type of increase and decrease species since the palm plantation was established and also the public perception related to it. The price data of the types of plants were obtained by economic valuation method by doing 3 types of approaches which are market value approach, replacement value approach and regulatory value approach.

In this study, respondents were indigenous Dayak people who were either related to oil palm cultivation or not with a total of people with a sampling unit of 1 person per family. This is because indigenous Dayak people are long-lived and settled in their residential areas around the oil palm plantations, so that whatever the type of work they do, they can still feel directly the impacts that result from the phenomenon of gain and loss of plant species.

The people in the village of Tajok Kayong and Nanga Tayap who are the respondents that entirely from the Dayak tribe and they accounted for 74 respondents. The Tajok Kayong Village has an area of 6.26% of the area of Nanga Tayap Sub-district, while Nanga Tayap Village has an area of 4.98% of Nanga Tayap Sub-district (BPS Ketapang [13]). The types of jobs do vary, with quite high amounts of income as it could be seen from the status of children.There are some who continue studying up to University level, so this proves that the economic prosperity of the people is quite high. Given that one of the most common tribe on the island of Kalimantan is the Dayak community, it is in accordance with the statement [14] that the Dayak people are Kalimantan island residents who occupy a region permanently for a long time This supports the idea that West Kalimantan Province has its own uniqueness towards the process of acculturation, because of the Dayak people could face some modern way of living syle in several aspects.

2.4 Data Analysis

The analysis method used to measure the interaction of the community with the diversity of plant species and the perception of the local community was conducted using Quantitatively Descriptive method through Likert Scale that were coded 0 = do not know, 1= strongly disagree, 2= disagree, 3= neutral, 4= agree, 5=strongly agree. It shows positive statements which become positive indication and negative form of negative statement becomes negative indication [15]. The distribution of responses from respondent each statement and each plants species that gain and loss/increased and decreased individually were diplayed in Tables.

3 Results and Discussion

3.1 List of gain and loss plant species based on community perceptions.

The existence of oil palm plantations around the village of Tajok Kayong and Nanga Tayap Village induce some changes in plant species diversity (either gain or loss). This phenomenon is felt by the community directly because they themselves feel that there are some types of plants that can never been found again, and also some types of plants that have only been discovered since the oil palm plantation was established. The types of plants listed in Table 1 below are the species identified by the community because they can well remember the species and its function at the time.

Table 1. List of types of loss and gain of plant species along with the average price based on the community.

No.	Loss Species	Average Price	Gain Species	Average Price
1.	*Shorea sp.*	2.400.000 IDR/m^3	*Diplazium esculentum*	2.000 IDR/bundle
2.	*Neolamarckia cadamba*	950.000 IDR/m^3	*Limnocharis flava*	1.500 IDR/bundle
3.	*Shorea sp.* (wood)	450.000 IDR/m^3	*Ipomea aquatica*	2.000 IDR/bundle
4.	*Shorea sp.*/Tengkawang (fruit)	3.000 IDR/Kg		
5.	Durian hutan "Pekawai" *(Durio kutejensis)*	55.000 IDR/Kg		
6.	Kusik (Red Durian)	100.000 IDR/Kg		
7.	*Momordica sp.*	3.000 IDR/Kg		
8.	*Artocarpus anisophyllus*	5.000 IDR/Kg		
	TOTAL	3.966.000 IDR		5.500 IDR

Price-related data is also obtained from the knowledge of the community who had used the types of plants, so that in the end the economic valuation can be done based on the price gained from the knowledge of the community. The purpose of the valuation is to know how big the existing value of biodiversity is, because the existence of natural resources itself is the initial factor in the future process of economic activity involving natural resources [16].

The species of plants which can not be found again by the community include *Shorea sp.*, *Neolamarckia cadamba*, *Shorea sp*/Tengkawang (wood and fruit), *Durio kutejensis* (pekawai), *Momordica sp* (Kusik), and *Artocarpus anisophyllus*. In the species of Meranti *(Shorea sp.)*, people use these woody plants for building and other needs. The selling price of this meranti reaches 2.400,000 IDR/m3. Such type of meranti cannot be be found any more by the community, but the community does not suffer significant losses because they feel the need for more energy and a much time to enter the area and harvest this timber. This is in accordance with the results of the study from Sunkar *et al.* [17] which showed that distance is a factor affecting the rate of encroachment and positive interaction. Thus, this is in harmony with the people who feel heavy to travel to the location of woody plants due to needed amount of time and energy. Then, people who have experienced more advanced life changes also make their habits to buy more building materials, so they prefer to spend money to build their own homes and other buildings such as public facilities nearby.

There is also a type of forest durian called "pekawai" by local people which can not be found anymore. This type of plant has an economic value with a selling price of 55.000 IDR / Kg. On the cultural side, there is a custom of the community that makes this type of pekawai a symbol of family heritage. Indeed, planting and caring for the tree for several generations train the young ones about the existence of pekawai so that the land has a marker linked with ancestors who planted the durian. This value is built by people who do have a high involvement in agricultural activities, so that the object used as a symbol will not be far from the crop itself. The system of knowledge possessed by society is traditionally a part of indigenous peoples culture and rural farmers [18]. So, even if the tradition of maintaining the pekawai has already faded away, it is important to know that it must be remembered for the future generation to grow, the culture can remain alive in the survival story in society.

In addition to the species that can not be found again, there are also new types of plants that can be found by the community with prices and benefits that can be taken from these types. The price and benefits of each species that emerged since the palm oil plantations become interesting, because it shows that the existence of oil palm plantations resulted in the gained of species biodiversity which also from the community has an economic value that can be utilized either directly or for sold to the market.

The types emerging from these plants include the Pakis (*Diplazium esculentum)* plant, with a selling price of 2.000 IDR / bundle, Genjer *(Limnocharis flava)* with 1.500 IDR/bundle, and Kangkung *(Ipomea aquatica)* with 2.000 IDR/bundle. These plants are used by the community because they are increasingly encountered since the existence of oil palm plantations, where previously the community had never found this type of fern. Kaplan et.al [19] explain that within the ethnobotany discipline there is a set of assumptions about the relationship between behavioral patterns with integrated socio-cultural arrangements in the language, cognitive system, rules and codes of local cultural ethics. This shows that the pattern of human behavior in the utilization of several types of plants into something that will affect the incidence of value to the type. Given the new plant species or new natural resources, humans will naturally find a way to make use of them. This happens because humans seek to utilize natural resources and the environment based on their experience and knowledge [20].

The existence of some of these new species have an effect on the existing biodiversity. The higher value of species diversity in a habitat, the equilibrium of the community will be higher [21]. The economic value derived from this newly discovered plant species is important because it indicates the presence of palm oil can provide the positive impact is also to be felt directly by the people who live nearby because those plants were naturally gained.

3.2 List of plant species that are decreased and increased individually based on community perceptions.

In addition to the phenomenon of the gain and loss of biodiversity, there are also some types of plants that have increased and also decreased individually (Table 2). This is seen from the knowledge of the local community that Dayak people who feel to find more of some certain species which were less and more encountered by the community, so that types of species can be identified along with the price and its usefulness.

Table 2. List of decreased and increased of plants individu along with the average price of the community.

No.	Decreased Species	Average Price	Increased Species	Average Price
1.	*Korthalsia flagellaris*	1.500 IDR/Kg	*Imperata cylindrica*	0
2.	*Cucurbita sp*	8.000 IDR/Kg	*Volvarariella volvacea*	12.500 IDR/Kg
3.	*Curcuma zanthorrhiza*	8.000 IDR/Kg	*Turnera ulmifolia*	0
4.	*Eurycoma longifolia*	40.000 IDR/Kg		
5.	*Tinospora cordifolia*	35.000 IDR/Kg		
6.	*Eusideroxylon zwageri*	16.500.000 IDR/m^3		
7.	*Shorea leavis*	Rp 9.750.000 IDR/m^3		
	TOTAL	26.342.500 IDR		12.500 IDR

There are several types of medicinal plants that have decreased the number of individuals based on the perception of the community, including species of *Curcuma zanthorrhiza*, *Eurycoma longifolia* and also *Tinospora cordifolia*, with total selling price of the three types include 83.000 IDR/ kg. Although they are perceived to be less in number by the community, they do not have a perceived influence because the level of dependence on the use of medicinal plants is very low, and also people realize that the use of generic medicine is clearly more efficient and it enables to save energy and time. In addition, the presence of general hospitals around their place make people change their habits and they now to prefer to come directly to seek doctors rather than plants in the forest that requires a lot of energy. This is in accordance with the results of Lovadi et al [22] who showed that harvesting plants in the forest is less effective to make people start thinking to plant or cultivate medicinal plants around the settlement or yard. So that changes in the pattern of community life was realized first before the Nanga Tayap oil palm plantation was built.

Furthermore, there are types of woody plants such as ulin (*Eusideroxylon zwageri)* and bengkirai (*Shorea leavis)* with selling price of 16.500.000 IDR/m3 and 9.750.000 IDR/m3 respectively. The perception of this species is not much different from the kind of woody plants that loss in the previous discussion, that the community feels no big influence on their lifestyle because of the existence of their habits that tend to buy directly the building materials rather than take and cut down trees from the forest which is considered as ineffective and a waste of time and energy.

Public perception about the use of medicinal plants is influenced by their style of life which is becoming more modern day by day, so this is in accordance with the statement by Waluyo [23] that the existence of modern life will change the original knowledge of ethnic Indonesians outside Java. Then, there are several types of plants such as tankos mushrooms (*Volvarariella volvacea)*, reeds (*Imperata cylindrica*), and bunga pukul delapan (*Turnera ulmifolia)* to be often found by the community. This suggests that the phenomenon of biodiversity gain is ecologically felt by society as well. The type of plants with selling value is tankos mushroom which reach 12.500

IDR/ Kg. This is interesting because in addition to the decreased or loss plant species, there are also increased plant individually such as the three species that are most remembered by the community.

3.3 Perceptions of Dayak People on plants

The presence of NTYE oil palm plantations surrounding the community caused some kind of changes in certain species of plants. It is important to know the perception of the community about loss and gain/decreased and increased of plants species because the Dayak community is known for its habits to use the type of natural resources that exist around it. The existence of plants has essentially been of great value to indigenous peoples, in accordance with archaeological evidence often used to show that at the beginning of civilization and human dependence on plants were used to survive by taking from natural resources for food [24].

The types of plants that previously existed around NTYE are woody plants including *Neolamarckia cadamba and Shorea sp.* Then, there are also fruit trees such as forest durian/*pekawai*, red durian/*kusik*, and forest pare *(Momordica sp).* Some types of plants are considered loss because they cannot be found again by the community. But not all societies fully feel that these lost species bring the negative impacts to them, this is seen with most (74.32%) people feeling that the existence of these species of plants is not considered as an alternative to their income (Table 3).

Table 3. Perceptions of Dayak Communities.

No.	Statement	Perception (%)			
		Do not know	Do not Agree	Doubt	Agree
1	Alternative Income Society	0.00	74.32	18.92	6.76
2	Oil Palm increase medicinal plants	17.57	75.68	0.00	6.76
3	Oil Palm increase food plants	2.70	85.14	8.11	4.05
4	Oil palm increase plants diversity	2.70	20.27	1.35	75.68
5	Easy to grow other plants	0.00	10.81	0.00	89.19

This happens because the existing community is not directly relying on economic activities involving plants that exist around them.

Considering medicinal uses, (78%) of the people felt that the medicinal plants became increasingly difficult to find, as did the food plants with as many as (85.19%) the community also found it increasingly difficult to find them. However, people who feel that the plants become more difficult to find do not feel lost because they are used to buy food in the market and to buy medicine at the nearest public health center. This is in accordance with the results from Purwanto [25] who showed that the traditional knowledge of society decreases due to socio-cultural changes that generally affect the social values, as the younger generation seek a more practical alternative. This lifestyle change in the modern direction is supported by the majority (89%) of the community who feels that with oil palm plantations they can still grow other plants around their homes, so that some types of crops can be planted and nurtured by the community and sometimes used as a source of income. According to Danoesatro [26] and Giono [27], the yard is usually planted with a variety of seasonal plant species for everyday purposes. Household restrictions are often referred to as living barns, live stalls or live pharmacies (Lovadi et al 2013). Then, considering the diversity of plant species, most people (76%) feel that in general, the existence of oil palm plantations around their places make the variety of plant species become increased. The types of plants referred by the community that considered as gained species consist of mushroom tankos (*Volvarariella volvacea)*, bunga pukul delapan *(Turnera ulmifolia)*, and also reeds *(Imperata cylindrica)*.

4 Conclusion

The loss of medicinal and food plants was felt by most people although they did not feel directly the loss because before the oil palm plantations were established, the pattern of society habits in fulfilling food needs and medicine had already changed to a more modern lifestyle. The existence of some lost economic value of some plant species that was decreased individually or lost does not have a significant impact on people's lives, because these species are not intensively used Indeed. The change of plant diversity does not really impact the Dayak community life pattern, on the other hand there are benefits that can be taken such as from the gain plant species and also opportunities to increase economic prosperity with the presence of oil palm plantations.

5 Suggestions

The research about gain and loss biodiversity based on community perceptions is still rare, so in the future it is necessary to conduct research related to it in order to obtain a symmetrical result both from the ecological and social side.

References

1. C.P. Soni, *Kearifan Lokal Masyarakat Adat Dayak Kanayaitn dalam Pengelolaan Hutan Adat (Marang) di Kampung Sidas Data Kecamatan Sengah Temila Kabupaten Landak Kalimantan Barat* [skripsi]. DIY (ID) : Universitas Negeri Yogyakarta (2012)
2. E.B. Walujo, *Keterintegrasian Ilmu Sosial dengan Ilmu-ilmu Lain di Indonesia*. Bunga Rampai Metodologi Penelitian. Departemen Pendidikan dan Kebudayaan, Dirjen Pendidikan Tinggi, Direktorat Pembinaan Penelitian dan Pengabdian Masyarakat. 173 – 176 (1992)
3. Direktorat Jendral Perkebunan. (2016). *Statistik Perkebunan Indonesia: komoditas kelapa sawit 2013-2015.* Jakarta(ID): Direktorat Jenderal Perkebunan, Kementerian Pertanian.
4. Budidarsono *et al, Oil palm plantation in Indonesia: the implications for migration. settlement/resettlement and local economics development,* (2013)
5. L.A. Budiasih, *Inventerisasi Jenis Tumbuhan yang Berkhasiat sebagai Obat di kawasan Hutan Baning Sintang Kalimantan Barat, Skripsi, Fakultas Pertanian Universitas Tanjung Pura Pontianak,* (2005*)*,
6. L.P. Koh, D.S. Wilcove, Cashing in palm oil for conservation. *Nature* 448: 993–994, (2007)

7. A. Zakaria , C. Theile , L. Khaimur, *Policy, practice, pride and prejudice: review of legal, environmental and social practices of oil palm plantation companies of the Wilmar Group in Sambas District, West Kalimantan (Indonesia).* Friends of the Earth Netherlands, Lembaga Gemawan, and KONTAK Rakyat Borneo, Amsterdam, the Netherlands, (2007).
8. B. Azhar, L.P. Chong, M. Zakaria, N. Hassan, M. Arif, Effects of monoculture and polyculture practices in oil palm smallholdings on tropical farmland birds. *Basic and Applied Ecology* **15:336-346**, (2014).
9. K. Obidzinski, R. Andriani, H. Komarudin, A. Andrianto, Environmental and social impacts of oil palm plantations and their implications for biofuel production in Indonesia. *Ecology and Society* 17(1): 25, (2012)
10. E.B. Fitzherbert, M.J Struebig, A. Morel, How will oil palm expansion affect biodiversity?, *Trends in Ecology & Evolution* 23: 538–545 (2008)
11. L.P. Koh, D.S. Wilcove, Is oil palm agriculture really destroying tropical biodiversity . *Conservation Letters* **1:60-64,** (2008)
12. Y. Santosa, A.Sunkar, Erniwati, I. Purnamasari, *Sejarah Perkembangan Status, Penggunaan Lahan, dan Keanekaragaman Hayati Kebun Sawit Indonesia* [laporan riset]. Bogor (ID): Badan Pengelola Dana Perkebunan Sawit dan Lembaga Penelitian Pengabdian Kepada Masyarakat IPB, (2016)
13. [BPS] Badan Pusat Statistik. (2017). *Kecamatan Nanga Tayap dalam Angka 2017.* Ketapang (ID): BPS Kabupaten Ketapang.
14. Darmadi. Dayak asal-usul dan penyebarannya di Bumi Borneo (1). **Vol.3, No. 2.** (Sosial Horizon, Pontianak, 2016).
15. A Afandi , *Perbandingan Persepsi Masyarakat Terhadap Hutan Rakyat Berbasis Karet Dan Kebun Kelapa Sawit* [skripsi]. Bogor (ID) : Institut Pertanian Bogor, (2011).
16. A.Fauzi, *Ekonomi Sumberdaya Alam dan Lingkungan : Teori dan Aplikasi.* (Gramedia, Jakarta, 2004).
17. A. Sunkar, Y. Santosa, S.H Hasibuan. Identifikasi Peubah Sosial Ekonomi Penentu Keberhasilan Konservasi Keanekaragaman Hayati, *Media Konservasi*, (2017).
18. S.B. Brush, *A non-market approach to proctecting biological research*. In: Greaves, T. (editor). (Intelectual Property Right for Indigenous People. Oklahoma City: Society for Applied Anthropology, 1994)
19. D. Kaplan and R.A. Manners, *Teori Budaya (terjemahan) Landung Simatupang. Pustaka Pelajar. Edisi III.* 294 hal, (2003)
20. E.B. Walujo, Sumbangan Ilmu Etnobotani dalam Memfasilitasi Hubungan Manusia dengan Tumbuhan dan Lingkungannya, *Jurnal Biologi Indonesia* **7** (2): 375-391 (2011).
21. A. Sunkar, E. Amzu, S. Nurjannah. Peran Kawasan Bernilai Konservasi Tinggi Bagi Pelestarian Keanekaragaman Hayati di Perkebunan Kelapa Sawit Provinsi Riau, *Risalah Kebijakan Pertanian dan Lingkungan,* (2016).
22. I. Lovadi, R. Linda, Meliki. Etnobotani Tumbuhan Obat oleh Suku Dayak Iban di Desa Tanjung Sari Kecamatan Ketungau Tengah Kabupaten Sintang, *Protobiont* **Vol 2** (3): 129 – 135, (2013).
23. E.B. Waluyo, Perkembangan pemanfaatan tumbuhan obat di luar Pulau Jawa. *Prosiding Pemanfaatan Tentang Tumbuhan Obat dari Hutan Tropika Indonesia,* (Institut Pertanian Bogor, Bogor,1991)

24. E.B. Walujo, Etnobotani: Memfasilitasi penghayatan, pemutakiran pengetahuan dan kearifan lokal dengan mengguna-kan prinsipprinsip dasar ilmu pengetahuan. *Prosiding Seminar Enobotani IV*. (Cibinong Science Center-LIPI, 18 Mei 2009, 2009).
25. Y. Purwanto, T.chikmawati, D.Setiadi, J.Batoro, Pengetahuan Tentang Tumbuhan Masyarakat Tengger di Bromo Tengger Semeru Jawa Timur, *WACANA Jurnal Sosial*, (2011).
26. Danoesatro, *Tumbuhan Obat Keluarga*, (Penebar Swadaya, Jakarta, 1980)
27. W. Giono, *Budidaya Tumbuhan Obat di Perkarangan*, (Agromedia Pustaka, Jakarta, 2004)

Tracing the causality between Livestock and poverty alleviation in the rural economy of Gilgit Baltistan-Pakistan

Gulnaz Hameed[1*], *Shagufta* Shaheen[1], *Abdul* Saboor[1], and *Khuram Nawaz* Sadozai[2]

[1]Department of Economics, UAAR, Pakistan
[2]Assistant Professor, Department of Agriculture and Applied Economics, UAP, Pakistan

Abstract. Livestock rearing and its production are deemed as a lucrative economic activity which provides livelihood and sustainable growth to rural population of Gilgit Baltistan . This sector not only delivers variety of food and non-food products but also contributes in income generation and subsequently diminishes the rural poverty. The fundamental aim of this research endeavor was to assess the poverty status and its determinants for livestock growers of the study area. The primary data was gleaned from 100 respondents through interview schedule following systematic sampling procedure. Uni-dimensional poverty estimates for sampled respondents employing Foster, Greer, and Thorbecke (FGT) poverty index are incorporated as dependent variable to model the binary probability expressions through Logit. Out of Explanatory variables used in model, benefit from the livestock, age, form size and gender of household head, showed a significant but negative relationship with poverty status whereas, household size and education of household head had a insignificant influence on poverty. This study recommends, that increase in farm size and youth's participation in livestock farming can elevate the household income. The capacity building of the livestock farmers through trainings and more women contribution in livestock sector should be encouraged to reduce poverty in the area which can play a remarkable role in sustainable economic growth.

1 Introduction

Agriculture is the lifeline for developing countries and contribution of livestock in agricultural sector has been 40 Percent globally and thirty percent in developing countries [19]. Two thirds of the world population kept domestic animals for their needs, from which 90 Percent were owned by rural smallholders [18]. Livestock provides food security, food nutrition and contributes to income generation which implies that growth in the livestock

* Correspondent Author: gulnazhameed@yahoo.com

sector offers a chance to alleviate/decrease poverty by providing food safety and improving the livelihoods of smallholder households [21, 32, 16] and thus plays a valuable role for sustainable agriculture especially in developing countries [42].

According to estimation the production of livestock globally valued about 883 billion dollars [22]. The economic benefits from livestock are further extended to upstream industry e.g. producers and transporters of feed, manufacturers of equipment and vaccines and seed companies etc. , downstream industries like butchers, milk processors, retailers , wholesalers etc. and allied services including veterinarians, dietary specialists etc. A variety of products (food and non-food) including milk, Dairy products, meat, eggs, wool, leather, bone products, pharmaceuticals, fats and industrial proteins are obtained from animals. Even animal wastes may not be wasted like intestines may further be utilized as crop fertilizer. Animal dung/manure is helpful not only in maintaining the land fertility but also used as fuel. Moreover, some types of cattle are used for on farm mechanical purposes (tillage etc.) and also for transport of folks and goods in regions with inadequate machine power [38, 52, 39]. So the production of livestock not only serve as the source of earning income but a major contributing factor to economic and food security and can be considered as cover/insurance in case of any hazard [19, 3]. For diversifying sources of income some of the farmers are reported to rear livestock as a tactic to lessen the chance of risks associated to climate, marketing conditions and other issues [34, 31].There are studies which also highlighted the social importance (besides economic value) of livestock in developing economies with rural poverty areas and is not restricted to nomadic or country side cultures only [1, 2].Even in developed nations rearing animals are crucial in maintaining the cultural values and connections of people to forefather's land and animals and so to whole community and it has been accepted even at low rate of return from ranching [6, 10, 46].

Due to extensive worldwide variation of cultivation practices in agriculture, the impact of producing meat on environment also varies across the regions. Some of the methods of meat production are intensive livestock farming, free range agribusiness, organic farming, fishing, and subsistence farming etc. Some of the negative effects of livestock production which have been reported are pollution due to fossil fuel, discharge of wastes and consumption of land and water. [20] reported this sector as a major stressing factor on ecosystems, source of greenhouse gases and also causes damage to biodiversity.

In spite of these effects one cannot deny the importance of livestock in economic activities, [51] analyzed the effectiveness of livestock based micro enterprises on poverty reduction in Sunsari district of Nepal using Cross-sectional data. Results regarding this study showed that livestock based micro enterprises are effective in generating income to reduce poverty. It was concluded the livestock is important for producing food, generating income and employment. Livestock sector has a significant share in agricultural GDP [49, 35] of the country. Livestock sector provides many products to rural people but there are certain threats such as lack of livestock market and social off-limits etc. [40].

World Bank [54] attempts to explore that the world's livestock division has changing rapidly through a strong as well rising demand so if it will have risen up commercially livestock can be helpful for reduction of poverty. Shugri [17] determined the various dimensions of agro-pastoral household as well as the poverty status and factors determining poverty. The cross-sectional data was used to measure poverty through FGT and logistic regression model was then employed for finding poverty correlates. The findings of binary logit model showed that number of livestock, farm size, expenditure, farm income, Oxen Ownership had a significant and negative effect towards the poverty. Based on empirical findings this study recommended to improve livestock market, improve infrastructure and remove inequalities among gender.

[23]The socio-economic factors affecting the household decision to rear livestock were identified. Three analytical techniques were used in this study. Heckman selection model, Ordinary Least Square and logistic regression model were used to analyze the data. The findings revealed that gender, age of household head, benefit from livestock and total family size has a significant effect towards probability to own cattle. Several recommendations were anticipated in this study such as to inspire youth to participate in livestock, eliminate inequalities among gender, encourage women in decision making and enhance the livestock provision institution.

In Pakistan, agricultural sector contributes importantly in economic growth as its share is nineteen point five three percent of Gross Domestic Product (GDP) with a growth rate of three point four six percent. About forty five percent of total labor force is engaged while livelihood of approximately sixty two percent of rural population depends directly or indirectly on this sector. In agriculture, the livestock sector accounts fifty eight point three percent, eleven point nine percent in GDP and it grew by three point four three percent as compared to last year growth rate of three point three six percent [27].

Rehman[7]highlighted the performance of livestock production including poultry product and explored the association between the livestock output (Products) and GDP in Pakistan over a tenure from 1980 to 2015. It was concluded that livestock production and GDP has a positive and significant relationship so the Government should provide incentives to producers and further schemes for development of livestock sector.

[43]It was analyzed that livestock is one of key source of income to local residence of economy and essential for rural livelihood. Sample of 137 respondents were taken whose livelihood depends on livestock in Pindsultani district of Attock. Qualitative analysis was used for livestock role in rural livelihood and its contribution to local residents. The results of this study showed that animal husbandry is significant for rural economies of Pakistan and is a source of livelihood poor rural people

The Northern Areas of Pakistan situated between Karakorum, Western Himalayas and Hindu Kush mountain ranges. Approximately one million of people living in northern areas of Pakistan are governed on livestock for annually income [9]. Gilgit Baltistan earlier a part of northern areas is a beautiful mountainous region of Pakistan in which an enormous number of households depends on agriculture as well as on livestock to uplift their income and livelihood standards. Surveys of farmer's reason for keeping livestock indicated domestic milk supply as being primary source and the second purpose is for dung production which is used as fertilizer for crops production [28].

Livestock sector's contribution is continuously increasing not only to rural economy but for the country's economy as well but there is still lack of critical analysis of its benefits and associated problems for future planning. So the current research is carried out to bridge the gap of this analytical deficiency especially in the area like GilgitBaltistan by exploring the socio-economic conditions of household of the area, assessing the impact of livestock on household poverty and filter the impact of livestock production on livelihood of respondents to give some policy recommendation for future.

2 Research Methodology

This section includes data collection technique, source of data and sample design. The conceptual and analytical framework is also given as a sub sections.

2.1 Data Collection

2.1.1 Locale

The study was conducted in Gilgit- Baltistan, Pakistan which comprises of eight districts. Out of eight districts, three districts Hunza, Nagar and Gilgit were selected purposively. Three villages from each district were chosen and surveyed. Misghar, Ispenj, Sher-e-baz; from Hunza, Thole, Chamiling and Gutas from Nagar and Nomal, Oshkhandas and Sultanabad were selected from Gilgit purposively.

2.1.2 Sample Size

The formula for sample size is called proportion sample [50](Yamane, 1967). The formula is

$$n = \frac{N}{1+N(e)^2} \quad (1)$$

Where N =Population Research, n =Sample Size, e =Level of Confidence
Our research population is 102078. So "N "is 102078 and e is taken 90 Percent. Then, a sample size of 100 household is selected.

2.2 Data Collection

Properly designed interview schedule including both open and closed ended questions was used to collect information from livestock growers regarding demographic characteristics production of livestock and its products, as well as problems related to livestock like lack of markets, feeding problems and many other problems facing by the farmers involved in livestock keeping. Questionnaire was design in English, translated and explained in local language. The purpose of survey was to gather information about the socio- economic opportunities and challenges faced by the farmer during keeping livestock and benefits/ gain from livestock etc.

3 Analytical Framework

This study includes data analysis in three phases. First phase involves descriptive analysis second phase estimates unidimensional poverty and in last phase logit model applied to determine the correlates of poverty in the study area.

3.1 Descriptive Statistic

It is a comprehensive description of the data and it converts the data in a sophisticated manner to establish the effects of quantitative data [45].

3.2 Poverty Estimation

In order to estimate Unidimensional poverty of sampled respondents of the research area first step is to set the poverty line or the threshold level under which the households are thought to be poor and it was set based on the Cost of the Basic Needs (CBN) method. According to Economic Survey of Pakistan, poverty line for year 2013-14 was Rs.3030 (CBN) which is inflated through Consumer Price Index to get poverty line of Rs. 3401.82 for the year 2016-17. Then poverty was calculated by applying FGT index (Foster-Greer-Thorbecke, 1984). Through FGT measure Head Count Ratio (HCR), Poverty Gap Ratio (PGR) and Squared Poverty Gap Ratio (SPGR) can be estimated by using formula:

$$FGT\alpha = \frac{1}{N}\sum_{i=1}^{N}\left[\left(\frac{Z-Y_i}{Z}\right)\right]\alpha \quad (2)$$

Where N is total number of respondents, Z is the poverty line as estimated and Y shows the consumption expenditure per person α may take the value of 0,1 and 2 for estimating HCR, PGR and SPGR respectively. HCR or incidence of poverty is used to quantify the percentage of poverty form over the entire population by dividing number of poor persons with total population. PGR shows how much households are far away from poverty line and figure out the mean distance under the poverty threshold means depth of poverty. While SPGR measure the average value of the square of depth of poverty intended for every household. It measure the severity of poverty by attaching more weight to the more deprived thus explains the inequality across the poor [8]. In addition the one is far from the poverty line, the larger the weight as well the severity of poverty [30].

3.3 Specification of Logistic Model

As in this study respondents are divided in poor and non-poor by estimating poverty so this binary relation converts the data limit in 1 or 0 forms. Distinctive method used to solve dichotomy variables is logistic regression. Major reason of using the logistic regression model is that the dependent variable is composed in the form of binary (0, 1) values [24, 13] while OLS regression method is inefficient in parameter estimations as well as heteroscedastic error correction. Consequently, the hypothesis testing and construction of confidence interval turn inaccurate and misleading [29]. So, to solve these problems and to generate relevant outcomes commonly used qualitative response a model is known as logit model [47]. In this model dependent variable takes the value of 1 if the individuals lie under the poverty line such as the poor through the probability as Pi, if not than value be 0 such as the non-poor by the probability is 1-Pi [33, 44, 26, 48].

In this study respondent's estimated poverty is a binomial quantitative dependent variable and takes the value of 0 and 1, 1 if household is poor and 0 if household is non poor. Model analyses probability of household poverty status with respect to different independent variables which could be numeric as well as nominal in the form of dummies. Relationship between the probability of dependent variable is equal to one as Y = 1, while the independent variables are determined through Logit function which is logarithms of odd Y = 1. It is assumed that the value of Y depends on the on value of $X_1....X_K$.

General Formula of logit Model is

$$\ln\left(\frac{Y_i}{1-Y_i}\right) = L_i=\beta_0 + \beta_1X_1+........... \beta_KX_K + \mu_i \quad (4)$$

Where $\frac{Y_i}{1-Y_i}$ is the Odd Ratio:

Y_i = Poor Household Probability, $1-Y_i$ = Non Poor Household Probability, $\beta_0.......\beta_k$ = estimated parameters, $X_i.....X_k$= independent variables, μ_i= error term/disturbance term

The specific model is written as:

$$L_i = \beta_0 + \beta_1 hhsize + \beta_2 age + \beta_3 benftLs + \beta_4 edu + \beta_5 gndr + \beta_6 landarea + u_i \quad (5)$$

Where, L_i is probability of being poor, β_0 = Intercept, X_1= Total number of family members, X_2= Age of household head, X_3= Livestock benefit of household, X_4= Education of household, X_5= Gender of household head, X_6= Landholding of Household, μ_i = Error term.

3.4 Maximum Likelihood Ratio (MLE)

In logit model coefficient are estimated through MLE because it maximize log likelihood rather to minimize the residual [24, 13].

3.5 Odds Ratio

Odds ratio is estimated to express the probability of occurrence of an event. One unit change in explanatory variable changes the odds which is estimated by exponentiation the resulted parameter e^{β} termed as odd ratio. If its value is greater than 1 it means that the independent variable increases the odds of occurrence of the outcome and vice versa [4, 5].

3.6 Marginal Effects

As in OLS, marginal effects are used as a basic measure for the description of result in logistic regression model. Both measures are similar with little contrast that is OLS measure the instant change in dependent variable due to unit change in independent variable while marginal effects calculate the distinct change in estimated probability due to a unit change in independent variable [53].

3.7 Multicollinearity Test

Before estimation of the logit model it is essential to check the existence of the multicollinearity amongst the continuous variables because the presence of the multicollinearity will affect the importance of parameter or estimates. If the multicollinearity goes out and shows to be significant than the instantaneous existence of the two variables can be weaken as well as strengthen the distinct effects of these variables so, by omitting the significant collaboration terms inaccurately will lead towards a specification biasness. In simple words, the coefficients of the interaction among the variables designate whether one of the double associated and related variables ought to be abolished from the model analysis [11]. Variance Inflation Factor (VIF) is a technique that is helpful to perceive the problems of multicollinearity for independent variables [14] .

$$\text{VIF}\ (X_j) = \frac{1}{1-R^2} \tag{6}$$

Here $R^2$is the coefficient of determination of the variable j which is regressed on the remaining quantifiable independent variables. An increase in the value of R^2 shows that the degree of multicollinearity tends to increase, which lead to rise in the variances as well as the standard errors of the OLS estimations. A value of VIF is greater than the 10 such as it will occur if R^2 is surpasses 0.90, so it is used by sign designed for presence of unadorned multicollinearity [15].

4 Results and Discussion

4.1 Poverty Estimation

The poverty estimates of the three districts namely Gilgit, Hunza and Nagar compared with estimated poverty line are highlighted in Table 1. The result of Hunza district shows that 74 percent households are absolute poor, depth of poverty is 37 percent and severity is 22 percent. For district Nagar incidence of poverty is 38 percent, depth is 19 percent while 10 percent people are severely poor. In the same way results of Gilgit district show 32, 19 and 12 percent incidence, depth and severity of poverty respectively. Table also shows that on average 44 percent people are living below poverty line in study area for the year 2016-17. Overall depth of poverty is 23 percent which means that and 13 percent SPGR for overall sampled areas of Gilgit Baltistan. It means 44 percent people on average are living below poverty line in research area while depth of poverty is 23 percent which means if resources are mobilized equal to 23 percent of the poverty line 3401.82 poor individuals can be transferred to the non-poor as it is the amount needed to bring each individual up to the poverty line and then at least in principle, poverty could be eliminated. For overall district severity of poverty 13 percent which indicates severe inequality among households of the rural area.

The comparison of poverty incidences across the districts of rural areas shows the proportion of rural households living in poverty being markedly highest in Hunza and Nagar respectively. According to the household survey results, the depth of poverty is higher in Hunza, followed by Nagar and Gilgit districts of selected rural areas, implying that more resource is required to bring the poor households out of poverty in Hunza than Nagar and Gilgit.

Table:1 Poverty Profile of Gilgit

Districts	HCR	PGR	SPGR
Hunza	0.74	0.37	0.22
Nagar	0.38	0.19	0.10
Gilgit	0.32	0.19	0.12
Total	0.44	0.23	0.13

4.2 Diagnostics of the Econometric Model

Before applying logit model multicollinearity of the independent variables was tested.

4.2.1 Multicollinearity Test

Annex-I shows the results of VIF and Tolerance and reflects that the designated data set has no problem of multicollinearity because of the value of each variable lies below the given criteria of the value 10 which recommended that the multicollinearity is not an issue. According to [12] multicollinearity is an issue if the value of any estimates is greater than or equal to 10.

4.2.2 Coefficient Estimation of Logistic Regression

The coefficient estimates of logistic regression model are shown in Table 2, which shows that benefit from livestock are statistically significant and reduce the chance of vulnerability toward poverty. This is understandable that livestock benefit means as a backbone for rural people livelihood. Livestock benefits are source of food, way of income, social security etc. More prominently, shock- absorbing ability of rural depends on livestock earning. Thus, those who own more livestock might be non-poor than otherwise. This is similar with the findings of [37] and[25] that the more the livestock the more the benefit and lower is the poverty level so are more likely to become better off.

Table 2: Estimated Measure of Logit Model

Covariate	Coefficient	Odd Ratio	Marginal Effect	X	Z-Value	P-Value
Constant	7.41				3.95	0.00
HH size	0.44	1.55	0.11	0.19	0.75	0.45
Age	-0.05	0.95	-0.01	44.06	-2.12	0.03
Benefit Livestock	-1.63	0.19	-0.38	0.81	-2.46	0.01
Education HH	-1.15	0.31	-0.24	0.02	-0.72	0.47
Gender	-1.42	0.24	-0.34	0.67	-2.28	0.02
HH Own Land	-.32	0.72	-0.08	9.39	-3.29	0.00
Pseudo R^2			0.22			
Likelihood Ratio			30.6(0.0000)			
Pearson goodness of fit test			94.8(0.19)			
Hosmer-Lemeshow goodness of fit test			5.35%			
Sensitivity Value			73.4%			
Specificity Value			74.5%			
Correctly Specified			74.0%			
Area Under ROC curve			0.80			
Iteration			6			

Source: Researcher's own estimation

Gender of household head is negative and significant towards poverty, this infers that male headed households are less likely to be poor than female headed households for the sake that male headed households are in better position of escaping out of poverty than the female headed ones. The possible explanation is that regarding farming experience males were in a better position than the female farmers. This is alike with the findings [36] that the female headed households are more likely to be poor. The findings of this study reflects that rise in household size has a positive relation with poverty this is because of the burden that large family size increase poverty level but family size has insignificant effect on poverty which shows that if there were higher numbers of household members in the family unit the higher the likelihood of such a family unit to own livestock. More household members reduced the costs for employing people to look after livestock since they provide family labor so increase in family size has no effect on poverty. Age of household head is negatively and significantly related to poverty which refers that older individuals have relatively more affluent experience in farming activities so are less poor. This is similar with the findings [41] that the age of the household head is negatively related with poverty. The land holding of the household is statistically significant and negative towards poverty and is similar to the result of [17] which means that more land provides more opportunity of farming practices for the rural people thus reduce the chances of poverty.

4.2.3 Odd Ratio Estimates of Logit Model

Odd ratio is a different way of explaining the results, as in our research value for poor is 1 and for non-poor is so less than 1 value is desirable. According to results given in Table 2 the value of odd in favor of being poor to non-poor is 0.94 for head age which means that head age on average decrease the odd of being poor by 6percent. However, the odd of education of household is 0.31 with insignificant statistics (P value greater than 0.005) which indicate that in 2016-17 education of household has no effect on poverty. The value of odd in favor of being poor to non-poor is 0.72 for owned land which narrates that land holding on average decrease the odd of being poor by 28 percent. Odd ratio of gender is 0.24 which indicates that the odd of being poor decrease on average by 76percentif male household head involve in income generating activities of house. Estimated odd of family size is 1.55 with insignificant value which shows that increase in number of family has no effect on poverty. The value of odd in favor of being poor to non-poor is 0.19 for the benefit from livestock asset which clearly explains that livestock benefit on average decrease the odd of being poor by 81 % for rural households. It shows that livestock are very important source for poverty reduction in the rural areas of Gilgit Baltistan.

4.2.4 Marginal Effects Estimates of Logistic Regression

The values of marginal effects shown in table 2 tell that one unit increase in the independent variable above a certain threshold level X will change the poverty by the value of estimates. The results suggest that if the age of household head increase by one year above approximately 44 years the probability of being poor to non-poor decrease by 1percent. In case of land holding of household, if land size increases by one unit above 9.39 kanals the probability of being poor decreases by8 percent. The estimated results of education of the household and household size are insignificant which show that these variables have no effect on poverty of the households in 2016-17. Benefits from livestock and gender are dichotomous variables so there is no worth of using this threshold interpretation.

4.2.5 Goodness of Fit and Accuracy Estimates of Logistic Regression

In logistic regression model goodness of fit has been tested by using Pseudo R^2 which shows that the correlates of poverty investigated in this study have explained 22 percent variation in the model and it is quite matches with any cross sectional analysis. P values for Pearson goodness of fit (0.19) and likelihood ratio (0.000) are less than 1 which indicates that at least one of the coefficient is non-zero and whole model is good fitted and results are highly significant.

4.2.6 Tests for Predictive Accuracy

Predictive Accuracy of the given model was tested by using Specificity and Sensitivity values and Receiving Operating Curve (ROC). Specificity measures the percentage of true negative correctly specified while Sensitivity measure that of positive. ROC is plotted with sensitivity and specificity on the vertical axis for each cut off point shown in Annex-II and III. Value of Specificity for logistic distribution is 74.5 percent and sensitivity is 73.4 percent. High value of both estimates is preferable and gives correct prediction about events and non-events. Model is correctly specified by 74 percent. These two measures collectively give the value of ROC. In present study area under ROC is 0.80 showing that logistic model is excellent and is indicating good predictive accuracy as shown in table 2.

5 Conclusion

Gilgit Baltistan is endowed with many natural base resources and rural population is directly and indirectly involved in livestock production. In this study the three common poverty estimates have been found to be 44 percent, 23 percent and 13 percent for head count, poverty gap and poverty severity, respectively for selected rural areas of Gilgit Baltistan. District wise poverty estimation reflects highest poverty in Hunza. The results of the econometric model indicated the influence of different variables on household poverty. A total of six independent variables were included in the model out of which four variables had showed the significant and negative effect towards poverty. Benefit from the livestock, age of the household and gender and holding land had a significant and negative association with poverty status. On the contrary, household size was found to be positive and education had negative but insignificant influence on poverty respectively. So the research concludes that livestock is essential for rural economy of study area as it plays a significant role in poverty eradication and for sustaining livelihood of deprived people.

5.1 Recommendations

The policy implications mentioned below are suggested as major recommendations on the basis of key findings of this research endeavor.

1. The study area was deprived of major veterinary services and proper market facilities. Therefore, establishment of veterinary service center and proper market in the study area can enhance the income level of the household thus descend poverty. Area specific interventions would be more beneficial.
2. Provision of microcredit to the female farmers for livestock rearing can stimulate the female contribution towards livestock farming and subsequently address gender inequality issue.
3. Livestock management based trainings in the study area can elevate the capacity building of livestock farmers.

In order to decrease any expected negative effects of livestock production on environment some improved environment friendly technologies are required to be introduced to reap the maximum socioeconomic benefits for sustainable economic growth.

References

1. A. Aleme, and Z. Lemma, Contribution of Livestock Sector in Ethiopian Economy: A Review, *Advances in Life Science and Technology*, **23**,1-13 (2015)
2. A. Ali, & M. Khan, Livestock ownership in ensuring rural household food security in Pakistan, *J Anim Plant Sci*, **23**(1), 313-8 (2013)
3. A. B. Johannesen, & A. Skonhoft, Livestock as insurance and social status: Evidence from reindeer herding in Norway, *Environmental and resource economics*, **48**(4), 679-694 (2011)
4. A. Field, *Discovering statistics using SPSS. 2nd eds* (Sage Publications, London, 124-189, 2005)
5. A. I. Abdelrahman, *Applying logistic regression model to the second primary cancer data*http://interstat.sta2010/articles/**1001006**.pdf.Searched on 30-12-2010 (2010)
6. A. M. McSweeney,& C. Raish, Social, cultural, and economic aspects of livestock ranching on the Santa Fe and Carson National Forests, *Rocky Mountian Research Station, Forest Service, US Department of Agriculture* (2012)
7. A. Rehman, L.Jingdong, A. A. Chandio, & I. Hussain, Livestock production and population census in Pakistan: Determining their relationship with agricultural GDP using econometric analysis, *Information Processing in Agriculture*, 4(2), 168-177 (2017)

8. A. Siddiqui, Economic Developments in the age of Globalization Sharpe Reference, a, b, c, *India and South Asia*, **419** (2006)

9. AKRSP, *An assessment of socio economic trend and impact in Northern Areas* (1991-1997): Findings from AKRSP's farm household income and expenditure survey Gilgit, Aga Khan Rural Support Program (2000)

10. B. J. Gentner, & J. A. Tanaka, Classifying federal public land grazing permittees , *Journal of Range Management*, 2-11 (2002)

11. C. R. Kothari, *Research Methodology: Methods and Techniques.* Second Edition (Washwa Prakashan Pub, New Delhi, India, 1990)

12. D. Asteriou,&S. G Hall, *Applied econometrics* (Palgrave Macmillan, New York, 2011)

13. D. N. Gujarat, (2005).*Basic Econometrics. Fourth Edition* (Mc. Grawhill Inc, New York, 2005)

14. D. N. Gujarati, *Basic Econometrics*, *Third edition* (McGraw-Hill Inc, New York, 1995)

15. D. N. Gujarati, *Basic Econometrics, 4th edition* (The McGraw-Hill Companies, New York, 2004)

16. DAAF.(2012). *The framework for the Zero Hunger Programme. Actuating The irrigated food security strategy for South Africa,* Department of Agriculture, Forestory and Fisheries (DAAF), South Africa.

17. F. F. Shugri, *Dimensions and Determinants of Poverty in Agro-Pastoral Households of Kabribayah District, Fafan Zone, Somali Regional State, Ethiopia* (Doctoral dissertation, Haramaya University, Haramaya) (2016)

18. F. J. C. Swanepoel, A. Stroebel, & E.Nesamvuni, Key Functions of livestock in Smallholder Farming Systems: A South African Case Study, *Animal Production Australia*, **24**, 237-240 (2002)

19. F. Swanepoel, A.Stroebel, & S. Moyo, T*he role of livestock in developing communities: Enhancing multifunctionality.* University of the Free State and CTA (2010)

20. FAO.(2006). *Livestock's Long Shadow, released by the Food and Agriculture Organization of the United Nations.*

21. FAO.(2009). *The State of Food and Agriculture Livestock in Balance*, http://www.fao.org/docrep/012/i0680e/i0680e.pdf, accessed 21.02.14.

22. FAOSTAT.(2013). *Statistical database of the Food and Agriculture Organization of the United Nations.*http://faostat3.fao.org/.

23. G. Mudzielwana, Determinants of cattle ownership and herd size in Vhembe District of Limpopo Province, South Africa (Doctoral dissertation, University of Limpopo) (2015).

24. G. S. Maddala, *Introduction to Econometrics.Third edition.*(Wiley India Ltd. p. 316-332, India, 2007)

25. G. Semer, Dimension and Determinants of rural household Poverty: The Case of Eastern Zone in Tigray National Regional State, An MSc Thesis Presented to the School of Graduate Studies of Haramaya University, Haramaya (2008).

26. G.S. Maddala, *Limited Dependent and Qualitative Variables in Econometrics* (Cambridge University Press, 1989).

27. GoP. (2016-17). *Over view of the economy.Ministry of Finance*. http://www.finance.gov.pk/survey/chapters_17/overview_2016-17.pdf.**10-11**.

28. J. Clemens, Human and economic issues associated with livestock production. Livestock, Fodder, Pastures and People: An Integrated Study in the Northern Areas of Pakistan, *International Centre for Integrated Mountain Development, Kathmandu, Nepal*, 35-56 (2005)

29. J. H. Aldrich, &F. D. Nelson, *Linear probability, logit, and probit models* (Vol. 45) (Sage, 1984).

30. J. Klugman, *A Sourcebook for Poverty Reduction Strategies:* Volume **2**: Macroeconomic and Sectoral Approaches (World Bank, Washington DC, 2002)

31. J. M. Kandulu, B. A. Bryan, D. King, & J.D. Connor, Mitigating economic risk from climate variability in rain-fed agriculture through enterprise mix diversification, *Ecological economics*, **79**, 105-112 (2012)

32. L. De Haan, & A. Zoomers, Development geography at the crossroads of livelihood and globalization, *Tijdschriftvooreconomischeensocialegeografie*, **94**(3), 350-362 (2003)

33. L.H. Dossa, B. Rischkowsky, R.Birner, & C. Wollny, Socio-economic determinants of keeping goats and sheep by rural people in southern Benin, *Agriculture and human values*, **25**(4), 581 (2008)

34. L.W Bell, & A. D. Moore, Integrated crop–livestock systems in Australian agriculture: Trends, drivers and implications, *Agricultural Systems* **111**, 1-12 (2012)

35. M. A. Iqubal, Role of livestock husbandry on rural transformation in north india: a case study, *Revijazageografijo* **5**(2), 83-94 (2010).

36. M. Abdikadir, *Dimensions and Determinants of Poverty in Agro-Pastoral Households of Jigjiga District, Somali National Regional State, Ethiopia*. An MSc Thesis Presented to the School of Graduate Studies of Haramaya University, Haramaya, (2013)

37. M. Hilina, *Dimensions and determinants of poverty in pastoral areas of Eastern Ethiopia: The case of Shinile zone in Somali national regional state.* An MSc Thesis, Haramaya University, Ethiopia, **125** (2005)

38. M. I. Meltzer, Livestock in Africa: The economics of ownership and production, and the potential for improvement, *Agriculture and Human Values,* **12**(2), 4-18 (1995).

39. M. Jodlowski, A. Winter-Nelson, K.Baylis, &P. D. Goldsmith, Milk in the data: food security impacts from a livestock field experiment in Zambia, *World Development*, **77**, 99-114 (2016)

40. M. M. Islam, S. Anjum, R.J. Modi, &K. N. Wadhwani, Scenario of livestock and poultry in india and their contribution to national economy. *International Journal of Science, Environment and Technology*, **5**(3), 956-65 (2016).

41. O. A. Omotesho, A Gender Analysis of Poverty among Rural Farming Households in North Central Nigeria, *European Journal of Social Science*, **28**(3), 366-376 (2012)

42. P. R. Lawrence, W. R. Meghan, K.A. Debra,& A. D Teresa, Importance of Animals in Agricultural Sustainability and Food Security,1-3 (2015).

43. Pirwahab, Livestock and Rural Livelihood in Pindsultani, Attock. The Exployer, *Journal of Social Sciences*, **1**(2), 51-53 (2015).

44. R. S. Pindyck, & D. L. Rubinfeld, *Econometric Models and Economic Forecasts*, (McGraw-Hill Book Co., New York, 1981).

45. S. L. Jackson, *Research Methods and Statistics*: A Critical Thinking Approach volume **3**. (Belmont, CA, Wadsworth, 2009).

46. S. Reist, F. Hintermann, & R. Sommer, *The livestock revolution: An opportunity for poor farmers.* Funded by Swiss Agency for Development and Cooperation.Switzerland, **1**–1 (2007)

47. T. Amemiya, Qualitative response models: A survey. *Journal of economic literature*, **19**(4), 1483-1536 (1981).

48. T. Liao, *Interpreting Probability Models: Logit, Probit, and other Generalized Linear Models.* Sage University Paper Series on Qualitative Application in the Social sciences.Thousand Oaks. CA: (Sage, 1994)

49. T. Serik, K. Musabekov, A. Yesbolova, S.Ibraimova, A. Mergenbayeva, Z. Sabdenova, & M. Seidahmetov, Issues in the Development of the Livestock Sector in Kazakhstan. *Procedia-Social and Behavioral Sciences*, **143**, 610-614 (2014)

50. T. Yamane, *Statistics, An Introductory Analysis,volume no.2,* (Harper and Row, New York, 1967)

51. UNDP, Livestock-based Micro-enterprise and Its Effectiveness on Poverty Reduction: A Study in Hansposa VDC of Sunsari District, **3**: 8 (2013)

52. V. Alary,C. Corniaux,,& D. Gautier, Livestock's contribution to poverty alleviation: How to measure it?. *World Development*, **39**(9), 1638-1648 (2011)

53. W. H. Green, *Econometrics Analysis.Fifth Edition.* (Dorling Kindersley, India, 2003)

54. WB, *World Bank prepared a book entitled, Livestock Development:* Implication for Rural Poverty, the Environment and Global Food Security (2001)

 https://doi.org/10.1051/e3sconf/20185200019

Annex-I

Variance Inflation Factor and Torelance

```
. vif
```

Variable	VIF	1/VIF
age	1.49	0.672010
gender	1.39	0.718571
livestockb~t	1.10	0.911794
education	1.05	0.948721
hhlandarea	1.04	0.963037
hh_size	1.03	0.969523
Mean VIF	1.18	

Annex-II

Sensitivity-Specificity Curve

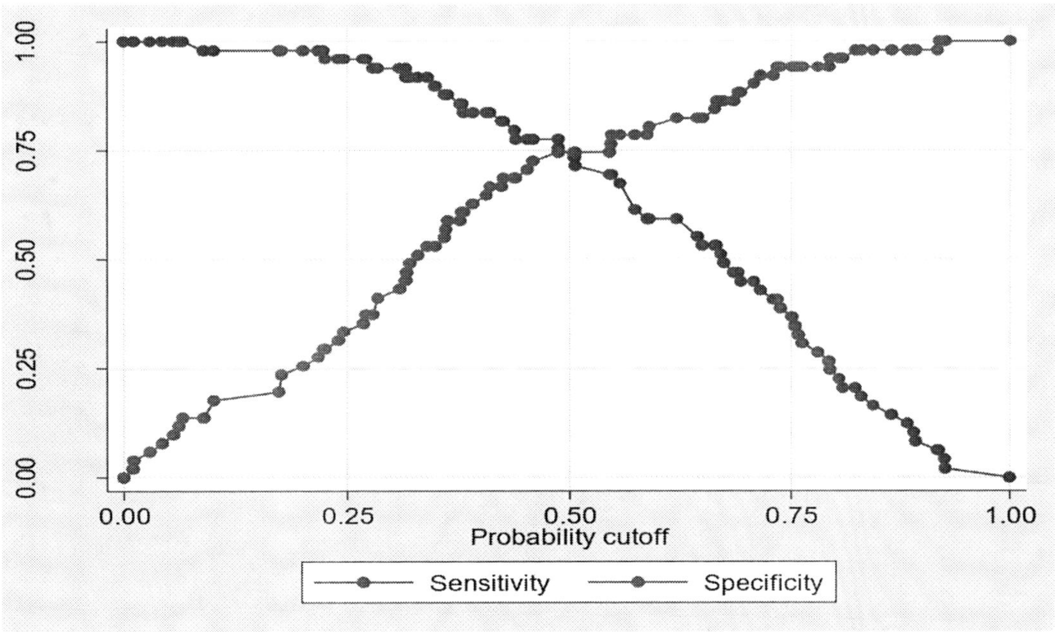

 https://doi.org/10.1051/e3sconf/20185200019

Annex-III

Receiving Operating Curve (ROC)

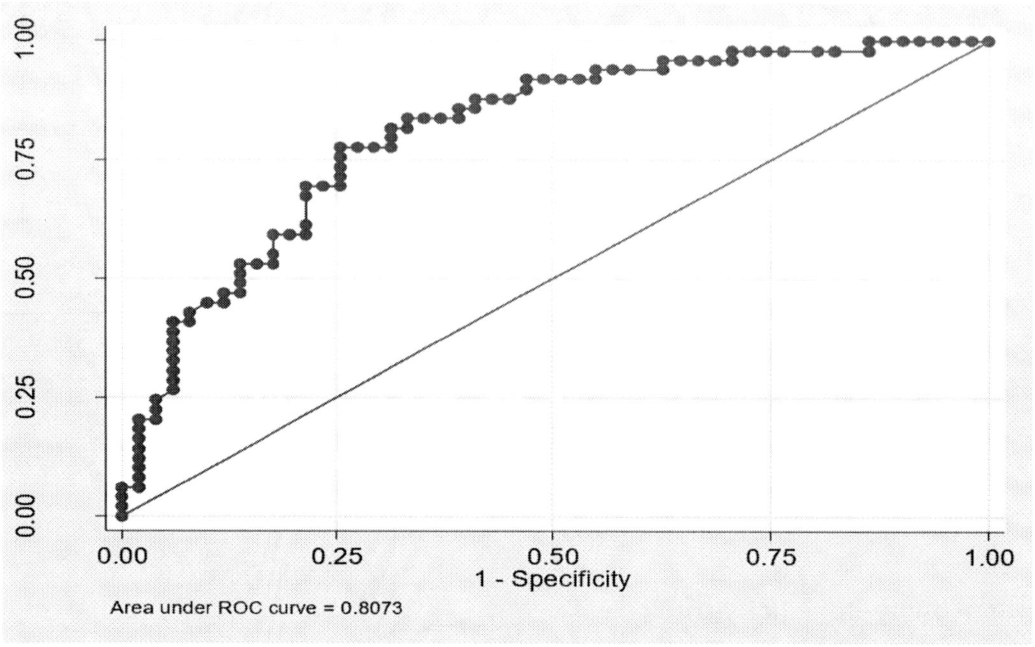

The approach of political ecology as an alternative method to overcome socio-economic problems

Husnul Isa Harahap* dan *Rosmery*

Department of Political Science Universitas Sumatera Utara (USU), Kota Medan, Indonesia

Abstract. There are many efforts that has been done by the society to meet their economic needs. But there are often obstacles in the processes. One among them an expensive food's prices. Is political ecological approach can be used as an alternative method to overcome the society social issues? It reliability is based on three reasons. First, so far the political ecological approach has a very small negative side effect for humans therefore it is environmentally friendly. Secondly, in some cases it is easy to develop at an affordable cost. Third, the results achieved is more valuable. The political ecological approach is thus a viable method developed and promoted through public policy.

1 Introduction

Society and socioeconomic problems are two unseparated things. When there is society there will be socioeconomic problems. This is because humans involve in both social and economics. Humans live in a society with primary (basic) and other needs that must be fulfilled. Socioeconomic problem arises when human has difficulty in fulfilling their need and when they finally use many methods to fulfil their needs.

One of the way to fulfil their needs is by increasing their economic income. As a result, it takes several efforts to be involved in the economic process. As for example, they breed animals to be sold for money, grows plants to increase their income. Economically, with the increased amount of income, their need would be fulfilled.

Nature resources can be used in economic processes. However, it cannot limit or prevent human behaviour against it. As a result, human beings use these natural resources to resolve their economic problems, even in ways that can cause harm to nature and for humans. In the end humans enter the circle of problems, especially when humans cannot resolve their economic problems and at the same time nature also loses resources to fulfil human needs.

Solution to the socioeconomic problems of human beings is done with an approach that is not environmentally based. Hence this solution often creates new problems. The result of the solution is often a controversy relationship between mankind and nature. For example, fertilizers are made to fertilize the soil. As an impact of this act, there will be degrading

* Corresponding author: husnul.harahap@usu.ac.id

upon the soil fertility. These condition makes us depend on a fertilizer. Similarly, among farmers, where they use inorganic to feed. As a result, people who are farming do not succeed to get the maximum profit cut of their efforts. The cause of the problem is that, cost of fertilizer is more expensive.

In accordance with this background, overall this study reveals three fundamental problems. First, environmental problems are lack of attention. Second, the old approach is not able to overcome environmental problems. Third, this old approach is also unable to overcome socio-economic problems. The question that arises, is there any alternative approach to solve this problem? This study offers a political ecology approach as an answer to this question. The political ecology approach is an alternative model for resolving socioeconomic issues.

2 Method

The Method that used in this study is qualitative. Data collection methods applied are literature and document studies. Literature materials and documents used are available both in online and offline. Method of analysis is with political ecology approach. According to Tim Forsyth who reviewed Piers Blaike's view that "environmental analysis and policy can be reframed towards addressing the problems of socially vulnerable people" [1].

The main idea on this study is elaborate relationship between environmental sustainability with the society welfare and vice versa. The society's efforts to increase their economic income can be done without damaging the environment, or efforts to utilize the available resources naturally turn out to increase the income of the society. Thus, environmental and social problems can be analysed simultaneously.

3 Alternative Model

Political ecology is a social science that emerged in the 1980s. This science is transdisciplinary. One of the most important things is that, it is between nature and the social world. As a transdisciplinary, it sometimes combines social science and the natural sciences. According to Paul Elliott Little "Some of the most important recent transformations in the ecological paradigm are the development of trans disciplinary syntheses between the social and natural sciences ..." [2].

This approach sees that socioeconomic and environmental problems can be solved simultaneously. The socioeconomic problem arises as a result of the process of producing certain subject that humans are distracted by the high cost, followed by the condition where the environment is contaminated. As a conclusion, the socioeconomic problem comes with environmental problems. In the case of farmers, there are cases where the price of poultry feed causes low profit to the farmer at the harvest time. Whereas the expensive poultry feed on the other hand pollutes the environment. The political ecology as an alternative method shows that if the utilization on natural resources be used to the fullest extent the social and environment issues will solve simultaneously.

For example an integrated farming system which integrated rice fields, ducks to support an organic farming [3]. In the early stages of this system succeed to decrease the dependence of farmer to chemical materials. The chemical materials is made from the fertilizer. Even though this fertilizer contains chemical materials, the cost is still expensive. This system has a potential to increase the income because the cost of the organic farming products are more expensive than an organic one. Therefore, this system can be develop easier.

From a socioeconomic perspective there are several problems arising from the use of chemicals in agriculture through fertilizers. According to I Wayan Supartha the first *"...terjadi penurunan produktivitas lahan pertanian akibat praktek-praktek penggunaan sarana produksi (pupuk dan pestisida) buatan yang berlebihan" [sic.]*[4] ["...there is a decrease in agricultural productivity due to the practice of excessive use of production of fertilizers and pesticides".] Secondly *"adanya ketakutan masyarakat konsumen terhadap produk-produk pertanian (konvensional) yang sarat dengan kandungan residu pestisida" [sic.]*[4] ["...there is a fear of society towards agricultural products (conventional) which are loaded with pesticide residue content".]

Another example of the alternative methods of ecological approach is integrated farming in Pekanbaru, Riau Province. According to the results of a study conducted by L. Siswati and R. Nizar with the theme of *Integrated Farming System for Horticulture and Beef Cattle to Improve Farmer Income*, there are several benefits that had been felt by the society. According to them there are three main benefits. The first benefit is empty land that can be productive. Average land area of 0.3 ha, cattle ownership of four to five animals [5].

The second benefit is integrated agriculture crop of horticulture and cattle can increase farmer's income. Horticulture income per planting period is 9,731,147 rupiah per year. Revenue from livestock business 9.345.328 rupiah -per year. The forms of mixed farming include vegetables, cattle and fruits such as papaya, melon, jackfruit, and sapodilla. It appears that this has a socioeconomic value.

The third benefit is this technique can reduce the use of fertilizer. As we know fertilizer are not good for the environment. Siswati explained that *"...penggunaan pupuk anorganik yang selama ini digunakan oleh petani apabila digunakan terus menerus dapat menyebabkan dampak negativ terhadap lingkungan, antara lain tanah menjadi padat akibat efek rekat (glueing effect) terutama pada pupk ammonium, bereaksi masam dan bila tercuci samapai ke air tanah bila air dikomsumsi dapat menimbulkan penyakit" [sic.]*[5] ["...the use of inorganic fertilizer that used by farmers continuously can cause negative impact to the environment such as, the soil becomes solid due to the gluing effect especially the ammonium fertilizer, react sour when it is washed to the ground, thus the water consumed can cause disease".] So this proved to be environmentally friendly.

Based on all the example, it appears that the alternative method provides considerable benefits, not only for humans but also for the environment. But the question is, if indeed the ecological approach is more profitable, why there are still many people who approaches methods that are not based on the environment? There are three explanations related to this. First, political ecology is an underdeveloped approach in the society. This has consequences many people didn't know about it.

Second, an organic models have been applied in society long time ago, thus it is easier to apply them, and supported by product advertisement that intense to socialize it. The equipment are available. The expert on it is easier to find. Similarly, the example that had been applied, is quite easy to find. Meanwhile, the approach of political ecology requires practice and integrated system. When this system is applied there will still be a doubt, such as will this system work? This is the sign that states there are many people who still do not fully believe in this system. So it can be summarized that there are still a problems of non-existent view of the system.

Third, the approach of political ecology actually requires an active role from the government. Therefore, the question is whether the government program is already based on political ecology approach to solve socioeconomic problems? In general, in many countries the political ecological approach has not become an issue of concern to the governments. In Indonesia there are several local governments that support this alternative method, such as in Bali.

Another question is, does the general government have or have not felt the need to popularize this method? Of course there is an opportunity to make it happen. But the thing that needs to underline is, in the end the course is not always enough without seriousness in the program's process. The success and failure of this program is strongly influenced by the seriousness of the implementation of the political ecology approach. Seriousness will obviously affect the results because without any seriousness the results achieved will not be maximized.

4 Conclusion

One of the causes of the emergence of social problems is the condition where society does not has the ability to fulfil the needs of the economy. This is inseparable from the fact that many people are trying to improve their economic needs. But some are successful and some are not. The lack of success is one of them based on circumstances that are considered not supportive. This situation arises from the choice of a very "instant" approach method to overcome the economic problem.

The political ecology approach provides an alternative method for resolving the social issues. It is based on three reasons. First, so far the ecological approach has a very small negative side effect for humans therefore it is environmentally friendly. Second, in some cases it is easy to develop with an affordable cost. Third, the results achieved have a more valuable quality. In some cases this alternative method has been successfully implemented. The political ecology approach is thus a viable method developed and promoted through public policy.

References

1. T. Forsyth, Political Ecology and the Epistemology of Social Justice, *Geoforum*, **39**, no. 2, pp. 756–764, Mar. (2008).
2. P. E. Little, Political Ecology as Ethnography: a Theoretical and Methodological Guide, *Horizontes Antropologicos* **3**, (2007).
3. M. C. T. Bintang, P. Marbun, and Yunilas, Pertanian Terpadu Sistem Integrasi Ladang, Sawah dan Ternak Itik untuk Mendukung Pertanian Organik, *Jurnal Abdimas TALENTA*, **1**, no. 1, pp. 97–100, (2016).
4. I. W. Supartha and N. L. Kartini, *Konsep dan Strategi Pengembangan Pertanian Organik di Bali*, presented at the Seminar Regional Pertanian Organik dan Prospek Pengembangannya di Bali, Fakultas Pertanian UNUD Denpasar, (2001).
5. R. Nizar and L. Siswati, Model Pertanian Terpadu Tanaman Hortikultura dan Ternak Sapi untuk Meningkatkan Pendapatan Petani, *Jurnal Peternakan Indonesia* **14**, no. 2, pp. 379–384, Jun. 2012.

Effect of monopalmitic content and temperature on precipitation rate in biodiesel-petroleum diesel blends

Indy Prasetya Girry[1], *Imam* Paryanto[1], *Tirto* Prakoso[2], and *Misri* Gozan[1*]

[1]Department of Chemical Engineering, Faculty of Engineering, Universitas Indonesia, Depok 16424, Indonesia
[2]Department of Chemical Engineering, Faculty of Industrial Engineering, Institut Teknologi Bandung, West Jawa, Indonesia.

Abstract The need for cleaner fuel increases alongside the necessity of reducing greenhouse gas emissions. Biodiesel is considered as one of best options for cleaner energy. However, it has disadvantages due to lower stability, lesser cold flow properties (CFP), and higher viscosity, which are mainly caused by fatty acid type and content. These works investigate CFP of various biodiesel blends B20 & B30, with the objective on how the monopalmytic as one of the saturated monoglycerides (SMG) content and Temperatures affect the precipitation rate. A set of 100ml separatory funnels containing various samples were put in refrigerators in controlled-temperature, i.e.15°C; 20°C; and room temperature. The rate of precipitation was observed every day for 2 weeks using 0.8μm membrane filter. Results indicated precipitation rate of B30 was higher than that of B20. As it was expected, the precipitate was taken at the same temperature, 20°C, resulting in B30 had more precipitate in the end (37.6 mg) compared to B20 (22.2 mg). The closer to cloud point also indicates the higher rate of precipitation as 4th day B20 at 15°C (25.4 mg) compared 4th day B20 at 20°C (19.3 mg). The change of crystal from less stable α-phase to more stable, less soluble β-phase is also proven to significantly affect the rate of precipitation.

1 Introduction

Biodiesel is a fast growing renewable energy that can be produced from animal and plant oil, as well as from oil waste such as cooking oil waste through esterification process (Altaie et al. 2016). The transesterification process transforms the fatty acid in oil into fatty acid methyl ester (FAME). The main composition of biodiesel itself are methyl ester consists of a different type; oleic, linoleic, palmitic. The properties of biodiesel largely depend on its

* Correspondent Author: Email: misrigozan@gmail.com

composition, and it mostly because of the oil sources. Fatty acid composition affects the cold flow properties (CFP) significantly. Biodiesel freezing point increases as atom carbon in the chain of carbon increases and decreases as double bond increased. Biodiesel that comes from saturated fatty acid has higher Cloud point (CP) and Pour point (PP). Thus Biodiesel from vegetable oil have CP and PP lower from animal oil [1]

Transesterification of triglycerides with alcohol produces alkyl esters (biodiesel) and glycerol. This reaction happened through series of transformation from triacylglycerol to diacylglycerol and monoacylglycerol to eventually become glycerol. In each step, methyl ester is made. In normal optimum condition yield of conversion very high, so that trace from triacylglycerol is very low, however this trace even in very minimum amount affect the biodiesel properties significantly [1]

Biodiesel has its own distinct advantages as it is more safety-renewable energy, nontoxic and better performance on a car than petroleum diesel, and has a lubricant effect [2]. However, until now only 20% biodiesel-80% Petroleum diesel blend (B20) has been allowed to be used in the car without modification. This is due to Cold Flow Properties (CFP) that causes precipitation to form at not just below cloud point (CP) temperature but also above CP [3]. This precipitate leads to the fuel filter and dispenser filter to eventually clogged. This made blend range that allowed only 5-20% as ruled in ASTM D9675.

The precipitation normally occurs because of following condition: (1) blending biodiesel with petroleum diesel, (2) Transesterification with branched-chain alcohol, (3) Winterization, (4) Chemical additives, (5) Fatty acid modification [4]. Winterization is the most common reason for precipitation. Mainly because of the temperature can go down below cloud point, however The precipitation also occurs above cloud point due to a huge gap between cloud point and final melting point. This happens on high saturated monoglycerides contain between 0.2-0.3%wt [5]

The formation of precipitate starts with a trace of monoglycerides that unsaturated become a cloud, or change in solution color into milky. At this point it will form metastable crystal. The next step crystal become polymorf and increases in number, bigger in size, and more stable. This stable crystal then creates precipitation in the bottom of the container [6].

This study focused on the precipitation occurs at temperatures above cloud point. The aim of this experiment is to know how the rate of precipitation happened during unstable period and stable period. Monopalmitic is chosen as variation as it affects the CFP significantly [7], and it is abundant in palm-oil feedstock.

2 Materials and Methods

2.1 Characterization of Biodiesel and Petroleum diesel

Biodiesel that was used are palmitic oil bought from Indonesia. The petroleum diesel was bought from Indonesia. The composition profile of this biodiesel and petroleum diesel was characterized by using Gas Chromatography/Mass Spectrometry.

Biodiesel was added volumetrically to petroleum diesel to obtain B20 and B30. To modify the monopalmitic content, a certain mass of monopalmitic is added to change the monopalmitic contain from around 0.4% to 0.6% and 0.8%. Magnetic stirer was used to homogenize the solution at 55°C and speed of 1000 rpm. Characterisation for biodiesel and petroleum diesel were done on the following ASTM properties: density, kinematic viscosity, cloud point, water and sediment, fame contain, acid number, oxidation stability, aromatic contain, monoglyceride contain, total glycerol, free glycerol.

In order to increase monopalmitic contain, standard equation for concentration are used as below:

$$m_a = \frac{C \times V \times \rho}{x} \quad (1)$$

m_a = mass of monopalmitic to add (g)
C = Concentration to increase (wt/wt%)
V = Volume of B100 (ml)
P = Density (g/ml)

2.2 Beaker Test

Biodiesel blend consist of B20, B30, and Monopalmitic variation were observed every day for the first week and on 10th and 14th day for the 2nd week. This biodiesel was stored under controlled-refrigerator at temperature of 15°C, 20°C, and room temperature. Room temperature was ranging between 28-30°C. The sample then was taken each day to undergo vacuum filtration with membrane filter pores 0,8µ. This method is a modification from Cold Soak Filter Test (CSFT) from ASTM D7501. The standard method from ASTM is for determination of fuel filter blocking potential of biodiesel. The initial method controled biodiesel around 4,5°C for 16 hour then filtered using vacuum filter 0,7µm. This work changed the temperature to observe how the biodiesel perform under various temperature that is higher than cloud point.

3 Results and discussion

3.1 Characterization result

Parameters that influence the performance of biodiesel have been analyzed . Tthe cloud point (CPs of B30 was higher than B20. Table 1 shows basic characterization as ruled in ASTM, for the petroleum diesel and biodiesel used in this experiment. Table 2 shows that Monopalmitic and Stearic as monoglycerides that dominant in the sample. As the precipitate contained low Monooleic, it was considered that monooleic has minimum effect on the precipitation.

Table 1 Parameter Analysis for Biodiesel and Petroleum Diesel

Paramater	Units	Petroleum diesel	Biodiesel
Density at 15C	kg/m3	842.5	856.8
Kinematic Viscosity	mm2/s	3.35	4.51
Cloud Point	°C	1.7	n.a
Water and Sedimen	% (wt)	n.a	0
Fame Contain	% (wt)	n.a	99.5
Acid Number	mg KOH/g	n.a	0.34
Oxidation stability	minutes	n.a	601.5
Aromatic Contain	% (wt)	2.54	n.a
Monoglyceride contain	% (wt)	n.a	0.4
Total Glycerol	% (wt)	n.a	0.1

Free Glycerol	% (wt)	n.a	0.003

n.a. = data not available

Table 2 Monoglycerides percentage contain

Monoglyceride	Mass Percentage
Monopalmitic	0.357
Monooleic	0.183
Monostearic	0.391

Table 3 Paramater Analysis on Biodiesel Blend

Parameter	Unit	Result		Petroleum diesel standard	Methods
		B20	B30		
Density at 15	kg/m3	841.1	849.1	815-870	ASTM D4052
Kinematic Viscosity	Mm2/s	3.75	3.82	2-4.5	ASTM D445
Cloud Point	°C	4.3	4.7	18 (Max)	ASTM D5773
Water	% (wt)	208.11	210.31	500 (Max)	ASTM D6304
Sediment	% (wt)	0	0	0.01 (Max)	ASTM D473
Fame Contain	% (vol)	19.9	29.9	20	ASTM D7806
Acid Number	mg KOH/g	0.089	0.101	0.6	ASTM D664
Oxidation stability	Hour	36.87	36.59	35 (Min)	EN 15751

3.2 Baker test result

Biodiesel sample was put in separator dunnel then vacuum –filtered on daily schedule . the results were put in g/100ml. Standard monopalmitin contain from biodiesel are around 0,4%. The result shown are for the initial monopalmitin contain and after modification

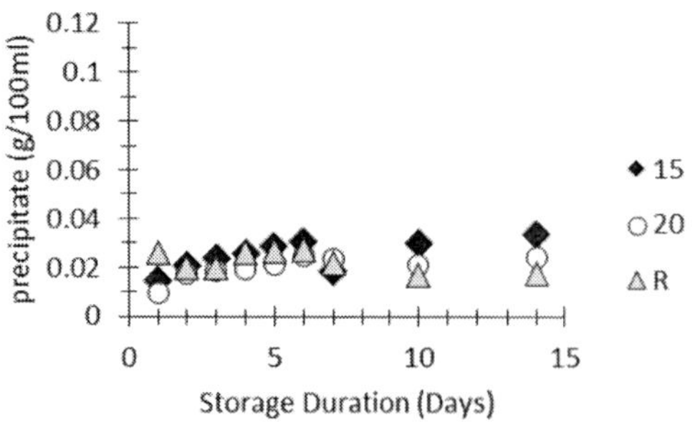

(a)

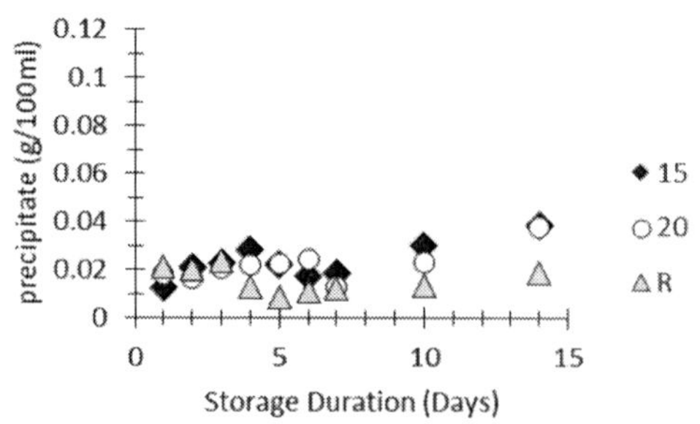

(b)

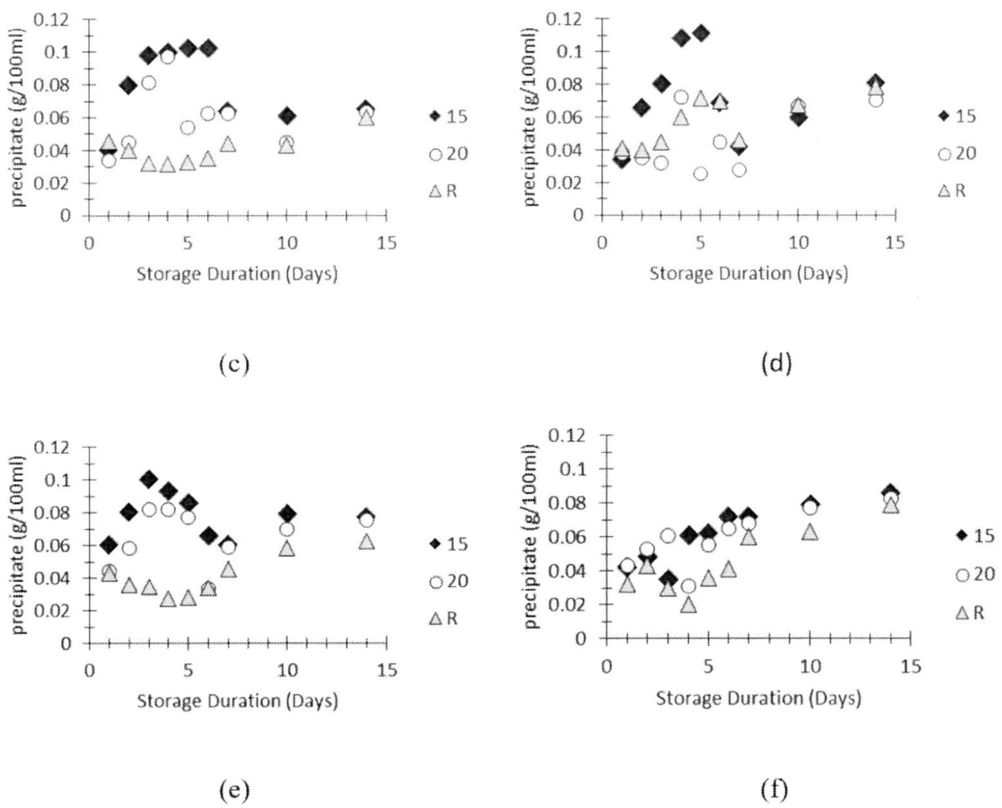

Fig. 1. The change of formation precipitate at :(a) B20, (b) B30, (c) B20 0,6% Monopalmitic, (d) B30 0,6% Monopalmitic, (e) B20 0,8% Monopalmitic, (f) B30 0,8% Monopalmitic

Between 5t[h] – 7th days at downeard fluctuation of precipitation weight

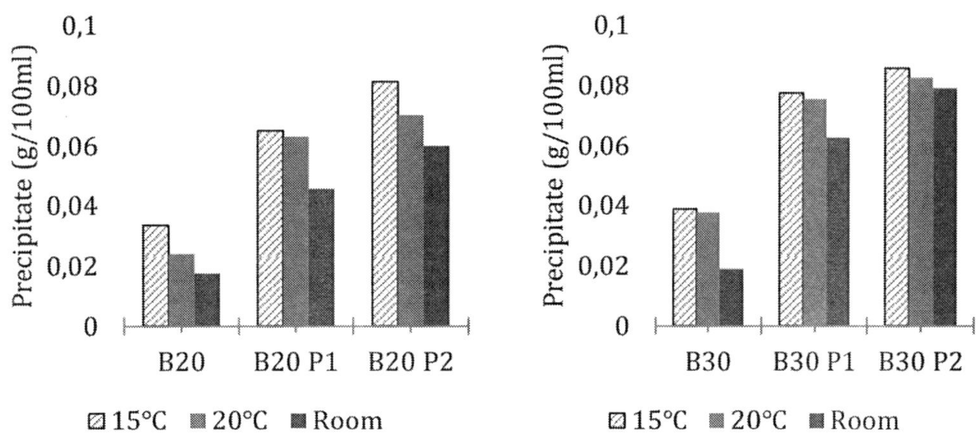

Fig. 2. Precipitate at the end of 14th day

and precipitation rate. This due to changes in the crystal stability.Differential scanning calorimetry can further explain that in biodiesel, at early start in a cooling process, metastable α-phase will change into its final form more stable but less soluble β-phase. This β-phase also has higher final melting points (FMT), which much effect the precipitate at the last day [8].

Moreover Chupka also found that at low concentration of monopalmitic,, all of the precipitate dissolved during the transformation process. Thus the decreasing of precipitate also affected by this.

Fig 1. Also explain how the change in temperature affect the rate of precipitation. B20 which stored at 15°C has higher rate at α-phase (0.003g/100ml/day) compared to B20 stored at 20°C (0.0026g/100ml/day). This due to the closer to cloud point (CP), in this experiment between 4-5°C.

Adding more monopalmitic increase the CP, as CP affect the precipitation rate, thus if we compare the precipitation rate at α-phase at the same temperature 15°C, B20 has lower precipitation (0.003g/100ml/day) compared to B20 0.8% Monopalmitic (0.0137g/100ml/day). Same result for B30 (0.0049g/100ml/day) compared to B30 0.8% Monopalmitic (0.0237g/100ml/day)

Moreover Monopalmitic has reverse micelles structure, this helps rate precipitation to increase as it can lower nucleation energy from monoglyceride. Monopalmitic can then assumed to work like catalyst (Dunn, 2009)

Fig. 5 shows the increae in Monopalmitate precipitate significantly. Adding 0,2% saturated monopalmitate increase almost 0.05 mg/100 of prexipitation. The B30 and B20 blends show less The B20 and B20 blends show less significant increase compared to the B20 0.4% Monopalmitic and B20 0.6% Monopalmitic.

4 Conclusion

This CFP study examined the effect of monopalmytic as one of the saturated monoglycerides (SMG) content and temperatures on the precipitation of biodiesel occurred even above cloud point in significant amount. The closer to cloud point the more precipitation produced. After several days, during the 5-7 days the amount of recipitation in the filter decresed, with the concentration on biodiesel-petroleum diesel blend.

5 Acknowledgment

Authors are grateful for the financial support provided by *Hibah Publikasi Internasional Terindeks untuk Tugas Akhir Mahasiswa* (PITTA) through Universitas Indonesia and research facilities provided by Univesitas Indonesia (2477/UN2.R3.1/HKP.05.00/2018).

References

1. M. Canakci, and H. Sanli, Biodiesel production from various feedstocks and their effects on the fuel properties. *J Ind Microbiol Biotechnol* 35:431–441 (2008)
2. U.S. Department of Energy, *Biodiesel Handling and Use Guide*, (2016)
3. G.M. Chupka, L. Fouts, J.A. Lennon, T.L. Alleman, D.A. Daniels, R.L. McCormick, Saturated monoglyceride effects on low-temperature performance of biodiesel blends. *Fuel Processing Technology* 118 302–309 (2014)
4. O. Edith,R.B. Janius, and R. Yunus, Factors affecting the cold flow behaviour of biodiesel and methods for improvement - A review. *Pertanika Journal of Science and Technology*, 20(1), 1–14, (2012)
5. G.M. Chupka, L. Fouts, and R.L. McCormick, Effect of low-level impurities on low-temperature performance properties of biodiesel, *Energy Environ. Sci.* 5 8734–8742, (2012)

6. E. Fredrick, I. Foubert, J. De Van Sype and K. Dewettinck,. Influence of monoglycerides on the crystallization behavior of palm oil, *Crystal Growth and Design*, 8(6), 1833–1839 (2008)
7. M.A.H. Altaie, R.B. Janius, Y.H Taufiq-Yap, U. Rashid, Basic properties of methyl palmitate-diesel blends. *Fuel* 193 (2017)
8. G.M. Chupka, J. Yanowitz, G. Chiu, T.L. Alleman, and R.L. McCormick, Effect of Saturated Monoglyceride Polymorphism on Low-Temperature Performance of Biodiesel. *Energy Fuels 2011* 25 (2010)

The concept of essential facilities in the efficiency of industrial natural resources

Kurnia[1,*], *Bramantyo* Suryodhahono[2], *R Ismala* Dewi[1], and *Putri* A Saragi[3]

[1]Faculty of Law, Universitas Indonesia, Depok, Indonesia
[2]Institute for Business Competition and Policy Studies Universitas Indonesia, Depok, Indonesia
[3]School of Environmental Science, Universitas Indonesia, Salemba, Indonesia

Abstact. The concept of 'essential facilities' is known as a tool to request an access of a vital facility to an undertaking having a dominant position in the market. The concept of 'essential facilities' can also be regarded as a code of ethics in the utilization of vital facilities because it regulates the ethics of utilizing it. According to the concept of 'essential facilities', the utilization of a vital facility must be fair and equitable among all parties and none of the parties can dominate it. This doctrine is of course applicable to maintain fair competition within a market. However, by opening an access to a vital facility in natural resources industry, the essential facility doctrine will also be a supporting factor to achieve green industry concept in natural resouces industry. In this paper, SWOT analysis will be conducted to explore the strengths, weaknesses, opportunities, and threats to essential facilities in the efficiency of industrial natural resources with regard to green industry concept, in order to realize a sustainable industry where there is efficiency, effectiveness, and also able to maintain environmental sustainability.

1 Introduction

Industrialization has now occurred in almost all parts of the world. Industrialization itself was started by the industrial revolution in the 18th century. Since the industrial revolution to date, the existence of industry has been overwhelming and scattered throughout the world, covering a huge number of sectors. China is one of the fastest growing industrialized countries. Industrialization in China and recent declines have greatly affected world commodity prices [1]. However, the existence of this increasingly large industry also affects the current state of the world environment, one of which is the mining industry. The existence of the mining industry has always been a problem because it has disrupted the ecological environment [2]. The existence of an increasing number of industries, if the production process does not pay attention to the existence of natural resources and environmental conditions, can cause serious problems for the world in the future.

Environmental problems caused by the development of industrial existence today has become the world's attention. The need for awareness for the world industrial sector to change the concept to be more environmentally friendly, both from the process of utilization of natural resources and production. Therefore, UNIDO together with the Philippine

* Corresponding author: k-toha@ui.ac.id

government held an International Conference on Green Industry in Manila in September 2009 [3]. UNIDO defines the green industry as an industry that encourages sustainable production and consumption patterns, i.e. energy and resource efficiency, low carbon and low waste, without pollution and safe, and environmentally friendly products [4]. One of the countries in Asia who participated in the declaration of Manila is Indonesia which then issued Law No. 3 of 2014 on the industry with the definition of green industry as an industry in the production process prioritize efforts to efficiency and effectiveness of resource use in a sustainable manner, aligning industrial development with the preservation of environmental functions and benefiting society [5].

The change of industry concept to green industry aims to realize a sustainable industry where there is efficiency and effectiveness in the use of natural resources so that in addition to providing benefits to the industry, the industry is also able to maintain environmental sustainability. In China, the green concept has been implemented before the Manila declaration, which since 2007 the development of the green mining industry and the construction of a green mine has been done to address the environmental problems of mining [2]. The concept of green industry can be realized in the use of equipment or facilities that are environmentally friendly or utilization of natural resources that are effective and efficient (which can be renewed better), so the application of green concepts can provide benefits to the industry as well as to the environment.

The term 'essential facilities' can actually be regarded as a doctrine of an economic principle because it relates to the use of essential facilities to meet the needs of each industry. The doctrine of 'essential facilities' requires that for a healthy and competitive business competition, competitors must have access to existing resources in order to start the competition from the beginning fairly [6]. The resources referred here are resources that can be utilized by anyone, not exception by large companies and industries that do require these resources to produce a product. Sometimes, the doctrine of 'essential facilities' also requires undertakings who monopolize the utilization of a essential facility must provide access to facilities deemed essential for the achievement of an effective and fair business competition. Provision of access to the intended facilities must be under the control of the monopolist [7].

The doctrine for the first time was known around 1970, when the US Supreme Court clearly provided a ruling that the monopoly could not refuse the competitor's access to vital assets to compete [8]. As in the formulation of the meaning of the doctrine Essential Facilities is a doctrine that prohibits the abuse of dominant position to deny access or use of essential facilities possessed or controlled by a business actor who has a dominant position, thus causing the competitor's business abilities to be hampered due to rejection access to these essential facilities [9].

The concept of 'essential facilities' is inseparable from the concept of business competition and production activities of a product by the company or industry, but the implementation of these activities must be sustainable and environmentally friendly. Environmentally friendly meant here is by avoiding the exploitation of existing natural resources and utilizing existing resources efficiently. The concept of environmentally friendly in business competition and production activities also means producing a useful product and service without causing environmental problems when the distribution of the products is made to various cities to the country. If a company or industry manages to achieve its production targets without damaging the environment, then they are considered to have successfully carried out the eco-efficiency activities in their business activities [10].

The facility is deemed vital if the facility is very important for the sustainability of the competition because the competitor cannot compete effectively on the relevant market without any access to the facility [11]. The form of a vital facility may be a transport terminal, infrastructure, other transportation facilities, financial, an intellectual property rights permit, or access to software code [12]. While the meaning of vital facilities in this study is a downstream distribution pipeline network that flows up to consumers. Thus, the use of the doctrine essential facilities is expected to be used to request access to monopolistic business actors in related industries. So that business actor must open access and cannot make refusal of cooperation (refusal to deal) [13].

In the law of business competition, monopoly garners the most attention. In fact, the monopoly itself is not a crime if it is obtained in a manner that is not illegal and fair. What is prohibited by business competition law is a monopolization that is the actions of a company that has a monopoly position to use its power in the relevant market. This monopolistic practice is the concentration of economic power by one or more business actors resulting in the control of the production and / or marketing of certain goods or services so as to create unfair business competition and may harm the public interest [14]. In other words, the monopoly position itself by companies is not something that is prohibited, but the potential for abuse or monopolization that these companies can do is against the law.

Various literature states that there are many negative impacts related to the monopolization by the perpetrator or a group of business actors that can harm consumers and other business actors, one of which is the entry barrier, where no other company is able to enter the monopoly market for a similar products, so that in turn small companies that are not able to enter the monopoly market will experience the dearth to develop fairly and eventually will go bankrupt. This is in line with the notion of monopoly in the Collins Economy dictionary that monopoly is: "one type of market structure that has properties, that one company with multiple buyers, a lack of substitution or substitution products and a barrier to entry, which cannot be entered by other business actors ". The entry barrier is something that degrades the desire, scope or pace of potential newcomers to enter the market [15]. Shepherd (1991) defines an entry barrier as a condition that makes it harder for new entrants to enter existing markets, where these obstacles can be any form of legal instrument such as licensing as well as a more general economic constraint or any strategic action of an existing company to inhibiting the desire to enter from newcomers [15]. According to the concept of 'essential facilities', the utilization of a vital facility must be fair and equitable among all parties and none of the parties can dominate it. This doctrine is of course applicable to maintain fair competition within a market, eliminating monopolization and abuses from the dominants. However, by opening an access to a vital facility in natural resources industry, the essential facility doctrine can be a key factor to achieve green industry concept in natural resouces industry as the industry will be more efficient by sharing the facilities among undertakings.

2 Material and method

The method used in this study is SWOT analysis method. SWOT analysis (SWOT alternative matrix) is an acronym for strengths, weaknesses, opportunities, and threats, i.e. structured planning methods within a company [16]. SWOT analysis in an industry is needed to know the purpose and facilitate the process to be run, so that industry can run effectively and efficiently. In this study, the SWOT analysis will be conducted to explore the strengths, weaknesses, opportunities, and threats to essential facilities in the efficiency of industrial natural resources. Based on identification using these four parameters (strengths, weaknesses, opportunities, and threats), the SWOT matrix determines which ones will be applied to identify four types of strategies, eg power-opportunity strategies (SO), weakness-opportunities (WO), power-threat strategies (ST) and weakness-threat (WT) [18]. The SWOT matrix is created in tabular form, and added by secondary data as support.

4 Results and discussion

In the case of the concept of 'essential facilities' on the efficiency of the industrial natural resources, there are strengths, weaknesses, opportunities, and threats that will be explained more clearly as follows.

4.1 Strength

There is competition between each industry in utilizing a facility that is essential or vital to run the business.

4.2 Weakness

Tend to feel 'afraid of not getting shares' considering all parties are entitled to access and utilize essential facilities.

4.3 Opportunity

There is dominance on one side in the utilization of essential facilities so that their needs remain fulfilled.

4.4 Threat

A decrease in production efficiency utilizing natural resources in industries, both the dominant and the dominated.
The following table presents the SWOT on legal protection cases for women environmental activists.

Table 1. SWOT matrix.

	S (Strength) Internal strength factors	W (Weakness) Internal weakness factors
O (Opportunity) External opportunity factors	SO strategy: create a strategy using force to take advantage of opportunities	WO Strategy: create strategies that reduce weaknesses by taking advantage of opportunities
T (Threat) External threat factors	ST Strategy: create a strategy using the power to deal with threats	WT Strategy: create strategies that reduce weakness and avoid threats

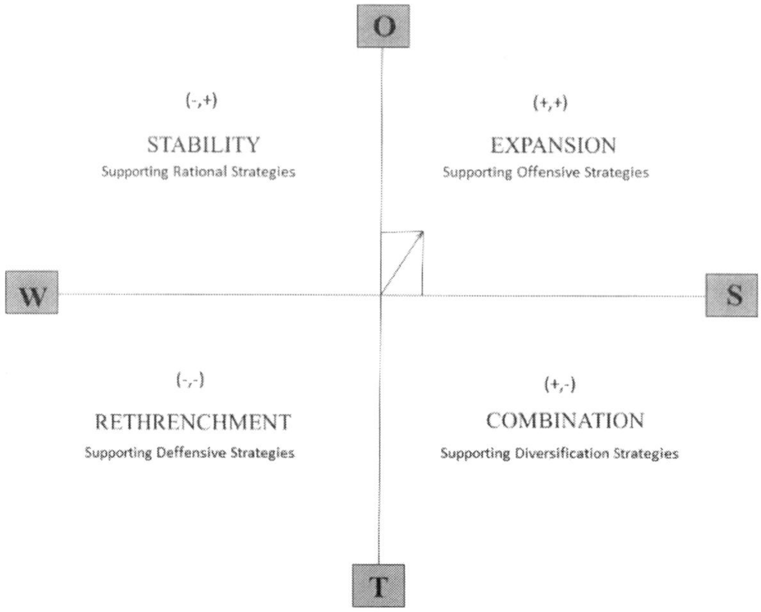

Fig. 1. Cartesian diagram.

In accordance with SWOT analysis results based on each strength, weakness, opportunity, and threat, then for this case SO strategy is used. In this case, it is necessary to create a strategy of using force to exploit opportunities. This strength if maximized can provide opportunities for the effective and sustainable use of natural resources. Essential facilities plays a role in the management and utilization of natural resources. Support vital essential facilities to be "heart" in a system of utilization and management of natural resources. Overall, the problems that have been faced are related to the availability and access of essential facilities.

Natural resources have an important role for human life and the sustainability of a society. Natural resources for various groups of people in Indonesia not only have economic value alone, but also social, cultural, and political meaning. Natural resources play an important role in the formation of civilization in human life. Therefore, every culture and ethnicity has its own distinct and diverse conceptions and views on the mastery and management of natural resources [15]. As from community communities in various places do not have adequate access to natural resources in the vicinity. This is not without reason, but is indeed because of the limitations that are owned especially in terms of downstream of the natural resources.

The public's ability to manage and utilize natural resources encourages outside parties to manage and utilize it. Communities who come from outside the community to manage and utilize the SDA are fully responsible for the sustainability of the utilization process. On the other hand, this SDA will generally be down-streamed through an industrial process to become a valuable product. In the process of industrialization of natural resources as inputs should be used as efficiently as possible. Therefore, the strength of these essential facilities has a very important central role as a force to be used as a reference in exploiting the availability of natural resources. Of course, this opportunity is closely related to the potential for availability and sustainability within the natural ecosystem. It becomes imperative because in the concept of sustainable development optimization of the benefits of natural resources and its integration with natural resources is the determinant of the success of sustainable development [15].

5 Conclusion

In an effort to realize sustainable development, essential facilities in the efficiency of industrial natural resources become one of the supporting factors. The complexity of the problems of industrialization as a downstream of natural resources is still being felt up to now. Based on the results of this study, it is known that to achieve the goal of sustainable development of the industrial world the concept of essential facilities must be applied. The concept plays a role in creating a strategy of using the inherent strengths therein to exploit law enforcement opportunities in the context of efficient use of natural resources in the industrial world. Thus, the future availability of natural resources and industrial sustainability can be realized.

Acknowledgements

We, the authors wish to thank the staffs for their time in formatting the manuscript and in supporting the editorial process. Special thanks are also extended, Intan Nurul Aini, and A W Utari for providing data and assisting us in the planning and production of the proceeding. Lastly we would like to thank the Ministry of Research, Technology, and Higher Education of the Republic of Indonesia as this research is funded by the Grant of Penelitian Dasar Unggulan Perguruan Tinggi-Kemristekdikti-Universitas Indonesia 2018 with contract number 271/UN2.R3.1/HKP.05.00/2018.

References

1. M. Stuermer. Industrialization and the demand for mineral commodities. *Journal of International Money and Finance*, **76**, 16-27 (2017)
2. X. Cheng, Q. Huang, S. Yang, J. Xu, Y. Fan, G. Xu, J. Yang, J. Yuan, W. Qi. Modes and experience of green mine construction in Yunnan, China: case studies. *IOP Conference Series: Earth and Environmental Science* **95** *042080*, (2017)
3. UNIDO, United Nations Industrial Development Organization, (2010)
4. Indonesia Ministry of Industry, http://www.kemenperin.go.id/download/6297/ Efisiensi-dan-Efektivitas-dalam-Implementasi-Industri-Hijau, (2012)
5. Undang-undang Republik Indonesia Nomor 3 tahun 2014, http://kemenperin.go.id/download/5181/Undang-Undang-No-3-Tahun-2014-Perindustrian.
6. P.A. McNutt, *Law, Economics, and Antitrust: Towards A New Perspective* (Edward Elgar Publishing Limited, Cheltenham, 2005).
7. T.F. Cotter. The essential facilities doctrine. *Antitrust Law and Economics*, (2010)
8. G. Vial. Routledge Taylor & Francis Group (2016)
9. C. Angelo, N. Antonio. Essential facility access in Europe: building a test for antitrust policy. *Review of Law and Economics* **3**, 1 (2007)
10. J.A. Pearce, R.B. Robinson Jr., *Manajemen strategis-formulasi, implementasi, dan* pengendalian *edisi 10 buku 1*. Terj. dari *Strategic management-formulation, implementation, and control 10th edition* (Bachtiar, Y., Christine: Penerjemah). (Salemba Empat, Jakarta, 2008)
11. S.W. Waller. Areeda, epithets, and essential facilities. *Wisconsin Law Review*, (2008)
12. L. Mays. The Consequences of search bias: how application of the essential facilities doctrine remedies google's unrestricted monopoly on search in the United States and Europe. *The George Washington Law Review* **83**, 2 (2015)

13. A. Firth, *EU Copyright law: the commentary.* (Edward Elgar Publishing Ltd., Cheltenham Glos, 2014)
14. A.F. Lubis. A.M. Tri Anggraini, Kurnia Toha, *Hukum persaingan usaha; antara teks dan konteks* (RDV Creative Media, Jakarta, 2009)
15. I. Widiyanto. Entry barrier antara persaingan potensial dan aktual. *Jurnal Bisnis Strategi* **5**, 2 (2006)

Local wisdom of indigenous society in managing their customary land: a comparative study on tribes in Indonesia

Maria

Faculty of Law, Private Law Department, Universitas Sumatera Utara, Indonesia

Abstract. The knowledge of indigenous people in Indonesia has developed for years in line with the development of human civilization. The development produces traditional knowledge and rules resulting from the adaptation process to its environment. The recent emergence of environmental crises has created a new awareness that this crisis can be resolved by returning to the local wisdom of indigenous peoples. The discourse on local wisdom has surfaced and is recognized as an important part of future development programs, including legal development. Local wisdom or environmental wisdom can be defined as the cultural knowledge possessed by a particular society that includes a number of cultural knowledge concerning models of sustainable use and management of natural resources. The legal relationship between the community and the land creates a right that gives the community a legal group, the right to use the land for the benefit of the society. In local communities, traditional wisdom manifests in the form of a set of rules, knowledge and skills as well as values and ethics that govern the social order of the communities that continue to live and evolve from generation to generation. This local wisdom is not only contained in the customary values and norms but also in the activities of indigenous people members in managing their lands.

1 Introduction

The knowledge of indigenous people in Indonesia has been developing for a long time in line with the development of human civilization. This development has emerged various new knowledge and traditional values as a result of the adaptation to their environment. In accordance with the basic human needs, one of the developing traditional knowledge is the knowledge about land use either for residential purposes or for commercial purposes [1].

In order to fulfill their needs, man will use everything provided in their surrounding environment. Man tends to lead themselves to Dutch Disease Syndrome which is human overexploitation of the natural resources for his own benefit without paying attention to the sustainability of those natural resources [2]. Therefore, man attempts to adapt in order to gain a balance and sustainability between the society and the environment, one of which is through establishing social control systems in the form of norms and customary law derived from that society. The current environmental crisis has raised a new awareness that such crisis can be resolved by going back to the local wisdom of the indigenous society. The idea of local wisdom has emerged to the surface and has been approved as an important part in the future development program including legal development [3]. Through this way, not only could the existence of indigenous people with all their wealth and local wisdom be saved, but also the ecological crisis caused mainly by the wrong perspective and behavior of the modern society [4].

The purpose of this article is to describe the local wisdom of indigenous peoples in Indonesia in managing their ulayat land. This discussion is then divided into several parts : (1) Understanding the meaning of local wisdom; (2) what is ulayat land; (3) local wisdom of indigenous peoples in Indonesia; (4) the relationship of UUPA and *tanah ulayat*.

2 What local wisdom is

Local wisdom can be simply defined as the cultural knowledge owned by a certain society which includes an amount of cultural knowledge concerning the models of sustainable natural resource utilization and management [5]. According to Law Number 32 of 2009, local wisdom is the noble values prevailing in the society's life to protect and manage sustainable living environment [6]. The Law also formalizes the government responsibility to acknowledge, respect, and protect the rights of the indigenous society, the traditional society, and the local wisdom over the coastal areas and small islands that have been used from generation to generation.

In addition, I Nyoman Nurjaya describes local wisdom in accordance with the Indonesian people mindset as follows:

> "Fundamentally, it originates from the values and religious systems that the community adheres to. Religious teaching and belief of the local community en soul, give color, and affect the environmental image in the form of attitude and behavior towards the environment. This implies that it provides guidance to humans to act in a harmony with the rhythm of the universe that it will create a balanced relationship between the human and the environment [7]."

Local wisdom contains local institution or system of norms embodying values, principle, structure, institution, mechanism and religion that grow, develop and are adhered to within the community. It serves as the instrument to maintain the order of interaction among the community (social order), the order of relationship with The Creator and spirits that are believed to have supernatural power (spiritual order), or the order of community's behavior towards their environment (ecological order) [7].

In relation to the utilization and management of natural resources and environmental sustainability, this indigenous people with their indigenous knowledge, their ability to uphold the customary law, their spiritual ability, and religion turn out to be wiser compared to other societies. The indigenous knowledge known as local wisdom grows and develops in the society as knowledge passed down from generation to generation as a part of their adaptation towards the environment.

3 *Ulayat* (communal) land

Van Vollenoven initially introduced the term "*Beschikkingsrecht*" for naming the *ulayat* rights; however, after the Indonesian independence, the communal land managed by this indigenous society was called *ulayat* land. The land ruled by indigenous society, also called customary land/*ulayat* land/*tanah hak pertuanan*, is a land under a legitimate reign of indigenous society [8]. Although the term of *ulayat* land differs for every region in Indonesia [9], there is a uniformity to call it *ulayat* land and the right over the land is called *ulayat* right.

The word *ulayat* itself means the environment of mastership and ownership according to indigenous literature [10]. The right of the indigenous society to manage *ulayat* land is called *ulayat* right. This right exists if there is a fellowship of the law "claims to have within a certain area the exclusive right to avail itself on land" [11]. Indonesia also recognizes the existence of *ulayat* right [12] proven bythe existence of various rules concerning *ulayat* right. Unfortunately, the implementation of this rule becomes unclear and, in fact, causes problems to the indigenous society [13]. There are many court cases where a group of people act on behalf of indigenous society to claim a freehold land to be the *ulayat* right of that indigenous society. The recognition of the existence of indigenous society in Indonesia [14] gives authority to organize their fellowship based on their customary law system as long as it is in accordance with the constitution of the Republic of Indonesia. For indigenous society, - considering the fact that Indonesia is an agrarian country – land has a very important function. Land plays an important role because of [15] (1) the nature of the land, land is the only natural resource that will stay in its original state under any circumstances, even becomes more useful. (2) the land fact, the reality of land is the residence of the fellowship of the law and all its members and gives a living for its owner, (3) the magic-religious aspect of land, as land is a unity which serves as a place for the burial of the land's owner and the ancestors of the fellowship of the law from the previous generations.

The religio-magic close relationship between indigenous society and the land makes the fellowship obtain the right to rule the land, use the land, gather the crop from plants and trees growing on the land, and hunt the animals living on that land [16]. Land is a place where indigenous society lives and a place that gives a living for them [17]. Hence, it can be said that "The ownership and use of land, especially among traditional producers in most developing countries, is not just a source of livelihood, but also a symbol of identity, dignity, solidarity, and peace" [18].

The land ruled by indigenous society, also known as customary land or *ulayat* land or*tanah hak pertuanan,* is land under the legitimate reign of indigenous society [8]. According to Boedi Harsono, *ulayat* right is the name given by the law and the jurists indicating a legal relationship between a certain indigenous society and a certain region which is the "*lebensraum*" [19] for its members forever [13].

The concept of customary law referring to *ipso facto*, occupation, placement, mastership, and cultivation is the beginning of the emergence of the right [20]. The relationship between a man and his land is highly determined by the intensity of *defacto*, man's utilization or cultivation over the land [21].

Ulayat land mentioned here is not always the land physically, it is also the forest, the pond, the river, the lake, and even the sea [22]. The utilization of the *ulayat* land also differs in each region depending on the creativity of the society in cultivating the land. For instance, in Juhar, Karo District, North Sumatra, the *ulayat* land is a pond that will be harvested once a year and enjoyed by all the members. These variations in indigenous land use arise in part from contextual and historical factors, as emphasized by many small-scale case studies of indigenous land use [23].

Iman Sudiyat named *ulayat* right as ancient right, which is the right owned by a certain clan/gans/stam, villages union (doper bond), or usually only one village to rule the whole land in that region [24]. *Ulayat* right is the highest mastership right in an indigenous society over the land that belongs to all its members [25].

Regarding the relationship between the indigenous society and their *ulayat* land, Ter Haar stated that this society has the right over their land and is allowed to apply that right inside and outside the land [10]. Applying it outside means the society is responsible for sustaining *ulayat* land and has the right to refuse or accept outsiders who want to enjoy their land based on the members' agreement. For instance, in indigenous society of Karo, an outsider who wants to enter and cultivate the *ulayat* land has tobe approved by the *runggun* [26] led by a chieftain. For Indonesia, such settlement earns support from cultural roots living and respected in the social interaction rules [27]. Meanwhile, applying it inside means the indigenous society organizes how the members carry out their right accordingly by limiting the allotment of personal rights and demand and withdrawing a certain part of the land from the right to enjoy it individually for the sake of public interest.

The legal relationship between society and their land creates a right that makes the society a legal group who can use the land for the society's benefit [13].

4 Local wisdom of indigenous society in Indonesia in managing *ulayat* land

Known as an archipelago, each tribe spread throughout Indonesia has their own local wisdom in managing their *ulayat* land. There are so many tribes with their own local wisdom inhabiting Indonesia. Nevertheless, we can pull out a connecting line as the common trait from the uniqueness of each tribe in managing their *ulayat* land. In local society, traditional wisdom manifests itself in a set of rules, knowledge, skills, and system of values and etiquette controlling social order in the community that keeps living and developing from generation to generation. This local wisdom is not only found in the customary norms and values but also in the activities of indigenous society in managing their *ulayat* land.

The local wisdom of the Malay indigenous societyis reflected in their proverb:

Kalau binasa hutan yang *lebat*
Rusak lembaga hilanglah adat [28]

The proverb above clearly shows a close relationship between the local society and their environment that they draw an analogy between deforestation and custom breakdown. It also shows a strong bond between Malay people and their environment.

For the Karonese indigenous society in Karo District, the concept of managing *ulayat* land is carried out with a ban on tree-cutting for commercial interest. This is controlled by all the members of the indigenous society in that region.

In the society of Nagari Kambang, Pesisir Selatan District, the society of Guguk Malalo, Tanah Datar District, and the society of Simanau, Solok District in West Sumatra, the concept of managing the forest is a communal conception based on the *ulayat* mastership over natural resources. This conception of *ulayat* is holistic, either for the land, forest, water, lake, orthe sea. This is in line with Minangkabau proverb:

Ka rimbo babungo kayu (Ke rimba berbunga kayu)
Ka ladang babungo ampiang (Ke ladang berbunga emping)
Ka sungai babungo pasie (Ke sungai berbunga pasir)
Ka lauik babungo karang (Ke laut berbunga karang) [29]

Article 5 of Regional Regulation Number 13 of 1983 formalizes the function and role of *KAN* (*Kerapatan Adat Nagari* = Council onTraditional Law and Customs) in preserving, sustaining and utilizing the natural wealth of district, *ulayat,* tribe and clan for the sake of public welfare that can be achieved by:

1. Conducting inventory of district's wealth and *ulayat*of tribe and clan.
2. Legalizing the wealth of district, *ulayat*, tribe, and clan done by *KAN*.
3. Utilizing the wealth of district, *ulayat*, tribe, and clan by society/district's member should be approved by *KAN* in a form of *KAN*'s legalization:
   a. The use of district's wealth should gain *KAN*'s approval in advance.
   b. The use of *ulayat* of tribe and clan should gain the chieftain's approval in advance.
4. Utilizing the wealth of district, *ulayat*, tribe, and clan by government for public and social interests should be approved by *KAN* in a form of *KAN*'s legalization:

   a. The use of district's wealth should gain *KAN*'s approval in advance.
   b. The use of *ulayat* of tribe and clans hould gain the chieftain's approval in advance.
5. Utilizing the wealth of district, *ulayat*, tribe, and clan by government for public and social interests is carried out under the members' agreement based on the principle of *"Adat diisi Limbago dituang/siliah jariah"*(the decision must be based on a consensus among the members of the society)
6. Utilizing the wealth of district, *ulayat*, tribe, and clan by investor/private, BUMN (State-Owned Enterprises), BUMD (Regional-Owned Enterprises), Cooperative, and Institution should be approved by *KAN* in a form of *KAN*'s legalization:
   a. The use of district's wealth should gain *KAN*'s approval in advance.
   b. The use of *ulayat* of tribe and clan should gain the chieftain's approval in advance.
7. Utilizing the wealth of district, *ulayat*, tribe, and clan by investor/private, BUMN (State-Owned Enterprises), BUMD (Regional-Owned Enterprises), Cooperative, and Institution is carried out under the members' agreement based on the principle of *"Adat diisi Limbago dituang / siliah jariah"* (the decision must be based on a consensus among the members of the society) in the form of:
   a. Inclusion of share
   b. Inclusion of Rights to Cultivate
   c. Inclusion of Rights to Build
   d. Inclusion of Rights to Use
   e. Inclusion of Rights to Lease
   f. Inclusion of Rights to Manage
   g. Inclusion in other forms [25]

The principle of *"Adat diisi Limbago dituang / siliah jariah"* (the decision must be based on a consensus among the members of the society) shows the local wisdom of Minangkabau society in preserving natural resources (*ulayat*) either in the forest, the land/field, the river, or the sea.

Unlike European people who lay the foundation of their legal culture in writing (literally), Indonesian people develop oral tradition culture [30]. A customary law mostly consists of unwritten laws that are understandable to the members. Hence, the way or method used to socialize the values, principles, and norms of the law is also carried out orally by using such instruments as: folklore, myth, chronicle, poem, song and proverb. The bad people are always described as being cynical and opportunistic but they will eventually be defeated by the good people. Meanwhile, the good people are described as being humble, polite and respectful that all their good deeds are always blessed by the Almighty God.

The indigenous society of Maluku living in small islands or coastal areas has a system called *"Sasi"* which is a prohibition against harvesting or taking away a certain natural resource (either from sea or land) for a certain time. *Sasi* is an attempt to maintain the quality and the population of the natural resources. The practice of *Sasi* covers both the natural resources in the sea and on land. On land, *Sasi* is applied for plants and fruits where at certain time the plants are not allowed to be harvested in order to get more gains that can fulfill the needs of the indigenous society. *Sasi* starts with a traditional ceremony called *"Tutup Sasi"* and ends with a ceremony called *"Buka Sasi"*. This shows that indigenous society and their environment have united that they value the nature as they value themselves [31].

Furthermore, the government also empowers the indigenous society in doing nature conservation. For instance, in managing Batang Gadis National Park, the indigenous society of Mandailing Natal has a local wisdom as follows: (1) *Harangan Rarangan* or a restricted forest, is a forest owned by a village that is not allowed to be used as farmland and the trees cannot be cut down neither for domestic nor commercial purposes. (2) *Lubuk*

Larangan is the belief in spirits, for instance the water and the fish in a river that are forbidden to take except for social and religious purposes. Taking them will reputedly lead to sanction and sickness. (3) *Bahasa Daun* means that the indigenous society of Mandailing Natal has the philosophy of interacting with the nature [32].

Traditionally, the community of Angkola and Mandailing in Tapanuli Selatan understand several basic concepts concerning spatial distribution and region mastership with its natural resources. Three of the concepts are *Banua*, *Huta*, and *Janjian*. *Banua* means one region, *Huta* means one settlement, and *Janjian* means the territorial fellowship of some *Banua* which is bound by one custom [3].

The legal protection for the rights of indigenous society of Baduy, particularly their *ulayat* right, is actually listed in the Regional Regulation of Lebak District Number 32 of 2001 on Protection of *Ulayat* Right of Baduy Society (National Gazette of Lebak District Number 65 Series C of 2001 as Supplement to the National Gazette of Lebak District Number 4) (hereinafter is called Regional Regulation of *Ulayat* Right of Baduy Society). The enactment of this Regional Regulation is done becausethe Local Government of Lebak District considers Baduy society as a society who has a strong bond with their customary law; therefore, there needs to be a legal protection for *ulayat* lands of Baduy society. The Regional Regulation of *Ulayat* Right of Baduy Society is one example of government concerns towards *ulayat* right of indigenous society where the government does not only recognize their right but also gives them a guaranteed protection.

5 UUPA (Basic Agrarian Law) protects *ulayat* land?

The enactment of UUPA (Basic Agrarian Law) is a fundamental change in land law (agrarian law). It is considered as the important milestone in the history of Agrarian in Indonesia in general and as the renewal of Agrarian Law in particular. This is because the Agrarian Law in Indonesia was previously dualistic in nature. It was rooted in the customary law and the Western Agrarian Law which was the result of the regulation in colonial era [33].

The recognition of customary law or customary rights particularly in the field of land is arranged in Law Number 5 of 1960 on the Basic Regulation of Agrarian Law and the rules underneath. The General Explanation III of Basic Agrarian Law [34] states:

"The Agrarian Law applicable to the land, water and airspace is *Adat* Law (customary law) as long as it does not contradict the national interest and the interest of the State, which are based on national unity, the Indonesian socialism, other legislations and religious law." The explanation of this article affirms that customary law is made as the foundation of the new law [35].

UUPA, as the law defending indigenous society's interest and the law embodying Indonesianationalism, also formalizes in Article 3 that *ulayat* right in villages can only be recognized and maintained as long as the society still exists when the UPPA is valid [36].

However, if one uses reverse logic [37], the government or the ruler at that time actually says that the indigenous society does not exist anymore. By using euphemism (substitution of expression for unpleasant word), the words "...as long as the society still exists..." mean the society is no longer existed as long as their existence is not yet proven. If the indigenous society as the element in establishing the State is not existed, then the State denies itself of being an existing state. The implementation is that the State itself is pessimistic towards the existence of the indigenous society.

The regulation of *ulayat* right means the recognition to guarantee and give legal protection for the rights of indigenous society along with its limitations. This regulation becomes a dilemma faced by *ulayat* right that it does not give legal protection and guarantee of law certainty of *ulayat* right following its purpose and foundation as a legal norm [8].

UUPA does not clarify the criteria regarding the existence of the *ulayat* right. Nevertheless, by referring to the fundamental understandings mentioned above, the criteria to determine the existence of *ulayat* right can be seen in the following conditions:

1. The presence of indigenous society that meets certain criteria, as the subject of the *ulayat* right;
2. The presence of land/region with certain borders as the *Lebensraum* and the object of the *ulayat* right; and
3. The presence of authority of indigenous society to carry out certain acts as explained above [38].

The practice of *ulayat* right, as long as it has been existing, is done by the indigenous society in accordance with their customary law. The *ulayat* right of the indigenous society is considered still in existence if: 1) There are still a group of people who feel attached to their customary law as one fellowship of a certain law, recognize, and carry out the laws in their daily lives. 2) There is a certain *ulayat* land as a living place for the fellowship of the law and the place to get their daily needs, and 3) There is a customary law regarding arrangement, mastership, and utilization of *ulayat* land that the fellowship abides by.

A research and determination for the existence of *ulayat* right are conducted by the Local Government involving the experts in customary law, the indigenous society in that region, Non-Governmental Organization, and the institutes managing natural resources. The *ulayat* land that still exists will be included in the base map of land registration by adding cartographic mark and if possible drawing the borders and including it in the land register. In the cases of land conflict, none of indigenous societies won the 1,400 cases of land dispute in West Sumatera Court. The same casea pplies to *Ulayat* Right of District where approximately 100 ha of the area has become the property of Ministry of Forestry [39].

The enactment of Regulation of the Minister of Agrarian Affairs Number 9 of 2015 on the Procedures of the Establishment of Communal Right over the Land of Indigenous Society and the Society in a Certain Region (*Permen ATR* 9/15) should be able to fulfill the needs of indigenous society regarding their rights over natural resources which have not yet been well protected. This makes the relationship between indigenous society and the natural resources become distant. The indigenous society has been forced to detach their bond with land, forest, water and other natural resources they own.

In actual practice, there are still pros and cons among the academicians regarding the enactment of *Permen ATR*. Some experts state that the regulation needs to be reviewed in terms of its lack of recognition of diversity in the character of the right of the indigenous society over their land and natural resources. Furthermore, the equalization of indigenous and non-indigenous society as the subject of communal right can cause an overlap of right between indigenous and non-indigenous society living in the same place and for a long time.

For *ulayat* right in particular, during the four decades of UUPA enactment, there is no law that implements such recognition [29]. There is no regulation that firmly recognizes the existence of *ulayat* right. In its implementation, *ulayat* right is not well identified that its status is often considered as the same as (included as) State's land. Overlap of *ulayat* land often happens when giving the Rights to Cultivate for farming and other purposes.

For instance, the land dispute between tribe of Dayak Bentian and HPH/HPHTI (Natural Forest Management Permit/Industrial Forest Plantation Permit) companies started when the forest which used to bea part of the tribe's territory was taken over by the companies [40]. The tribe was troubled by the companies making buildings on their *ulayat* land and disturbing the river ecosystem.

6 Local wisdom in the management of *ulayat* land: past and present

Conflicts of the management of natural resources happen because of the overlap of claim on territory. The indigenous society who feel they have the highest right over their territory are often ignored by the government who gives concessions to companies for using customary land or making it as a conservation area. Actually, the participation of indigenous society in preserving *ulayat* forest in the conservation area is very important since they have lived there from generation to generation long before the conservation area was made and they have had intense interaction with the area. This intense interaction surely enables them to be the reliable guardians of the conservation since they already have their own values or principles of conservation (local wisdom) in their souls. Basically, the ones who can protect environment are those who have conservation awareness.

However, as the time changes, the local wisdom of managing *ulayat* land does not only collide with regulations but also with the sweet talk from the plantation owners. Many chieftains of the indigenous societies were deceived by the plantation owners to hand their *ulayat* land for money, resulting in reducing and even losing the *ulayat* land.

7 Conclusion

The recognition of *ulayat* land under the UUPA is the recognition of the right of indigenous society, on condition that the *ulayat* right still exists and does notcontradict the higher purposes. This means there is a limitation to the existence of the right which is put on a weak position, even very possibly to be eliminated.

The regulation of rights of indigenous society for natural resources is still general and abstract. It seems like there is still a limitation to the rights of the indigenous society thus legislation initiative only rewrites the rights of indigenous society listed at the higher regulation. Therefore, clear justification and regulation are needed in determining the rights of the indigenous society to eliminate the obscurity of its existence.

As Hooker said, "The basic issue at the root of all Indonesian land system is the relationship between the community and individual interest."

References

1. N. Kosmaryandi. Kajian penggunaan lahan tradisional Minangkabau berdasarkan kondisi tanahnya (Study of Minangkabau traditional landuse based on its soil condition). *Media Konservasi* **X**, 2 (2005)
2. A. Satria. Penguatan kapasitas masyarakat desa dalam akses dan kontrol terhadap sumber daya alam. Seminar and Workshop Toward the 2030 Village, at the Management and Business Master's Program of IPB. (2007)
3. Z.B. Lubis. Menumbuhkan (kembali) kearifan lokal dalam pengelolaan sumberdaya alam di Tapanuli Selatan. *Antropologi Indonesia, Indonesian Journal of Social and Cultural Anthropolgy* **29**, 3 (2005)
4. S. Keraf. *Etika Lingkungan* (Kompas, Jakarta, 2002)
5. Y.R. Zakaria. *Hutan dan Kesejahteraan Masyarakat* (Penerbit WALHI, Jakarta, 1994)
6. Law Number 32 of 2009 on Environmental Protection and Management Article 1 Paragraph 30
7. S. Irianto. *Hukum yang Bergerak: Tinjauan Antropologi Hukum* (Yayasan Pustaka Obor Indonesia, Jakarta, 2009)

8. D. Samosir. *Hukum Adat Indonesia Eksistensi Dalam Dinamika Perkembangan Hukum di Indonesia* (Nuansa Aulia, Bandung, 2013)
9. *Taneh Kesain* (Karo), *Ulayat* (Minangkabau), *Popah* (Sakai)
10. S. Soekanto. *Kamus Hukum Adat* (Penerbit Alumni, Bandung, 1982)
11. Mahadi. *Uraian Singkat Tentang Hukum Adat Sejak RR Tahun 1854* (Penerbit Alumni, Bandung, 1991)
12. Basic Agrarian Law, Environmental Management Act, Plantation Law, Water Resources Law, and other laws
13. A.S. Hutagalung, *Tebaran Pemikiran Seputar Masalah Hukum Tanah* (Lembaga Pemberdayaan Hukum Indonesia, Jakarta, 2005)
14. The recognition on indigenous society is regulated in Article 28 Paragraph I (2) of the 1945 Constitution which reads: The cultural identity and rights of indigenous society are respected in line with the times and civilizations.
15. I.G.A.B. Wiranat. *Hukum Adat Indonesia* (PT. Citra Aditya Bakti, Bandung, 2005)
16. T. Setiady. *Intisari Hukum Adat Indonesia (Dalam Kajian Kepustakaan)* (Penerbit Alfabeta, Bandung, 2008)
17. S. Soekanto, S.B. Taneko. *Hukum Adat Indonesia* (Penerbit Rajawali, Jakarta, 1986)
18. Egemi 2006a, 2006b; El Hadary 2010; Komey 2009 in Yasin Abdalla Eltayeb ElHadary and Franklin Obeng Odoom.*Conventions, Changes, andContradictions in Land Governance inAfrica: The Story of Land Grabbing in North Sudan and Ghana*, Africa Today, Journal, p. 59.
19. Lebensraum is the right of a nation's living space to ensure its welfare and security. Retrieved June 29, 2017 from http://www.organisasi.org/1970/01/ajaran-lebensraum-autarki-dan-pan-region-oleh-karl-haushofer-orang-jerman-pengertian-dan-arti-defenisi-ilmu-sejarah-dunia.html, *Ajaran Lebensraum, Autarki dan Pan-Region oleh Karl Haushofer Orang Jerman- Pengertian dan Arti Defenisi –Ilmu Sejarah Dunia*
20. O.K. Saidin, *Mencari dan Menjadi Hukum Indonesia, Refleksi Pemikiran Prof. Mahadi* (PT. Raja Grafindo Persada, Jakarta, 2016)
21. S. Wignjosoebroto. Perbedaan konsep tentang dasar hak penguasaan tanah antara apa yang dianut dalam tradisi pandangan pribumi dan apa yang dianut dalam hukum positif Eropa. A Paper presented in the Seminar on Agrarian Law in the Commemoration of the 30th Birthday of UUPA (1960-1990), (1990)
22. In Karo Regency, for example, *tapin* (a fountain) becomes ulayat land managed by indigenous society.
23. C.L. Gray, R.E. Bilsborrow, J.L. Bremner, F. Lu. Indigenous land use in the ecuadorian amazon: a cross-cultural and multilevel analysis. Springer Science + Business Media, LLC 2007, Published online: 19 October 2007, Hum Ecol (2008) 36:97–109, Journal, p. 98
24. I. Sudiyat. *Hukum Adat Sketsa Asas (*Liberty, Yogyakarta, 1978)
25. H. Panuh. *Pengelolaan Tanah Ulayat Nagari Pada Era Desentralisasi Pemerintahan di Sumatera Barat (*PT. Raja Grafindo Persada, Jakarta, 2012)
26. *Runggun*can be defined as an institutional process of formal deliberation and decision making through consensus by a group of people bound by such kinship as *senina*, *anakberu*, and *kalimbubu*, which forms a unity (*sangkep si telu)*. Herman Slaats and Karen Portier, *Traditional Decision-Making and Law* (Gadjah Mada University Press, Yogyakarta, 1992)

27. Maria SW Sumardjono, et al. *Mediasi Sengketa Tanah*, Jakarta: Kompas, 2008, p. 5.
28. T. Effenddy. *Tunjuk Ajar Melayu* (*Butir-Butir Budaya Melayu Riau*) (Adicita Karya, Yogyakarta:, 2004)
29. K. Warman, at al. *Nasib Tenurial Adat Atas Kawasan Hutan* (Huma dan Qbar, Jakarta Selatan, 2007)
30. D. Rato. *Hukum Adat di Indonesia* (Laksbang Justitia Surabaya, Surabaya, 2014)
31. Kuwati, dkk. Konservasi Berbasis Kearifan Lokal (Studi Kasus: Sasi di Kabupaten Raja Ampat), Proceedings of Raja Ampat National Seminar WAISAI 12 – 13 August 2014, Universitas Kristen Satya Wacana, (2014)
32. G. Lumbantoruan. Kearifan Lokal Masyarakat Hukum Adat Mandailing Natal dalam Pengelolaan Taman Nasional Batang Gadis Sumatera Utara, Retrieved August 18, 2017 from http://lumbanlaw.blogspot.co.id/2013/01/kearifan-lokal-masyarakat-hukum-adat.html
33. Indonesia had been colonized by the Dutch for almost 350 years; consequently, there are still many colonial state regulations still in force today in accordance with Article 1 of the Transitional Constitution of the 1945 Constitution.
34. The new Agrarian Law must correspond to the legal consciousness of the masses, since the Indonesian people are largely subject to customary law, then the new agrarian law will be based also on the provisions of customary law, as the original law, which is completed and adapted to the interests of society in a modern state in relation to the international world and adapted to Indonesian socialism. The customary law in its growth is inseparable from the political influence and colonial society which is capitalist and self-governing society
35. M. Koesnoe. *Catatan-catatan Terhadap Hukum Adat Dewasa Ini* (Airlangga University Press, Surabaya, 1979)
36. AP. Parlindungan. *Hak Pengelolaan Menurut Sistem UUPA* (Mandar Maju, Bandung, 1989)
37. Reverse logic is the opposite action to the command "as long as it has existed..." its implication should have no longer been in existence."
38. M.S.W. Sumardjono. *Kebijakan Pertanahan Antara Regulasi dan Implementasi* (Kompas, Jakarta, 2001)
39. Emil Kleden and the results of the research by Asep Yunan Firdaus Jawahir Thontowi, et al. Aktualisasi masyarakat hukum adat (MHA): Perspektif hukum dan keadilan terkait dengan status MHA dan Hak-hak Konstitusionalnya. Jakarta: Pusat Penelitian dan Pengkajian Perkara, Pengelolaan Teknologi Informasi dan Komunikasi Kepaniteraan dan Sekretariat Jenderal Mahkamah Konstitusi Republik Indonesia, (2012)
40. M.S.W. Sumardjono. *Mediasi Sengketa Tanah* (Kompas, Jakarta, 2008)

Characterization of cellulase from *E. coli* BPPTCC-EGRK2

Mas Gunawan Haryanto[1], *Siswa* Setyahadi[2] , *Muhammad* Sahlan[1], *Masafumi* Yohda[3], *Yosuke* Fukutani[3], *Eko Agus* Suryono[4], and *Misri* Gozan[1*]

[1] Chemical Engineering Department, Faculty of Engineering, Universitas Indonesia, Depok 16424, Indonesia
[2] Center for Bioindustrial Technology, Agency for Assessment and Application of Technology, 15314 Serpong, Indonesia
[3] Department of Biotechnology and Life Science Tokyo University of Agriculture and Technology 2-24-16 Naka-cho, Koganei, Tokyo Japan 184-8588
[4]Faculty of Biology, Gadjah Mada University, Indonesia

Abstract. Cellulase enzymes are widely used in various industries such as detergent industry, bioethanol, animal feed, textile and paper. This research focused on characterization of cellulase produced from Eschericia coli BPPT-CC EgRK2, which is a recombinant that can produce protein enzymes endo- β-1,4-glucanase. Eschericia coli BPPT-CC EgRK2 was cultured in 1 litre liquid medium Luria Bertani. Because the bacteria is intracellular, sonication is needed for cell disruption to get the cellulase enzyme. The enzyme activity was then analyzed by CMC substrate at different concentration. The protein content analysis was carried out using Bradford method; the molecular weight analysis was done using SDS-PAGE; while the enzyme kinetics was plotted using Michaelis-Menten model. Our results showed the highest enzyme activity was 2.403 U/ml and the protein concentration was 5.352 mg/ml. The Michaelis-Menten constant (Km) and maximum velocity (Vmax) for CMC substrate hydrolysis were 0.07 μmol/ml and 2.49 μmol/ml/sec, respectively. The cellulase molecular weight was 58 kDa using SDS-PAGE with 7.5% stacking gel. The results indicated that Eschericia coli BPPT-CC EgRK2 is a promising renewable source for cellulase production for industrial application.

1 Introduction

Enzymes are proteins that function as biocatalysts in a chemical or biological reaction [1]. The enzyme is a protein biomolecule that acts as a catalyst (a compound that speeds the reaction process without reacting) to *an organic chemical reaction. Substrate as the initial molecule will be accelerated its conversion into another molecule called product. The type of product to be produced depends on a condition / substance, called a promoter. The work of enzymes is influenced by several factors, mainly substrate, temperature, acidity, cofactor

* Corresponding author: misrigozan@gmail.com

and inhibitor. Each enzyme requires different temperatures and pH (acidity levels) because different enzymes are proteins, which can change shape if temperature and acidity change.

One of the most widely used enzymes in the industry is the cellulase enzyme. Cellulase can convert cellulose, the main polymer in biomass, into glucose which can be further fermented to ethanol. Cellulase is currently the third largest industrial enzyme worldwide [2]. Cellulase is widely used in biotechnology applications, such as fiber modification in the paper and textile industries, but cellulase also has excellent capabilities in the industrial development of ethanol production from lignocellulose. Cellulase can be used for biomass fermentation into bioethanol [3]. A specific enzyme that is used to degrade cellulose is cellulase enzyme. Enzymatic process in hydrolyzing cellulose is a better and more beneficial way because the surrounding condition for enzyme's activity could be controlled [4].

The conventional method to produce cellulase is by using lignocellulosic fungi such as Aspergillus sp., Trichoderma sp., and Neurospora sp. [5-9]. However, cellulase can also be secreted by bacteria, such as Bacillus sp. Bacteria has a faster growth rate and is genetically easier to recombine compared to fungi [10]. This study examined an *E. coli* EgRK2 recombinant strain. The recombinant strain was used in this experiment because the result of the previous research showed the wild type has low activity , the recombinant strain grows faster on inexpensive media, ease of isolation and purification of proteins excreted in comparison with the wild type [11].This study examined the effects the enzyme's hydrolysis performance on Carboxymethyl Cellulose (CMC) as the substrate. The aim of the present research is to describe the biosynthesis, isolation and characterization of a cellulase from *E. coli EgRK2* isolate.

2 Materials and Methods

2.1 Sample Collection

The research was performed with an inoculum of *E. coli* EgRK2 recombinant strain which obtained from BPPT (Agency for the Assessment and Application of Technology Culture Collection) Laptiab, Puspiptek, Serpong,Tangerang, Indonesia.

E. coli EgRK2 is a recombined product of a Bacillus sp. RK2 isolate. The gene that expressed endo-β-1.4-glucanase (gene attB-EgRK2) in Bacillus sp. RK2 was isolated and amplified by the Polymerase Chain Reaction (PCR) method.In this study we used GatewayR (Invitrogen) method and two primers such as egrk2F: 5'- AAA AAG CAG GCT CGA TGA AAC GGT CAA TYT Y -3'; egrk2R: 5'- AGA AAG CTG GGT ACT AAT TKG KTT CTG WTC CC -3' and attB1 adapter: 5'- G GGG ACA AGT TTG TAC AAA AAA GCA GGC T -3'; attB2 adapter: 5'- GGG GAC CAC TTT GTA AA GAA AGC TGG GT -3'[11].

These design primers were used for amplifying endonuclease EgRK2 gen from chromosomal of B amyloliquefaciens BPPTCC RK2. pDONR221 as a donor vector with Kanamycin resistance and pDESTI4 as a destination vector with ampicillin resistance. E. coli DH5α for host cloning and E. coli BL21 Star for expression gen. It was then recombined and expressed in the pEXP-EgRK2 vector, which was transformed to E. coli DH5α [11]. The isolate was inoculated to Luria-Bertani (LB) slants of agar, and incubated at 37°C for 24 h and used as a culture stock.

2.2 Cellulase Enzyme Production

A starter medium was made by inoculating 1 loops of culture stock into sterile 1L LB medium, which was previously autoclaved at 121°C, and 1.2 atm for 2 hours. After autoclaved the LB medium, we put on the rotary shaker at 37°C and 150 rpm for 24 hours. Then we centrifuge the medium at 4°C and 5000 rpm for 25 minutes. After the pellet and supernatant separate, we add tris-HCl buffer to the pellet and mixed until homogen. At the end of the process, the culture medium was sonicated for 15 min to break the cells, and centrifuged at 4,000 rpm for 15 min, to obtain the crude extract that was meant to be used as the enzyme source.

2.3 Enzyme Assay

The assay of enzyme activity was done by the DNS (3.5-Dinitrosalicylic Acid) method. The cellulose existence was identified by determining reducing sugar content that was formed as a result of hydrolysis. The enzyme activity calculated using Eq. (1):

$$\text{Activity (U/ml)} = \frac{mg\ glukosa\ x\ 1000}{Mr\ glukosa\ x\ 30\ min\ x\ 0.1\ ml} \quad (1)$$

DNS method determined the amount of reducing sugar which was formed (mg) as the sugar content; The molar mass of glucose (180 g/mol) as Mr reducing sugar, the hydrolysis period of cellulose (30 min), and the amount of crude extract analyzed as sample (0.1 mL). Cellulase activity was measured in U/mL. The amount of enzyme required to liberate 1 μmol of reducing sugar per minute under the assay condition is defined as 1 unit.

The protein content was measured using the Bradford protein assay. An enzyme sample of 2 μL was mixed with 795 μL aquades and 0.2 mL Bradford reagent. After that measuring the sample absorbance at 595 nm wavelength.

2.4 Enzyme Characterization

2.4.1 Gel Electrophoresis

The determination of cellulose enzyme molecular weight used sodium deodecyl sulphate poly acrylamide gel electrophoresis (SDS-PAGE) with 7.5% stacking gel at constant current of 20 mA. The following protein calibration kit was used: ovotransferrin (78 kDa), albumin (66.25 kDa), ovalbumin (45 kDa), carbonic anhydhrase (30 kDa). The molecular weight of crude enzyme was determined after gel staining with Comassie Brilliant Blue G-250.

2.4.2 Substrate Specificity and Kinetic Parameters

To investigate the kinetic paramaters of CMC as cellulose enzyme substrate, hydrolysis experiments were conducted. Experiments were carried out on CMC (1; 1.25; 1.5; 1.75; and 2.0% wt). The kinetics of CMC degradation by cellulase was evaluated using the Michaelis–Menten model, with linearization using the Lineweaver-Burk approach shown in Eq. (2).

$$\frac{1}{Vo} = \frac{Km}{Vmaks}\frac{1}{[S]} + \frac{1}{Vmax} \quad (2)$$

[S] is the concentration of carbon source (mg/mL), Vmax is the maximum rate of reaction at infinite reactant concentration (μmol/mL/sec), and Km is the Michaelis constant (μmol/mL). The pH condition is 7 and temperature 40°C.

3 Results and Discussion

3.1 Cellulase Enzyme Activity

Our study showed that the highest enzyme activity on CMC hydrolysis is 2.403 U/ml which obtained from 2% CMC concentration as shown in Fig 1.

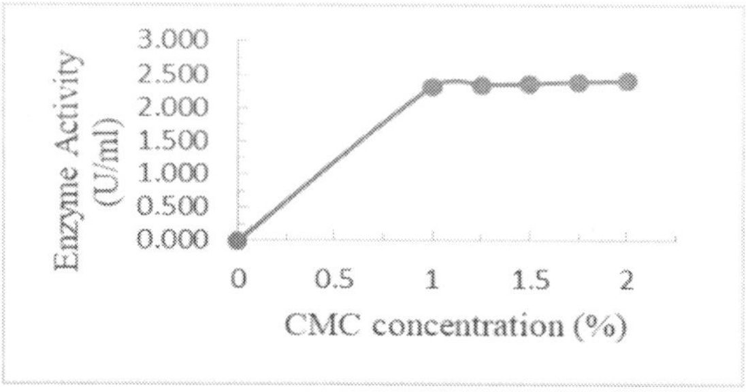

Fig. 1. Cellulase Enzyme activity over different CMC concentration

From this figure we know that reducing sugar concentration gradually increasing during the the substrate is higher. The result of enzyme activity not significantly different from the previous work with CMC as substrate [11] and the optimum condition in this research is 40°C for the temperature and 7 for pH based on the previous work [4] [11]. This result was slightly higher compared to cellulase activity from several studies as shown in figure 2.

Table 1 The variation of microorganisms for cellulase production

Microorganism	Carbon Source	Enzyme activity	Reference
E. coli EgRK2	CMC	Cellulase - 2.49 U/mL	This research
E. coli EgRK2	OPEFB CMC	Cellulase - 2.665 U/mL Celluase - 2.5 U/mL	[11]
*Bacillus sp.*RK2	OPEFB	Cellulase - 2.498 U/mL	[11]
Bacillus subtilis	CMC	Cellulase – 0.9 U/mL	[12]
Bacillus pumilus EB3	OPEFB CMC Cellulase Powder	Endoglucanase – 0.076 U/mL Endoglucanase – 0.063 U/mL Endoglucanase – 0.025 U/mL	[13]
Bacillus pumilus	CMC	CMCase – 1.9 U/mL, cellobiase – 1.2 U/mL	[14]
Neurospora crassa	Wheat Straw	FPase – 0.667 U/mL	[15]
Chaetomium globusum	Avicel CMC	FPase – 0.11 U/mL, CMCase – 1.05 U/mL, β-glucosidase – 0.5 U/mL FPase – 0.18 U/mL, CMCase – 2.00 U/mL, β-glucosidase – 1.5 U/mL	[16]
Trichoderma ressei	Cellulose	β-glucosidase – 4.54 U/mL	[17]
Phenerocheate	Rice straw	FPase – 1.438 U/mL	[18]

chrysosporium			
Trichoderma harzianum	OPEFB	FPase – 5.1 U/g	[7]
Trichoderma ressei	Straw	FPase – 0.041 U/mL	[9]
Trichoderma koningii	Cellulose	FPase – 1.4 U/mL	[8]
Trichoderma ressei	Corn stover	CMCase activity 2596.3 IU/g dry culture	[19]

The cellulase activity values were 2.665 U/mL using *E. coli* EgRK2 and OPEFB as substrate, 2.5 U/mL using CMC as substrate [11]; 0.041 U/mL using *Trichoderma reesei* and straw [9]; 2.00 U/mL using *Chaetomium globusumn* and CMC [16]; 1.9 U/mL using *Bacillus pumilus* and CMC [14]; 0.063 U/mL using *Bacillus pumilus* EB3 using CMC [13]; and 0.9 U/mL using *Bacillus subtilis* and CMC [12].

3.2 Enzyme Characterization

3.2.1 Molecular Weight and Protein Assay

The crude enzyme from the culture broth was analyzed with SDS-PAGE to know the molecular weight of this protein. Because this crude enzyme was not purified yet, there was still more than one protein band. From the SDS-PAGE we know the molecular weight is 58 kDa. After that we analyzed the protein concentration of this broth. The concentration result is 5.352 mg/ml. This is not concentration of cellulose enzyme, because it is no purified, we can't measure the enzyme concentration.

3.2.2 Enzyme Kinetics on CMC Hydrolysis

Cellulase enzyme from *Eschericia coli* EgRK2 was used to hydrolysis of CMC. CMC degradation experiments showed the Vmax values were 2.49 μmol/mL/sec. The Km values for CMC hydrolysis were 0.07 μmol/mL for the recombinant strain. Previous research on Km and Vmax values which used CMC as substrate have been conducted by using same isolate, *Acinetobacter anitratus and Branhamella sp*. as shown in Figure 3 [11][20]. In CMC degradation, cellulase from the *E. coli* EgRK2 had higher Vmax values (2.49 μmol/mL/sec) compared to two other strains (0.48 and 2.56 μmol/mL/sec for *A. anitratus and Branhamella sp*., respectively. But had lower Vmax values compared to the previous research with same EgRK2 isolate (2.543 μmol/mL/sec). Additionally, Vmax values were also higher than cellobiose degradation by using A. anitratus (0.24 μmol/mL/sec) and Branhamella sp. (0.34 μmol/mL/sec). Thus the cellulase produced in this research had a better performance value than that produced in other research [20]. The Km values of cellulase from the *E. coli* EgRK2 recombinant in CMC degradation (0.07 μmol/mL) were smaller than the cellulase from same EgRK2 isolate in the previous research, *A.anitratus* (0.497 μmol/mL) and *Branhamella sp*. (0.790 μmol/mL). Thus in our research the enzyme affinity was better and the carbon source needed to reach its half maximum rate was lower.

Table 2 Values of cellulase Vmax and Km from various carbon source and microorganism

Carbon Source	Microorganism	Maximum enzymatic rate, Vmax (μmol/mL/sec)	Michaelis-Menten Constant, Km	Reference
CMC	*E. coli* EgRK2	2.49	0.07 μmol/mL	This research
CMC	*E. coli* EgRK2	2.543	0.097 μmol/mL	[11]
OPEFB	*E. coli* EgRK2	1.75	0.26 μmol/mL	[20]
CMC	*Acinetobacter anitratus*	0.48	4.97 mg/ml	[20]

	Branhamella sp.	2.56	7.90 mg/ml	
Cellobiose	*Acinetobacter anitratus*	0.24	0.32 mM	[20]
	Branhamella sp.	0.34	2.50 mM	[20]

4 Conclusion

Our research shows that cellulase production by the *E. coli* EgRK2 had Vmax value for CMC degradation were 2.49 µmol/mL/sec, while Km values were 0.07 µmol/mL. The molecular weight was 58 kDa and the protein concentration wa 5.352 mg/ml. This research demonstrated a possible way to develop an efficient hydrolysis system for CMC using *E. coli* EgRK2. The result of this research have known that *E. coli* EgRK2 become a promising renewable source for cellulase enzyme for industrial application.

Acknowledgements

Authors are grateful for the financial support provided by Japan Student Service Organization (JASSO), research facilities provided by Tokyo University of Agriculture and Technology and conference funding from USAID through SHERA program- Center for Development of Sustainable Region (CDSR). Authors also gratefully acknowledge for the Center of Bioindustrial Technology, the Agency for Assessment and Application of Technology (BPPT), Indonesia, for their kind technical support.

References

1. H.S. Fogler, *Elements of Chemical Reaction Engineering.* **4th ed**, pp. 394-418. (Prentice-Hall International, Inc, New Jersey, USA ,2006).
2. M. Gozan, M. Edita, D. H. Park, and P. Bambang Cellulase Immobilization Using Reversible Soluble-Insoluble Polymer. *International Journal of Pharma and Bio Sciences*. ISSN 0975-6299 (2011)
3. S. Coseri, Cellulose: To depolymerize… or not to?. *Biotechnology advances*, 35(2), 251-266, (2017)
4. M. Gozan, M.B. Agung., S. Siswa. Variation of C/N Ratio and Fermentation Time In Respone Surface Methodology For Cellulase Production from *Bacillus sp*. BPPT CC RK2. *International Journal of Pharma and Bio Sciences*. ISSN 0975-6299 (2013)
5. M. Romero, J. Aguado, L. Gonzalez, and M. Ladero. Cellulase production by Neurospora crassa on wheat straw. *Enz. Microb. Technol*. **25**: 244-250 (1999).
6. S. Wen, C.Z. and W. Liao. Production of cellulase/β-glucosidase by the mixed fungi culture of Trichoderma reesei and Aspergillus phoenicis on dairy manure. *Proc. Biochem*. **40**: 3087- 3094 (2005).
7. M.Z. Alam, A.A. Mamun, I.Y. Qudsieh, S.A. Muyibi, H.M. Salleh, and N.M. Omar. Solid state bioconversion of oil palm empty fruit bunches for cellulase enzyme production using a rotary drum bioreactor. *Biochem. Eng*. J. **46**: 61-64 (2009).
8. Z. Wang, H.X. Ong, and A. Geng Cellulase production and oil palm empty fruit bunch saccharification by a new isolate of Trichoderma koningii D-64. *Proc. Biochem*. **47**: 1564-1571 (2012).

9. V. Rosyida, A.W. Indrianingsih, R. Maryana, and S.K. Wahono. Effect of temperature and fermentation time of crude cellulase production by Trichoderma Reesei on straw substrate. *Energy Procedia*. **65**: 368-371 (2015).
10. M. Maki, K.T. Leung, and W. Qin. The prospects of cellulase-producing bacteria for the bioconversion of lignocellulosic biomass. Int. J. Biol. Sci. 5: 500-516 (2009).
11. S.Z. Amraini, P.A. Lina, H. Heri., S. Siswa., F.R. Siti., D.H. Park., G. Misri. Production and Characterization of Cellulase From *E. coli* EgRK2 Recombinant Based on Oil Empty Fruit Bunches. *The Korean Society for Biotechnology and Bioprocess Engineering and springer*. **22**:287-295 (2017).
12. S. Sethi, A. Datta, B.L. Gupta, and S. Gupta. Optimization of cellulose production from bacteria isolated from soil. *ISRN Biotechnol*. 2013: 985685 (2013)
13. H. Ariffin, N. Abdullah, M.S. Kalsom, Y. Shirai, and M.A. Hassan. Production and characterization of cellulose by *Bacillus pumilus EB3*. *Int. J. Eng Technol*. **3**:47-53 (2006).
14. O.S. Kotchoni, O.O. Shonukan, and W.E. Gachomo. Bacillus pumilus BpCRI 6, a promising candidate for cellulose production under conditions of catabolite repression. *Afr. J. Biotechnol*. **2**: 140-146 (2003).
15. M. Romero, J. Aguado, L. Gonzalez, and M. Ladero Cellulase production by *Neurospora crassa* on wheat straw. *Enz Microb. Technol*. **25**:244-250 (1999).
16. U. Kalsom, M.S., A.B. Ariff, and H.S. Zulkifli. The treatment of palm empty fruit bunch fibre for subsequent use as substrate for cellulose production by chaetomium globosum kunze. *Bioresour. Technol*. **62**: 1-9 (1997).
17. C. Li, Z Yang., R.H.C. Zhang, D. Zhang, S. Chen, and L. Ma. Effect of pH on cellulose production and morphology of Trichoderma reesei and the application in cellulosic material hydrolysis. *J. Biotechnol*. **168**: 470-477 (2013).
18. M.M.H. Khan, S. Ali, A.F. Razi and M.Z. Alzam. Use of fungi for the bioconversion of rice straw into cellulose enzyme. *J. Biotechnol*. 168: 470-477 (2007).
19. H. Chen, Q. He, and L. Liu. Cellulase production from the corn stover fraction based on the organ and tissue. *Biotechnol. Bioproc. Eng*. **16**: 867-874 (2011).
20. M.M. Ekpergin. Preliminary studies of cellulase production by *Acinrtobacter anitratus and Branhamella sp.Afr. J. Biotechnol*. **6**: 28-33 (2007).

Smallholder Plantations Palm Oil Productivity: a Comparative Study in Two Sub-district Areas in Aceh Province and North Sumatera Province

*Mawardati** and *Munawar* Khalil

Department of Agribusiness, Faculty of Agriculture, Universitas Malikussaleh, 24355, Reuleut Main Campus, North Aceh, Aceh Provice, Indonesia

Abstract. This study was conducted on smallholder plantations in two locations: in Sub-district of Kuta Makmur, District of Aceh Utara, Province of Aceh and Sub-district of Kualuh Hulu, District of Labuhanbatu Utara Province of Sumatera Utara in January to March 2017. This study aims to analyze palm oil productivity differences of the smallholder plantations in both districts based on the age group of plant and location. The analytical method used is the analysis of variance (ANOVA). The results of analysis shows that there are differences in palm oil productivity significant smallholder plantations based on research locations. Furthermore, the results describes, palm oil productivity based on age group of plants only in Sub-district of Kualuh Hulu showing the significant, whereas in Sub-district of Kuta Makmur did not significant. This is due to the differences in the level of application of technology in managing palm oil cultivation by smallholders at both sites.

1 Introduction

Palm oil is a very pledge trading commodity. In the future, palm oil will believe not only for producing the downstream industries needed by humans such as cooking oil, butter, soap, cosmetics and others, but it can also be a substitution of petroleum fuels that are currently mostly filled with petroleum. Moreover, petroleum source is not renewable (non-renewable), while palm oil is a source of vegetable oil that can be updated (renewable). The development and expansion of Indonesian palm oil plantation by both large plantations and the community from year to year continues to increase rapidly even very astonishing. The expansion of plantation area the production of palm oil in the form fresh fruit bunch or in the form of Crude Palm Oil (CPO) also significantly increase. Undertaking increase production through area expansion are still necessary as long as the available potential and feasible. However, the most important is increasing productivity and quality of production.

The largest area of palm oil development in Indonesia is in the Island of Sumatera, especially Province of Riau and Sumatera Utara. Palm oil plantations also developed almost in various other provinces in Indonesia, one of the several province is province of Aceh.

* Correspondent Author: mawardati@yahoo.com

Sumatera Utara Province and Aceh Province the two provinces contiguous to natural conditions, such as climate and soil fertility are also not much different. The different types of cultivated plants are common including palm oil. Although the development of palm oil production in Sumatera Utara Province faster than the province of Aceh, but the average productivity of palm oil in both these areas is still relatively low, both at the provincial level and at the district level even at the sub-district level.

The low productivity of palm oil overall is due to the low productivity of palm oil by smallholder plantations. In fact, under optimum conditions, palm oil production can reach 20-25 ton TBS/hectare/year or about 5-6 ton palm oil/hectare/year [1]. Furthermore, [3] added that the production of palm oil crops with seeds from plant isolation method reached between 7 to 9 ton/hectare /year or 32 to 40 ton TBS/hectare/year. The data obtained from related officials in both regions shows that the average productivity in District of Labuhanbatu Utara is 13.03 tons of TBS hectare/ year and Sub-district of Kualuh Hulu, District of Labuhanbatu Utara, Sumatera Utara Province is 12.45 tons TBS/hectare/year [1]. Meanwhile, the average productivity in District of Aceh Utara is 9.50 tons of TBS/hectare/year and Kuta Makmur Sub-district, Aceh Utara District, Aceh Province is 4.59 tons TBS/hectare/year [2].

The result shows that the average productivity of palm oil smallholder plantations in Sub-district of Kuta Makmur, Aceh Province is lower compared to Sub-district of Kualuh Hulu, Sumatera Utara Province. The high and low production of palm oil in research areas are depends on maintenance including the application of technology to different age groups of plants. This is due to the differences productivity of palm oil in different age groups of plants. The research results of [5] showed that at 3-7 years plants are the average production of TBS per hectare continued to increase, and at the age of 8-13 years plants average production was stable and highest while the age crops above 13 years there is a decrease in the production of TBS at various levels of land suitability. The reality in the research locations shows that there are various composition of plant age in palm oil smallholder plantations in both Aceh Province and Sumatera Utara Province. The purpose of this research is to compare the gap of palm oil productivity of smallholder plantations based on research location and age group of plants.

2 Methodology

This research was conducted by using survey method. The population of this research is all farmers of palm oil smallholder plantations in research area. The location of this research is purposive, in Kuta Makmur Sub-district, Aceh Utara District, Aceh Province and Sub-district of Kualuh Hulu, Labuhanbatu Utara District, Sumatera Utara Province with consideration of both districts have the composition of palm oil plantations with various age groups. Sampling was done by grouping the palm oil farmers based on the age of the plants in each selected clusters as the sampling frame. The sample was selected by proportionally in age group of plants. The number of sample in this study were 84 respondents from Sub-district of Kuta Makmur and 95 respondents from Sub-district of Kualuh Hulu. Overall, there were 179 farmers as sample.

The analysis technique used Analysis Of Variance (Anova), that is to test the differences of productivity based on location and age group of plants with general model as follows:

$$Y_{ij} = \mu + t_i + \varepsilon_{ij}$$

Description: Y = TBS Productivity, μ = overall average, t_i = the mean of the i group, I = the i-th group, j = jth sample, ε_{ij} = Error term

3 Result and discussion

3.1 Productivity of palm oil smallholder plantations based on research location

The results show the average productivity of palm oil plantations based on location are presented in Table 1.

Table 1. Productivity of palm oil (ton tbs / hectare) based on research location year 2016.

Size	Sub-district of Kuta Makmur, District of Aceh Utara, Aceh Province	Sub-district of Kualuh Hulu Utara, Labuhanbatu District, Sumatera Utara Province
Average value	9.241,70	16.764,76
Maximum Value	13.900,00	29.408,00
Minimum Value	4.000,00	6.344,00
Standard Deviation	1.813,37	6.098,43

Table 2 shows the differences of average productivity of palm oil smallholder plantations in the research location. The average productivity of palm oil in Sub-district of Kualuh Hulu is higher than Sub-district of Kuta Makmur. The result of *analysis of variance* (Anova) shows the value of F-test 118.495 with probability (sig.) 0.000. This means that there is a significant difference in productivity between Kuta Makmur Sub-district, Aceh Utara District, Aceh Province with Kualu Hulu Sub-district, Labuhanbatu Utara District, Sumatera Utara Province. In other words, the productivity of palm oil in Sub-district of Kualuh Hulu is higher compared to Kuta Makmur Subdistrict, statistically also has significant result.

The low productivity of palm oil in Sub-district of Kuta Makmur is related to plant treatment or treatment of plant technical cultivate such as the utilization of production factors and application of technology under the Kualuh Hulu Sub-district. In addition, the productivity of palm oil crops is also determined by different land characteristics in each area of development. The productivity of each area is different according to the land suitability class [4].

3.2 Productivity by total age group and total productivity

According to [5] the average weight of TBS respondents to the age of the plant and the suitability class of the land. In the plant age 3 to 7 years the average production of TBS per hectare continues to increase, and at 8 to 13 years of age the average production is stable and achieves the highest productivity whereas the age of the plant above 13 years decreases the production of TBS at various levels of conformity land. The high productivity of palm oil in the form of TBS in a location iare influenced by the composition of plant age in the area. More on palm oil productivity by plant age and total productivity in the study area is shown in Table 2.

Table 2. Productivity of palm oil at various age group of plants in sub-district of Kuta Makmur, Aceh Province, year 2017.

Size	Productivity of TBS (ton / hectare) Age 3 -7 yr	Productivity of TBS (ton /hectare) Age 8-13 yr	Productivity of TBS (ton / hectare) Age> 13 yr

Average value	9,17	9,33	8,89
Maximum Value	12,00	12,63	15,00
Minimum Value	7,20	7,50	4,00
Standard Deviation	1,36	1,23	2,29

Table 2 shows that the average productivity for the 8-13 year age group is much lower than that of [5] that in the lowest grade of land suitability the average productivity of TBS in the plant age group of 8-13 years can reach an average of 22 ton TBS/hectare/year. This shows that the productivity of palm oil in research area is still very low, not only in the age group of young plants and old plants but also occurs in the age group of adult plants.

Table 2 also shows that although there is a differences in average productivity between plant age groups, there is no high variation. The results of the analysis of the productivity differences between age groups of plants by using *analysis of variance* (ANOVA) F-test values obtained at 0 .460 with probability (sig.) of 0, 633 . There is no significant differences in productivity between age groups of plants in various age groups of plants in Sub-district of Kuta Makmur Aceh Utara District Aceh Province.

The non-productivity of palm oil according to the age group of plants in the study area may be caused by plant maintenance/ maintenance factors, including the application of technology and the knowledge level of farmers. The low knowledge of farmers on palm oil cultivation techniques is the most dominant factor, because it impacts on its ability in plant maintenance [6].

In addition, the significance of productivity by age group is also due to the high productivity differences within the same age group. The high differences in productivity within the group, statistically causing non-significant productivity between age groups of plants.

According to [7] the dosage of fertilization on oil palm plants yields different according to plant age. However, based on observations and interviews with farmers found that most of the respondents did not know how many fertilizer doses should be given according to the age of their palm oil crops. According to the pruning of leaves/ repletion, although some of the farmers respondents do this method, but generally they do not understand exactly how many optimal leaves that should be left in the trees that correspond to the age of their plants. Overall, productivity of palm oil in the study area is also low.

Table 3. Average productivity of palm oil at various age group of plants in sub-district Kualuh Hulu, Sumatera Utara Province, year 2017.

Size	Productivity of TBS (ton / hectare) Age 4-7 yr	Productivity of TBS (ton / hectare) Age 8-13 yr	Productivity of TBS (ton / hectare) Age> 13 yr
Average value	9,84	15,99	21,00
Maximum Value	14,91	26,95	29,41
Minimum Value	6,34	6,94	11,50
Standard Deviation	5,59	3,87	7,34

Table 3 shows that the average productivity with the highest plant age with the highest productivity is found in the age group> 13 years. In addition, there is a high productivity variation between the highest productivity with the lowest productivity and between the age groups of the plant. This condition is also reinforced by the result of analysis of differences of average productivity between age group of plant with value of F-test equal to 29,22 with probability (sig.) Equal to 0, 000 . This shows that there are significant differences in palm oil productivity between plant age groups in smallholder plantations in Sub-district of Kualuh Hulu, Labuhanbatu Utara District, Sumatera Utara Province.

Theoretically, differences in plant age group will cause different productivity of *fresh fruit bunch* from palm oil crops. However, these result are not fully consistent with the findings of [5] which conclude in addition to the differences in productivity between the age groups of oil palm trees, the highest productivity was found in the age group of 8-13 years.

Based on the results of the analysis on the two research areas showed that the average productivity of palm oil in Sub-district of Kualuh Hulu, Sumatera Utara Province is higher than the average productivity of palm oil in Sub-district of Kuta Makmur, Aceh Province. Nevertheless, the average productivity is still below the average of optimum productivity.

4 Conclusion

There is significant difference of average productivity of palm oil plantation between Kuta Makmur Sub-district, Aceh Utara District, Aceh Province and Sub-district of Kualuh Hulu, Labuhanbatu Utara District, Sumatera Utara Province. Overall, the average of palm oil productivity in Sub-district of Kualuh Hulu, Sumatera Utara Province is higher than Kuta Makmur Sub-district, Aceh Province. There is an average difference of productivity based on plant age group in Sub-district of Kualuh Hulu, Sumatera Utara Province. While in Kuta Makmur Sub-district of Aceh Province. Besides the lower average productivity there is also no significant difference based on the age group of the plants caused by the lack of knowledge and ability of farmers in the treatment of crops, especially the level of application of technology in various age groups of palm oil crops.

The maximum productivity of palm oil plantations overall is strongly influenced by maximum productivity in each age group. To achieve the maximum productivity as much as possible local government especially related offices in order to improve the ability of farmers through counseling and cultivation techniques, especially the application of appropriate technology advice for each age group of plants. Further research is needed to find out more about the cause of productivity gaps either based on age group of plants or by region / location by including soil type factor and climate factor in analysis.

References

1. Directorate General of Plantation. Land Area, Production, and Productivity Smallholder Plantations. Sumatera Utara Province. Jakarta (2015)
2. BPS. *Aceh Utara Dalam Angka.* Biro Pusat Statistik Kabupaten Aceh Utara (2015)
3. Fauzi, Y, Yustina E.W, Iman. S Rudi. H. *Kelapa Sawit. Budidaya Pemanfaaatan Hasil dan Limbah, Analisis Usaha dan Pemasaran.* (Penebar Swadaya. Jakarta, 2003)
4. Harahap, I. Y, Pangaribuan, Y and Listia, E. Initial Gaze Growth and Productivity Potential Various Varieties of Palm Oil Planted with High Population. *Jurnal Penelitian Kelapa Sawit* **14**(1) : 1-10 (2006)
5. Mangoensoekarjo.S and Semangun.H. *Manajemen Agrobisnis Kelapa Sawit.* (Gadjah Mada University Press, Yogyakarta, 2005)
6. Mawardati. *Dibalik Rimbunnya Kelapa Sawit Rakyat.* (Unpad Press, Bandun, 2010)
7. Pahan. I, *Panduan Lengkap Kalapa Sawit. Manajemen Agribisnis dari Hulu Hingga Hilir*. (PT. Penebar Swadaya, Jakarta, 2006)

Monostearin effects on the formation of precipitate in palm oil biodiesel and petroleum diesel blends with various storage temperature

Mohamad Aufar Ghaizani, *Imam* Abdurrosyid, *Imam* Paryanto, and *Misri* Gozan*

Department of Chemical Engineering, Faculty of Engineering, Universitas Indonesia, Depok 16424, Indonesia

Abstract. Biodiesel is one of the renewable energy forms that is on demand in Indonesia. In the biodiesel production process, impurities such as Saturated Monoglycerides (SMG) in the form of monostearin that precipitate at low temperature are commonly formed. This is caused by high Final Melting Temperature (FMT) of SMG. Formation of solid deposits when it reaches a temperature above Cloud Point (CP) is then unavoidable under these conditions. The use of palm oil biodiesel and petroleum diesel blends (BXX) with biodiesel blend ratio of 10% (B10), 20% (B20) and 30% (B30) accelerates precipation process which renders clogging on fuel filters. These works examined the effect monostearin content and temperature on the precipitation rate. Investigation is carried out at 15°C, 20°C, 25°C, and room temperature (30-33°C) with varying content of monostearin (0.4%, 0.7%, and 0.9%). Early results show that for 0.4% monostearin content in 100 ml B20, the amount of precipitate formed at temperature 15 oC was 31.3 mg. This value was higher than that at room temperature (6.3 mg) after 2 weeks storage time. This value was higher compared to B20 (21.5 mg). This indicates that as the biodiesel ratio in BXX become higher, the amount of FAME will increase which effect monostearin solubility in BXX at low temperature.

1 Introduction

The energy crisis is the biggest problem we faced today, where the energy crisis will bring a great impact on the community. Demand and needs of the energy alone are increasing each day while energy resources declining and soon it will run out. This suggests that renewable energy should be of more concern because even though fossil fuels still be the primary option for energy in the world but eventually increase CO2 levels and cause greenhouse effects. To maintain environmental and the sustainability of energy resources, alternatives of fuel source is needed.

Biodiesel is a liquid-form fuel produced from biomass from various fields of agriculture, forest products, and biodegradable factory waste [3]. Biodiesel consists of long chains of methyl or ethyl esters generally produced by transesterification of short chain monohydric alcohols. In

* Corresponding Author; misrigozan@gmail.com

biodiesel production process formed impurities such as saturated monoglycerides (SMG) and precipitate at low temperature operating conditions. This is due to high final melting temperature (FMT) or melting point of SMG and will form solid deposits above cloud point (CP). The use of biodiesel mixed with petroleum diesel will enhance precipitation at low temperatures and will cause clogging of fuel filters in a vehicle [4].

Bondioli et al (2008) examined the occurence of precipitation in FAME production of palm oil and soybean oil from the final abstertion and formed insoluble white solids. Tang et al (2008) examined solid residue content from SME (soybean methyl ester), FAME-cotton seed, and animal fat FAME and mixed with USLD 2 using GC and FT-IR. The results of FT-IR and GC analysis shows that precipitate on SME biodiesel mostly contain Free Steryl Glucoside (FStG). In addtion, the precipitates formed in animal fat biodiesel mostly contain saturated monoglycerides in the form of monopalmitin and monostearin while in cotton seed biodiesel the precipitate mostly contain FStG and saturated monoglycerides.

The soluble precipitate will again become impurities [2]. Impurities, either derived from raw materials or from other components resulted in biodiesel production, in small concentrations, may affect the performance of biodiesel at low temperatures. In the biodiesel production process often found residue produced by imperfect biodiesel conversion and appears in the final product of biodiesel. Insoluble impurities in biodiesel become important parameter because it will cause clogging in fuel filter in a vehicle [4].

Currently there are addiotional parameters on ASTM in the form of Cold Soaking Filtration Test (CSFT) and Filter Blocking Test (FBT) which will become mandatory for biodiesel that will be blend with petroleum diesel. FBT is a dimensionless unit that defines filter blocking or plugging tendency of particulate, taken from the volume needed to achieve pressure difference of 105 kPa or 300 ml of volume at a velocity of 20 ml/min. CSFT combine cooling temperature and time which in this case 4°C and 16 hours for ASTM (cold soaking).

2 Materials and Methods

2.1 Fuels and Monostearin

Biodiesel used in this study was bought from palm oil biodiesel from Wilmar Nabati Indonesia CO., meanwhile the petroleum diesel from Shell Indonesia that has aromatic content 2.54%. These biodiesel and petroleum diesel were characterized by using Gas Chromatography/Mass Spectrometry as shown in the Table 1.

Table 1. Biodiesel and Petroleum Diesel Characterization

Parameter	Units	WNI	Shell
Density at 15°C	kg/m^3	856.8	842.5
Cloud Point	°C	n.a	1.7
FAME Content	%	99.5	n.a
Oxidation Stability	minute	601.5	n.a
Aromatic Content	%	n.a	2.54
Monoglycerides Content	% vol	0.4	n.a
Total Glycerol	% vol	0.1	n.a
Free Glycerol	% vol	0.003	n.a

The monostearin is known to have high influence in the precipitate formation [4]. Monostearin used in this study were produced by TCI with 63.4% purity. Monostearin were added to this TCI-biodiesel, so the total volumetric content of monostearin became 0.4%,

0.71%, and 0.92% (v/v). Ratio of blending in biodiesel and petroleum diesel also varied and characterized with methods listed in the Table 2.

Table 2. Characterization of BXX WNI-S

Parameters	Units	Results			Standard Solar 48		Methods
		B10	B20	B30	Min	Max	
Density at 15°C	kg/m^3	848.7	841.1	849.1	815	870	ASTM D4052
Cloud Point	°C	3.4	4.3	4.7	-	18	ASTM D5773
Sediment	% mass	0	0	0	-	0.6	ASTM D473
FAME Content	% vol	10.1	19.9	29.9	-	20	ASTM D664
Oxidation Stability Rancimat Methods	Hour	44.22	36.87	36.59	35	-	EN 15751

The results of characterization of biodiesel-petroleum diesel blends for blending ratio B10, B20, and B30 without monostearin content modification in Table 2 shows that all cloud points are below the lowest temperature used in this study (15°C) with the highest value came from B30 blend, that is 4.7°C. This indicated that all the blends were qualified to be used to quantify the formation of precipitate above the cloud points.

2.2 Beaker Test

In ASTM D7501 (Standard Test Method for Determination of Fuel Filter Blocking Potential of Biodiesel (B100) Blend Stock by Cold Soak Filtration Test (CSFT)), 300 ml of biodiesel (B100) stored at temperature 4.5 ± 0.5°C for 16 hours, warmed up to 25 ± 1°C (standard room temperature), then filtered with vacuum filter through 0.7 μm fiber glass filter on 70-85 kPa vacuum level. ASTM D7501 method is inefficient when used for large samples and it doesn't show the effect of different storage temperature to the formation of a precipitate. In this study, the temperature and soaking time were modified. 100 ml of BXX (Biodiesel and Petroleum Diesel Blend) stored in cooling chamber with a temperature of 15°C, 20°C, 25°C, and room temperature (range from 30°C to 33°C) for 2 weeks. These variation is based on real operational condition and storage of fuel in lower temperature area.

3 Results

Increasing in blending ratio will results in cloud point rise. The cloud point of palm oil biodiesel is higher than petroleum diesel. This study used variation of blending ratio, that was B10, B20, and B30. The results are shows in Table 3.

Table 3. Precipitation Formed from Beaker Test (mg/ml)

BXX - Monostearin Content	Temperature			
	15	20	25	Room Temperature
B10 - 0.4%	0.192	0.141	0.100	0.070
B10 - 0.71%	0.372	0.238	0.111	0.109
B10 - 0.94%	0.540	0.445	0.277	0.279
B20 - 0.4%	0.313	0.215	0.182	0.063
B20 - 0.71%	0.853	0.481	0.226	0.254

B20 - 0.94%	1.302	1.109	0.726	0.627
B30 - 0.4%	0.358	0.282	0.185	0.127
B30 - 0.71%	0.853	0.481	0.226	0.254
B30 - 0.94%	2.187	1.711	1.096	1.049

The results in Table 3 show that the effect of blending ratio at temperature 15°C, 20°C, and 25°C were significant compared to the effect of blending ratio at room temperature (±30°C). Precipitate formed in B20 was significantly lower at 15°C. The difference on the formation of precipitate in B20 and B30 was not significant above temperature 15°C. The results also showed that precipitate was more likely to form as the percentage of monostearin increase. In B10, the effect of temperature on the precipitate formation was not significant. The temperature affected precipitation more on biodiesel with higher blending ratio.

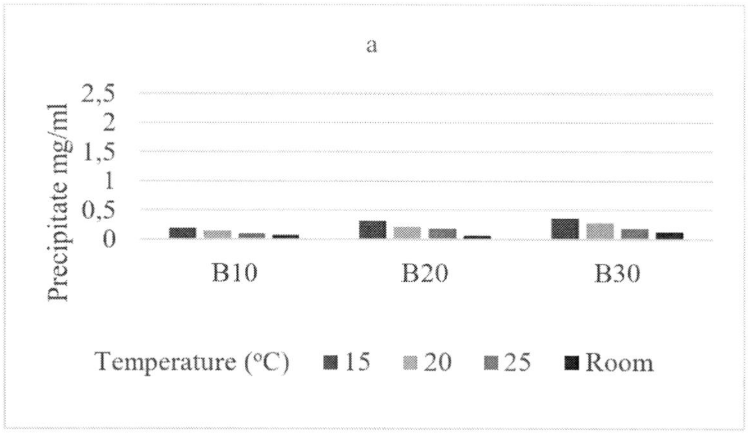

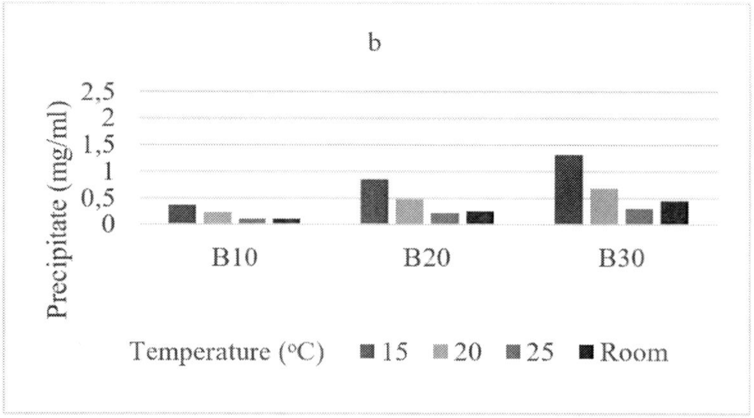

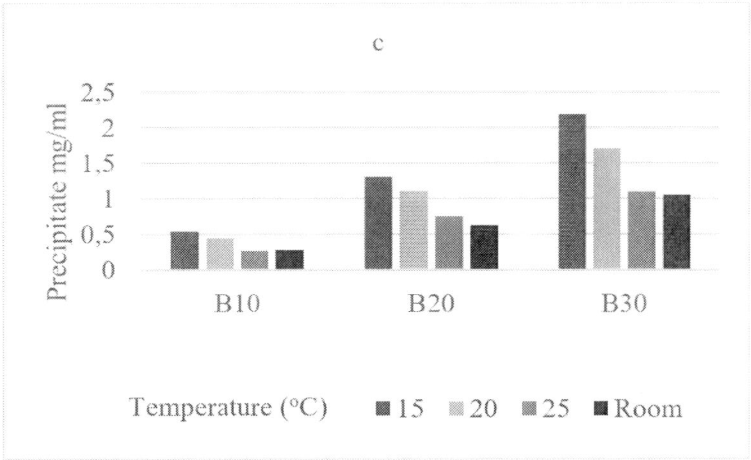

Fig. 1. Comparison of precipitate formed in BXX (Biodiesel and Petroleum Diesel Blend) resulted from beaker test with (a) 0.4% Monostearin Content, (b) 0.71% Monostearin Content, and (c) 0.92% Monostearin Content

JAMA (Japan Automobiles Manufacturers Association) conducted soaking test on B20 with high and low monostearin content on various temperature (5°C, 15°C, 20°C, and 25°C). The result shows that the precipitation in low monostearin content was not found at all temperatures. Meanwhile, in high monostearin content the precipitate was formed at temperature 5°C, 15°C, and 20°C only (GAIKINDO, 2015). Those results corresponding with this study where on cold soaking at temperature above 20°C and blending ratio below B20 did not affect precipitate formation significantly.

4 Conclusion

This study has shown that monostearin contents and temperature affected the formation of precipitate on biodiesel-petroleum diesel blends after 2 weeks storage period time. The lower temperature and higher monostearin content produced more precipitate. The higher percentage of B100 in blending ratio also promotes the formation of precipitate.

Acknowledgments

Authors are grateful for the financial support provided by *Hibah Publikasi Internasional Terindeks untuk Tugas Akhir Mahasiswa* (PITTA) through Universitas Indonesia and research facilities provided by Univesitas Indonesia (2477/UN2.R3.1/HKP.05.00/2018).

References

1. P. Bondioli, N. Cortesi and C. Mariani, Identification and quantification of steryl glucosides in biodiesel, *European Journal of Lipid Science and Technology* 110(2):120–6, (2008)
2. G. Chupka, L. Fouts, J. Lennon, T. Alleman, D. Daniels, R. Mccormick, Saturated monoglyceride effects on low-temperature performance of biodiesel blends. *Fuel Processing Technology*; 118:302–9, (2014)
3. A. Dufey, *Biofuels Production, Trade, and Sustainable Development: Emerging Issues. London*, (International Institute for Environment and Development, 2006)

4. RO. Dunn, Effects of minor constituents on cold flow properties and performance of biodiesel. *Progress in Energy and Combustion Science*;35(6):481–9. (2009)
5. GAIKINDO. Bio Diesel B20 Movement (Technical Concern). (2015)
6. H. Tang, S. Salley, K. Simonng, Fuel properties and precipitate formation at low temperature in soy-, cottonseed-, and poultry fat-based biodiesel blends. *Fuel* 87(13-14):3006–17, (2008)

Microsatellite analysis of genetic diversity and structure of *Bruguiera gymnorrhiza* and *Kandelia obovate*

Mohammad Basyuni[1,2*], *Shigeyuki* Baba[3], and *Hirosuke* Oku[4]

[1]Department of Forestry, Faculty of Forestry, Universitas Sumatera Utara, Jl. Tri Dharma Ujung No. 1 Medan, North Sumatera 20155, Indonesia
[2]Mangrove and Bio-Resources Group, Center of Excellence for Natural Resources Based Technology, Universitas Sumatera Utara, Medan North Sumatera 20155, Indonesia
[3]International Society for Mangrove Ecosystems, Faculty of Agriculture, University of the Ryukyus, 1 Senbaru, Nishihara, Okinawa 903-0213, Japan
[4]Molecular Biotechnology Group, Tropical Biosphere Research Center, University of the Ryukyus, 1 Senbaru, Nishihara, Okinawa 903-0213, Japan

Abstract. Microsatellite loci were used for estimating genetic diversity and structure for three populations of *B. gymnorrhiza* and *K. obovata* (Rhizophoracea) in Okinawa, Japan. Thirty propagules of individual samples representing the population of both species were genotyped at five microsatellites. The level of observed heterozygosity (H_O) was observed for several population, overall loci, ranged 0.422-0.800 with an average 0.627 for *B. gymnorrhiza* and 0.477-0.822 with an average 0.665 for *K. obovata*, indicating both species had relatively low genetic diversity. Both species showed low levels of allelic diversity, 3-5 and 3-5 alleles per locus, respectively. Gene diversitywas also maintained within populations (H_S: 0.741 and 0.954). Additionally, an analysis of molecular variance (AMOVA) based on the immeasurable alleles model (*F*-statistics), for *B. gymnorrhiza* and *K. obovata* found that most of the variation resided within individuals in the total populations, i.e. 79.78 % and 69.90 % respectively, and among individuals within populations, i.e.14.30 % and 27.95 % respectively. There was little variation between populations, i.e. 5.92 % and 2.15 % for *B. gymnorrhiza* and *K. obovata*, respectively. The high-level genetic differentiation within individuals and populations both species may be due to the geographic range of the species, mating system, and environmental factors.

1 Introduction

Mangroves are typically tropical and sub-tropical coastal vegetation. In Japan, natural mangrove forests mainly distribute around Ryukyu Islands group, growing well in Okinawa,

* Correspondent Author: m.basyuni@usu.ac.id

Iriomote Island, Ishigaki Island, and Amami Island in its northern Japanese limit on southern Kyushu [1]. The mangrove area at Okukubi River and Katabaru River, Okinawa are inhabited by four significant species, including *Bruguiera gymnorrhiza*, *Kandelia obovata*, *Rhizophora stylosa*, and *Lumnitera racemosa*. *B. gymnorrhiza* is denser in population, growing taller, and preferring to grow under covered conditions, while *K. obovata* and *R. stylosa* need more sunlight [2].

One of the important characterizations of mangroves Rhizophoracea is the production of viviparous offspring. Their propagules develop into seedling while they are still attached to the mother tree. In viviparous mangrove species, the embryo ruptures the pericarp, grows beyond it before dispersals such as *B. gymnorrhiza, K. obovata* and many other species of Rhizophoracea [3-4]. However, the effect of this important dispersal mechanism on the level genetic variation within and among populations is not much concerned.

In addition, the rate of gene flow through seed dispersal of a species also plays an essential role in determining population genetic structure. The restricted mangroves area in Japan, therefore the conservation and management resources have become an urgent task. Reliable information on the distribution of population genetic structure and mating system of mangrove tree species is essential to establish an active network of conservation populations [5].

Therefore, the present study was aimed to investigate population genetic diversity and structure of two mangrove species, *B. gymnorrhiza* and *K. obovata*. Information of genetic variation parameters will not only clarify the pattern of population genetic structure but also contribute to its protection, management, and utlization of mangrove tree genetic resources.

2 Materials and methods

2.1 Study site and sample collection

This study site was carried out in Okukubi River and Katabaru River, Okinawa, Japan. Thirty propagules per tree as materials for progeny arrays was used. While 30 propagules each species representing as population (Okukubi left side, Okukubi right side, and Katabaru) as shown in Fig. 1. Collected samples were kept at -80 °C until required for DNA extraction.

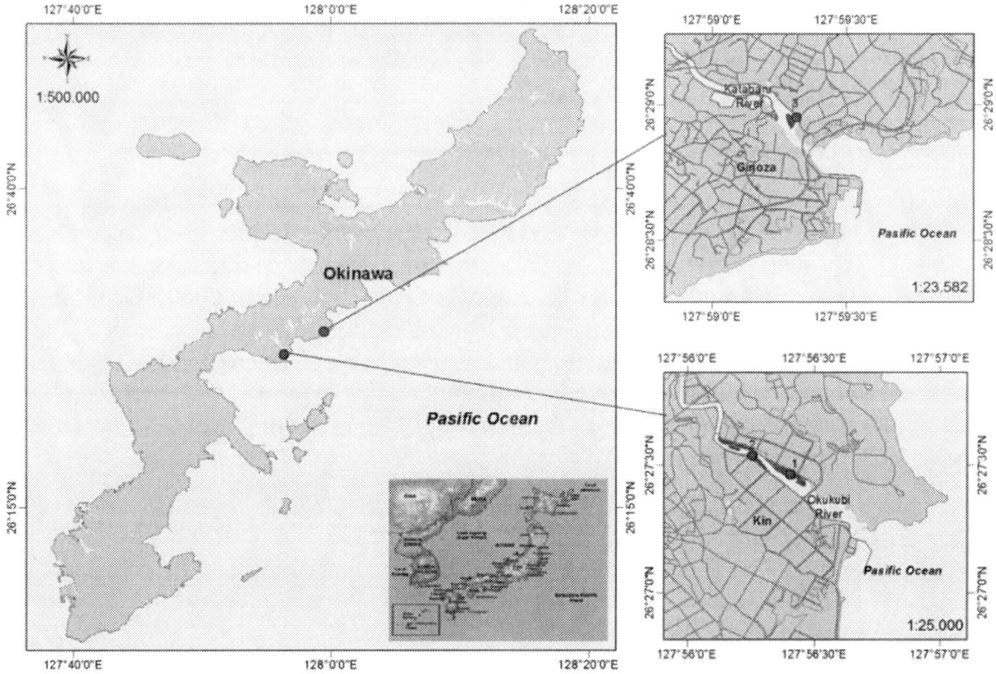

Fig. 1. Study site of Okukubi River and Katabaru River, Okinawa, Japan

2.2 DNA extraction and microsatellite analysis

Total DNA was extracted from the propagules as progeny arrays using modified CTAB procedure [6]. Five primer pairs, MSBgy025, MSBgy027, MSBgy030, MSBgy031 and MSBgy188 [7] were used in *B. gymnorrhiza* as well as five primer pairs (Kcan004, Kcan005, Kcan009, Kcan011, and Kcan034) of *K. obovata* according to Sugaya et al. [8]. PCR mixtures condition (10 µl) contained 10X Ex Taq, 1.5mM $MgCl_2$, 0.2 mM of each dNTP, 0.1 µM of each primer (fluorescent forward and reverse), 10 ng genomic DNA and 1.0 U of Taq polymerase. PCR amplification was performed for 3 min at 94°C for, continued by 35 cycles of 1 min at 94°C, 30 s at the improved annealing temperature and 30 s 72°C, with a final extension of 5 min at 72°C using a PCR System Model PC-806 (ASTEC). The amplicons were detached using an ABI PRISMR 310 Genetic Analyzer and genotypes were analyzed by GeneScanTM and ABI PRISMTM Genotyper software version 2.0. (PE Applied Biosystems).

2.3 Data Analysis

2.3.3 Genetic diversity

Genetic diversity overall populations and all microsatellite loci were estimated as the mean number of alleles per locus (A), mean expected heterozygosity (H_E), mean observed heterozygosity (H_O) and the inbreeding coefficient (F_{IS}) using GDA, version 1.1 [9]. Genetic differentiation between populations was calculated according to Nei's gene diversity within a population (HS), gene diversity between populations (D_{ST}), the coefficient of gene diversity (G_{ST}) [10], and calculating of allelic richness (A_R) implemented by FSTAT version 2.9.3 [11].

F-statistics (F_{IS}, F_{IT}, and F_{ST}) were computed every locus. The possibility of the F-statistics being more significant than zero was detected by bootstrap analysis using 1000 repeated, with a 99% confidence distance, using GDA, version 1.1 [9].

2.3.4 Genetic structure

Determination of genetic structure was done using the unbounded alleles model [12]. Analysis of molecular variance (AMOVA) [13] to quantity variance components and their importance between populations, among individuals within populations, and within populations were computed using the ARLEQUIN version 2.000 [14], in which significance levels for the overall values were decided after 1023 permutations.

To estimate gene flow, multilocus calculates the mean effective number of migrants (Nm) between population using private allele methods of Slatkin (1985) and were adjusted for sample size as provided in Barton & Slatkin [15], implemented by program GENEPOP version 3.1b [16].

3 Results and discussion

3.1 Genetic diversity

The microsatellite loci of *B. gymnorrhiza* and *K. obovata* were lowly polymorphic with 3-5 and 3-5 alleles per locus, with five microsatellite loci respectively (Table 1). In *B. gymnorrhiza*, allele sizes ranged from 200 to 257 reappearances for locus MSBgy025, from 50 to 79 repeats for MSBgy027, from 50 to 70 for locus MSBgy030, from 160 to 200 reiterations for locus MSBgy031, and from 139 to 167 replications for locus MSBgy188. At the species level, the mean observed heterozygosity (H_O), mean expected heterozygosity (H_E), mean gene diversity within the population (H_S), and mean coefficient of gene differentiation (G_{ST}) were 0.627, 0.744, 0.741, and 0.045 respectively (Table 1). At the population, the mean observed heterozygosity (H_O), mean expected heterozygosity (H_E) and mean inbreeding coefficient (F_{IS}) was 0.627, 0.926, and 0.144 respectively (Table 2).

In *K. obovata* allele sizes diverged from 390 to 480 repeats for locus Kcan004, from 350 to 400 recurrences for Kcan005, from 285 to 350 repeats for locus Kcan009, from 200 to 246 replications for locus Kcan011, and from 75 to 170 duplications for locus Kcan034. At the species level, the mean observed heterozygosity (H_O), mean expected heterozygosity (H_E), mean gene diversity within the population (H_S), and mean coefficient of gene differentiation (G_{ST}) were 0.665, 0.966, 0.954 and 0.014 respectively (Table 1). At the population, the average observed heterozygosity (H_O), average expected heterozygosity (H_E), and mean inbreeding coefficient (FIS) was 0.665, 0.966, and 0.298, correspondingly.

Table 1. Profile of microsatellite loci overall populations in (a) offspring *B. gymnorrhiza* and (b) offspring *K. obovata*

Locus	A	A_R	H_E	H_O	H_S	D_{ST}	G_{ST}
(a) *B. gymnorrhiza*							
MSBgy025	3	2.89	0.707	0.678	0.710	0.045	0.064
MSBgy027	5	3.56	0.790	0.433**	0.759	0.033	0.041
MSBgy030	4	3.07	0.829	0.422**	0.795	0.036	0.043
MSBgy031	4	3.50	0.791	0.800	0.740	0.053	0.067
MSBgy188	4	3.66	0.755	0.800	0.748	0.008	0.010
Mean	4	3.34	0.744	0.627	0.741	0.035	0.045

Locus	A	A_R	H_E	H_O	H_S	D_{ST}	G_{ST}
(b) *K. obovata*							
Kcan004	3	2.11	0.985	0.759	0.980	0.005	0.005
Kcan005	4	2.91	0.943	0.545**	0.933	0.013	0.014
Kcan009	5	3.27	0.976	0.822	0.953	0.023	0.023
Kcan011	4	3.90	0.971	0.722	0.969	0.005	0.005
Kcan034	5	3.47	0.952	0.477**	0.932	0.022	0.023
Mean	4.2	3..3	0.966	0.665	0.954	0.014	0.014

A: number of alleles detected, A_R: allelic richness, H_E: expected heterozygosity, H_O: observed heterozygosity, H_S: gene diversity within the population, D_{ST}: gene diversity between populations, G_{ST}: coefficient of gene differentiation. The deviation from Hardy-Weinberg equilibrium was calculated using a χ test, ** $P<0.01$

Table 2. Descriptive statistics over all loci for three populations of *B. gymnorrhiza* and *K. obovata*

Population	*N*	*A*	H_E	H_O	F_{IS}	Outcrossing rate
(a) *B. gymnorrhiza*						
Katabaru	30	9.2	0.754	0.673	0.108	0.805
Okukubi left	30	5.8	0.671	0.680	-0.014	1.028
Okukubi right	30	9.6	0.790	0.527	0.337*	0.495
Mean	30	8.2	0.738	0.627	0.144	0.776
(b) *K. obovata*						
Katabaru	29.6	26.6	0.931	0.704	0.241	0.612
Okukubi left	29.4	24.8	0.955	0.618	0.351*	0.480
Okukubi right	29.6	30.6	0.958	0.673	0.302	0.536
Mean	29.5	27.3	0.948	0.665	0.298	0.543

N: sample size over all loci, *A*: number of alleles detected, H_E: expected heterozygosity, H_O: observed heterozygosity, F_{IS}: inbreeding coefficient, [a] [31]. The deviation from Hardy-Weinberg equilibrium was assessed using a χ test, *$P<0.1$

Both species of the three populations, one involved critical departures from Hardy-Weinberg equilibrium (HWE); Which correspond to that population with substantial levels of inbreeding (F_{IS}) (Table 3). In each case, deviations from HWE were due to homozygote overabundance.

Table 3. Estimates of *F*-statistics for each locus and over all loci all populations. *F*- statistics bootstrap analysis obtains bootstrap confidence intervals *F*-statistic, using 1000 replicates and 99 % confidence interval (CI)

Locus	F_{IS}	F_{IT}	F_{ST}
(a) *B. gymnorrhiza*			
MSBgy025	-0.026	0.071	0.094
MSBgy027	0.429	0.464	0.061
MSBgy030	0.469	0.502	0.063
MSBgy031	-0.087	0.021	0.099
MSBgy188	-0.069	-0.056	0.012
Overall loci	0.153	0.209	0.066
Bootstrap CI			
Upper	0.461	0.495	0.097
Lower	-0.076	-0.031	0.031
(b) *K. obovata*			
Kcan004	0.226	0.288	0.008
Kcan005	0.415	0.427	0.021
Kcan009	0.138	0.168	0.035
Kcan011	0.232	0.238	0.007
Kcan034	0.488	0.506	0.034

Locus	F_{IS}	F_{IT}	F_{ST}
Overall loci	0.298	0.313	0.021
Bootstrap CI			
Upper	0.475	0.459	0.035
Lower	0.193	0.174	0.008

F_{IS}: inbreeding coefficient, F_{IT}: overall inbreeding coefficient, F_{ST}: fixation index

3.2 Genetic structure

To measure the degree of population distinction the infinite alleles model (F_{ST}) was used. The total value of genetic differentiation for this model resulted different values and significantly different from zero ($P<0.1$): F_{ST} *B. gymnorrhiza* and *K. obovata* were 0.066 and 0.021 respectively, indicates significant structure population exist within populations. The overall values of F_{IS} both were 0.153 and 0.298 respectively, suggesting little to no breeding. The overall values of F_{IT} both were 0.209 and 0.313 respectively (Table 3), indicating a moderately high correlation between alleles within individuals relative to the allelic array of the total populations.

For F_{ST}, the greater of all pairwise comparisons were significantly higher than zero ($P<$ 0.005). Using Wright's manual [17], for F_{ST} all populations pairs in *K. obovata* and one population pair in *B. gymnorrhiza* in the limit 0.00-0.05 and two population pairs in *B. gymnorrhiza* in the range 0.05-0.15 (Table 3), suggesting moderate and little genetic diversity.

Table 4. Hierarchical analysis of molecular variance (AMOVA) of three populations of *B. gymnorrhiza* and *K. obovata*

Source of variation	d.f.	Sum of squares	Variance component		P
			Absolute	%	
B. gymnorrhiza					
Among populations	2	18.909	0.11774		<0.001
Among individuals				5.92	
Within populations	90	193.903	0.28423		<0.001
Within individuals	93	147.500	1.58602	14.30	<0.001
Total	185	360.312	1.98799	79.78	
K. obovata					
Among populations	2	12.033	0.05079		<0.001
Among individuals				2.15	
Within populations	87	258.350	0.65977		<0.001
Within individuals	90	148.500	1.65000	27.95	<0.001
Total	179	148.883	2.36056	69.90	

d.f.: degrees of freedom, P: P-value

Both species of these populations also had to estimate of the number migrants (Nm)>1, indicates some gene flow. For *B. gymnorrhiza* predicts of the number of migrants between Katabaru and Okukubi right, Katabaru and Okukubi left, Okukubi right and Okukubi left were 2.659, 1.788, and 1.189 respectively. Whereas for *K. obovata* estimates of the number of migrants between Katabaru and Okukubi right, Katabaru and Okukubi left, Okukubi right and Okukubi left were 5.128, 3,486, and 3.999 respectively. Overall populations and loci, both species the number of migrants using the private alleles method based to Barton and Slatkin [15] were 1.826 and 4.681 respectively, indicating relatively high of private alleles in these populations with the mean frequency of private alleles p(1) = 0.049 and

0.029 respectively. It is noteworthy, however, that these values of *N*m correspond to a historical average of the number of migrants per generation.

The genetic structure was also tested by analysis of molecular variance (AMOVA) using on the infinite alleles model (Table 4), both species showed that most variation was among individuals within population (14.30% and 27.35% respectively) and within individuals in the total populations (79.78% and 69.90% respectively), indicating significant genetic differentiation within individuals. There was littlevariation between populations (5.92% and 2.15% respectively).

For *B. gymnorrhiza,* the levels of genetic variation, the mean of the number of alleles per locus (A), and the average of gene diversity (H_E) were comparatively high with 3-5 and 0.744 respectively than those reported previously using allozymes marker [18] observed the genetic diversity values (A=1.10 and H_E=0.035±0.025). The levels of heterozygosity (H_O) detected for every population, overall loci (Table 3), indicate that these populations had relatively high genetic variation, these calculations of heterozygosity were much higher than those descrbed prior using allozyme to estimate H_O=0.009 [18]. Similarly, Sugaya et al. [7] also reported the genetic variation values of *B. gymnorrhiza* (A=2-5, H_O=0.031-0.500) using microsatellite marker.

Meanwhile, *K. obovata* also showed the high level of genetic diversity (A=46-73, H_E=0.966). The observed heterozygosity (H_O=0.665) was much higher than those reported previously using allozyme marker [19] found the gene diversity (H_E) and the number of alleles per locus were 0.033 and 1.11±0.04 respectively. Huang et al. [20] reported the values of genetic diversity (A=1.2, H_O=0.04, H_E=0.049) in *B. gymnorrhiza,* Southern China. Meanwhile, Takeuchi et al. [18] demonstrated the low level of genetic variation in *K. obovata*, Southwest of Japan (H_E=0.012±0.012). However, using microsatellite marker, Sugaya et al. [8] reported high values (A=3-9, H_O=0.250-0.938) in *K. obovata*. Factors such as isolation, small populations and gene flow, all of which directly affect useful population sizes, may have a significant influence on the levels of genetic diversity observed within and among populations of woody species [21]. In addition, the hierarchical analysis of molecular variance (AMOVA) of both species also showed that most of the diversity was separated among individuals within populations and within individuals in the total populations. There was little variation among populations. Differences in the breeding system may result in dissilimarity in population genetic structure. Hamrick et al. [21] pointed out that in outcrossing species, the majority of the total genetic diversity (H_E) located within populations (H_S), whereas in selfing species is more distributed among populations (F_{ST}). In this study, the genetic diversity of both species maintained within populations with the relatively high of H_S, 0.741 and 0.954 respectively (Table 4).

The comparable results were obtained for other mangroves tree species. Partitioning of variation was detected within the populations of *B. gymnorrhiza* (70%), and for *B. sexangula* 60% of the difference was observed among populations [22]. Jian et al. [23] reported the variation within populations of *Heritiera littoralis* from Australia and China and within China were 46.9 % and 72.2 respectively. Basyuni et al. [6] observed in *Rhizophora mangle* from Ecuador that the variation was partitioned among individuals within populations (33.3 %) and among individuals (66.6 %) in the total populations. Meanwhile, the genetic variation of *Sonneratia alba* in China was maintained within populations, 79.39 [24].

Also, assessment of genetic variation of population levels throughout the worldwide range of *A. marina* showed that most variation was found between populations (41-71%) and within individuals in the total populations (31-49%) [25]. Hamrick et al. [21] reported that studies on genetic diversity within populations of woody species showed that a high level of genetic diversity mostly resides within populations.

The discrete subpopulation model indicated little and moderate levels of genetic differentiation using Wright's interpretive guidelines [17] (Table 6). The extent and the pattern of genetic diversity in forest trees are strongly regulated by their mating system and gene flow [26]. Indirect estimates of gene flow using private alleles resulted relatively high for overall population of both species (*N*m=1.826 and 4.681, respectively). If *N*m is less than 1, then changes in allele frequencies resulting from a genetic drift of neutral alleles can occur. Such changes are not likely if *N*m is greater than 1 [27]. Moreover, viviparous seed dispersal is potentially a much greater source of gene flow in mangroves [20].

As measured by electrophoretically distinguishable alleles, which are often considered to be neutral variations, the value for *N*m in a range of forest tree species indicates fairly high levels (> 1) of gene flow. This is to be expected because most trees have high outcrossing rates, and there is a definite association between outcrossing rate and the level of gene flow [28]. The high value of *N*m for both species is comparable to estimates reported for other mangroves tree species, such as in *R. mangle* in Pacific coast, *N*m=3.174 [29] and *R. mangle* in Ecuador, *Nm*=3.778 [6].

Gene flow switch of gametes or genes among distributed trees, and it is reciprocally correlated to population adjustment. It occurs through the forward motion of pollen and seed, which can be incoherent by a broad varied of non-living and living mechanisms. Understanding how populations are dispread, preserved, or restricted by gene flow is indispensable for control precise tree populations in their primary environment [30].

The structure of genetic diversity within and among populations of species is an inherent feature of the evolution of mangrove forest, that may be correlated to evolutionarily oxidosqualene gene and salt tolerance gene [31-32], and must, therefore, be considered in developing any conservation strategy. Our results suggested supporting the conservations efforts of the mangrove ecosystem in Okinawa. The protection implies of the available data of these studies on genetic variation and structure are that, in the conditon of outcrossing species, some populations in Okinawan mangroves, especially *B. gymnorrhiza* and *K. obovata* may sustain much of the genetic diversity.

Acknowledgements

A part of this study was funded by an International Research Collaboration 2016-2018 from the Directorate for Research and Community Service, Ministry of Research, Technology and Higher Education, Republic of Indonesia.

References

1. M. Basyuni, H. Sagami, S. Baba, H. Iwasaki, H. Oku, Diversity of polyisoprenoids in ten Okinawan mangroves, Dendrobiology **75**, 167-175 (2017).
2. K. Kathiresan, S. Baba, Facts about mangroves of Japan. Seshaiyana 7:3-6 (1999).
3. M.Z. Hussain, Silviculture of Mangroves. Unasylva 181 (46):36-42 (1995).
4. M. Basyuni, H. Sagami, S. Baba, H. Oku, Distribution, occurrence, and cluster analysis of new polyprenyl acetones and other polyisoprenoids from North Sumatran mangroves, Dendrobiology **78**, 18-31 (2018).
5. L. Graudal, E. Kjaer, A. Thomsen, and A.B. Larsen, Planning national programmes for conservation of forest genetic resources. *Danida Forest Seed Centre. Technical Note* **48**: 9-25 (1997)
6. M. Basyuni, S. Baba, H. Oku, IOP Conference Series Materials Science and Engineering **180**, 1, p. 012243 (2017).

7. T. Sugaya, T. Takeuchi, H. Yoshimaru, and M. Katsuta, Development and polymorphism of simple sequence repeat DNA markers for Kandelia candel (L.) Druce. *Molecular Ecology Notes* **2**:65-66, (2002).

8. T. Sugaya, H.Yoshimaru, T.Takeuchi, M.Katsuta, K.Fujimoto,and S.Changtragoon, Development and polymorphism of simple sequence repeat DNA markers for Bruguiera gymnorrhiza (L.) Lamk. *Molecular Ecology Note* 3:88-90, (2003).

9. P.O. Lewis, and D. Zaykin, Genetic Data Analysis: Computer program for the analysis of allelic data, version 1.1. Free program distributed over the internet from GDA homepage at http://hydrodictyon.eeb.uconn.edu/people/plewis/software.php (2002).

10. M. Nei, *Molecular evolutionary genetics*, (Columbia University Press, New York. 512pp, 1987).

11. J. Goudet, FSTAT, a program to estimate and test gene diversities and fixation indices (version 2.9.3). (2001)

12. M. Kimura, and J. Crow, The number of alleles that can be maintained in a finite population. *Genetics* 49:725-738, (1964)

13. S. Schneider, D. Roessli, and L. Excoffier, *ARLEQUIN version 2.000: A software for populations genetics data analysis*, (Genetics and Biometry Laboratory, University of Geneva, Switzerland, 2000)

14. M. Raymond, and F. Rousset, GENEPOP (Version 3.1b): Population genetics software for exact tests and ecumenicism. *Journal of Heredity* **86**:248-249 (1995)

15. S. Wright, The interpretation of population structure by F-statistics with special regard to system of mating, *Evolution* **19**:395-420 (1965).

16. T. Takeuchi, T. Sugaya, A. Kanazashi, H. Yoshimaru, and M. Katsuta, Genetic diversity of Kandelia candel and Bruguiera gymnorrhiza in the Southwest Islands, Japan. *J.For.Res.* **6**:157-162, (2001).

17. M. Sun, K. C. Wong, and J. S. Y. Lee, Reproductive biology and population genetic structure of Kandelia candel (Rhizophoraceae), a viviparous mangrove species. *American Journal of Botany* **85**(11):161-1637 (1998).

18. S. Huang, Y. Chen., and H. Fan. Genetic variation on mangrove forest, Kandelia candel (L.) Druce (Rhizophoraceae) in Southern China. The Symposium on the Phylogeny, *Biogeography and Conservation of Fauna and Flora of East Asian Region*, 177-188, (1997).

19. J.L. Hamrick, M.J.W. Godt, and S.L.Sherman-Broyles, Factors influencing levels of genetic diversity in woody plant species. *New Forest* **6**:95-124, (1992).

20. P.D. Abeysinghe, L. Triest, B.De Greef, N. Koedam, and S. Hettiarachi, Genetic and geographic variation of the mangrove tree Bruguiera in Sri Langka. *Aquatic Botany* **67**:131-141, (2000).

21. S. Jian, T. Tang, Y. Zhong, and S. Shi, Variation in inter-simple sequence repeat (ISSR) in mangrove and non-mangrove populations of Heritiera littoralis (Sterculiaceae) from China and Australia. *Aquatic Botany* **79**: 75-86 (2004).

22. H.S. Li, and G. Z. Chen. Genetic diversity of Sonneratia alba in China detected by inter simple sequence repeats (ISSR) analysis. *Acta Botanica Sinica* **46**(5):515-521. (2004)

23. T.L. Maguire, P. Saenger, P. Baverstock, and R. Henry. Microsatellite analysis of genetic structure in the mangrove species Avicennia marina (Forsk.) Vierh. (Avicenniaceae). *Molecular Ecology* **9**:1853-1862 (2000).

24. K.S. Bawa, Breeding systems of tree species of a lowland tropical community. *Evolution* **28**:85-92 (1974).
25. M. Slatkin, Gene flow and the geographic structure of natural populations. *Science* **236**:787-792 (1987).
26. D.R. Govindaraju, A note on the relationship between outcrossing rate and gene flow in plants. *Heredity* **61**:401-404 (1988).
27. J. Nunez-Farfan, C.A. Dominguez, L.E. Eguiarte, A. Cornejo, M. Quijano, J. Vargas, and R. Dirzo, Genetic divergence among Mexican populations of red mangrove (Rhizophora mangle): geographic and historic effects. *Evolutionary Ecology Research* **4**:1049-1064, (2002).
28. K.S. Bawa, D. R. Perry, and J. H. Beach. Reproductive biology of tropical lowland rain forest trees. I. Sexual systems and incompatibility mechanism. American *Journal of Botany* **72**:331-345 (1985).
29. M. Basyuni, R. Wati, Bioinformatics analysis of the oxidosqualene cyclase gene and the amino acid sequence in mangrove plants, *Journal of Physics : Conferece Series* **801** 012011 (2017).
30. M. Basyuni, Sumardi, Bioinformatics approach of salt tolerance gene in mangrove plant Rhizophora stylosa, *Journal of Physics : Conference Series* **801** 012012 (2017).

Development of stingless beekeeping projects in Malaysia

Mohd Mansor Ismail [1,*] and *Wan Iryani Wan* Ismail[2]

[1]Institute of Agricultural and Food Policy Studies, Universiti Putra Malaysia, Putra Infoport, Jalan Kajang-Puchong 43400 UPM Serdang, Selangor, Malaysia
[2]School of Fundamental Science, Universiti Malaysia Terengganu, 23100 Kuala Terengganu, Terengganu, Malaysia

Abstract. Malaysia is in need of new high income project to meet the targeted high income nation by 2020. One such potential project in agriculture is stingless beekeeping. The project could supplement the existing beekeeping projects from *Apis spp.* such as *Apiscerana* and *Apis mellifera* in terms of honey production and pollination services. The introduction of stingless beekeeping in 2004 by MARDI was expected to provide new alternative species that was free from diseases and required simple farm management practices where adopters can operate the project as a part time job. The remaining labour hours can be used to earn off farm income. The project alone could generate a monthly income of up to RM5000 for the adopters with available capital and land, and this will enhance additional income and dietary consumption of the stingless bee products. The project is hypothesized to provide positive relationship with income and nutritional quality. For those farmers with land, government assistance is needed in providing cheaper credit to purchase stingless bee colonies which are the major portion of initial investment. In summary, the project is expected to reduce rural and urban poverty and generate a new source of wealth to private investors in Malaysia.

1 Introduction

Beekeeping has long been ignored by many developing countries as a source of additional income and foreign exchange. The industry has started as early as 1980s. Two main commercial species in Malaysia are *Apiscerana (*local bees*)* and *Apismellifera (*imported bees). But this species was plagued with diseases and was easily attacked by birds, pests and mites. The stingless bees of Trigona spp. is a new beekeeping activity that could complement natural honey production and pollination services. The aim of this study is to justify the development of stingless beekeeping in Malaysia. The basic valuation of the project mainly based on previous study on the feasibility of stingless bee farming, financially and economically. The financial study will provide the return on investment of stingless beekeeping project.

* Corresponding author: mkk@upm.edu.my

 https://doi.org/10.1051/e3sconf/20185200028

2 Honey trade

With plentiful nectar resources, surprisingly, Malaysia produces a paltry estimate of less than 5% (98mt /year) of her honey needs in 2000 but the country manages to increase its self-sufficiency level to 284% in 2010 (Table 1). This is a great achievement for the beekeeping industry and for Malaysia as a whole. According to [1] there are more than 700 stingless beekeepers in Malaysia, consisting 58, 593 colonies. The honey production of this stingless bee was estimated at 138, 580.71kg amounted RM 16, 988, 308. But, the country still imports its honey from abroad, which is reflected in prolong honey trade deficit after 2010 (Figure 1), mainly for industrial uses. The rest of the honey industrial requirement and for re-export purposes are imported from the temperate countries such as Australia and New Zealand. Imported honeys from China and Iran are much cheaper than the local honey but Malaysian prefers local honey [2]. This is reflected in the premium prices of all local honey (*Apis and Trigona spp.*) sold in Malaysia.

Malaysia imported low price honey for industrial use and exported high price honey for table consumption. In industry, honey is mainly used as ingredients or sweetener or additives in the confectionary, bakery, medicinal and pharmaceutical industries. One of the major honey exporters in the country is Summer Pacific Sdn. Bhd. located in Bintulu, Sarawak. The company produces about 100 mt of Acacia honey every month. Unlike in Europe and the America, Malaysians, especially the Malays, consumed honey mainly for medicinal in nature, rather than as food supplements and, that is why, the per capita consumption in Malaysia is very low (estimated about 100g/year).

Table 1. Export, import and balance trade of honey in Malaysia, 2000 – 2015.

Year	Export		Import		Balance of Trade	
	Quantity (Ton)	Trade Value (US$ '000)	Quantity (mt)	Trade Value (US$ '000)	Quantity (mt)	Trade Value (US$ '000)
2005	88.44	213.23	2,167.35	4,058.36	(2,078.91)	(3,845.13)
2006	238.72	486.03	2,482.82	4,246.53	(2,244.10)	(3,760.50)
2007	2,263.40	3,387.17	4,935.51	6,781.79	(2,672.11)	(3,394.62)
2008	4,572.63	7,223.92	6,744.85	8,016.33	(2,172.22)	(792.40)
2009	7,382.46	10,826.25	8,232.90	9,340.22	(850.44)	1,486.03
2010	13,453.57	14,445.94	7,913.57	9,491.50	5,540.00	4,954.44
2011	2,508.56	6,934.71	3,159.62	13,672.35	(651.06)	(6,737.64)
2012	1,578.91	4,946.04	2,207.09	14,319.65	(628.18)	(9,373.61)
2013	696.28	2,283.02	1,783.79	11,562.52	(1,087.51)	(9,279.50)
2014	801.88	3,199.47	3,156.85	11,810.79	(2,354.98)	(8,611.32)
2015	3,521.27	5,592.28	4,913.35	14,022.56	(1,392.08)	(8,430.28)

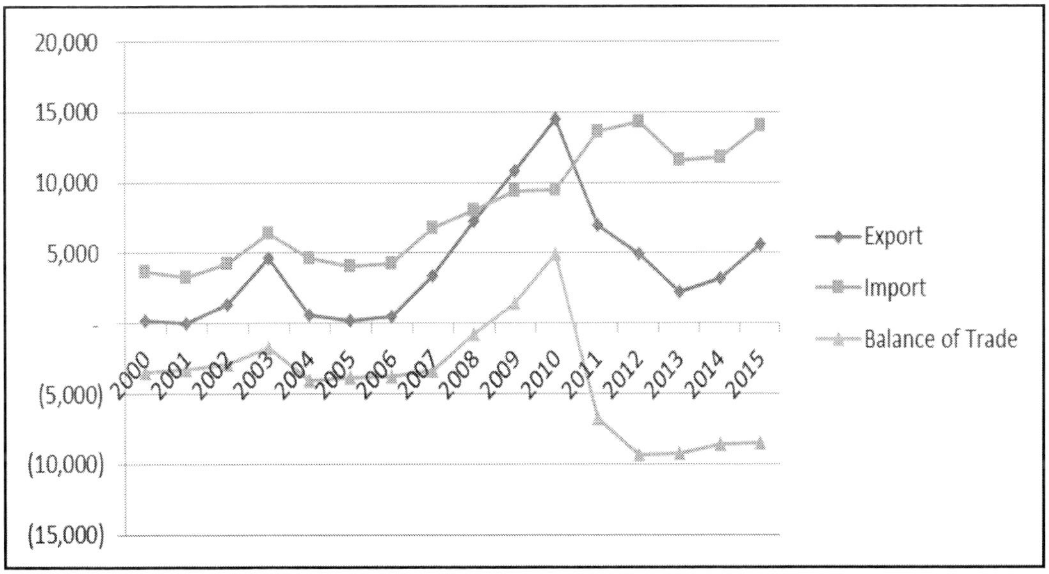

Fig. 1. Export, import and balance of trade for honey in Malaysia, 2000 – 2015 (US$).

Note: USD1 is equivalent to RM4.00, 2018

3 Natural factor endowment

Malaysia possesses abundant natural resources that can sustain beekeeping activity producing bee products and by products that can be converted into high value-added health food, rich in both enzymatic and non-enzymatic antioxidants. Beekeeping plays a major role in socio-economic development and environmental conservation in Malaysia. It is a source of food (e.g. honey, pollen and brood), raw materials for various industries (e.g. beeswax candles and lubricants), medicine (honey, propolis, beeswax and bee venom) and as additional sources of income. It is estimated that the beekeeping industry is capable of generating millions of revenues each year from sales of bee products and by-products. It is an important income generating activity with high potential for improving incomes especially in the fruits and pineapple plantations and to rural farmers' bordering tropical forest reserves.

Johor, Selangor and Melaka and the bees are from the species of *Apiscerana* (the local bee) and, Apismellifera (the imported bee). There are about 700 beekeepers in 2010 (10000 colonies) compared to 946 in 1988 (7000 colonies). The yield from local bees is about 2-5 kg per year and *A. mellifera* (the imported bee) is expected to produce up to 50 kg per colony per year but prone to pests and diseases. The problems in rearing *Acerana* and *A mellifera* prompted the industry to search for new species and found the *Trigona spp*. This species are quite diverse with estimated 50 species in Malaysia but only two species are expected to have commercialization potential. The research and development on the technical, financial and economic aspects of commercializing *Trigona spp*.is urgently needed to complement the existing production of natural honey in the industry.

Apiaries in Malaysia have aspects for improvement which can lead to higher yields and to a sustainable industry while exposing urgent technical problems. As a longer term objective, selections of new species like the *Trigonaspp* need to be started to obtain aspects like pests tolerant strains. By products such as beebread and propolisare still unexplored at the moment. Postharvest drying of honey and downstream processing are still remained untouched. All these technical aspects must be combined with financial and economic aspects before any suggestion to commercialize the stingless bee production.

These aspects of beekeeping practices which we have identified in its technical perspective and futuristic vision can be improved via a larger coordinated programme involving collaborations among many disciplines of agriculture. It is expected that the improvement in research and development on the technical aspect of stingless beekeeping will eventually increase the production of local natural honey.

If natural honey production from *Trigona spp.* is feasible, then other by products can also be harvested such as beebread, propolis and pollination services, which are associated with the stingless beekeeping industry. The demand for honey and other by-products and services from the stingless beekeeping activities are to be explored and connected into the supply-chain concept of management. Other related industries, such as, probiotics (beverages), cosmetics (beauty soaps) and ecotourism can be linked and connected into a commercial critical-mass for the bee supply-chain network. The pollination services from honeybees in the USA is estimated to worth more than USD $ 200 billion annually and it worth more than the honey production. No estimates are available for Malaysia at the moment. The potential of *Trigonaspp.*is enormous and what is urgently needed is the research on technical and financial feasibility of this activity and if the production systems possess comparative advantage, the *Trigona* farming is internationally competitive meaning the natural honey from *Trigona spp.* can be one of the Malaysia's export items in the future.

4 Profitability

In order for an industry to be sustainable, a project must be consistently profitable. Without profit, any companies will have to exit the industry. Since there are no technological constraints, any farmers can participate in stingless beekeeping. All they need is to follow a short course in stingless beekeeping that is organized by many research entities including private sectors, government agencies and University Putra Malaysia. A standard return on investment is presented in Table 2.

Table 2. Benchmark of Return on Investment (ROI) for 200 stingless bee colonies.

Items	RM	RM
Tax	0%	25%
Revenue	64,800.00	6,4800.00
Cost	1,800.00	1,800.00
Depreciation	1,280.00	1,280.00
EBT	61,720.00	61,720.00
Tax	0.00	15,430.00
EAT	61,720.00	46,290.00
Monthly Income	**5,143.33**	**3,857.50**
Monthly Cash Flow	5,250.00	3,964.17
Total Investment	86,400	86,400
ROI	**71.44%**	**53.58%**

The Stingless beekeeping project can be considered as a high income project since it can provide the participant with more than RM5000 income per month (with zero tax). More incentives are also given by the Malaysian government in the promotion of food production.

5 Conclusion

Stingless beekeeping projects are a new venture in beekeeping in Malaysia. Since the beekeeping management in *Apis spp.* is plagued with many farm management issues, the stingless beekeeping is preferred currently. The project is expected to provide more than 70% return on investment and a monthly income of more than RM5000. This is in line with the government intention of increasing the number of high impact agriculture project to the people. The stingless beekeeping project can be treated as a potential commercial project or as a rural and urban poverty eradication project in Malaysia.

Acknowledgements

The authors thankfully acknowledge the financial support received from the MOE RACE research grant.

References

1. Department of Agriculture, Putrajaya. Statistics of Stingless Beekeepers in Malaysia. Unpublished Report. Putrajaya, Malaysia, (2017)
2. M. Mansor. Honey marketing in Malaysia. Agricultural Marketing Issues of Selected. Commodities, Editor Fatimah Mohamed Arshad and Amin Mahir Abdullah, (2012)

Standard contracts in bank credit agreements

Montayana Meher* and *Ningrum Natasya* Sirait

Doctoral Progam in Law, Graduate School, Universitas Sumatera Utara, Medan, Indonesia

Abstract. The application of standard contracts in bank credit agreements have triggered many legal issues in questions. In banking practice, each bank provides a form of credit agreement that has been prepared in advance. The type of agreements known as standard agreements. Using standard agreement is often problematic. Problems encountered in the standard agreements are the validity of the standard agreements and the making of clauses or provisions which are unreasonably burdensome to the parties, especially the debtor, which is called the exoneration clause. The existing credit agreements in Indonesian banking system are so diverse, means there is no uniformity. Each bank produce or prepare its own rules with certain clauses that may impair the customers. There is a need for a legislation that regulates credit matters and it is expected that it contains rights and obligations of the parties in preparing the agreement.

1 Introduction

Agreement is prepared based on the principle of freedom of contract between parties where parties seek to reach necessary consensus through a process of negotiation between them. These days there is an increasing tendency that many contracts in business transactions existed not through a sensible negotiation process between the parties. However, the agreement existed because one party has prepared the standard terms or clause on printed form and presented to the other party for approval with literally no opportunity to negotiate on the terms. Such contracts are define as standard contracts.[1]

Freedom of contract is based on an assumption that parties to the contract have an equivalent bargaining position. In reality, the parties do not always have equal bargaining position. As a consequences, party with stronger bargaining position tends to control the other party with weaker bargaining position.[2] Party with weaker bargaining position will accept the contents of the contract under pressure. If the person tries to propose bargain alternative or option of the content in the contract, the person is likely to lose what he/she needs obtain from the agreement. Therefore, there are only two options for the party with weaker bargaining position, either to accept or to refuse the offer (take it or leave it). [3]

In accordance with the main principle of an agreement, namely the principle of freedom of contract as implied in Article 1338 of the Civil Code, the parties that bind themselves in a credit agreement can refer either to the provisions existing in the Civil Code or to the mutual agreement. This means that in the case of mandatory provisions, the agreement must

*Corresponding author: montayanameher@yahoo.com

accord with the provisions of the Civil Code, whereas in the case of non-compulsory provisions, it is entirely left to the parties' mutual consent. [4]

The law recognizes a person's autonomous right to freely make an agreement with any person and determines the content of the agreement. In addition to the principle of freedom of contract, the principle of binding force states that all legally-made agreements shall apply as laws to those who prepared them, as set forth in the provisions of Article 1338 of the Civil Code.

The other fundamental principle of contract law is the principle of a consensus. In addition to the above principles, there is also a principle of equilibrium stating that the principles of contract law in particular or the legal instruments contained in the Civil Code shall be established based on the legal values and norms. [5]

The phenomenon of imbalance position in contracts as stated earlier can be observed from a number of contract models, especially standard contracts with unfair clauses. For example such as credit provision in banking sector, where we may find clause on responsibility transfer of business actors and or consumer's compliance to new, additional, new or subsequent regulations, and/or further changes made unilaterally by the business actors.

In banking practices, each bank provides a form of credit agreement that has been prepared in advance. This credit agreement form shall be submitted to the debtor for approval without giving any opportunity to the other party to negotiate on the terms provided. [6] The use of standard agreements is problematic. The problems encountered in the standard agreement are the validity and the making of clauses or provisions which are unreasonably burdensome to the other party, especially the debtor.

The exoneration clause usually contained in the agreement as an additional clause is usually disadvantageous for the consumer who generally has weaker position compared to the producer. The existence of such clause place a burden to the consumer. The application of certain clauses by a party with a stronger position which is harmful is usually known as the misuse of the condition (*misbruik van omstadigheden*). [7]

2 Problem

The problem question in this study is why are the validity of the standard agreement and the making of its clauses or provisions unreasonably burdensome to the other party, especially the debtor?

3 Research methods

This research is normative legal study. The data used was in this paper obtain from secondary data consisting of (a) primary legal materials in the form of legislation, (b) secondary law materials in the form of writings, both books and articles containing comment or analysis related to the subject matter, and (c) tertiary legal materials in the form of dictionaries. The data collected using a literature review and document analysis and using qualitative method.

4 Discussion

4.1 Legal relationship between the bank and its customers

As a financial institution, banks play a strategic role in the life of a state's economy. The institution is intended as an intermediary body between parties with surplus of funds and parties with lack of funds. Banks will engage in as credits provider in various services to its customers. Banks serve financial assistance as well as smoothening the payment system mechanism for all economic sectors. [8]

The word credit comes from "*credere*" which means trust. This means that the creditor give their trust to the debtor that the credit provided will certainly be repaid in accordance with the agreement. As for the recipient of the credit, this means accepting the trust, so the borrower has an obligation to repay the loan in accordance with the time period. Therefore, to make sure that the customer is truly trustworthy, the bank usually conducts credit analysis before the credit is given. Credit analysis includes the background of the customer or the company, its business prospects, the collateral provided and other factors. The purpose of the analysis is for the bank to be confident that the credit provided is completely secure. [9] According to Law Number 10 of 1998 on Banking Service, credit is the provision of money or equivalent claims, based on a loan agreement between the bank and another party requiring the borrower to repay the debt after a certain period of time with added interest.

Any credit approved and agreed between the creditor and the debtor must be set forth in a written credit agreement. This credit agreement should be carefully considered by both, the bank as a creditor and the customer as a debtor. The credit agreement has very important function in the provision, management and administration of the credit.

The relationships between the bank and its customers can be categorized as the relationship between a creditor and a debtor, trust relation (*fiduciary relation*) and *confidential* relation. The relationship between bank and customer are accompanied by *prudential relation* principle. These four relationships underpin the relationship between the bank and its customers.

Symons Jr argued that the relationship between bank and customer is a debtor-creditor relationship which suggests a determination of obligations in a narrow sense. The term refers to an unconditional assurance by the debtor to pay a certain amount of money at a certain time to the creditor that has provided the money. It further indicates that the debtor has no other obligations, unless otherwise specified, especially with regard to the use of the borrowed money. In a bank deposit for example, the bank may use the money freely at will. However, the relationship between a bank and a customer is not solely a debtor-creditor relationship, it also serves as a fiduciary relation. [10]

4.2 Validity of the standard agreement

In the contract law, there are three principles which are interrelated to each other, namely the principle of consensus, the principle of freedom of contract, and the principle of binding force. The principle of consensus was born at the initial moment of the agreement, namely at the time when the parties reach a mutual agreement. The parties enter the stage of freedom of contract when determine the rights and obligations as well as other matters related to the substance of the agreement, Under this principle, the parties can determine the form and content freely as long as they can be held accountable. Reciprocal approval of the form and content of the agreement becomes the recognition of a valid will to the content of the agreement. As a result, the agreement is binding for both parties and should be performed in a good faith. [11]

Under the principle of freedom of contract, an individual possess the right to have the freedom to sign a contract, to determine the content of the contract, to determine the form of the contract, and to choose the law applicable to the contract concerned. If the principle of consensus relates to the commencement of a contract, the principle of binding force relates to the legal consequences and the principle of freedom of contract relates to the content of the contract.

Freedom of contract is one of the most important principles in the agreement. Freedom of contract is the byword of free will and substantiate human rights. This freedom of contract is a base in the ideology of individualism found in Greek era, continued by the *Epicurists* and thrived in the *Renaissance Age*. The principle among others, convey through famous intellectual such as Hugo de Groot, Thomas Hobbes, John Locke and Rosseau. The uttermost of its development achieved after the French revolution.

In the nineteenth century, as the economic doctrine of *laissez faire* prevailed, freedom of contract became a general principle in favor of the free competition. Freedom of contract became the embodiment of the principle of free trade law. Any state intervention to the contract is against the principle of the free trade. Freedom of contract has become a new paradigm in the contract law that is highly exalted by philosophers, lawyers and courts.[10]

Freedom of contract may only achieve fairness if the parties have equal bargaining power. If the bargaining power is unequal, then a contract tends to be or become *unconscionable*. In addition, although the principle of equilibrium and conformity of the parties' position exists, the implementation often results in an imbalanced and unsuitable contract (inappropriate and unfair).

A credit agreement is a standard contract in which the content or clauses of the credit agreement has been standardized and set forth in a written form but not bound in a particular form (*vorn vrij*). A prospective debtor signs only if he/she is willing to accept the content of the agreement. The bank does not give an opportunity to a prospective debtor to discuss further details about the content or clauses proposed by the bank. In this agreement, the position of the prospective debtor is weak, so he/she can only accept all the conditions proposed by the bank, otherwise, the prospective debtor will not get the requested credit.[11]

Principally, the parties to a contract are free to self-govern the contract in accordance with the principle of freedom of contract, as defined in Article 1338 Paragraph (1) of the Civil Code. Article 1338 Paragraph (1) specifies that all legally- agreed contracts pertain as laws to those who bound them.

In an agreement where the terms, conditions or contents set forth in the clauses have been standardized, the legal position of one party is more dominant than the other party's. This has made a party to be restricted or constrained in expressing his/her will because he/she being offered an agreement with predetermined terms. In standard agreement, the rights of one party are more prominent than other's and this has consequences to the rights and obligations between the parties to be imbalanced.

Jurists such as Sluijter and Pitlo, disagree the existence of the standard agreement. Sluijter argues that this standard agreement is not accepted as an agreement, because the entrepreneur's position in the agreement is similar to private law maker (*legio particuliere wetgever*). The conditions specified by the entrepreneur in the agreement are laws and not considered as agreement. While Pitlo maintains that the background to the growth of the standard agreement is the social and economic need, large corporations, semi-governmental enterprises or government enterprises entering into cooperation for their interests in determining certain conditions unilaterally. By using this standard agreement, the entrepreneur will obtain the efficient use of cost, labor and time which describes a picture of a pragmatic society. [12]

However, there are also jurists who are proponent to this standard agreement. Stein argues that a standard agreement can be accepted as an agreement based on the willingness and trust that parties are bound to the agreement. If the debtor accepts an agreement, it means he or she freely and voluntarily agrees on the content of the agreement.

Asser-Rutten argues that everyone who signs an agreement is responsible for the content that he/she has agree and signed a. If somebody signs a standard agreement form, the signature raises the expectation that the undersigned knows and wants the content of the signed form. It is impossible for a person to sign what he/she does not know nor understand. [13]

Another jurist, Hondius insists that the standard agreement has binding power based on customs prevailing in the community and trade transactions. While an Indonesian jurist, Mariam Darus argues that although theoretically a standard agreement does not meet the provisions of the law and is rejected by some jurists, but in reality, the needs of the society require the existence of this standard agreement. [14]

4.3 The exoneration clause in a credit agreement

The phenomenon of imbalance in a contract as mentioned above can be observed from a number of contract models, especially standard contracts with clauses of biased content. In the practice of credit provisions in the banking sector, for example, there is a clause that obliges the customer to comply with any bank guidance and regulation, either existing or to be regulated later, or a clause that exempts the bank from customer's loss as a result of the bank action. [15]

The standard agreement brings about negative impact in the sense that a party with a strong bargaining position can impose his/her will on the weak, and the strong benefits from the action. In the development of various jurisdictions, the state intervenes to protect the weak party either through judicial decisions or by issuing regulations made by the legislature. In that connection, some basic rules that must be observed for an unsigned written agreement and contain the standard terms such as the principle of *duty to read* prevailing in the United States prior to the 1960s for signed documents, the principle of *public policy* and the principle of *unconscionability*.

In 1854, the British parliament introduced a law that attempted to limit the freedom of the seller to make clauses in a standard contract to protect the purchaser of goods/services, beginning with the law on the channel and train transport. In the nineteenth century, other laws pertaining to certain contracts were also enacted in England as in other industrialized countries. For example, in 1893 the *Sale of Goods Act* was enacted. Later in the United States, similar provisions, namely the *Uniform Sales Act*, in 1906, were accepted by the *National of Commissioners on Uniform State Laws*. [16]

Therefore, the arrangement of the content of an agreement is not solely left to the parties, but it should be monitored by government institution especially against the standard agreement. [17] The government as the bearer of the public interests should keep the balance of individual and public interests. Allowing the making of an agreement to be in the sole mechanism of freedom of contract would create the imbalance and unconformity of the contracting parties. Therefore, the principle of freedom of contract requires restrictions on its operation by determining prohibited or required clauses in the form of laws.

5 Conclusions

5.2 Summary

Despite the existence of standard credit agreements which poses some problems in terms of the construction of clauses, especially the exoneration clause that incriminates the debtor, such agreements are still needed. Therefore, the arrangement of the content of an agreement is not solely left to the parties, but it should be monitored especially against the standard agreement. The government as the bearer of the public interests should keep the balance of individual and public interests.

5.2. Recommendation

The existing credit agreements in the banking system in Indonesia are so diverse that there is no uniformity. Every bank creates its own rules with certain clauses that may harm customers. Therefore, there is a need for legislation to regulate the credit agreement in terms of the rights and obligations of the parties.

References

1. S.R. Sjahdeni, *Kebebasan berkontrak dan perlindungan yang seimbang dalam perjanjian kredit bank di Indonesia*, (Institut Bankir Indonesia, Jakarta, 1993).
2. R. Khairandy, *Itikad baik dalam kebebasan berkontrak*, (Fakultas Hukum UI, Jakarta, 2004).
3. Shidarta, *Hukum Perlindungan Konsumen Indonesia*, (Grasindo, Jakarta, 2004).
4. M. D. Badrulzaman, *K.U.H. Perdata Buku III tentang Hukum Perikatan dan Penjelasannya*, (Alumni, Bandung, 2005).
5. H. Budiono, *Kumpulan tulisan hukum perdata di bidang kenotariatan*, (Citra Aditya Bakti, Bandung, 2008).
6. J. Ibrahim, *Bank sebagai lembaga intermediasi dalam hukum positif*, (CV. Utomo, Bandung, 2004).

A. Miru & S. Yodo, *Hukum perlindungan konsumen*, (Raja Grafindo Persada, Jakarta, 2004).

7. M. Djumhana, *Hukum perbankan di Indonesia*, (Citra Aditya Bakti, Bandung, 2003).
8. Kasmir, *Dasar-dasar perbankan*, (Raja Grafindo Persada, Jakarta, 2000).
9. R.S.H. Bako, *Hubungan bank dan nasabah terhadap produk tabungan dan deposito (Suatu tinjauan hukum terhadap perlindungan deposan di Indonesia dewasa ini)*, (Citra Aditya Bakti, Bandung, 1995).
10. T. Kamello, Karakter hukum perdata dalam fungsi perbankan melalui hubungan antara bank dengan nasabah, *Pidato Pengukuhan Jabatan Guru Besar Tetap dalam Bidang Ilmu Hukum Perdata*, (Fakultas Hukum Universitas Sumatera Utara, Medan, 2006).
11. C. Elliot, and F. Quinn, *Contract law, Fourth Edition*, (Pearson Education, England, 2003).
12. R. Usman, *Aspek-aspek hukum perbankan di Indonesia*, (PT Gramedia Pustaka Utama, Jakarta, 2001).
13. Sluijter, *De standaardcontracten, de grenzen van de particuliere wetgever*, (Kluwer-Deventer, 1972).
14. Asser-Ruttem, *Handleiding tot de beofening van het Nederland burgerlijk,* (W.E.J. Tjeenk-Willink Zwole, 1968).
15. M.D. Badrulzaman, *Aneka hukum bisnis*, (Alumni, Bandung, 2005).
16. A.Y. Hernoko, *Hukum perjanjian: Azas proporsionalitas dalam kontrak komersial*, (LaksBang Mediatama Yogyakarta Bekerjasama dengan Kantor Advokat "Hufron & Hans Simaela", Yogyakarta, 2008).
17. M. Fuady, *Perbandingan hukum perdata*, (Citra Aditya Bakti, Bandung, 2005).

Efficiency analysis of cotton production: a case study of gassol and lau local government of Taraba State, Nigeria

Muhammad Bala[1*], *Mad Nasir* Shamsudin[2], *Alias* Radam[3], dan *Ismail Abd.* Latif[4]

[1,2,4]Department of Agribusiness and Information System, Faculty of Agriculture, Universiti Putra Malaysia, UPM Serdang Malaysia
[3] Department of Management and Marketing, Faculty of Economics and Management, Universiti Putra Malaysia, UPM Serdang, Malaysia
[1]Department Agricultural Economics and Extension, Faculty Agricultural Sciences, Taraba State University Jalingo

Abstract. This Study analyzed the efficiency of cotton Production in Gassol and Lau Local Government Areas of Taraba State. Data were collected with the aid of structured questionnaires administered to 81 randomly selected farmers in the study areas and analyzed using descriptive statistics and stochastic frontier production function analysis. The result revealed that majority of farmers (67.9%) were young, mostly males (86.42%) and educated (86.41%). The result of the budgeting technique showed that the total revenue (TR), gross margin (GM), net Farm income (NFI) and Return on Investment (ROI) per hectare were ₦55, 108.70, ₦21, 128.70, ₦20, 008.70 and ₦0.62 respectively. The result of the stochastic frontier production analysis revealed that the variance parameter indicates positive and significant relationship at 5% level of probability for agro-chemicals and family labour at 1% level of probability for fertilizer, labour and farm size. Mean technical efficiency index was 0.77 while the minimum and maximum technical efficiency were 0.37 and 0.97 respectively. The study also identified that low price of cotton seed and high cost of inputs are the major challenges faced by cotton producers. The study recommended that inputs should be subsidized by the government and be channeled through the community leaders.

1 Introduction

Cotton *(Gossipium SPP)* is a plant grown in warm countries that has a soft white hair around its seed used in textile industries, and of great importance to international trade Korcher, [1], David and Adamu, [2], Helfand, S.M. [3]. Cotton remains the most important natural fibre of the 20th Century that contributed to the development of the Nigeria economy. It is commonly grown under favorable condition like soil type and fertilizer requirements. Nigeria has experienced its peak period in cotton production in 2003/2004 with about 536,400 metric

* Correspondent Author: moh39dbala@gmail.com

ton output. Since then production has been in decline due to price fluctuation, low farm income, pest infestation and primitive techniques of production, Muhammad, [4].

The Nigeria's output of cotton in 2006/2007 was 631,500 bales, CBN [5]. This shows that compare to that of 2003/2004 a decline in cotton production is being experienced in the country. To blow-away the future occurrence, the Federal Ministry of Science and Technology set up council known as the Raw Materials Research and Development Council, RMRDC [6] to work out ways of implementing the production of some specific crops; and cotton is one of them. According to the council report, it recommended that a collaboration effort has to be made with Afcott Nigeria Plc Yola, Adamawa State, to improve on the cotton seed variety so that high yielding and pest resistant variety would be available to the farmers. Prior to council recommendation, cotton production in Afcott Nigeria Plc, which was established in 1986, has been on the increase. As Mshelia [7] opined, there has been an increase in the average output of farmers as follows: 1.230 tones/ha in 1988, 1.391 tones/ha in 1989 and 1.530 tones/ha in 1990, which in each case compared favorably with the National average of 250kg/ha. Further decline in the output of cotton in the recent past made the then Nigerian's president Obasanjo's administration to come up with a new policy on revitalization of textile industries in Nigeria with a capital production taken 20 billion Naira through Cotton Development Committee with Hon. Minister of Agriculture and Rural Development as Chairman of the, CDC [8].

Investigation has shown that there was a decline in cotton production in Nigeria after every four years. This endangered the future prospect of the Nigeria textile industries, which resulted into falling of the country's foreign exchange. The decline may be attributed to the oil boom in seventies, where it becomes prominent and took over from the agricultural sector in terms of relative importance in almost all aspect of Nigeria economic life. Subsequently, the agricultural sector becomes dormant and neglected, Argbokan [9], Aminu, [10]. Prior to the oil boom, cotton was one of the main sources of foreign exchange and the second largest employer of labour after the Government sector, Gbadegesin and Uyoubisere, [11]. Cotton output, in 2010, was almost equivalent to the requirement of the textile industry, USDA, [12]. However, eight years later, cotton production in the country could only account 49 percent of the requirement of the Nigeria textile industry, while the remaining 51 percent was imported, USDA, [12].

Even though, Nigeria is a bonafide member country for international cotton adversary committee ICAC, [13], the quantity of cotton lint required by our spinning industries is always increasing in contrast to the supply, Mshelia, [7]. For Nigeria to maintain a stable domestic need as well as surplus for exportation, effort has to be made towards increasing output and efficiency in cotton production. The decline in the annual growth rate from 61.7% in 1995 – 1998 was 15.3% in 1999 – 2002 to 4.24% in 2003 – 2006 and to 3.98% in 2007 – 2010 should be tackled by the government. This is a matter of the utmost concern, which, if the situation persists, the demand for our local textile industries can hardly be met. Therefore, there is a need to further investigate and analyze the efficiency in cotton production by small-scale farmers as 80 percent of the cotton production in the country is done by them, Ibrahim, [14]. The specific objectives of the study were to:

a) Describe the socio-economic characteristics of cotton farmers in the study areas,
b) Determine the profitability of cotton production,
c) Examine the technical efficiency of factors employed in the production,
d) Analyze the influence of some socio-economic characteristics of farmers on their technical efficiency, and
e) Describe the constraints affecting cotton production in the study areas

2 Methodology

The study was conducted in Gassol and Lau Local Government Areas of Taraba State. The state is located in north-eastern part of Nigeria. It is roughly between latitudes $6^0 30'$ and $9^0 36'$N of the equator and longitudes $9^0\ 11'$ and $11^\circ\ 50'$E of the Greenwich Meridian (Taraba State Government diary, [15]. The state has land area of about 60,291km^2 with a population figure of 2,300,736 people; comprising 1,199,849 males and 1,100,887 females respectively, NPC, [16]. The data was collected from 81 cotton producers, using multi-stage sampling techniques. The sample's respondent was obtained by dividing the multiple of cotton farmers in each district and the total number of the respondent (81) by the number of cotton growers (1350) in the study area. This can be seen in table 2 by using the formula adopted by Adebayo and Olayemi, [17].

Table 1 Distributed Questionnaires in Eight Villages

Districts	Population of Cotton Growers	Questionnaires
Abbare	255	15
Lallami	230	14
Lau	192	12
Kunini	185	11
Maihula	195	13
Donadda	150	9
Kurmo	83	4
Gassol	60	3
Total	**1350**	**81**

Source: Statistical records of Affcott out growers scheme, 2001.

$$\mathbf{S = P/p \times Q/1} \qquad \mathbf{(1)}$$

Where:
S = Total Number of respondents sampled.
P = Number of cotton farmers at each location.
p = Total population of cotton farmers in the study area.
Q = Total number the questionnaires to be administered.

2.1 Method of Data Analysis

The data collected was subjected to descriptive and inferential statistics. The descriptive statistic was used to achieve objectives one and five of the study. Also, the descriptive statistics, like frequency and percentage was used to describe variables and their occurrences among respondents. Mean was used as a measure of central tendency. Gross Margin analysis gives the difference between the gross income while the total cost of production was used in determining the profitability of cotton production among farmers. The inferential statistics involved the use of stochastic frontier production function to determine the technical efficiency of cotton farmers in the study areas. The gross margin was explicitly stated thus;

$$\mathbf{GM = GFI - TVC} \qquad \mathbf{(2)}$$

Where:
GM = Gross Margin (₦/Ha).
GFI = Gross Farm Income (₦).
TVC = Total Variable Cost (₦/Ha).

$$\textbf{Net return per output} = \textbf{TR} - \textbf{TC (TFC + TVC) Q} \quad (3)$$

$$\textbf{NFI} = \textbf{GM} - \textbf{TFC} \quad (4)$$

Where:
NFI = Net Farm Income (₦/Ha).
TFC = Total Fixed Cost (₦/Ha).
Q = Total Quantity of Cotton (Kg).

2.1.1 The Stochastic Frontier Production Function

The use of the stochastic frontier production function has some conceptual advantage in that, it allows for the decomposition of the error term into random error and inefficiency effect rather than attributing all errors to random effects, Xu and Jerry, [18], Ojo, [19].

It is specified as:

$$Y_i = f(x_i;\beta) + (v_i - u_i).i = 1, 2 \quad (5)$$

Where:
Y_i = an output of the ith farm
x_i = Vector of input quantities of the ith farm
β = Vector of the estimated parameter
v_i = Assumed to account for random factors such as weather, risk and measurement error.

2.2 The Empirical Stochastic Production Model

The stochastic frontier production model used was specified as follows: -

$$\text{Log } Y_i = \beta o + \beta_1 \log X_1 + \beta_2 \log X_2 + \beta_3 \log X_3 + \beta \log X_4 + \beta_5 \log X_5 + \beta_6 \log X_6 + \beta_7 \log X_7 + \beta_8 \log X_8 + vi - ui \quad (6)$$

Where;
Log Y_i = output (Kg of seed cotton of ith farmer)
X_1 = Total land area under cultivation (Ha)
X_2 = family labour used in production (in Main days)
X_3 = hired labour used in production (in Main days)
X_4 = Quantity of Agro-chemicals (in litres) per Ha
X_5 = Quantity of fertilizer (in Kg) per Ha
X_6 = Expenses on ploughing (Animal traction) in naira per Ha
X_7 = Quantity of cotton seeds (in Kg) per Ha
X_8 = depreciation on fixed cost items (in Naira)
V_i = Random noise (white noise) which are the N $(0, \sigma^2{}_v)$

 https://doi.org/10.1051/e3sconf/20185200030

Ui = are inefficiency effects which are non-negative, half normal distribution N (σ^2_u)

Note: The variable X_8 (depreciation on fixed cost items) was included because straight line method was used in computing fixed farm assets.

The corresponding cost frontier as used by Ogundari *et.al*;[21] can be derived analytically and written in general form as;

$$\mathbf{C = f(\rho, Y_i, \gamma_i) + (V_i - U_i)} \tag{7}$$

Where;
C = Minimum Cost Used in the production of Y_i.
ρ= Vector variable of input prices.
f = suitable function form.
Y_i = Value of output in Kg.
$\propto_i$ = parameter to be estimated.

Using Sheppard's Lemma, Bravo-Ureta and Rieger, [22]; the minimum cost input demand equation is obtained as;

$$\frac{\delta c}{\delta p} = X(P, Y, Q) \tag{8}$$

By substituting equation (3.6) and (3.7) into (3.8) yield the economic efficiency of input vector X_e. The technically efficiency input vector (X_t) and economically efficient input vector can be used to compute the cost of technically efficient (X_t, P) and economically efficient (X_e, P) input combinations with the firms observed output. The cost of farm's actual operating inputs combination is given as (X_a, P). These three cost measures are the basis for computing the following: technical, economic and allocative efficiency indices as explained by Bravo-Ureta and Rieger, [22].

$$TE = E_t \frac{P}{X_e} \rho \tag{9}$$

$$EE = X_e \frac{P}{Xa} \rho \tag{10}$$

$$AE = \frac{EE}{TE} = X_e \frac{P}{Xt} \rho \tag{11}$$

Where: TE, EE and AE are technical efficiency, economic efficiency and allocative efficiency respectively.

2.3 The Inefficiency Model

It is assumed that the technical inefficiency effects are independently distributed and arises by truncation (at zero) of the normal distribution with mean, and variance , where is defined by:

$$u_{ij} = \lambda_0 + \lambda_1 z_{1ij} + \lambda_2 z_{2ij} + \lambda_3 z_{3ij} + \lambda_4 z_{4ij} + \lambda_5 z_{5ij} \text{ ....} \textbf{(12)}$$

Where;
U_{ij} = the technical inefficiency of the farmer
Z_1 = Years of the farming experience
Z_2 = Years of formal education
Z_3 = Extension contact (number of meeting)
Z_4 = House hold size
Z_5 = other occupation (where one indicate farming and zero otherwise)
λ_0-λ_5 = unknown parameters to be estimated)

The maximum likelihood estimates (MLE) of β and γ coefficients was estimated simultaneously using computer program frontier 4.1, where the variance parameters are expressed in terms of $\sigma_s^2 = \sigma_u^2 + \sigma_v^2$ and $\gamma = \sigma_u^2/\sigma_s^2$ (Coelli, [23], Ajibefun, [24] .

3 Results and Discussions

3.1 Socio-Economic Characteristics of the Respondents

Table1.1 reveals that 67.9% of the cotton farmers are below the age of 41years. The mean age of the cotton farmers was 32 years. The preponderance of farmers in the cotton farming profession means that their productivity is expected to be high since they are expected to be active and energetic. This result conformed to Ibrahim [25] who noted in his findings that farmers in their active years are productive and can easily adopt agricultural techniques. The result also shows that most farmers are males (86.42%) as against (13.58%) females. This indicates that men participate more in cotton production than their female counterparts. The low percentage of women in cotton production may be explained by socio-cultural factors affecting women and not as a result of technical and managerial inefficiencies. The finding reveals that 86.42% of farmers have family sizes of 1-10 people in their household while 13.58% have more than ten (10) people. The mean of the family size was 6 people. The respondents' large family size is above the recommended average of four per family in Nigeria. The large family size is relevant to cotton production because family labour constitutes the highest form of labour supply in cotton production in Nigeria [14].

The study shows that most farmers (86.41%) know how to read and write, while only 13.59% did not attain any form of education. The mean age of schooling is 9 years meaning that most of the farmers have attained education above primary school level. It can be inferred that adoption of farming techniques may be easier as most of them would find it easy to read the specification labeled on the innovations. Consequently, this will help them to access relevant information that will aid in their production, Alabi and Aruna, [26]. The respondents' farming experience indicates that majority (50.62%) had 11 years and above, whiles those below 10 years constituted only (49.37%). The mean years of experience is about 15 years which is high indicating that cotton farmers are experienced and might know the good practices involved. It is a general opinion that experience farmers would be more efficient, have a better knowledge of climatic conditions and market situations and are expected to run a more efficient and profitable enterprise, Oluwatayo et al, [27].

Table 2 Socio-economic characteristics of the respondents

Socio-economic variables	Frequencies	Percentages
Sex		
Male	70	86.42
Female	11	13.58
Age (Years)		
< 30	20	24.69
31-40	35	43.21
41-50	11	13.58
≥ 50	15	18.52
Mean	32	
Household Size		
1-5	49	60.49
6-10	21	25.93
≥ 11	11	13.58
Educational Level		
Non-formal education	11	13.58
Primary school	20	24.69
Secondary	25	30.86
Tertiary	25	30.86
Experience in Years		
1-5	21	25.92
6-10	19	23.45
≥ 11	41	50.62
Total	**81**	**100**

Source: Field Survey, 2012.

3.2 Gross Margin Analysis

The result in table 1.2 shows that the average variable cost/ha was ₦32,860.00 which represent 96.70% of the total farming cost while average fixed cost was ₦1,120 which represent (3.3%) of the total cost of farming. Thus average selling price/kg was ₦58.00 and the total cost of production was ₦35.76 which is the ratio of total cost (33,980.00) to total output per hectare (950.15), i.e. $\left(\frac{33980}{950.15} = 35.76\right)$. The Total Revenue (TR), Gross Margin (GM) and Net Farm Income (NFI)/ha of cotton were, ₦55, 108.70 [58*950.15], ₦22,248.70 [55,108.70-32,860.00] and ₦21,128.70 [22,248.70-1,120] respectively. From these results, it can be deduced that cotton farming is a profitable venture as the return on investment is ₦0.62. This finding is in conformity with the finding of Mshelia [7], Ibrahim [25] and Ibrahim [14].

Table 3 Average cost and returns to cotton farmers

Variables	Value (N)
A. Variable cost	32,860.00
B. Fixed cost	1,120.00

Total cost of production	33,980.00
C. Returns	
Total average output (kg)	950.15
Average price/kg	58.00
Total revenue	55,108.70
Gross margin (TR-TVC)	22,248.70
NFI (GM-TFC)	21,128.70
Return on naira invested (R.O.I.)	0.62
Source: Field Survey, 2012.	

3.3 Results of The Stochastic Frontier Production Functions Analysis.

Table1.3 shows the maximum likelihood estimates of the respondents. The results show that there is positive relationship between variable inputs with output variable. The relationship is also statistical significant at 10% level for land, labour and for fertilizer while agrochemicals is significance at 5% level respectively. Fertilizer coefficient has the highest value of 0.47, followed by family labour and farm size with 0.32 each, and indicated that if more of these variable inputs were use there will be more than proportionate increase in the variable output of cotton. Since fertilizer has the highest coefficient, it means that more fertilizer should be use by the farmers as 1% increase of it will lead to 0.47 increase in the variable output.

The inefficiency model, as in Table1.3 also, shows that the coefficient of the entire variable had negative sign. Farming experience and visitation by extension agent were statistically significant at 5%, while literacy level was statistically significant at 1% level. The negative and significant relationship of the variables in the inefficiency model suggests that inefficiency in production was less among cotton farmers. The signs and coefficients in the inefficiency model are interpreted in the opposite way such that a negative sign decrease inefficiency and positive sign increase inefficiency. This shows that some socio-economic characteristics have influence on the inefficiency of farmers' output. Inefficiency parameters established the fact that inefficiency of cotton production decrease with increases in farming experience. The estimate of variance parameter Sigma Squared ($\sigma^2 = 0.82$) is statistically significant at 1% level of probability and significantly different from zero, indicating a goodness of fit and correctness of the distributional form assumed for the composite error term.

Gamma ($\gamma = 0.56$) is statistically significant at 1% shows the amount of variation resulting from the technical inefficiency of the farmers. This means that 56% of the variation in farmers' output is due to difference in technical efficiency, implying that the ordinary least squares (OLS)) estimate will not be adequate in explaining the inefficiencies on cotton farming. The predicted technical efficiency varies across the respondents, ranging between 0.70 and 0.90 on the scale of maximum one with mean technical efficiency of 0.97. This implies that cotton producers are 97% efficient in the use of their input. The return to scale was 1.15 which is greater than one, suggested that the producers of cotton are operating at stage I of the production curve. At this stage, marginal physical product of cotton is greater than the average physical product (MPP >APP), meaning that the rate of inputs used in the production is increasing and it reaches its maximum at the end of stage I. Although, this is a rational zone of production yet the point of diminishing returns or efficiency has not been attained. The farmers can only attain that point where Marginal Physical Product (MPP) equals Average Physical Product (APP) (MPP=APP) with elasticity equal to one, i.e. a 1% change in input will bring about 1% change in output. This is the point where farmers use optimum amount of variable inputs.

Table 4 Maximum Likelihood of Parameters of Stochastic Productions

Variables	Parameters	Coefficient	t-value
Constant	β_0	0.43*	0.365
Land (x_1)	β_1	0.32*	0.298
Labour (x_2)	β_2	0.32*	0.206
Agrochemical (lit)/h (x_3)	β_3	0.04**	0.213
Fertilizer (kg)/h (x_4)	β_4	0.47*	0.396
Inefficiency model			
Constant	δ_0	0.27*	-0.298
Farming experience	δ_1	-0.26**	-0.242
Literacy	δ_2	-0.61*	-0.393
Extension agent contact	δ_3	-0.47**	-0.264
Variance parameters			
Sigma squared	σ^2	0.82*	0.24
Gamma	γ	0.56*	0.37
Minimum TE	0.37		
Maximum TE	0.97		
Mean TE	0.77		
Returns to scale	1.15		

Source: Survey Data, 2012. Note: *,**, and *** denote significance at 10%, 5% and 1% levels.

3.4. Frequency Distribution of Technical Efficiency of Cotton Farmers

Table1.4 shows the frequency distribution of technical efficiency of cotton farmers. The result shows that 80.25% of the cotton farmers fall to the range of 0.60 and above while, 19.75% of them fall below the range of 0.60. The best farmer has a technical efficiency of 0.97 while the worst farmer has a technical efficiency of 0.37. The mean technical efficiency is 0.77, meaning that farmers (on average) were able to obtain 77% optimal output from a given quantity of production inputs. This implies that farmers were not efficient as their observed output was 23% less than the maximum output. This can be increased by 23% through improved resource allocation with no additional cost. The magnitude of a mean technical efficiency of the farmers is a reflection of the fact that most of them carry out cotton production under technical condition involving the use of local inputs. The mode of the technical efficiency was 0.97 meaning that majority of the farmers had technical efficiency of 0.97.

Table 5 Technical efficiency score of cotton farmers

Efficiency Score	Frequency	Percentage (%)
0.30-0.39	4	4.94
0.40-0.49	5	6.17
0.50-0.59	7	8.64
0.60-0.69	25	30.86
0.70-0.79	17	20.99
0.80-0.89	17	20.99
0.90-0.99	6	7.41
Minimum 0.37		

Maximum 0.97		
Mean 0.77		
Mode 0.97		
TOTAL	**81**	**100**

3.5 Constraints associated with Cotton Production.

From Table1.5, we can see that low price of cotton seeds ranked first as one of the major constraints faced by cotton farmers in the study areas. This represented 21.88%, pest and disease infestation ranked second with 21.09%, while high cost of inputs, inadequate funds, shortage of labour and insufficient extension agents occupied the third, fourth and fifth positions with 19.53%, 15.63%, 11.72% and 10.16% in that order. This finding is in conformity with the work of Dittoh (1980) who worked on output and hectare response to price of cotton and wheat in North-eastern Ghana using time series data and observed that statistical estimate of supply response has been too low because of incorrect formulation of price to which the farmers react. In the same vein, the finding is in line with that of Mshelia [7], Aminu [10] and Ibrahim [25] who, also revealed that cotton farmers respond positively to price of seed of cotton in Adamawa State by cultivating more hectare of land the following year. The high cost of inputs in the study areas may be attributed to low supply of inputs used, especially fertilizer and insecticide which directly affected yield negatively as compared to yield of 1500kg-2000kg/ha by intensified inputs used by its out growers scheme in Adamawa State, Mshelia, [7].

The research also agreed with the findings of Yilmaz and Ozkan [28], who discovered that labour is the most critical factor in cotton production as its insufficient supply in critical period could result in total failure in farming business. The inadequate funds may probably be the reason why farmers could not purchase more of fertilizer and insecticides because of vicious cycle of poverty experienced by most of the peasant farmers in Nigeria. The insufficient extension agent is prevalent in most community of Nigeria today as one hardly see extension agent in villages. Even their demonstration plot or SPAT cannot longer be seen these days instead they are concentrated in towns and cities, where they can enjoy social amenities. This is a threat to agricultural growth and development as most farmers cannot supply correctly some agricultural innovations on their farms that require the attention of the extension agents.

Table 6 Constraints associated with cotton production.

Constraints	**Frequency**	**Percentage**	**Rank****
Low price of seed of cotton/Lint	28	21.88	1
Pest and Diseases infestation	27	21.09	2
High cost of input	25	19.53	3
Inadequate credit facilities	20	15.63	4
Shortage of labour	15	11.72	5
Insufficient extension agent	13	10.16	6

Total	128 *	100	

Source: Field Survey, 2012.
* Multiple responses used.
**Ranks in descending order of magnitude.

5 Conclusion

The study concludes that the stochastic frontier production function model is the best fit for the data. the study investigated the mutual effect of input on the output variance, whereas the profitability of cotton and technical efficiency were estimated. Results from the analysis recorded that one-naira investment in cotton production will yield 0.62 kobo. The results from the analysis also recorded that technical efficiency enhances the variability of cotton production in the study area with some socioeconomic characteristic influence. The results from the study shows that farmers in the study areas face some constraints in their production processes.

6 Recommendations

Based on the farmers' technical efficiency the study recommends that the conventional input factors such as, farm size, fertilizer, agrochemicals and labour are essential in the development of cotton production as they increase output positively in the production process. Therefore, an efforts and actions should be taken to expand efficiency performance at which the cotton farmers operate. Alongside cotton farmers themselves, government should play a fundamental role to improve efficiency through the extension agent to education the farmers on the use of new farming techniques as policy implications to be use by the government to achieve the set objective. Based on the findings of this study, it is recommended that policy action could be taken in order to help enhance cotton output, eradicate technical inefficiencies and tone down its effect in the production process.

References

1. S.C, Korcher, *Tropical Crops, a text book of economic Macmillan Publishers*, (New Delhi, India, 1981)
2. G. David and J. Adamu, *Crops of the Drier Region of Tropics*, (Longman Press. 1991)
3. Helfand, S.M.: Farm Size and the Determinant of Production Efficiency in the Brazilian Centre-West Proceeding of the 25th International Conference of Agricultural Economics, Durban, South Africa, 16th-22nd August, 2003, Pp 605 – 612, (2003)
4. Muhammad, B. Economic Analysis of Cotton Production in Gassol Local Taraba State. Unpublished PGD Thesis Agricultural Economics, Federal University of Technology, Yola, (2007)
5. CBN: Statistical Bulletin Vol. 18, December (2007)
6. RMRDC, (2004): Report on Survey of selected Agricultural Raw Materials in Nigeria Federal Ministry of Agriculture FCT Abuja, Nigeria.
7. S.I. Mshelia, Economic Analysis of Cotton Production in Nigeria. A Case Study of Afcott Out-growers Scheme in Adamawa State. Unpublished M.Sc. Agric Economics Thesis. University of Ibadan. Pp 15 – 18 (1991)

8. C.D.C.: Minutes of industrial Meeting of the Cotton Development Committee held at the Ministry's Conference Room, Federal Ministry of Agriculture and Rural
9. Development, Area II Garki, Abuja on Tuesday 30th August 2005.
10. B.E. Argbokam, Resuscitating Agricultural Production (Cocoa, Cotton, Groundnut, Palm oil and Rubber for Export. A paper presented at the 10th Annual Conference of zonalresearch units of the Central Bank of Nigeria, On the theme, Endowment, growth and Macro-Economic Management in Nigeria: held in Oweri: June 4th - 8th (2001)
11. S. Aminu, Economic Analysis of Cotton Production in Numan Local Government Area of Adamawa State, Unpublished B. Agric Tech., Agricultural Economics and Extension, Federal University of Technology, Yola, Pp 45 – 46 (2004)
12. R.A. Gbadegesin and E. Uyoubisere, Cotton Production in Nigeria. Extension Bulleting No. 14, Agricultural Extension and Research Liaison Services, Ahmadu Bello University, Zaria, Nigeria (1994)
13. USDA: United State Department of Agriculture, (2018).
14. ICAC: Supply and Distribution of Cotton Worldwide 2**(1)**: 1, (2002)
15. E.J. Ibrahim, Economic of Sole Cotton Production in Lau Local Govt. Area of Taraba State, Nigeria. Unpublished M.Sc. Agric Economics Thesis, Federal University of Technology, Yola. Pp 29 – 40 (2008)
16. Taraba State Official Dairy. Ministry of Education, Taraba State, Nigeria, (2010).
17. National Population Commission, *Republic of Nigeria Official Gazette,* Printed and Published by the Federal Government Printer, Lagos, Nigeria. No. 24 Vol. 94, 15th May, (2007)
18. E.F. Adebayo and J.K. Olayemi, Application of Goal Programming to Resource Allocation Behaviour of Dairy Producers in Adamawa State, Nigeria. *Journal of Sustainable Development and Environment* 1: 60 – 69, (2005)
19. X. Xu and S.R. Jeffery, Efficiency and Technical Progress in Traditional and Modern Agriculture Evidence from Rice Production in China Agricultural Economics 18: 157 – 165, (1998)
20. S.O. Ojo and P.B. Imoudu, Productivity and Technical Efficiency: A Comparative Analysis of oil Palm Farms in Ondo State, *Journal of Agriculture, Forestry and Fisheries, 1: 40-46,* (2000)
21. G.E. Battese and T.J. Coelli, Frontier Production Functions, Technical Efficiency and Panel Data". *Journal of Econometrics* **38**: 387 – 399, (1992)
22. D.K. Orgundari, S.O. Ojo and I.A. Ajibefun, Economics of Scale and Cost Efficiency in Small Maize Production. Empirical Evidences from Nigeria. *Journal of Social Sciences* 13**(2)**: 131 – 136, (2006)
23. B.E. Bravo-Ureta and L. Reiger, Dairy farm Efficiency Measurement using Stochastic frontier and Neo-Classical Duality. *America Journal of Agricultural Economics* **73**: 421 – 428, (1991)

24. T.J. Coelli, A Guide to Frontier Version 4.1: A Computer Programme for Stochastic Frontier Production and Cost Function Estimation. Department of Econometrics, (University of New England, Armadale, NSW 2351, Australia, 1995)
25. I.A. Ajibefun, Investigation of Technical Inefficiency of Production of Firms Under the National Directorate of Employment in Ondo State, Applied Tropical Agriculture: **3**: 15 – 28, (1998)
26. M.S. Ibrahim, Resource use efficiency in small-scale cotton production in Adamawa State, Nigeria. Unpublished M.Sc Agric Economics Thesis, University of Maiduguri.Pp 50 – 55, (2002)
27. R.A. Alabi and M.B. Aruna, Technical efficiency of Family Poultry Production in Niger–Delta, Nigeria. *Journal of Central European Agriculture,* 06 **(4)**: 531-538, (2006)
28. I.B. Oluwatayo, A.B. Sekumade and S.A. Adesoji, Resource Use Efficiency of Maize Farmers in Rural Nigeria: Evidence from Ekiti State. *World Journal of Agricultural Sciences*. 4(1): 91-99 (2008)
29. I. Yilmaz and B. Ozkan, Economic Analysis of Land Tenure Systems in cotton Production in Turkey, *International Journal of Agriculture and Biology* 6**(60):** 1023-1025, (2004)

Settlement of Indonesian disorientation of democracy: perspective of legal culture

Muhammad Yusrizal Adi Syaputra* and *Mirza* Nasution

Doctoral Program in Law, Graduate School, Universitas Sumatera Utara, Medan, Indonesia

Abstract. The orientation of Indonesian democracy is the establishment of an Indonesian rule of law based on Pancasila. The disorientation of Indonesian Democracy caused the unstable political situation, uncontrolled freedom of press, uneven law enforcement. This situation has resulted in the low quality of democratic implementation in Indonesia compared to ASEAN countries. This research uses a normative juridical method with qualitative. The theory of "law is the spirit of the volk" which pioneered by von Savigny and the theory of legal system by Friedman will be the theoretical analysis in this research. Based on theory law is the spirit of the volk by Savigny, the current Indonesian state administration is not in accordance with the culture and history of the Indonesian nation. The cause of disorientation of democracy in Indonesia, first, the existence of cultural degradation of society. Second, the destruction of the legal and political system. Third, uncontrolled freedom of press. Pancasila as the *volkgeist* (spirit of the nation) of Indonesia should be the basic of state administration. Based on the legal system theory, the revitalization of Pancasila's values and the legal culture education is the solution to resolve disorientation of Indonesian democracy.

1 Introduction

Better political, economic, legal, social and cultural change is at the heart of the reform movement of 1998. All the components of the student-driven nation bring democracy to build an open society life and increasingly visible public participation.[1]Post-New Order (era reform) in 1999 marked with the spirit of democratization and uphold human rights (HAM), but also colored euphoria of freedom and spirit of resistance to all things that smell of new order, so impressed irrational and emotional[2].

The reform movement was born as the answer to the crisis that hit various aspects of life in the political crisis, economy, law, and social crisis which in turn gave birth to the crisis of trust. The demands of the reform movement in 1998 were the transformation of the political system from authoritarian to democratic, the amendment of the 1945 Constitution, the freedom of press, the protection of human rights, the protection of constitutional rights of citizens and the implementation of elections.

* Corresponding author: yusrizal.1986@gmail.com

The reform wave of 1998 has ushered Indonesia into three political choices: first, the continuation of the reform movement. Second, the occurrence of the Revolution. Third, the ongoing involution. The choice happens to continue the path of reform.

During the reign of President B.J. Habibie, there was little a change. For the sake of Indonesia's economic recovery, the government continues to cooperate with the International Monetary Fund (IMF), granting freedom to the media, freedom of expression such as the liberalization of political parties, freedom of expression and the annulment of subversive legislation, and issuing human rights law[3]. In addition to the transition of political power from the new order to the reform order, it turns out that Indonesia's reforms provide a record of human rights violations. Some brief notes on human rights violations are to include:

First, there is a case that is not yet known who the perpetrators are, although there are indications that the perpetrator has a power of national security. This is evident from the case of Marsinah and the killing of Udin journalists. *Secondly*, even though the perpetrators are known and allegedly tight is the security apparatus, the perpetrator was never filed by the court. For example, the case of the DPP-PDI office raid on the Diponegoro road, Jakarta on July 27, 1996. *Third*, the existence of qualifiers certain areas as military operation area (DOM), in Timor-Timor (now Timor Leste), Irian (Papua), and Aceh. This has the potential to encourage the use of violence that leads to human rights violations, since such acts can be qualified as political violence so that the institution must be positioned as a responsible party. *Fourth*, even though the security authorities acknowledge their involvement in a case, the acknowledgment is deconstructed to merely a person's fault. Or the qualification of errors is denied merely a procedural error, so the punishment being handed down becomes light, a court decision unacceptable to common sense. For example, in the case of the Trisakti student shootings (May, 12th 1998), also in the case of kidnapping numbers of pro-democracy activists during 1997 to 1998.

The 1999 general election was an important political breakthrough of the President B.J.Habibie government in order to form a new government that gained popular power. The amendment of the 1945 Constitution and the direct presidential and vice-presidential elections also became the agenda, marked by the formation of a National Team of Reform towards the Civil Society to design a political, economic and legal reform program[4]. One of the working groups formed is the Law and Legal Reform Group, which is responsible for drafting the proposed amendments to the 1945 Constitution and direct presidential elections. This group produces two things. The main point is to proposed amendments of the 1945 Constitution, and the direct presidential election.

The principle of 1998 reform in Indonesia is the strengthening of the supremacy of civil society in Indonesia. This is triggered by the development that occurred in the international world. Renewal ideas such as freedom of the press, freedom of expression, political freedom, equality and justice enter through communication channels such as mass media, public discourse, and increasingly sophisticated global information technology.

Economic, political, and legal reforms will not produce long-term result if not seen as well as a cultural reform. The reformers should be at the forefront of enlightenment that builds the national awakening. The lack of enlightenment of democracy and human rights and the environment will only make the reformists suffer from myopia which in turn will only promote the short-term interests of the individual, family, class, more than the long-term interests of the whole nation. The lack of a cultural perspective, will only make reforms into temporary phenomena and keratonism will be control back in a period time. The deficient in cultural reform will make corruption, collusion and nepotism as a norm and the reform as an exception at the other side. It could proven as the Indonesian experiences during the Old Order and the New Order[5].

The escalation of complexity due to modernization brings the universal problems experienced by various societies, especially in Indonesia, namely the issue of social integrity and social solidarity amidst the facts of pluralism and relativism of values. Democracy in Indonesia, which became the spirit of the goal of the 1998 reform process, becomes invaluable if its structure is not based on the legal, social, economic, and political reform context and especially the morality of the Indonesian people.

In fact, currently Indonesia has experienced degradation of the system in the implementation of life in society, nation and state. The available law is increasingly confusing and fraudulent in its formation in the House of Representatives. Political parties no longer carry out their role as agents of social change of society, but rather to the means of achieving political power alone. The state economy is controlled by the owners of capital and the dependence of foreign loans. There is a disorientation of democracy that the 1998 reform movement aspires to, especially in the press or media. This is evident from the spirit of media coverage that is no longer based on the news to build and increase social awareness of Indonesian society or more precisely said that the press/media increasingly wild and can not be controlled by the government in providing information to the public. Indonesian society culture in the use of information technology increasingly unlimited change the behavior of society becomes worse because of easy to receive information that not yet known truth.

2 Formulation of the Problem

The problem to be answered in this research is; What is the form of disorientation of democracy in politics, law, government and media in Indonesia?

3 Research Method

This is a legal research that uses normative legal research methods. Normative legal research is a study that examines the study of documents, which uses various secondary data such as legislation, court decisions, legal theory, and can be the opinions of scholars. This research will uses qualitative analysis by explaining the existing data and theory that related to answer the problems.

4 Literature Review

The Hegel concept of the "universal idea" (universal spirit) has disturbed the romantic-historical German writers. To them, there is no such thing as a "universal soul". There is only "the soul of the nation". The universal soul can not negate the soul of the nation at the local level. The soul of the nation is an independent unit and is not subject to the universal value unit as Hegel states. The privilege of this nation's soul was initiated by Gustav Hugo (1764-1861) and fully developed by Frederick Carl Von Savigny (1779-1861).[6]

According to Savigny that to the organic relationship between the law and the character of a nation. Law is a reflection of Volkgeist. Savigny stated that the law is embedded in culture. The law is understood as the soul of the nation (*Volksgeist*), which is closely related to the common language[7].

Savigny's thesis states that the law from the beginning of history has attached national features. As with language, customs, and constitutions all of which are peculiar to the people. The law does not arise by chance but is born from the inner consciousness of the people. Seen as a form of cultural tradition, history and express a sense of cultural unity of a nation. According to Savigny there is no law that transcends time and space. Laws are always contextual and historical.

According to Savigny, the development of law and society through three stages. *The first* is development through the formation of political elements that are legal principles not found in the law, but are part of the belief and spirit of the society (*volksglauben*). *The second* is to transform political elements into legal technical elements of juristic skill. In this period the community is at the top of a legal culture and is a good time to codify the law. *The third* is marked by the decline in the existence of a society/nation. At this stage the law is no longer the breath and pulse of the life of a society, but only becomes an asset and hegemony of jurists. If this situation is present, then the nature of the society has lost its identity and by itself the law no longer has a meaningful role[8].

Savigny's view was then reinforced by his disciple G. Puchta. Puchta formulated that the law grows together with growth, and becomes strong with the strength of the people, and in the end it dies when the nation loses its nationality.[9]

The state organization which based on law and politics always gives the color that law and politics are two inseparable parts. The state organization is based on law, while politics is the driving force for the country to work to meet the needs of its people. Starting from the economic, social, administrative, service, legal and other needs. The implementation of the state in the form of a democratic government as it happens in Indonesia. Democracy is the choice of the will of the Indonesian people, adapted to the circumstances of Indonesian society. Adjustment of the implementation of democracy in Indonesia with the legal system is a step to become Indonesia becomes a strong legal state and make as a prosperous country.

The occurrence of synchronization to all areas of state administration is not limited to the mismatch between the law alone. But it relates to the whole sub-system of a legal system applicable in Indonesia. The Indonesian legal system is a sequence of parts of the overall forming of a complex national legal system and an interplay of entities.

The pattern of democratic implementation based on the law in Indonesia can not be separated from the implementation of state law. the implementation of positive law in Indonesia is based on the legal system adopted by Indonesia. Talking about the legal system can not be separated from a view of Lawrence M Friedman. Friedman is a legal thinker who argues that a legal system within a country can work well and must meet its elements entirely.

Friedman states that the three elements of the legal system consist of[10]:

1. Substance rule of the law, inside it covers all the rules both written and unwritten, both material law and formal law.
2. The legal structure, encompassing legal institutions, legal apparatus and law enforcement system. The legal structure is closely related to the judicial system implemented by law enforcement officers, in the criminal justice system, law enforcement applications carried out by investigators, prosecutors, judges and advocates.
3. Legal culture, is the emphasis of the culture in general, habits, opinions, ways of acting and thinking, which direct social forces in society.

The placement of Friedman's view of the legal system against Indonesia then, in principle, the Indonesian legal system can also be categorized as comprising three elements of the friedman's legal system.For the substance of the law, the implementation of government is based on the constitution and legislation available in various forms of written norms, and unwritten norms.

The legal structure of the legal system in Indonesia is also fulfilled because in Indonesia there are legislative bodies, legislative bodies, executive and judicial bodies (judiciary) institutions or legal institutions such as courts, prosecutors and others. For legal culture that seen from the characteristics of Indonesian society in implementing law and government.

The legal system of Indonesia is an open system, meaning among the elements that make up the mutual legal system. In addition, the Indonesian legal system receives influences from the environment, both in the form of information and in the pressure groups of the political elites who suppress and impose the will on bodies given the legislative and executive councils or the judiciary.[11]

The influence of the legal system in the implementation of government democracy in Indonesia is seen in all elements of Indonesian legal system. Indonesian democracy is the embodiment or implementation of the legal system within the framework of government. Democracy is a typical government that puts the will of the people and is based on the constitution. Democracy without constitution will be brutal and create chaos.

5 Result and Discussion

The rule of law is a very important tool in the administration of a democratic government. The development of Indonesia's increasingly dynamic government with a wide spectrum. The paradigm shift in the implementation of a centralized centralized government of Indonesia into decentralization also requires legal prerequisites in accordance with the paradigm. The demands of the 1998 reforms to make the government of Indonesia a democratic government have come into being with the development of a governance system. The spaces of government are filled with democratic colors, including law, politics, economy, social and others.

Today the democratic system is considered the best system among all the bad. That's why 90-95% of countries in the world claim to adhere to democracy. Although in practice its application depends on the interpretation of each country and the rulers of the so-called democratic country.[12]one of the most commonly expressed weaknesses is that democratic systems rely too much on majority voices in accordance with the doctrine of "one man one vote". The principle of democracy is according to Jimlly Asshiddiqie synonymous with the capitalist system in the business world that prioritizes the principle of "one share one vote". Who has the most shares is the one who determines the decision. Though the majority of votes do not necessarily reflect the truth and justice.

To compensate for the inherent weakness of the democratic system, the decision-making process in the dynamics of state power must be balanced with the principle of justice, nomocracy (rule of law). True democracy is a democracy based on law (certainty), order, fair and equality of the parties. According to Asshiddiqie that the idea of democracy and nomocracy must be aligned and in line to be said as a democratic state as well as a state law.

Modern democracy in line with the rule of law has the following basic criteria: first, the freedom that is expressed in opinion and the association set forth in the constitution. Second, public participation in politics. Third, the existence of commonality before the law and government on every citizen. Fourth, protection of human rights.

Historically, since the early movements of Indonesian independence, the leaders and predecessors of the Indonesian nation such as H.O.S.Tjokrominoto, Soekarno and Mohammad Hatta have emphasized the importance of popular sovereignty. Among others, it is because of the favor and the nature of that vocation, Sarekat Islam (SI) until the first bellah of this century appeared as the only symbol of resistance against the colonists. When later born various political parties as a container of the struggle to achieve independence, the sovereignty of the people is always included as ideals that will be realized if Indonesia gained independence. The spirit of upholding the sovereignty of the people that causes the founders of the Indonesian nation prefer the form of the state of the Republic rather than the form of a royal state.

People's sovereignty which is the basis of the implementation of the system of governance in Indonesia is set in the 1945 Constitution. It is a derivation of Pancasila as the foundation of the Republic of Indonesia. The five precepts of Pancasila are Belief in the One God; Just and civilized humanity; The unity of Indonesia; Democracy led by the wisdom of thoughts deliberations of representation and Social Justice for all the people of Indonesia

5.1 Politics

During the New Order period, the growth of political parties was limited as a result of the ongoing instability of parliamentary democracy in the early 1950s. In this case, the existence of political parties is more regarded as a problem. So the space is limited and its existence is only intended as a tool of the legitimacy of the then ruling regime of President Soeharto[13]. At that time the political party could not carry out its role and function, other than just as a means of mass mobilization in elections and a mere power channel.

2004 was the first year of the presidential and vice presidential elections in Indonesia directly by the people. People are given the freedom to choose a leader, and it is regulated in the 1945 Constitution. Until now three times the president and vice president of Indonesia directly elected by the people. The political process in the presidential and vice-presidential elections in Indonesia keeps the political party into a vehicle for the implementation of elections. Candidates for president and vice president are proposed by political parties either themselves or any combination of political parties with the provisions regulated by law. Political parties remain the main actors in the political system in Indonesia. Political parties become channels to occupy candidates for government authority either in the center or in the region. Slightly different from the presidential and vice-presidential elections, regional heads as leaders in the region are also directly elected by the people to be proposed by political parties or independently.

According to Huntington, stability, party hooliganism and party system will be highly dependent on the degree of institutionalization and participation. Broad participation coupled with a low degree of institutionalization of political parties will result in anomic politics and violence. Participation without an organization will degenerate into a mass movement, while organizations that do not engage in participation tend to lead to personal clicks.[14]

Political parties have a special place in the political system of Indonesia. But the political part has not contributed the most in the effort to increase and significant change of decision making in government. Political parties have lost the direction and purpose to give and accommodate the interests of society, so that the current political party becomes a place of transaction interests of the political elite who dominate power in Indonesia. Political parties turn into transactional, party cadres no longer really come from people who are fully part of political parties. Not a few senior party officials who served previously are members of other political parties, then move to another party, when he no longer get the original position of the party.

Based on these conditions, it is difficult to expect the political party will carry out its function maximally as an institution for the formulation of political ideals, articulating political interests, providing enlightenment, and conducting constructive public mobilization and conducting guidance and political recruitment. The plurality of post-reform political parties does not address the political demands of society as a whole. The large number of political parties is not proportional to the ability of political parties to aggress and articulate public interest as well as pluralistic.[15]

5.2 Law

In the field of law became a sharp spot in the 1998 reformation. The demand for legal reform is the absolute desire of the reform movement. As a result, all areas of law are subject to change both in the process of formation and implementation up to their supervision. In this research. Issues of law enforcement and formation into spaces that must be filled with change. Legal establishment (legislation) by House of Representative and President is explicitly regulated by the 1945 Constitution. The People's Representative Council together with the President has a legislative function that is the function to formulate law.

Based on Friedman's legal system theory that the implementation of a legal system influenced by legal structuring components, legal substance and legal culture in the Indonesian context becomes something that can be seen clearly;

The presence of legal institutions as part of the implementation of the legal system such as the legal establishment (the People's Representative Council with the President), and the implementing institutions of law such as the judiciary, the Police Service, the Prosecution Service, the Penitentiary, the Corruption Eradication Commission (KPK) affect how the law works within the Indonesian legal system.

Corruption is a form of abuse of power for personal, group or group interests. Corruption today is a global problem for every country including Indonesia. Eradication of corruption if guided by Friedman's legal system theory then all components of the legal system must be considered its role. From the start of the legal structure, its substance and its legal culture must be implemented well and in accordance with their respective functions. Improvements in the legal structure by optimizing the performance of legal institutions, eliminating rivalry in the handling of corruption cases between the police, prosecutors and the commission to eradicate corruption is one way to combat corruption in Indonesia. Building cooperation among law enforcement agencies and mutually shaping commitment responsibilities is the first step in war against corruption.

Culturally, the people of Indonesia have a soul that is embedded in the nation moral values and ethics that can be used as guidance of life in society, nation and state. The social character of Indonesian society with the spirit of *gotong royong* must be returned to the real purpose so as not to get that character or nature that becomes the basis of corruption, collusion and nepotism.

The appreciation of Pancasila values as Volksgeis (soul of the nation) to all levels of society is a form of reform of the culture of Indonesian society. Pancasila is supposed to be a life guide and guidance in running the law and government of Indonesia.Pancasila as the spirit of nationalism.

5.3 Government

Democratic principles embodied in legislation are reflected in the institutional structures and mechanisms of state and government to ensure the establishment of the legal system and the functioning of democracy. From an institutional perspective, democracy is usually organized through a system of separation of powers or distribution of powers. Separation of powers tends to be horizontal in the sense that separation of powers into functions reflected in checks and balances of state institutions. In a vertical sense, the power of state administration is passed down to state institutions under the constitution.[16]

The implementation of presidential government system in Indonesia is regulated in Article 4 of the 1945 Constitution. The mechanism of election of the head of government in the presidential system is through direct election by the people, while the head of government in the parliamentary system is not through direct election by the people, but through mechanism in parliament so that the basis of legitimation politics comes from parliament. Consequently, the head of government (prime minister) in the parliamentary system is accountable to parliament.[17]The basis of presidential legitimacy in the presidential system is politically derived from the people. The basis of legitimacy is obtained by the president through direct electoral mechanism by the people with a fixed term of office. Therefore, politically the president is responsible to the people not to the parliament. The consequence is a permanent presidential term, president and vice president can not be dismissed amid his term of office for political reasons.

The implementation of Indonesian presidential system in multiparty system can be seen that the application of presidential system with multiparty is not suitable. Currently,

Indonesia is implementing an extreme multiparty system, should the implementation of a presidential system be implemented with a limited multiparty system. Presidentialism prevailing today needs to be refined in order to produce a strong, stable and effective system of government on the one hand, and to avoid authoritarianism on the other . The effectiveness of presidential government can, among other things, be achieved through the simplification of party system and political grouping in parliament, while authoritarianism can be avoided through the institutionalization of oppositional forces.

5.4 Media

In the context of democracy, mass media or the press is expected to become an institution that contributes to strengthening democracy through the coverage of political life that can educate public readers to understand public issues so as to provide quality political support. Therefore, efforts at the institutional level have been made through the creation of laws that create a democratic media system, but unfortunately there have been no serious attempts to reform the media workers.[18]

Habermas emphasized the importance of the availability of "undistorted communications" channels, which are the primary means of liberation for participation in the public sphere, the place where individuals can interact with society at large. It is undeniable that the provision of objective and partial channels will help minorities to engage more actively. In other words, "undistorted communication" will empower minority groups in demanding position; also demand equal rights.[19]

The reality of the existence of the press as a pillar of democracy in Indonesia today no longer plays its role well. Some electronic media or print media that no longer be a tool of people's voice, but has become the owner of political elites. There are several private television stations in Indonesia where the owners or high-ranking officials are plunged into practical politics, and use the television media as a tool of personal political campaigns, groups or groups. Media press in Indonesia, currently no longer stands on independence and neutrality in the news. The media has changed its function, becoming a tool of particular elite political control or ruler in order to provide information to the public.

But as a comparison in the United Kingdom, starting in 2001 the type of digital media, which uses internet, mobile cellular, and special television (digital) has been used for public service, as well as commercial. Sir Christopher Bland, head of the BBC, in 2005 the digital media system was used extensively to complement the conventional broadcast role both for television and radio. In 2001 the BBC in his country began to enjoy the digital television era with BBC Three, BBC Four, and BBC Children Channel. As for radio, digital systems are used for BBC Network X, Network Y, Network Z, BBC Sport (BBC Radio 4 and 5), and BBC Asean Network. While outside the BBC who already use the digital system is ITV, Channel 4, Chanel 5, and S4C. But for Indonesia, this type of media is still very limited, because it is only an alternative media.[20]

Media commercialization in Indonesia and the suburings of mass media business in Indonesia make new land the political means of power. Weak legal substance that regulates media/press in Indonesia to make the media or press can easily inform the news that is not necessarily the truth.

6 Conclusions

6.1 Summary

Implementation of democracy in Indonesia in the field of politics, law, government, and the field of a press in Indonesia has not yet realized the criteria in accordance with the ideals of democracy of Indonesian society. The implementation of democracy in politics, law, government, and Press does not work well. Political parties have not implemented their roles maximally, law enforcement is not fair, there are government policies that do not side with the people, and freedom of the press that violates the law and ethics.

6.2 Recommendations

Firstly, in the political field is to carry out the function of political parties as a pillar of democracy and instill political behavior or culture in the political elite and society in accordance with the values of Pancasila.

Secondly, In the field of law, it is recommended to revitalize the process of formation, development and law enforcement in Indonesia. The formation of law in accordance with the values of Pancasila and the interests of the people of Indonesia with the standardization of the process of formation will result in good, sustainable, equitable, fair and beneficial law for all Indonesians. Legal development is done by establishing a centralized legal institution in order to avoid disorder, miscommunication in carrying out the responsibility of developing the national law. Law enforcement must be resolved through procedural rules and implemented by competent legal institutions. The existence of synergy between law enforcement agencies such as Police, Attorney, Corruption Eradication Commission, and Court and Correctional Institution is the effort to restore the democratic orientation of national law.

Thirdly, in the field of government, it is suggested to the Government of Indonesia to realize policies that prosper the people of Indonesia.

Fourthly, in the press field, it is advisable to limit press freedom to law and ethics, maintain media neutrality by separating the political interests of media owners with journalistic professionalism.

References

1. F. Hamzah, *Demokrasi transisi korupsi : orkestra pemberantasan korupsi sistemik,* p. 2 (Yayasan FAHAM Indonesia, Jakarta, 2012)
2. Wahyuno, S. Arinando, dan N. Triyanti (Eds.), *Menggugat superiorisasi mahkamah konstitusi dalam menyelesaikan sengketa PEMILU, memahami hukum: dari konstruksi sampai implementasi,* p. 254 (Rajawali Pers, Jakarta, 2009)
3. A. Priyadi, A.E. Priyono, dan U. Hamid (Eds.), *Reformasi di mata keluarga korban pelanggaran HAM, merancang arah baru demokrasi : Indonesia pasca reformasi,* p. 9 (Kepustakaan Populer Gramedia, Jakarta, 2014)
4. V.S. Subekti, *Menyusun konstitusi transisi: pergulatan kepentingan dan pemikiran dalam proses perubahan UUD 1945,* p. 63 (PT. RajaGrafindo Persada, Jakarta, 2008)
5. A.A. Nugroho, *Rakyatisme : dan esai-esai lain,* p. 152-153 (Kompas, Jakarta, 2017)
6. B. L.Tanya, Y. N. Simanjuntak, M.Y. Hage, *Teori hukum : strategi tertib manusia lintas ruang dan generasi,* Cetakan III, p. 103 (Genta Publishing, Yogyakarta, 2010)
7. R. Cotterrell, *Law, culture and society : legal ideas in the mirror of social theory,* p. 103 (Ashgate Publishing Limited,England, 2006)
8. L. Pospisil, *Anthropology of law: a comparative theory*, p. 142 (Willey, USA, 1971)
9. Rahardjo, *Ilmu hukum,* Cetakan VIII,p. 316 (PT. Citra Aditya Bakti, Bandung, 2003)

10. L. M. Friedman; *The legal system; a social scince prespective*, p. 12-16 (Russel Sage Foundation, New York, 1975)
11. B. Mustapa, *Sistem hukum Indonesia terpadu*, p. 74 (PT. Citra Aditya Bakti, Bandung, 2003)
12. J. Asshiddiqie, *Pokok-pokok hukum tata negara pasca reformasi*, p. 146 (PT. Bhuana Ilmu Populer, Jakarta, 2007)
13. B. Winarno, *Sistem politik indonesia era reformasi*, Cetakan II,p. 100 (Media Pressindo, Yogyakarta, 2008)
14. S.Huntington, *Tertib politik pada masyarakat yang sedang berubah (trans.)*, p. 476-477 (Rajawali Press, Jakarta, 2004)
15. A. Sanit, *Demokrasi, kekuatan masyarakat dan strategi alternatif*, in M. MD dan A. MWK, *Reformasi politik dan kekuatan masyarakat*, p. 130 (LP3ES, Jakarta, 2002)
16. A.M. Fatwa, *Potret konstitusi pasca amandemen UUD 1945*, p. 11 (Kompas, Jakarta, 2009)
17. H. Y. AR, *Presidensialisme setengah hati dari dilemma ke kompromi*, p. 17 (PT. Gramedia Pustaka Utama, Jakarta, 2010)
18. I.G.N. Putra, Demokrasi dan Kinerja Pers Indonesia, *Jurnal komunikasi*, **3**, **2**, 132 (2004)
19. Centre for Innovation Policy and Governance (CIPG), Tantangan media dan demokrasi: seri 4 rangkaian CREAME (critical research and methodology), http://cipg.or.id/wp-content/uploads/2015/05/CREAME-4.-Media-Demokrasi-2012.pdf
20. M. Solihat, Diversifikasi media massa dan demokrasi di Indonesia; penguatan peran media massa serta masyarakat dalam mewujudkan demokrasi, *Jurnal Ilmu Politik dan Komunikasi*, **2**;101 (2015)

Triangular concept of legal pluralism in the establishment of consumer protection law

Muldri Pudamo James Pasaribu* and *Ningrum Natasya* Sirait

Doctoral Program in Law, Universitas Sumatera Utara, Indonesia

Abstract. The new paradigm in legal pluralism is closely related to the phenomenon of globalization. Laws of various levels move into limitless territories. There is a strong contact and adoption between international, transnational, national and local laws. Such circumstances make it impossible for mapping that a certain law (international, national, local) is separated from other law systems. This is a normative legal research with a comparative law approach. Law as a global phenomenon has the common values throughout the world, namely ethical moral values, social values and formal values of the state. The same values apply to the consumer protection law in Indonesia. Law Number 8 of 1999 on Consumer Protection (UUPK) in Indonesia was developed on the basis of legal pluralism. The data were analyzed using a Triangular Concept of Legal Pluralism developed by Werner Menski. In conclusion, UUPK is a form of legal pluralism. It is enacted based on the community needs, legitimized by the state and based on values and ethics.

1 Introduction

Legal pluralism is not a new phenomenon in Indonesia. Customary law used to be the original Indonesia's legal system. However, in its development, Indonesia adopted the Islamic Law System, and then, due to a long period of Dutch colonialism, it adopted the Dutch legal system rooted from the Roman law system. Therefore, examining the Indonesia's legal system cannot be done by determining to which cluster the legal system is oriented, either to a civil law system or to a common law system. This is the evidence that the pluralism of the legal system has existed in Indonesia since a long time ago.

In addition, it has been proved that in some parts of the archipelago there was once a common law legal system, e.g. in Kalimantan under the British colonies and in Papua under the Dutch colonies. Moreover, geographically, these two areas are very close to the countries adopting Anglo-Saxon legal system, such as Papua with Australia, Kalimantan with Malaysia, Brunei Darussalam and Singapore. After its independence, the Republic of Indonesia enacted a legal unification referring to a Civil Law legal system.[1]

If the historical reality mentioned above is used as a reference or benchmark in categorizing Indonesia legal system orientation, it can be said that it lies on the both sides. This means that Indonesia adopts a mixed legal system or Mixed System. The tendency of

* Corresponding author: muldripasaribu73@gmail.com

adopting a mixed system is also inevitable due to the continuous recognition of the original law of Indonesia (Adat Law/Customary Law) as a legal source. In addition, the continuous adoption of a legal codification as a civil law system shows that Indonesia adopts a legal pluralism.[1]

Until recently, there have been many concepts and attributes about legal pluralism proposed by the experts. The legal pluralists of the 1970s proposed the concept of legal pluralism, although somewhat varied, as the existence of more than one legal system within the same social field. In line with this idea, Sally Engle Merry argues, "legal pluralism is generally defined as a situation in which two or more legal systems coexist in the same social field".[2]

The legal pluralism should always consider the state laws on the one hand and the people's laws on the other. Such people's laws include customary law, religion, customs or other social conventions that are viewed as law. In addition, in the current era of globalization, it is necessary to take into account the presence of international law in the arena of legal pluralism. In the empirical reality, especially in the field of economy and human rights, the effect of international laws is very visible.[3]

The consumer law and the consumer protection law are two areas that are difficult to separate and draw the boundaries. Essentially, the consumer protection law is part of the consumer law. The consumer law serves as the overall principles and rules governing relationships and problems of supply and use of products (goods and/or services) between providers and consumers in social life.[4]

The scope of consumer protection law as a special part of the consumer law is the overall principles and rules governing and protecting consumers in the relationships and problems of supply and its users in social life. For consumers, it takes legal principles that prevent dishonest acts in business, or the existence of market dominance with various business practices. It is possible that such business practices hamper the entry of new companies or harm other companies in unnatural ways. As for the end-user, it is necessary to have legal rules that guarantee the safe conditions of every product for human consumption, supplemented with correct, clear, honest and responsible information.[4]

Based on the above explanation, it can be defined that consumer law and consumer protection law in Indonesia are established based on three elements that include the normative element as the basis of its applicability, the sociological element as the element of society needs and the value element needed as the ethics in its enactment.

Werner Menski offers a conceptual approach called "Triangular Concept of Legal Pluralism". According to Menski, law as a global phenomenon has the common values throughout the world, namely ethical moral values, social values and formal values of the state. Essentially, a law comprises a number of cultural variations and always requires plurality. Facing the globalization era, modern legal experts have abandoned three classical legal approaches – normative (positivism), empirical (sociological, anthropological, psychological and others), and values and moral (philosophical) approaches – that tended to be narrow by using only one type of approach. However, the theory of triangular concept of legal pluralism uses all the three approaches.[5]

2 Formulation of the Problem

The problem of this research is formulated as: how is the formation of Consumer Protection Law in Indonesia viewed from the triangular concept of legal pluralism, and how is the possibility of that concept uses in a different legal system?

3 Research Method

This type of research is an analytical descriptive legal research. The data sources were secondary data comprising primary, secondary and tertiary legal materials.[6] Normative legal methods refer to legal norms in various laws and regulations.[7] Data collection is done through library research study, i.e. document data collection, literature and studying the provisions of related legislation. The qualitative data analysis then will be used through a comparative law approach which is oriented to the legal theory approach as the base of thinking. Comparative approaches are used by comparing legal systems with one another, which has more or less similarities, so that the similarities and differences between the legal systems were found.[8] Triangular concept of legal pluralism theory by Menski then tested to get a new concept related to the problems to be answered in this research.

4 Literature Review

In a global context, law has similarities all over the world, in the sense that everywhere the law is made up on the basis of ethical values, social norms, and rules made by the state, although later in reality, there are many more specific cultural variations. The Menski's legal pluralism approach uses the three main types of legal approaches: laws created by the society, laws created by the state and laws that arise through values and ethics.[5]

According to Gerald Turkel, to understand the legal system comprehensively there are three approaches that can be used in, that is: (1) The moral approach to law, which focuses his attention on the moral basis of the law and its legal validity is the legal consistency with ethics external or moral values, (2) The jurisprudence approach to law, which is focused his attention to the independence of the law, and his legal validity is internal law consistency with the rules of the norm, and principles of the law itself; and, (3) The sociological approach to law, which focuses attention on law and social action, where the validity of the law is consequences law for its people.[9]

The theory and approach advanced by Gerald Turkel later developed further by Werner Menski by combining the three approaches so that a new theory is emerging known as the theory of triangular concept of legal pluralism.

Werner Menski offers a new legal interaction that would serve as the solution, through the idea of a new legal pluralism, with a combination of state law (positivism), society (socio-legal approach), and natural law (moral/ethic/religion).[5]

The law as a tree, then we cannot look at parts of the tree partially, but totally. We should not look at any of the woods only in function as "roots" (which in law are the societies and cultural values) that absorb food from the soil, we should not look at any of the woods only in the form of stems that reinforce the tree (the law is the law positive), and we should not look at any of the woods only as branches of the branches that extend into the sky and in various directions to inhale the heavenly scent (in law are moral, religious, and ethical values).[10]

5 Results and Discussion

Consumer interests have long been a concern in all countries across the world, and consumer protection issue is one of the most important issues in the world today. The reason is that all members of the society are consumers who need to be protected from the quality of goods or services provided by the producers to the public. In reality, consumer is the decisive party in the fostering of capital to drive the wheels of the economy.[11]

The consumer protection movement in Indonesia is becoming popular from the similar movement in the United States. The Indonesian Consumers Foundation (YLKI), popularly known as a pioneer of consumer advocacy in Indonesia, was established on 11 May 1973.

YLKI together with BPHN (National Legal Development Board) formulated the Consumer Protection Bill. In the late 1990s, the Consumer Protection Law was not only championed by consumer agencies and the Ministry of Commerce, but it was also strictly required by the international financial institution (IMF/International Monetary Fund).[12] Another important development was that in 1962 President John F. Kennedy sent to the so-called Special Message on Protecting the Consumer Interest to the American Congress, where he established four rights that every consumer possessed: the right to safety, the right to be informed, the right to choose, the right to be heard.[13]

In addition, the factors that influence the formulation of UUPK in Indonesia are the emergence of several cases that harm consumers and end with a solution that does not satisfy the consumers. In the case of the Republic of Indonesia against Tan Chandra Helmi and Gimun Tanno [14] known as the case of toxic biscuits, for example, the consumers' claims were viewed only from the criminal and administrative aspects. As a result, the victims or consumers were not compensated on the basis of civil liability. Another case involved Janizal et al. against PT Kentamik Super Internasional, known as the Perumahan Narogong Indah Case. In this case, the developer was won, and the developer even sued back the consumers who were considered to have committed defamation. [15]

In this case, there is a group in the society whose role is very important to note in causing legal movement. Epistemic community is an entity in which academicians, experts, scientists or legal drafters interact across state boundaries to generate the concept of global norms used as the solution to a certain problem. The definitive conceptual framework of an epistemic community is a network of professionals with recognised expertise and competence in a particular domain and an authoritative claim to policy relevant knowledge within that domain or issue-area.[16]

Furthermore, UUPK was issued by the Government of Indonesia on 20 April 1999. The UUPK came into effect on April 20, 2000. This period was the beginning of the recognition of consumer protection and since then it has become legitimately formal as a means of legal force for the consumers and as a demand for the business actors as the providers/producers.

In its current development, there is a tendency to expand the scope of the consumer protection law, as done by The Economic Law and Improved Procurement System Project (ELIPS), which proposes 9 (nine) items of consumer protection law formulation: (1) inequalities in bargaining power; (2) freedom of contract versus justice in contract; (3) requirements to provide information to consumers, including general announcement law and financial announcement law; (4) regulations on the seller's behaviour/conduct, including false guidance/direction and cunning in the trade; (5) regulations on product quality, including product warranty and safety; (6) access to credit (reporting, credit, non-discrimination); (7) limits of the right to terminate the guarantee period; (8) price regulation; and (9) rectification. [17]

Consumer protection in the era of globalization becomes very important because consumers do not only have universal rights but also very specific rights both related to its situation and condition. Thus, in addition to universal values, the legislation should also contain specific values in accordance with the values that grow in the society. Changes in the social structure always require changes in the law. [18]

Carefully studying the articles in UUPK reveals that some of the provisions contained in the law are in accordance with the values of Islamic business ethics. Although written in different wordings, the substance and purpose of such provisions are just the same – to protect consumers. For example, the provisions contained in Article 7 letter a) a, b, c, d, e regulate the necessity of having good intentions in doing business. In addition, Article 8 paragraph (2, 3, 4) defines the obligation to sell goods with good quality. The prohibition of concealing defective goods in Article 8, not playing oaths (providing incorrect advertisement information) and not rivalling friends by fooling consumers with gifts are substances containing ethical elements.[19]

Related to the Triangular Concept of Legal Pluralism approach, these three facts above, can be described:

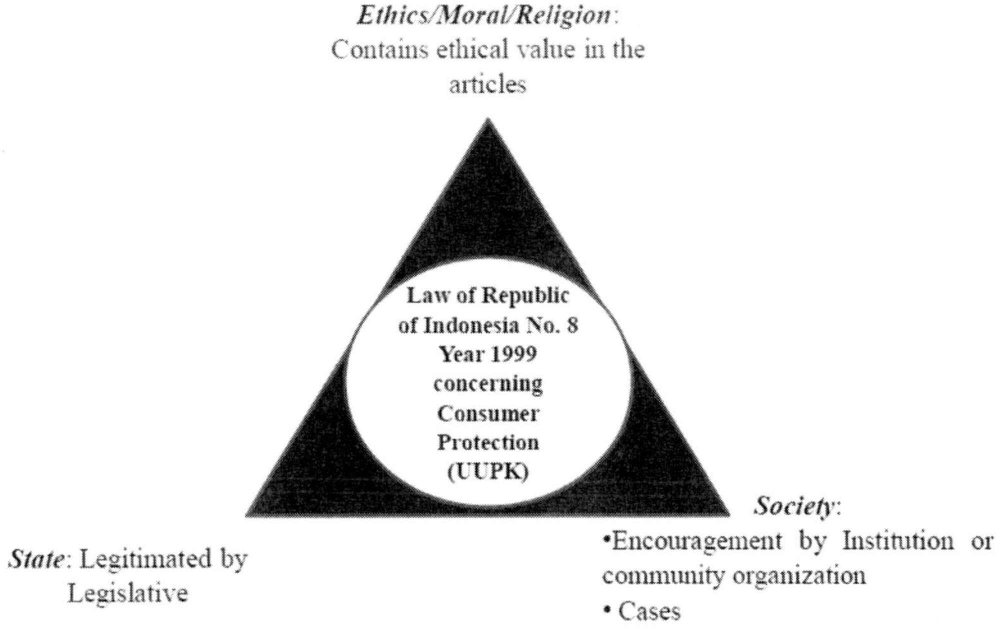

Fig. 1. Establishment of Indonesian Consumer Protection Law with Triangular Concept of Legal Pluralism approach

Figure 1 above shows that the establishment of the UUPK cannot be separated from the development of the three regions around the circle. This UUPK area is a legal pluralism. If viewed from the Triangular Concept of Legal Pluralism, regulations in the field of consumer protection will continue to grow. The area of pluralism in the triangle will continue to expand. One of the examples that will expand such area is a series of endorsements of the establishment of the World Trade Organization (WTO) which has been ratified by Indonesia through Law Number 7 of 1994 on Endorsement of Agreement Establishing the World Trade Organization. This endorsement enables products from other countries to meet the Indonesian market, which has implications for the increasing interest of Indonesian consumers to purchase.

Today, the growth of internet users increasing rapidly. The internet is an effective medium for businesses to introduce and sell goods or services to potential customers from all over the world. The practice of online transactions, namely e-commerce, causes many problems that tend to harm consumers. One of them is a legal problem in conducting transactions. E-commerce transactions are also a sale and purchase contract that is the same as conventional transactions. The difference is the media used. In e-commerce transactions, the media used is electronic media, namely the internet. So that the agreement or contract created is through online. In making an obligation, it must be clear in terms of national legal aspects through consumer protection law, as well as aspects of international law through international agreements or harmonization of laws.

Strengthening consumers through consumer education will also greatly affect the changes in consumer behaviour. This development is what makes the need to do amendment of UUPK to protect the development of modern consumers' needs, such as this online shopping behaviour.

The amendment of UUPK can be done by considering the eight principles under Fuller's principles of legality, (1) a system of governance operates through general norms, and all or most of the norms partake of the following properties: (2) they are promulgated to the people who are required to comply with them; (3) they are prospective rather than retrospective; (4) they are understandable rather than hopelessly unintelligible; (5) they do not contradict one another and do not impose duties that conflict; (6) they do not impose requirements that cannot possibly be fulfilled; (7) they persist over substantial periods of time, instead of being changed with disorienting frequency; and (8) they are generally given effect in accordance with their terms, so that there is a congruence between the norms as formulated and the norms as implemented.[20]

One of the conditions is found in the number 7 (seven) that is consistent at all times. According to Fuller, the stipulation that the law must be constant/ consistent at all times is not absolute, since the law must respond to changes occurring at any time (Introducing such frequent changes in the rules). The law should be able to respond to changes that occur at any time, meaning that the regulations in the field of consumer protection should be able to accommodate all kinds of cases or problems such as online shopping behaviour that arises from the development of the times in order to avoid the legal vacuum.

If this concept of legal pluralism from Menski related to the concept of the three elements of the legal system introduced by Lawrence M. Friedman, then we can say that the plurality of laws concerns not only the substance or structure but also the higher plurality of elements of "legal culture" which includes the plurality of existing habits, the plurality of opinions that exist, the plurality of existing beliefs, as well as the plurality of ways of thinking and acting in the field of law.[21]

The differences which exist between civil law and common law should not be exaggerated. It is also important to note that differences on many issues exist both among civil law and among common law countries. The differences between civil law and common law systems are more in styles of argumentation and methodology than in the content of legal norms. By using different means, both civil law and common law are aimed at the same goal and similar results are often obtained by different reasoning. The fact that common law and civil law, despite the use of different means arrive at the same or similar solutions is not surprising, as the subject-matter of the legal regulation and the basic values in both legal systems are more or less the same.[22]

Even though it be admitted that the civil law and the common law started from opposite extremes, it is sometimes said that as a result of the movements each has made in the direction of the other, there is no longer much difference between them. The same social needs, and similar economic and technical conditions, have led to the adoption of similar solutions for their legal problems.[23] The similar between the civil law system and the common law system is that both are open to the development of the times.

6 Conclusions

Globalization is not only indicated by borderless state but also by borderless law. The laws of a certain region are able to penetrate into other limitless territories. International and transnational law can penetrate into the territory of any country, even into the local area down to the grassroots level. On the contrary, local laws and principles are also adopted in part or in whole and may become an international law. This is called the legal pluralism.

The establishment of UUPK can be studied through Triangular Concept of Legal Pluralism approach, where the social pressure of local and international community encourages the state to legitimize it. The interaction between moral values, ethical values and religious values with the state and the society in the UUPK ultimately creates a new legal pluralism. The national market is also increasingly open through the process of economic globalization. Starting with the 1994 World Trade Organization agreement. In this situation the government must further increase legal efforts to ensure legal certainty and people prosperity. In line with the development of modern consumer society's needs, it is necessary to do amendment of legislation that would fit the development and needs of the society.

A legal system is a part of the life of the people in every country. Different characteristics influenced the culture of the people. Each legal system may have some advantages and deficiencies. If a foreign legal system has some advantages, there is a possibility to incorporate them in the domestic legal system. In that way the resulting convergence of the two legal systems can only contribute to their common goal of creating a fair and just legal system which can provide legal certainty and protection to all citizens and legal persons.

References

1. N. Qamar, *Perbandingan sistem hukum dan peradilan*: *Civil law system dan common law system*, p. 3-4 (Refleksi, Makassar, 2010)
2. S. E. Merry, Legal pluralism, *Law Soc. Rev.* **22:** 869-896 **(**1988)
3. S. Irianto, Sejarah dan perkembangan pemikiran pluralisme hukum dan konsekuensi metodologisnya, *Wacana Antropologi* **1** (1998)

a. Z. Nasution, *Hukum perlindungan konsumen: Suatu pengantar*, p. 21-22 (Diadit Media, Jakarta, 2006)

4. W. Menski, *Perbandingan hukum dalam konteks global: Sistem Eropa, Asia dan Afrika*, Translated by M. Khozim, p. 228-242 (Nusamedia, Bandung, 2008)
5. Amiruddin and Z. Asikin, *Pengantar metode penelitian hukum*, p. 118 (Raja Grafindo Persada, Jakarta, 2004)
6. S. Soekanto and S. Mamuji, *Penelitian hukum normatif*, p. 10 (Raja Grafindo Persada, Jakarta, 2004)
7. Soeroso, *Perbandingan hukum perdata,* p. 6 (Grafika, Jakarta, 2005)
8. G. Turkel, *Law and Society Critical Approachs*, p. 10 (Allyn and Bacon, USA, 1996)

a. Ali, *Menguak teori hukum dan teori peradilan*, p. 159 (Kencana, Jakarta, 2009)

9. J. Sidabalok, *Hukum perlindungan konsumen di Indonesia*, p. 23-24 (Citra Aditya Bakti, Bandung, 2014)
10. Sudaryatmo, *Memahami hak anda sebagai konsumen*, 1st Edition, p. 23 (PIRAC, Jakarta, 2001)
11. Z. B. Carroll, *Busines and society*, in K. Bertens, *Pengantar etika bisnis*, p. 248 (Kanisius, Yogyakarta, 2013)
12. Putusan PN Tangerang No. 30/Pid.B/1990/PN/Tng
13. Putusan PN Jakarta Pusat No. 3138/K/Pdt/1994/PN.Jkt.Pst. Putusan Kasasi MA No. 3138/K/Pdt/1994.
14. P. M. Hass, Introduction: eepistemic communities and international policy coordination, *Int. Organ.*, special issue: Knowledge, power, and international policy coordination, *Cambridge Journals*, **46:1**; 1-35 (1992)
15. Z. Nasution, Sekilas hukum perlindungan konsumen, *Majalah hukum dan pembangunan*, Tahun XVI, **6**, 70 (1986)
16. Watson, *Comparative law and legal change*, Lecture at King's College on 6 February 1978, and the author wishes to dedicate it to Otto Kahn-Freund.

17. N. S. Imaniati, Perlindungan konsumen salah satu upaya penegakan etika bisnis pada masyarakat islam dalam era globalisasi, *Jurnal Mimbar,* **4 Th.XVII** (2001)
18. L. L. Fuller, *The morality of law*, 2nd ed., chap. 2. (Yale University Press, New Haven, Conn., 1969)
19. D. Johnston, *Roman law in context*, p. 6 (Cambridge University Press, United Kingdom, 2004)
20. Pejovic, Civil law and common law: two different paths leading to the same goal, Victoria University of Wellington Law Review, **32**;817-842 (2001)
21. J. Dainow, The civil law and the common law: some points of comparison, *The American Journal of Comparative Law,* **15**:**3**; 419-435 (1966)

E3S Web of Conferences 52, 00033 (2018) https://doi.org/10.1051/e3sconf/20185200033
CSSPO 2018

The prevention of child trafficking crimes and its legal reform

Nanci Yosepin Simbolon* and *Madyasah* Ablisar

Doctoral Program in Law, Graduate School, Universitas Sumatera Utara, Medan, Indonesia

Abstract. The news about human trafficking especially children in Indonesia is increasing both domestically and across borders. The crime also includes in the form of labor smuggling, immigrant smuggling, slave trade, women and children. One of the most serious and very complex issues is the impact it has on and directly related to the child's fate. In 2016, the Ministry of Women Empowerment and Child Protection reported, there were 943 victims of human trafficking reported in 65 the media. Separately, the Indonesian Child Protection Commission officially identifies 307 child trafficking victims. It remains unclear whether these two processes lead to an investigation or provision of protection to victims. This research uses normative research method to find out the prevention of child trafficking crime and form of legal renewal about child trafficking. This study states that the prevention of child trafficking victims is by prevention, prosecution to traffickers, and protection measures against victims. In addition, there needs to be a renewal of criminal law which should also focus on the protection of victims of trafficking in persons so as to provide opportunities for victims to gain their rights.

1 Introduction

The case of child trafficking is also linked to the praxis of globalization driven by cross-border business practices, transnational corporations, and international economic and financial institutions. In this context, globalization is driven by the exploration of the business sector that continually seeks new areas for production, distribution, and markets most profitable for the process of capital and profit accumulation. The existence of likage between child trade issues. Globalization and the transnational business of globalization pressure has led to child labor and forced labor systems.

The news about human trafficking especially children in Indonesia is increasing both domestically and across borders.[i][1] This can be seen from various crimes committed by individuals and corporations within the borders of an increasingly diverse country. The

* Corresponding author: yosepinn@yahoo.com

crimes also include among others the smuggling of labor, immigrant smuggling, slave trade, trafficking of women and children.

The trafficking of children in Indonesia has threatened the existence and dignity of humanity that endanger the future of the child.[2] Globally, child trafficking is an organized crime that extends beyond the border of the country so it is known as transnational crime. Indonesia is listed and declared as one of the countries of international child trafficking and transit, especially for the purpose of prostitution and child labor in the world.[3]

This unity of commitment to the abolition of child trafficking was known through the Palermo protocol in 2001. The deal on the abolition of child trafficking as a global issue, in line with the scope of the deal of eliminating terrorism, arm smuggling, ilegally drug trafficking and psychotropic, money laundering, smuggling people and child trafficking. Indonesia has ratified the Palermo protocol for the abolition of transnational crime.[4]

The Convention on the Rights of the Child (CRC) is one of the conventions governing matters relating to the protection of the rights of the child. Article I states that y is meant by child is every human being below the law. Under this provision it is further determined that there is a requirement for the State to pay attention to all forms of violence against children.[5]

Specifically, for child trafficking, it is contained in Article 35 which states that the parties would take all appropriate national, bilateral and multilateral measures to prevent the abduction of the sale of or sale in children of welfare.

In order to increase the protection to victims of trafficking in persons, it is carried out by increasing the accessibility of services through the establishment of Integrated Service Centers at Public Hospitals owned by Central, Provincial and District/City Government as well as Central Police Hospital and Bhayangkara Hospital in the region.[6] In addition, many NGOs and community organizations established women's crisis centers, Drop In Centers, or shelters.[7]

Strengthening the commitment of the Government of the Republic of Indonesia in the elimination of trafficking in persons is reflected in Presidential Decree No. 88/2002 on the National Plan of Action on the Elimination of Trafficking of Women and Children and the Act of the Elimination of Human Trafficking National Action on Combating Trafficking in Persons 2015-2019.[8]

2 Problem

The problem of this research is formulated as: How is the prevention of child trafficking and its legal reform?

3 Research methods

In accordance with the problems above, this research material is included in qualitative research through normative juridical approach by looking at the rules, both the primary legal material (the Law on the Elimination of Crime of Trafficking in Persons, the Criminal Code) as well as secondary legal materials (textbooks, and others), and tertiary legal materials (dictionaries, encyclopedias, etc.).

4 Discussions

Theory of Legal Protection against children presented by Philipus M Hadjon, which focuses on legal protection in the field of child rights. According to Barda Nawawi Arief, child protection theory can be interpreted as an effort to protect the law against various freedom and fundamental rights of children (children rights and freedoms of children) and various interests related to the welfare of children. Child protection in general is a way of life and livelihood of the child that can guarantee its growth and development properly, spiritually, physically and socially.[9]

Legal protection is all efforts are made consciously by all individuals and government agencies, the private sector aimed at seeking security, control and compliance welfare in accordance with the rights that exist as stipulated in Act Number 39 of 1999 concerning Human Rights. Every Indonesian child is a valuable asset of the nation, the future generation and human resources of Indonesia that will be the determinant of the future of the nation and state. The State is obliged to create a sense of security and provide legal protection to every Indonesian child in order for them to grow and develop properly and participate in development. Legal protection of children is the legal protection of various fundamental rights and freedoms of children and various interests related to the welfare of children.[10]

Child Protection is all activities to guarantee and protect children and their rights in order to live, grow, develop and participate optimally in accordance with human dignity and values and to be protected from violence and discrimination. Child protection is a joint activity aimed at securing the security, procurement and fulfillment of the spiritual and physical well-being of the child in accordance with his or her interests and rights.[11]

Child protection is a manifestation of justice in a society. Law is a guarantee for child protection activities. Legal certainty should be endeavored for the sake of continuity of child protection activities and prevent misconduct that brings unintended negative consequences in the implementation of child protection.

Human trafficking is an act of recruitment, transportation, shelter, transfer, transfer or acceptance of persons with threats of violence, the use of harassment, abduction, capture, falsification, fraud, abuse of power or vulnerable positions, debt bondage or paying or benefits, persons exercising control over such persons, whether committed within States or between States, for the purpose of exploitation or exploitation of persons.

Similarly, child prostitution is not always a victim of trafficking, because to be categorized as a trafficking victim the child must be recruited by a network of syndicates, or the child is deceived with incorrect information about the work to be performed.[12] Child trafficking may be for various purposes, including for prostitution, for pornographic objects, for forced labor, for organ transfusion, for drug traffickers, for adoption and so forth.

The anti-trafficking unit of the National Police's Criminal Investigation Unit reported 110 case investigations during 2017, this number declining compared to 2016 with 221 cases.10 Figures reported in 2016 may include cases of forced marriage and human organ trafficking. The Supreme Court reported 256 rulings, up from 119 decisions in the previous year, with a verdict of up to 7 years in prison.[13]

In 2017, the Ministry of Women Empowerment and Child Protection reported a partnership with a communications company to collect open information on 943 victims of trafficking reported in 65 print, online and broadcast media as an effort to extend the method of identifying victims. (Children, 2016) Separately, the Indonesian Child Protection

Commission officially identifies 307 child trafficking victims.[14] It remains unclear whether these two processes led to the investigation or protection of victims.

In Indonesia, women, men, and children are exploited for forced labor in the fishing, fish processing, construction sectors; in plantations, including oil palm plantations; and in mining and manufacturing.[15] Many women are exploited in the domestic sector and the sex trade. Victims are often recruited with the lure of a job offer in a restaurant, factory, or a domestic worker but is actually used as a commercial sex worker (CSW).[16]

Debt is very common for trafficking victims. Women and adolescent girls were used as prostitutes in mining operations in Maluku, Papua and Jambi provinces.[17] Child sex tourism is commonly found in the Riau Islands bordering Singapore, and Bali is also a destination for Indonesian tourists involved in child sex tourism.

Trafficking of children for the purpose of prostitution or some of the so-called prostituted children is one of the worst forms of child labor, this can be seen from the ILO Convention 182 which has been ratified by the Indonesian government as a law. Law No. 1 of 2000 on Ratification of ILO Convention 182 on the Prohibition and Immediate Action of the Elimination of the Worst Forms of Child Labor.[18]

The worst forms of child labor can be categorized as child slavery or similar to slavery such as child trafficking, childhood as security of debt or forced labor, the use or offer of a child for prostitution or for the production of pornography or pornographic performances, the use or offer of a child for the activity dark / forbidden especially for the production and sale of drugs, workplace or natural environment that endanger the health, safety and morals of children.

Addressing the issue of child trafficking, the Government of the Republic of Indonesia issued a policy in the field of criminal law to prevent and combat trafficking in persons including minors. This is followed by the issuance of Law Number 21 of 2007 on the Eradication of Crime of Trafficking in Persons and the Draft of the National Criminal Code where there are several articles regulating the crimes of trafficking in persons. This legal tool is a way to overcome the crime of trafficking of minors by implementing policies that include: Prevention, prosecution to traffickers, and protection measures against victims.[19]

In the National Coordination Meeting for the Crime of Human Trafficking (NCMCHT) of the Ministry of Women Empowerment and Child Protection , which took place on 28 August in Bali. Whether it's from self-help communities, academics, business world and mass media. Cooperation is needed to overcome the problems of people's trade from upstream. In this meeting also recommended some things for human trafficking.

The reported human traffficking data is very small compared to the actual data. Based on data obtained from the Ministry of Foreign Affairs from 2012 to May 2016, there are 1,223 human trafficking cases involving Indonesian citizens abroad. Therefore, the government is committed to ending human trafficking with the support of all relevant stakeholders and the community.

For this issue it is strongly recommended to increase efforts in investigating, prosecuting and punishing labor agencies, brokers and corrupt public officials involved in trafficking; develop and implement procedures to identify potential victims among vulnerable groups, including repatriation of migrant workers, prostitution workers, and fishing vessel crews.

In addition, it also provides training for the Ministry of Marine Affairs and Fisherie staff and supervisors related to the identification of victims and referral procedures; conducting training for judges, prosecutors, police and social workers on anti-trafficking laws; take steps to eliminate recruitment fees imposed by labor suppliers to prospective workers; proactively provide identified victim reintegration services; promoting official and safe migration with efforts to prevent trafficking in persons; increase the human resources of anti-trafficking task force (co-tasks) and improve coordination among ministries; create a data collection system to track anti-trafficking efforts at all levels of law enforcement; train hospital staff and other healthcare providers in terms of ensuring government-funded care for victims of trafficking in persons; as well as creating a national protocol explaining the responsibility for prosecuting trafficking cases outside of the victim's home province.

The Ministry of Foreign Affairs assisted 478 victims of trafficking of persons from Indonesia abroad through their consulates and embassies; an increase of 413 victims in the previous reporting period. (Indonesia 2017) In 2016, the Ministry of Foreign Affairs sent 13,714 Indonesian citizens, and other governments deported 27,855 people, compared to 9,039 people discharged and 85,490 deported in 2015. Of these figures, the Ministry of Foreign Affairs identified 602 trafficking victims (compared to 541 victims of 2015) and assisted in the immediate repatriation of 460 victims, compared with 306 victims in 2015. The Ministry of Foreign Affairs spent 240,398 US dollars to compensate victims, providing them temporary shelter and providing other services upon their return to Indonesia. Kemenlu also handed them over to local government institutions for further handling.

The government continues to rely on international organizations and NGOs in identifying victims-notably victims of foreign nationals in Indonesia-as well as efforts in complementing government-funded protection services. While endorsing the ILO Maritime Labor Convention in September and establishing protocols for the inspection of crew casualties in 2015, government institutions do not uniformly comply with both mechanisms during the reporting period.[20]

The government continues to work with NGOs in identifying trafficking victims among crews of ships secured or drowned as part of a 2014 moratorium on illegal fishing, but there is no data on the amount until the reporting period. Throughout the reporting period, the Government has initiated a new mechanism to improve the facilitation of victim protection services, but it does not say how often the government uses applications connecting users with the nearest Indonesian Ministry of Foreign Affairs and Embassy hotline and consulate abroad.

The renewal of the criminal law is an attempt to reform and reorient the criminal law in accordance with socio-political, socio-philosophical and socio-cultural values of society underlying criminal policies. The criminal law policy is part of the criminal policy. In the perspective of criminal policy, criminal law policy can be interpreted as crime prevention policy by using criminal law. Viewed from a political point of view, criminal law can be interpreted as a policy to realize better legislation in society to achieve what is desired.[21]

According Harkristuti Harkrisnowo there are several articles in the Criminal Code that can be used to capture some deeds of trafficking people although not out of various weaknesses. Article 297 of the Criminal Code specifically regulates the trafficking of women and boys under age. Seen from the angle of the corbann, almost all cases found victims are women and children under age, including infants.[22]

Another problem relating to Article 297 of the Criminal Code is about the age limit of immature (under age) for trafficked boys. As is known, there is no single provision that explicitly gives the age limit of immature or adult age. In the articles that stipulate that the victim should be underage, there are also articles specifically mentioning the age of 12 years, 15 years, 17 years so there is no clear benchmark for this age problem.

Regarding this age limit there must be a strict stipulation that there is only one understanding. In addition to Article 297 of the Criminal Code, Article 324 can also be used to capture some of the trafficking of persons because this passage prohibits acts that can be categorized as human trafficking, but the object is specifically mentioned as a slave so that the validity of this article becomes very narrow. In addition to the Criminal Code, Act Number 23 of 2002 concerning Child Protection may also be used to encompass traffickers as provided for in Article 83 that any person who trades, sells or kidnap a child for himself or for sale is punished with a maximum imprisonment of 15 (fifteen) years and a short 3 (three) years and a fine of not more than Rp 300,000,000 and a minimum of Rp 60,000,000. but in this law, the scope is limited to children so that traffickers of persons who are not children are not subject to this law.

5 Conclusions

Human trafficking is an act of recruitment, transportation, shelter, transfer, transfer or acceptance of persons with threats of violence, the use of harassment, abduction, capture, falsification, fraud, abuse of power or vulnerable positions, debt bondage or paying or benefits, persons exercising control over such persons, whether committed within States or between States, for the purpose of exploitation or exploitation of persons. Similarly, child prostitution is not always a victim of trafficking, because to be categorized as a trafficking victim the child must be recruited by a network of syndicates, or the child is deceived with incorrect information about the work to be performed. Child trafficking may be for various purposes, including for prostitution, for pornographic objects, for forced labor, for organ transfusion, for drug traffickers, for adoption and so forth.

The renewal of the criminal law is an attempt to reform and reorient the criminal law in accordance with socio-political, socio-philosophical and socio-cultural values of society underlying criminal policies. The criminal law policy is part of the criminal policy. In the perspective of criminal policy, criminal law policy can be interpreted as crime prevention policy by using criminal law. Viewed from a political point of view, criminal law can be interpreted as a policy to realize better legislation in society to achieve what is desired. Several articles in the Criminal Code that can be used to capture some deeds of trafficking people although not out of various weaknesses. Article 297 of the Criminal Code specifically regulates the trafficking of women and boys under age. Seen from the angle of the corbann, almost all cases found victims are women and children under age, including infants. Relating to Article 297 of the Criminal Code is about the age limit of immature (under age) for trafficked boys. As is known, there is no single provision that explicitly gives the age limit of immature or adult age.

The presence and performance of the Task Force at both the provincial and local levels should be maximized, the need for special studies for all districts / municipalities in the case of trafficking in children, and the organization of productive economic enterprises should be evaluated because of the tendency of some inadequate targets and the possibility of creating programs - continued empowerment programs, and the importance of re-registration of trafficking cases recorded in relevant and strategic agencies such as the police, the Social Service, the Office of Manpower, and Non-Governmental Organizations in order that case handling can be more targeted.

References

1. H. Harkrisnowo, *Laporan Perdagangan Manusia Di Indonesia.* p. 69 (Universitas Indonesia, Jakarta, 2003)
2. F. Astrid, Pemberantasan Perdagangan Orang Dalam Surat Kabar Elektronik di Lima Negara Asean. *Jurnal Komunikasi KAREBA*, **3**, p.3 (2011)
3. Farid, *Kejahatan Seksual Terhadap Anak, Implementasi Konvensi Hak Anak Di Indonesia Dan Agenda Aksi dari Stockholm.* p. 10 (Jakarta, 1997a)
4. Farid, *Kejahatan Seksual Terhadap Anak, Implementasi Konvensi Hak Anak Di Indonesia Dan Agenda Aksi dari Stockholm.* p. 10 (Jakarta, 1997b)
5. J. Collins, The Convention on The Rights of The Child: Is Thailand To Blame? *Childen's Legal Rights Journal*,**35,** p 85 (2015)
6. N. Fadilla, Upaya Perlindungan Hukum Terhadap Anak Sebagai Korban Tindak pidana Perdagangan Orang (The Legal Efforts of Child as a Criminal Victim in Human Trafficking). *Jurnal Hukum dan Peradilan,* **5**, p. 181-194 (2016)
7. J. Weaner, The Cambodian Women's Crisis Center: Safety, Shelter, Training...And Then? *McMaster School For Advancing Humanity*, **7**. p.7 (2008)
8. A. T. Wulang, J. Nawawi, & Nurlimah. Analisis Penerapan Kebijakan Pencegahan dan Penghapusan Perdagangan Perempuan dan Anak di Kota Makasar, *Government: Ilmu Pemerintahan*,**6**, p. 14 (2013)
9. A. Syamsuddin, *Tindak pidana khusus*, p. 193 (Anak Negeri Publisindo, Jakarta, 2014)
10. B. N. Arief, *Beberapa Aspek Kebijakan Penegakan Dan Pengembangan Hukum Pidana,* p. 156 (Kencana, Jakarta, 2008)
11. Arif Gosita, *Masalah Korban Kejahatan*, p. 18 (Bhuana Ilmu Poluler, Jakarta, 2004)
12. S. Agustina, Perdagangan Perempuan dan Anak Sebagai Kejahatan Transnasional, *Jurnal Hukum Projustitia*,**24**, p. 48 (2006)
13. A. S. Nugroho, Bongkar Kasus Hukum Trafficking, Satgas Tindak Pidana Perdagangan Orang Polri Sudah Ciduk, Jakarta: Republik.co.id (2017)
14. R. A. Susilo, Kebijakan Perlindungan Hukum Bagi Anak Korban Trafficking, *Peraturan Perundang-undangan Di Indonesia*, **8,** p. 6. (2016)
15. H. Handjani, R. Relawati, & E. Handayanto, Gender Role In Traditional Fisherman Family and The Implication on Women Empowerment Model in Coastal Area of South Malang, *Jurnal Perempuan dan Anak*, **1** p. 21 (2015)
16. T. Marlina, Kajian Tentang Perlindungan Perempuan Korban Trafficking dalam Perspektif Kesehatan Jiwa, p. 18 (2012)
17. L. Rahman, E. Noerdin, E. Aripurnami. & R. L. Yuningsih, Representasi Perempuan Dalam Kebijakan Publik Di Era Otonomi Daerah. *Woman research Institute*, **1**, p. 43 (2005)
18. Y. M. Simarmata, Implementasi kebijakan Pelarangan dan Penghapusan Bentuk-Bentuk Pekerjaan terburuk Bagi Pekerja Anak di Kota Bandung. *Fakultas Hukum universitas Atma Jaya*, p. 11 (2015)

19. S. Y. Ardianto, Perlindungan Hukum terhadap Anak Sebagai Korban Dari tindak Pidana Perdagangan Orang di Kota Pekanbaru. *Jurnal Ilmu Hukum*, **1, p.** 4 (2016)
20. Y. Indriaswari, A. Prantiasih, & R. A. Atok, The Role Of Integrated Pusat Pelayanan terpadu Pemberdayaan Perempuan dan Anak (PPTPPA) in Accompanying Child Abuse Victim in Malang regency, p. 4 (2017)
21. B. N. Arief, *Bunga Rampai Kebijakan Hukum Pidana.* P. 33 (Citra Aditya Bakti, Bandung, 1996)
22. H. Harkrisnowo, *Laporan Perdagangan Manusia Di Indonesia,* P. 56 (Universitas Indonesia, Jakarta, 2003)

Effectiveness of international fund in the implementation of child rights: CSO perspective

Ningrum Natasya Sirait* and *Rosmalinda*

Doctoral Program in Law, Graduate School, Universitas Sumatera Utara, Medan, Indonesia

Abstract. The fulfilment of child rights in Indonesia has been implemented by various parties including Community Social Organization (CSO). This study discusses the effectiveness of international fund in the implementation of child's rights. Article 4 of the Convention on the Rights of the Child (CRC) stated that state party will fulfil the rights of the child through international cooperation. CSOs with headquarters located in Medan, who manage the international funds in conducting children programs are acting as informants to this research. Depth interviews and Focus Group discussions were used as tools to obtain relevant information. The study figured out that CSOs had cooperation in implementation the child rights in Sumatera Utara Province. They cooperated with international agencies namely IOM, ACT-Japan and Catholic Relief Service. CSOs also manage to access international funds through (a) call for proposals and (b) establishing intensive communications with donor agencies. As conclusion, the use of international fund is proven to be effective based on (1) able to implement CSOs vision and mission, (2) CSOs are able to achieve program output, for example: palm oil company raise it awareness on child right and the impact to the children to access their education right especially in palm plantation area which is usually located in remote and isolated area.

1 Introduction

The study discussed on the efforts made by the Community Social Organization (CSO) in accessing international funds to implement child rights. In specific, the area of study is focused in the province of Sumatera Utara, Indonesia. Furthermore, the discussion will draw the effectiveness of the use of international funds to fulfil the child rights. The concept of child rights in this study is referred to Convention of the Child Rights (CRC).

There are 4 categories of child rights as stated in CRC namely; the right to survival which includes the right to preserve and sustain life and the right to the highest standard of health and the best possible care. Protection Rights, it is covering protection from discrimination, exploitation, violence and neglect. Right to development which included right to education and right to attain a standard of living by physical, mental, spiritual, moral and social development. Right of participation is the right to express opinion in all matters affecting the child. [1]

* Correspondent Author: ningrum.sirait@gmail.com

Article 4 CRC states that the exercise of the rights of the child as mentioned in CRC shall be undertaken by the States parties and known as the state's obligations. CRC, as well as other human rights instruments such as International Covenant on Civil and Political Rights (ICCPR) and International Covenant on Economic, Social and Cultural Rights (ICECSR) mentioned four obligation of state parties namely to respect, to promote, to fulfil and to protect.[2] Mohammad Farid mention this obligation as a generic obligation, that is the obligation of State parties who ratified CRC must ensure the right to human beings in its jurisdiction without discrimination in any form.[2] This obligation is done in order to meet the standards of CRC.

The CRC is an international legal instrument approved by the UN General Assembly on November 20, 1989 in New York. CRC aims to achieve the best interest of children as the principles of international standards in concern of the children. Indonesia is a State party to the CRC based on ratification through Presidential Decree (*Keputusan Presiden*) namely Keppres No 36 year 1990. Article 4 CRC mentioned that, in the context of the fulfilment of economic, social and cultural rights, States Parties shall take appropriate legislative, administrative and other measures from their available sources and, where appropriate, within the framework of international cooperation. [2]

The principles of international cooperation as stated in article 4 of CRC referred to articles 55 and 56 of the United Nations (UN) Charter which mentioned that All members pledge themselves to take join and separate action in co-operation with the organizations for the achievement of conditions of stability and well-being which are necessary for peaceful and friendly relation among nations based on respect for principle of equal rights and self-determinations of people.[3] This international cooperation can be done through the elaboration of agreement, projects or programs taking a variety of forms, including technical assistance from UN agencies, bodies and organs, bilateral and multilateral aid treaties of aid and loans. [3]. Meanwhile, in Indonesia the international cooperation as mentioned in article 4 CRC regulated in Law No 37 year 1999 concerning Foreign Affair and Law No 24 year 2000 concerning International Agreement. Both laws are the basis rules for international cooperation which involved foreign governments, international organizations, UN agencies and International Non-Government Organization (I-NGO).

I-NGO has a status as an inner consultative status Committee on Non-Governmental Organizations at Economic and Social Council (ECOSOC) and United Nations Educational, Scientific and Cultural Organization (UNESCO).[4] There were 109 registered I-NGO through Forum Registration Meeting *(Forum Rapat Registrasi)* facilitated by Directorate of Social and Culture and International Organization of Developing Countries, Directorate General Multilateral Ministry of Foreign Affairs *(Direktorat Sosial Budaya dan Organisasi Internasional Negara Berkembang, Direktorat Jenderal Multilateral Kementerian Luar Negeri)* in the middle of 2011. Some I-NGOs which concern on child rights are *Care International, Catholic Relief Service* (CRS), Child Fund Indonesia (CFI). *Inter church Organization for Development Cooperation* (ICCO), *Lutheran World Relief* (LWR), Oxfam GB, *Save the Children* (SC), Terre des Hommes, the Asia Foundation and World Vision (WV).[4]

A cooperation undertaken by the Indonesian government in order to meet with the standard of CRC is the use of the Child Protection System (SPA). This project is implemented by the Ministry of Women Empowerment and Child Protection and is supported by United Nations Children's Fund (UNICEF). CSO as an organization which is working directly with the community also conducts foreign relations as stated in the article 1 of law No 37 year 1999 concerning Foreign Affair, Therefore, the government is not the only institution cooperating with the UN agency in the fulfilment of children's rights.

There was a movement of UN to decrease poverty number internationally. It was the Millennium Development Goals (MDGs) which conducted in 1990 to 2015. . It targeted eight key issues namely poverty, education, gender equality, infant mortality, maternal health, disease, the environment, and global humanitarian cooperation. One strategy to achieve the MDGs is with international cooperation as international fund .[5] Post 2015, UN continues the development movement known as *Sustainable Development Goals* (SDGs), this movement was supported by 193 states.[6] As Indonesia is a member of UN, it committed to the development issues including poverty.

Poverty is a major factor of violation of children's rights.[6] Goal No 17 of SDGs mentioned about partnerships to achieve goals, strengthen implementation and revitalize global partnerships for sustainable development.[6] This global partnership involves governments, donor agencies such as UN agencies and I-NGO for sustainable development and not limited to information sharing but also financial support. International donor agencies serve as distributors of developed country funds to developing countries as a form of their humanitarian responsibility. [5]

International cooperation, especially related to funding has always been a serious issue, especially in developing countries with bureaucratic issues. [5] Several issues of international fund management can be found in several literatures discussing international fund management in post-tsunami Aceh and earthquake. Joe Leitman, the World Bank's international fund manager for example stated that the ineffectiveness of international fund management is seen when only one per eight of the funds are absorbed from the total US$ 12.8 billion.[7, 8] The second issue relates to the ineffectiveness in fund management related to the concept of the program. Some program implementations do not consider to local wisdom such as the issue of children, schools and orphanages, as an impact the projects do not fitted with the needs of local residents. [8, 9].

The effectiveness issue in managing the international funds should not induce the number and amount of the international fund. It must be existed. The international aids are needed by some developing countries. [5] Donors should consider expanding relationships with non-governmental organizations or CSOs that will mediate the distribution of aid to proper targets. This strategy will reduce the cost of bureaucracy and the risks. The Information Technology (IT) revolution must also be proclaimed in every process of assistance. The monitoring and evaluation process becomes easier and the risk of abuse can be suppressed. In addition, donors should be more selective towards programs that are a form of aid implementation. If necessary from the donor party, firstly design program criteria as detailed as possible. So the cost prediction can be compiled early. [5]

CSO is also well known as *Non Government Organization* (NGO). This institution is a collection of more than one person formed to achieve certain goals; most CSOs are formed by people who work voluntarily to achieve a good goal that is for the welfare of society. [10] The existence of this CSO exists at local, national, regional and even international level. They work on special issues within a particular area. [11]

There is a regulation concerning CSOs namely Law No 17 year 2013 concerning Community Organization. This rule also regulates CSOs working for the fulfilment of children's rights under CRC. Further their existence is set within Government regulation (*Peraturan Pemerintah*) No 58 year 2016 concerning Implementation of Law No 17 year 2013 concerning community organization. A CSO which concerned on children issues are available in Sumatera Utara Province since 1987, almost three years before the CRC was ratified by the Indonesian government. Meanwhile, two others were established several years after CRC has been ratified by Indonesia. The operational systems of these CSOs are the factors to know the ability of CSOs in managing international funds in order to fulfil child rights in accordance with the CRC.

2 Research question(s)

This research is intended to examine the main problem of the effectiveness of international funding managed by CSOs in the implementation of child rights in Sumatera Utara province.

This issue was developed into three research questions related to the implementation of child rights which referred to CRC such as: what programs are performed by CSOs in Sumatera Utara province for children; how can CSOs access International funding to support the implementation of child programs and the question on CSO's accountability system.

3 Research Methodologies

This research is qualitative descriptive research. Primary and secondary data were obtained by conducting literature review and field studies in 4 months. The implementation of the study is limited only in the issue of the effectiveness of International Funding in the implementation of the child rights in Indonesia in focused of Sumatera Utara province. Specifically, the location of research in Medan is based on two reasons. First, Medan is the capital of the Sumatera Utara province where all bureaucracy and administration are centred. Secondly, Medan is the domicile for CSO main offices that have managed international funds in the implementation of children's programs in Sumatera Utara province. Moreover, all data collection activities that literature review and interviews conducted in Medan. [13, 14] Informants are from CSO which has been working for more than 10 years in child rights in Sumatera Utara province and have visions, missions and programs related to fulfil child rights in Sumatera Utara province. In detailed, there were two leaders and two senior CSO staffs.

4 Results and discuss

There are three CSOs which concerns on child rights and implementing child rights programs in Sumatera Utara province. These three CSOs are involved as informants in research on the effectiveness of international funds to fulfil child rights. Firstly, *Yayasan Kelompok Kerja Sosial Perkotaan* (YKKSP), Education and Information Centre for Child Rights. YKKSP is a Non-profit organization which established in1987. [14]

Secondly, Centre for child study and Protection (*Pusat Kajian dan Perlindungan Anak*) or YPKPA. It was established on 21 October 1996. The organization was registered in the Department of Justice and Human Rights of the Republic as stated in certificate Number AHU -4047 .AH.01.02. in 2008 and *Badan Kesatuan Bangsa dan Politik* in Sumatera Utara Province. [15]

Thirdly, *Yayasan Pusaka Indonesia* (YPI). YPI is an organization concerned with child protection especially in Sumatera Utara province. YPI was established on 10 December 2000. [16] These three institutions are institutions that are concerned with the fulfilment of children's rights and are headquartered in Medan City.

Table 1. Vision and Mission of the CSOs.

No	Name of CSOs	Vision	Mission
1	KKSP Foundation	to be able to take part in the realization of healthy, skilled, creative and independent children in preparing themselves for maturity and spurring	1. Develop models of education, health services and social welfare and provide protection for children 2. Develop a network of studies and information about children in improving children's rights

No	Name of CSOs	Vision	Mission
		increased public participation in providing protection, education and health services to children.	3. Maintain the paths of structural reform that have been taken in the struggle for democracy and justice
2	PKPA Foundation	For the implementation of the best interests of the child	Advocacy for possible policy changes for child well-being and protection and implementation of the rights of the child.
3	Pusaka Indonesia Foundation	Create an environment that respects and protects the Rights of the Child	Provide services and protection of children through legal aid, information access, policy advocacy and community empowerment to respect and protect the rights of children.

The three CSOs are headquartered in Medan and engaged to fulfil child rights even before CRC was ratified by Indonesia through Keppres No 36 year 1990. This situation is based on the child's condition in Sumatera Utara province before 1990. There were some children labour issues namely street children, agriculture such as chocolate, rubber and palm plantation child labours. International Labour Organization (ILO) as a body which concerns with labour issue focus on special case namely Jermal issue.[17] Jermal is a large platform constructed off the ocean's coastline for fishermen, who use it to bring in the catch that they sell for their living. In particular the program of elimination of children working on the Jermal was conducted between years of 1990 to 2000 and involved two CSOs which involved in this research namely YKKSP and YPKPA.

Sumatra Utara province is known for its Palm Oil plantation industry which involves environmental and social issues. Currently, the situation and condition of child labour in Sumatera Utara province is still a concern, one of them is the condition of children in palm oil plantations. [18] It is a violation of the rules of Roundtable on Sustainable Palm Oil (RSPO). One CSO who became an informant had a program related to the issue of child labour in oil palm plantations. The program is supporting by an I-NGO in Indonesia namely ICCO. However, the program does not provide direct services for children; PKPA advocated the palm oil plantation to implement child rights issues such as should not use children as employees since it is prohibited by national and international law. The next description will discuss about the CSO programs that will be seen related to the fulfilment of 4 basic rights of children in the CRC namely the right to life, the right to grow and develop, the right to protection and the right of participation. [1, 2]

4.1 Programs on child protection

In general, the programs of CSOs which involved in this study provided in their official website, as below:

1. KKSP Foundation
   a. Giving basic rights of children and adolescents namely right to life, growth, protection and participation.
   b. Providing protection for children from exploitation, other rights violations and violence.
   c. Empowering communities, government and private groups that are naturally and strategically linked with children to develop greater independence, views, opinions and community participation in all aspects of life including upholding the rights of the child.
   d. Developing an assessment centre and information network for child protection at local, national, regional and international levels.

2. PKPA Foundation
   a. Researches and studies about child and woman issues
   b. Education and training for children and women
   c. Litigation and non litigation advocacy for children and women
   d. Publication and socialization of the rights of the child and woman
   e. Networking for children and women
   f. Child protection programs in emergency situations
3. Pusaka Indonesia Foundation, there are 6 main programs namely;
   a. Access for legal services including legal assistance to children in conflict with the law.
   b. Disaster respond and mitigation such as developing a safety culture through environment program.
   c. Health; promoting tobacco control movement through preventing more number of early-age smokers and the suppression of the spread of infectious diseases
   d. Inclusive education; to reduce and eliminate barriers to access and participation for who are discriminated against and marginalized to access their education right.
   e. Entrepreneurship; to develop various business units and public fundraising in support of organizational independence that is able to contribute to community empowerment and child protection.

Programs to fulfil the child rights as committed by the three CSOs are strategies to improve the welfare of children as regulated in Law No 4 year 1979 concerning Child welfare especially child labour.[18] These CSO programs are structured as part of the CSO's role as members of in Indonesia society under article 20 Law No 35 year 2014 concerning amendment of Law No 23 year 2002 concerning Child Protection which stated “State, Government, Local Government, Community, Family, and Parents or Guardians shall be responsible and responsible for the organization of Child Protection”. Another reason that should be considered is CSO programs are also a form of lack capacity of Indonesian government in handling the increase number of children's problems. [19]

Since 2016 to now, YPKA has cooperation with ICCO to advocate the rights of children who work in palm oil plantations. Based on in-depth interviews with PKPA staff, it was found that child protection programs in palm oil plantations have produced short story films to advocate palm oil plantation companies on the rights of children. They can use the movie to advocate other palm oil plantations in Indonesia to aware about child rights. There is a warning for plantation labour to should not involve their children to work in the palm oil plantation. PKPA also produced a book containing tools kits discussing the business principles and rights of children in the oil palm plantation sector. [20] This book can be guidance for the palm oil plantation to have an SOP concerning children right in palm oil plantation. This project aims to encourage the business sector to have a concern and awareness that the business has a responsibility to fulfil the rights of children especially in oil palm plantations. The PKPA staff further stated that if the oil palm company does child protection in accordance with the regulation then it is a social investment. The company not only has "prospective workers" but also "well-educated and skilled workers" in developing sustainable palm oil plantation development. It means that the palm oil companies need to invest in fulfilment of child rights.

4.2. CSO efforts in accessing the international fund

There are three stages when a CSO implemented its program since it is a community empowerment program. The stages are the approach with training, the attitude change and the skill improvement phase. [21] In regard to support their operational, CSO access International funding(s). Commonly, there are some strategies which can be used by CSOs in fundraising for examples apply the call proposal via official website of the donors. Nowadays, some CSOs worked together with their *stakeholders* (SHs) and communities in planning and implementing the programs including the financial aid issue.

Ben Davis mentioned in a research report concerning sustainability and diversity of fund that the financial sustainability of CSOs is a major challenge to ensure the creation of an effective Indonesian NGO sector. [22] The effectiveness of CSOs in managing international funds is part of CSO's ability to build financial sustainability and diversify. The sustainability funding is the ultimate objective to ensure the impact of the work of institutions especially CSO which concerns on child rights. [22]

International best practice shows that in order to survive, CSOs need long-term financial support from various sources. Some possible sources of funding form the basis of the CSO category, i.e. limited or unlimited funds (indication of degree of flexibility) and short-term or long-term funds (indication of sustainability). Further funds that are accessed by CSOs are: (a) grants (project funds or core funds) from international sources, such as international NGOs or international donors or domestic sources, such as government; (b) donations (endowment, one-time or regular donations, and fund-raising) from individuals, corporations or foundations; (c) self-fund (input from service activities and sales of products or services such as training, sales or technical assistance) from the government, the general public, domestic interest groups or international donors; and (d) in-kind (non-money) contributions (for example, office space or free work).[21]

Based on the results of interviews with the three CSOs regarding accessing the international funds, as below:

a. CSOs access international funds from UN agencies such as UNICEF, UNHCR and UNFPA. Governments of foreign countries such as US Aid, Aus-Aid and ACT-Japan. *International agencies* such as IOM and I-NGO such as *Catholic Relief Service*.
b. CSOs access the international funds through various strategies for examples through *call for proposal*, invitation from the donor agency, sending a mini proposal to the intended donor. Furthermore, in applying a proposal to international donors, the CSOs do it in joined application of two or more, a CSO coalition.

The strategy of accessing international funds which figure out by the three CSOs who became informants was a strategy which has been reviewed by Ben Davis's as stated above.[22] Currently, CSOs should develop independence in funding by fundraising by diversifying fund raising methods through *Corporate Social Responsibility* (CSR). In accessing the international fund, all CSOs provided proposals concerning child rights. In line with the proposal, they attached the logical frameworks of the child rights projects. Furthermore, in implementation of the project, they used their SOP and Children Protection Policy (CPP). These tools must be used to ensure that the projects were fitted with child rights.

One lesson learned from the CSOs that CSR was a new strategy in child rights issue. CSO advocates the oil palm plantation business by implementing child right programs. This program advocates the CSR of oil palm companies to develop child protection programs. The company developed child friendly community in the village of oil palm workers. The company facilitated the infrastructure for education in the area of oil palm plantations. However, CSR as a new fundraising strategy may also have a negative side. CSR can be a new excuse for the government in the implementation of its responsibilities as a state party of CRC which stated the state obligation to fulfil the rights of citizens including children. [23]

4.3 Monitoring and evaluation of the international Fund

All three CSOs access international funds directly to donor agencies domiciled abroad and within the country, Indonesia. Standard Operational Procedures (SOPs) covering administration and finance are the instruments that guide the control of the effectiveness of the use of international funds. The three CSOs mentioned that they always discuss with the donors about the administration and financial SOP in order to fit their SOP with the donors standards. They reviewed their SOP annually in the annual program meeting. The issue of sustainability is an important and fundamental issue. With that purpose all three CSOs perform social audits and not merely financial audit but also cover program audit. Program and financial audits are conducted internally and externally by teams that have been contracted by CSOs. [24] The three CSOs explained in the in-depth interview that evaluation and monitoring were conducted regularly. External audits are conducted at least once a year involving teams from outside CSOs. Audit results are further reported to donor agencies and / or the general public. After they conducted the monitoring and evaluation, they will report it to the donors periodically while for the board committees they report it annually.

5 Conclusions

CSO as a community group with a goal to build a community has an important role as an agent that fulfils the rights of children in Indonesia. The study shows that three CSOs headquartered in Medan, Sumatera Utara province have gained access to international funds from UN agencies, specialized bodies of the UN, foreign governments, I-NGOs and even multi-national corporations in the form of CSR. The study also shown that the international funds were effective managed by all three CSOs by using logical frame work of projects, children protection policy (CPP) and SOP in implementation the child rights program which collaboration between the CSOs and the donors. Furthermore, the organizations achieved its vision and mission as stated in their established certificate through achieving the project outputs as stated in the MoU between the CSOs and donors. Last but not least, CSO need to continue and harmonized their child program with the current condition of both environmental and social issues related to society in fulfilment the children's rights in specific child labour issue in the oil palm plantation sector.

Acknowledgements

We would like to thank the Research and Service Institute of Universitas Sumatera Utara who supported the authors to implement research program through TALENTA research scheme of Fiscal Year 2018 based on the Rector Letter Number: 433 / UN5.1.R / SK / PPM / 2018.

References

1. *Convention on the Rights of the Child*
2. Mohammad Farid, *Kewajiban Negara dan Upaya Implementasi Konvensi Hak Anak,* (to be published)
3. Mervat Rishmawi, *Article 4, the Nature of States Parties' Obligation,* Martinus Niijhoff Publisher, Leiden-Boston, 36 (2006)

4. Direktorat Sosial Budaya Dan Organisasi Internasional Negara Berkembang Direktorat Jenderal Multilateral Kementerian Luar Negeri, *Direktori Organisasi Internasional Non-Pemerintah (I-NGO) di Indonesia,* 1-20 (2011)
5. YP2SU, *Menilik Efektifitas Peran Lembaga Donor dalam Pembangunan Manusia (2017)*
6. Ishartono & Santoso T. Raharjo, Sustainable Development Goals (SDGs) dan Pengentasan Kemiskinan**,** *SWJ,* **6** (2), 154-272 (2016)
7. W. Fengler, A. Ihsan, *Hope High for Acehnese to Emerge from Poverty,* (The Jakarta Post, Indonesia, 2006)
8. R.A. Felayanti, Efektivitas Bantuan Luar Negeri di Aceh selama 2004 – 2010, setelah Tsunami Samudra Hindia tahun 2004, *JHI,* **Tahun IX No.1**, 31-48 (2016)
9. B. Philips, *An Assessment of the Impact of the 26th Dec 2004 Earthquake and Tsunami on Aquaculture in the province of Aceh and North Sumatera, UN Food and Agriculture Organization,* Indonesia (2005)
10. M.P. Karns, K.A. Mingst, K.W. Stiles, *International Organization; The Politics & Processes of Global Governance,* 241 (2015)
11. R.N.N. Sirait, *CSO Networking; A Strategy To Achieve Asean Vision,* (to be published)
12. B. Sunggono, *Metode Penelitian Hukum*, (PT.Raja Grafindo Persada, 103-104, 1998)
13. Dragan Milovanovic, *A Primer in the Sociology of Law,* (Harrow and Heston Publishers, 2nd Edition, 1-21, 1994)
14. http://www.kkspfoundation.org/page-about-us
15. http://www.pkpaindonesia.org/index.php/tentang-kami/profil-pkpa
16. http://pusakaindonesia.or.id
17. https://news.detik.com/berita/1005086/sejumlah-anak-ditemukan-masih-bekerja-di-jermal-di-sumut
18. I. Sukadi, Tanggung Jawab Negara Terhadap Anak Terlantar Dalam Operasionalisasi Pemerintah Di Bidang Perlindungan Hak Anak, *DJJSH,* **5** (2), Desember 2013, 117-133 (2013)
19. A.S Karokaro, *Kala Anak-anak jadi Buruh Harian Pemanggul Sawit*, MSBL, 17 Desember (2014)
20. A. Sofian, M. Lubis, Rosmalinda, P.S. Maharani, M.W. Lubis, S. Sucihati, Sagita, *Toolkits, prinsip-prinsip bisnis dan hak anak di sektor Perkebunan Kelapa Sawit*, PKPA, (2017)
21. K.S. Purba, H. Gurning, *Pemberdyaan, Partisipasi dan Peranan LSM dalam Pembangunan Masyarakat,* (USU Press 23-35, Medan, 2011)
22. B. Davis, *Keberlanjutan Finansial dan Diversifikasi Pendanaan : tantangan bagi LSM Indonesia,* (NSSC Publication, Research Series #2, Indonesia, 2015)
23. P. Patra, A. Behar, *Implications of Corporate Social Responsibility on Civil Society in India,* State of Civil Society reports 2015 (2015)
24. Manoj Fogla, *Social Accountability Standards for Voluntary Organisations,* Financial Management Service Foundation (2007)

A cointegration analysis of crude palm oil price in Thailand

Patchaya Songsiengchai[1,*] , *Shaufique F.* Sidique[2,3], *Marcel* Djama[3,4], and *W.N.W.* Azman-Saini[2]

[1]Faculty of Management Science, Nakhon Ratchasima Rajabhat University, 30000 Nakhon Ratchasima, Thailand
[2]Faculty of Economics and Management, Universiti Putra Malaysia, 43400 UPM Serdang, Malaysia
[3]Institute of Agricultural and Food Policy Studies, Universiti Putra Malaysia, 43400 UPM Serdang, Malaysia
[4]Cirad, UMR Moisa - Marchés, Organisations, Institutions et Stratégies d'Acteurs, 34398 Montpellier Cedex 5, France

Abstract. Reliazing the pass-through effects of global commodity prices on domestic prices, this study develops a vector error correction model (VECM) to test for the determinants and direction of causality between global prices and crude palm oil (CPO) price in Thailand. Malaysian crude palm oil, world soybean oil and world crude oil prices were investigated as factors affecting the Thai CPO price. Using the Johansen cointegration test, the result unveils a presence of long-run relationship among the determinants. This long-run relationship, proposes that CPO price flows in Thailand are positively influenced by the Malaysian CPO price and the error correction term suggests that approximately 35 percent of total disequilibrium in Thai CPO price was corrected in the following month. Moreover, the findings show Granger causality from each of the Malaysian CPO price and the world soybean oil price for the Thai CPO price. Information flow regarding the price movements of the Malaysian CPO and soybean oil affect the Thai CPO price and vice-versa. Whereas, the evidence for a causal relationship that runs from the world crude oil price to the Thai CPO price is found, but not in reverse.

1 Background

Palm oil is an essential product that is regularly consumed in households and is used as a raw material in industrial production for numerous products. Many countries are concerned about the price of palm oil used in industries and commercials because if the price of palm oil rapidly increased, it will cause inevitable rising in the living standards and lifestyle [1-3]. As the global prices of food and commodities are now widely believed to be co-integrated [4], there could be pass-through; at least partly, between the global prices and final consumer prices. When there is a rise in global prices, it eventually affects the domestic product prices in one way or another [5]. Domestic policies that aim to protect the

* Corresponding author: patchaya.s@nrru.ac.th

domestic prices may not be able to prevent the effects of global price changes in the long-run [6].

In the case of Thailand, the production volumes of Thailand may be small compared to that of leading countries, Indonesia and Malaysia. It can sufficiently produce and supply palm oil for the domestic market and export the surplus to its neighbouring countries [7]. The trend of palm oil supply and demand in Thailand is continuously increasing. Several projects were supported by the government to expand oil palm cultivation areas [8]. Its demand increased because the palm oil price is affordable compared to other edible oils and is used as an input to other industries and biodiesel production. High demand for palm oil and greater yield at a lower cost in the production of oil palm makes palm oil industry worthy of attention. Therefore, to create a stable and sustainable industry for the long-run, the effects of global price changes on palm oil price are essential to study. This study employed a cointegration analysis of price transmission from global prices to crude palm oil price in Thailand. An excellent understanding of the price mechanism and its determinants will encourage and support the productive production [9].

There is numerous literature on the effect of crude oil price pass-through to the domestic economy and agricultural commodity prices. Applanaidu et al. [10] showed the prices of biofuel feedstock, such as palm oil, soybean oil, rapeseed oil and maize are inevitably following crude oil prices. Jiménez-Rodríguez [11] showed the evidence of oil price shock on inflation and domestic products. Brown and Yücel [12-13] demonstrated that increases in oil price is reflected in reducing output efficiency as well as the rising cost of production. Lardic and Mignon [14] presented the consumption and investment effects after oil price shock. Also, Balach et al. [15] revealed the impact of oil price increase in soybean oil price.

Contrary to the above studies, Yu et al. [16] stated that the shocks in crude oil price have no significant effect on the variation of the edible oil prices. Chuangchid et al. [17] showed fairly weak dependence between crude oil and palm oil prices. Additionally, Hameed and Arshad [19] provided evidence of the strong relationship between petroleum and palm oil prices in the long-run but not in the short-run. These contradictory findings of the previous empirical studies raise a need for further studies to clarify the conceptions.

Moreover, there are a number of evidence from previous studies showing the relationship between crude palm oil (CPO) and soybean oil prices, and the impact of Malaysian CPO price to the Thai CPO price. According to Abdullah and Wahid [20] soybean oil price is a common factor affecting CPO price, which has been widely accepted in previous studies and concludes that they are two close substitute commodities. Also, Rungreunganun et al. [21] found that besides soybean oil price, Malaysian CPO price affects the Thai CPO price. Hence to study the price mechanism of crude palm oil in Thailand, this study used world crude oil, Malaysian CPO, and world soybean oil prices as the determinants affecting Thai CPO price.

The remaining part of the paper is organized as follows: Section 2 presents the methodology used in the study, Section 3 illustrates data and empirical model, Section 4 shows empirical results, Section 5 discusses the findings, and Section 6 concludes the paper and suggests policy implications.

2 Methodology

2.1 Cointegration and vector error correction model

According to Engle and Granger [19], two or more series of non-stationary data may exist as a stationary linear combination if they are cointegrated. Such that, the linear combination can be written as:

$$\mu_t = y_t - \alpha_0 - \beta_i x_i \tag{1}$$

where μ_t is stationary, and y_t and x_i are set of non-stationary data. This combination also represents the relation of long-run equilibrium among the variables.

To estimate the long-run relationship of the above equation, Hoffman and Rasche [23] suggested using a vector error correction model (VECM) that offers a long-run relation and provides efficient coefficient estimates. The estimating procedure of VECM model consists of four steps. First, testing for unit root tests. This step is a requirement before processing to cointegration analysis, to check whether the data series are stationary and examine the integration properties of the data. The widely used Augmented Dickey-Fuller (ADF) and Phillips-Perron (PP) unit root tests are applied in this study. Next, identify and estimate a vector autoregressive (VAR) model of the integrated series. Then, selecting an appropriate lag order of a VAR model using model selection criteria such as Sequential Likelihood Ratio (LR), Final Prediction Error (FPE), Akaike Information Criterion (AIC), Schwarz Information Criterion (SIC) and Hannan-Quinn Criterion (HQC). Finally, employing Johansen [24] cointegration test and Johansen and Juselius [25] maximum likelihood for indicating the rank of cointegration equation and estimating a VECM model. The VECM model can be written as:

$$\Delta y_t = \Pi y_{t-1} + \sum_{i=1}^{k-1} \Gamma_i \Delta y_{t-i} + \mu + \epsilon_t \tag{2}$$

where y_t is a (n x 1) vector of the n variables. Δ is a first difference operator. Π is a (n x n) coefficient matrix. Γ denotes a (n x (k-1)) matrix of short-run coefficients, ϵ_t represents a (n x 1) vector of white noise disturbances. The reduced rank of matrix Π ($0 < r < n$) can be separated into a (n x r) matrix of coefficients α and a (n x r) matrix of cointegrating vectors β. Such that, the reduced rank of matrix Π can be written as $\Pi = \alpha\beta'$ where α represents the importance of the cointegration relationships in the individual equations of the speed adjustment to disequilibrium and β indicates the long-run relationship.

2.2 Causality test

To estimate the short-run deviation from their long-run relationship, Granger [26] suggested an error correction term in the stationary model as:

$$\Delta y_t = \alpha_1 + \gamma(y_{t-1} - \beta x_{t-1}) + \sum_{i=1}^{k} \delta_i \Delta y_{t-i} + \sum_{i=1}^{m} \eta_i \Delta x_{t-i} + \epsilon_{it} \tag{3}$$

where $(y_{t-1} - \beta x_{t-1})$ is the error correction term (ECT_{t-1}). x_t is said not to Granger cause y_t if x_t cannot help to predict the future of y_t.

3 Data and empirical model

An econometric approach has been adopted in this paper to examine the determinants of CPO price in Thailand. The model as the following will be estimated:

$$\text{TCPO}_t = f(\text{MCPO}_t, \text{WSB}_t, \text{WCO}_t) \tag{4}$$

where TCPO refers to the crude palm oil price in Thailand, MCPO is Malaysian crude palm oil price, WSB is the world soybean oil price and WCO is world crude oil price. All data are compiled from the World Bank database, except Thai crude palm oil price is collected from the Department of Internal Trade, Ministry of Commerce, Thailand. The unit of crude palm oil prices and soybean oil price series was in USD/ton, while the world crude oil price series was in the unit of USD/barrel. The monthly data over 1996 to 2015 period, total 237 observations were obtained to the study.

4 Results

Table 1 illustrates the summary statistics for the Thai CPO price and its determinants. During the period of 1996 to 2015, the average price of world soybean oil was the highest at 689.67 USD/ton, with the maximum of 1,414.42 USD/ton and the minimum of 321.43 USD/ton. The average of Thai CPO price was the second highest price series in the study, but is quite volatile with a standard deviation of 283.54, a maximum of 1,903.19 and a minimum of 192.54 USD/ton. The difference between the average value of Thai CPO and Malaysian CPO is around 60 USD/ton. Except Thai CPO price, all price series are positively skewed, with kurtosis coefficients less than three or thin tails. These are confirmed by the result of Jarque-Bera test statistics that all price series data are not a normal distribution. Therefore all price series data are transformed into natural logarithm.

Table 1. Summary statistics of Thai CPO price and its determinants

	TCPO	MCPO	WSB	WCO
Mean	640.11	580.41	689.67	56.57
Median	581.75	509.47	586.78	50.18
Maximum	1,903.19	1,248.55	1,414.42	132.72
Minimum	192.54	185.07	321.43	9.82
Std. Dev.	283.54	250.67	284.35	35.43
Skewness	1.00	0.68	0.72	0.46
Kurtosis	4.83	2.66	2.44	1.83
Jarque-Bera	72.61***	19.41***	23.85***	21.94***

Table 2 presents unit root test results at a level and first difference. Augmented Dickey-Fuller (ADF) and Phillips-Perron (PP) unit root tests were conducted for stationarity test. Both unit root tests showed that at the level all the series data are non-stationary (p-value > 0.05), implying that all the variables are not $I(0)$. For the first difference, the unit root tests were also conducted. The computed values of statistics are all significant at one percent level (p-value < 0.01), indicating the rejection of the existence of a unit root for each price series in their first difference. Thus, all the variables are integrated of order 1 or $I(1)$.

Table 2. ADF and PP unit root tests results

Variable	ADF Test				PP Test			
	Level		First Difference		Level		First Difference	
	t-value	*p*-value	*t*-value	*p*-value	*t*-value	*p*-value	*t*-value	*p*-value
Intercept:								
LTCPO	-1.991	0.291	-11.206	0.000	-2.086	0.251	-11.430	0.000
LMCPO	-1.828	0.366	-6.262	0.000	-1.728	0.416	-10.499	0.000
LWSB	-1.578	0.492	-10.601	0.000	-1.518	0.523	-10.624	0.000
LWCO	-1.564	0.499	-12.487	0.000	-1.548	0.508	-12.481	0.000
Intercept and trend:								
LTCPO	-3.261	0.076	-11.177	0.000	-2.878	0.172	-11.396	0.000
LMCPO	-2.707	0.235	-6.250	0.000	-2.082	0.553	-10.477	0.000
LWSB	-1.927	0.637	-10.587	0.000	-1.856	0.674	-10.612	0.000
LWCO	-1.804	0.700	-12.511	0.000	-1.708	0.745	-12.506	0.000

Estimation results of the VAR model for determining an optimum lag order is given in Table 3. The all criteria Sequential Likelihood Ratio (LR), Final Prediction Error (FPE) and Akaike Information Criterion (AIC) called for three lags, except the Schwarz Information

Criterion (SC) and Hannan-Quinn information criterion (HQ) that called for two lags. Therefore, this study decided to choose lags three as an optimum lag in the model.

Table 3. VAR lag order selection criteria

Lag	LogL	LR	FPE	AIC	SIC	HQ
0	68.043	NA	6.72e-06	-0.5593	-0.4994	-0.5351
1	1204.488	2223.263	3.78e-10	-10.3449	-10.0450	-10.2239
2	1251.417	90.169	2.88e-10	-10.6150	-10.0752*	-10.3972*
3	1273.041	40.793*	2.75e-10*	-10.6641*	-9.8844	-10.3496

Note: * denotes significance at 5% level

As all variables in the model are integrated of order one, the Johansen cointegration test is selected to identify long-run relationships The results are provided in Table 4. Assuming there is no deterministic trend in data and no intercept or trend in the cointegration equation, the results of both Trace statistics and Max-Eigen Statistics suggest there are two cointegrating equations at the 5% significance level. These indicate that there are long-run relationships between Thai crude palm oil price and its determinants. The long run relationships estimation are showed in equation (5).

Table 4. Johansen cointegration test

H_0: rank = r	Eigen value	Trace Statistic	0.05 Critical Value	*p*-value	Max-Eigen Statistic	0.05 Critical Value	*p*-value
r = 0*	0.162	72.651	47.856	0.000	41.031	27.584	0.001
r ≤ 1*	0.093	31.620	29.797	0.031	22.797	21.132	0.029
r ≤ 2	0.029	8.823	15.495	0.382	6.807	14.265	0.512
r ≤ 3	0.009	2.016	3.842	0.156	2.016	3.842	0.156

Note: * denotes rejection of the hypothesis at the 5% level.

$$LTCPO_t = \underset{(7.4712)}{1.1440}\, LMCPO_t - \underset{(-1.4777)}{0.3055}\, LWSB_t + \underset{(1.7749)}{0.0808}\, LWCO_t \qquad (5)$$

Note: Numbers in parentheses are t-statistics.

The above equation indicates that in the long run Malaysian CPO price and world crude oil price are positively impacting Thai CPO price, while the world soybean oil price negatively induces it. The estimated elasticity of the determinants of Thai CPO price is 1.1440, 0.0808 and -0.3055, respectively. However, only the Malaysian CPO price seems to be significant factor.

Since the result of Johansen showed cointegrating vectors are existing, we estimate the vector error correction model to indicate the speed of adjustment to long-run changes. The VECM model is determined by the maximum likelihood method, and the optimum lag is decided by the AIC criterion to be 3. The robustness of the VECM model is evaluated by using the Portmanteau autocorrelation test and the autocorrelation LM test. The Portmanteau autocorrelation test shows that the Ljung-Box *Q*-statistics and Adjusted *Q*-statistics are equal to 152.6996 and 158.1062, respectively. Therefore, we cannot reject the null hypothesis of no serial correlation up to lag 12. Moreover, the autocorrelation LM test presents the LM statistics of 16.0111 at 12th lags which the statistics cannot reject the null of no serial correlation indicating that there is no serial correlation in the model up to lag 12. Hence, we can continue to interpret the result of VECM model.

The result in Table 5 revealed the error correction term (ECT) in each equation, however, this study only focuses on the Thai CPO equation. The ECT in the Thai CPO price equation showed a coefficient of -0.3492 and significant at the 1% level. This implies that the speed of adjustment towards the equilibrium takes place by 34.92% per month. In other words, if there is a shock occurred in the Malaysian CPO market, a unit increases in Malaysian CPO price that increases 1.14 units in Thai CPO price is corrected by a factor 34.92% per period. It will revert to its long-run equilibrium within four months of correcting by 80% of the deviation and will take almost seven months of adjusting 95% of the deviation. This result confirms the effect of global prices pass-through to the domestic price and clearly demonstates how important could be the effect of shocks to crude oil and Malaysian CPO prices on the Thai CPO price.

Table 5. Error correction model

Error Correction:	**D(LTHCPO)**	**D(LMYCPO)**	**D(LWSB)**	**D(LWCO)**
	-0.3492***	-0.1407***	-0.0796**	0.1397**
Cointegration equation	(0.0598)	(0.0434)	(0.0338)	(0.0579)
	[-5.8376]	[-3.2438]	[-2.3573]	[2.4150]

Note: ** and *** denote significance at 5% and 1% levels, respectively. Numbers in parentheses are standard errors and numbers in brackets are t-statistics.

This paper also checked the robustness of the error correction model in Thai CPO price using the *Q*-statistic of the Ljung-Box statistic autocorrelation test and the autocorrelation LM test as well as the heteroskedasticity ARCH test. The diagnostic results are provided in Table 6. The results indicate that the single error correction model of Thai CPO price does not suffer from the serial correlation problem since the *Q*-statistics and the statistic of the Breusch-Godfrey Serial Correlation LM test cannot reject the null hypothesis of no serial correlation; meaning that the residuals in the model are not correlated with one another. Besides, the chi-square statistics of ARCH test for heteroskedasticity is not significant, implying that the residuals are homoskedasticity.

Table 6. Residuals diagnostics of the estimated error correction model

Diagnostic tests	**Thai CPO price**	
	Statistics-value	*P*-value
Q(4)	1.6653	0.7970
Q(8)	5.9212	0.6560
LM(2)	0.0425	0.9790
LM(4)	6.3521	0.1743
ARCH(1)	0.6075	0.4357

Note: $Q(p)$ is the *Q*-statistics of standardized residuals at p^{th} order that test for serial correlation. LM(k) is LM test for serial correlation up to k^{th} order. ARCH(m) is autoregressive conditional heteroskedasticity test in m lag order.

Table 7 reveals the results of Granger causality, showing the bidirectional causality between Thai CPO price and Malaysian CPO price, and world soybean oil price and Thai CPO price, but only a unidirectional causality running from world crude oil price to the Thai CPO price.

Table 7. Results of granger causality

Variable	**Chi-Square Test**	**df**	***p*-value**	**Causality**
TCPO → MCPO	14.5388	3	0.0023	Yes
MCPO → TCPO	8.8838	3	0.0309	Yes
TCPO → WSB	7.3678	3	0.0611	Yes
WSB → TCPO	6.4220	3	0.0928	Yes
TCPO → WCO	0.2297	3	0.9727	No
WCO → TCPO	7.9662	3	0.0467	Yes

5 Discussion

The findings uncover that Thai CPO price is positively related to Malaysian CPO and world crude oil prices, while negatively related to world soybean oil price. These findings are consistent with Rungreunganun et al. [21] in the view that Malaysian CPO and soybean oil prices are the factors affecting Thai CPO price. Unfortunately, the sign of soybean oil price is found to be contrasted with the study. Furthermore, the results are in line with findings provided by Applanaidu et al. [10] that showed the positive relationship between crude oil and vegetable oils.

The causality analysis reveals that bidirectional causalities are running between the Malaysian CPO and world soybean oil prices to the Thai CPO price, while unidirectional causality is running from world crude oil price to the price of Thai CPO. These causality findings are similar to those obtained by Nazlioglu and Soytas [4] and Hameed and Arshad [19]. This is in contrast with Yu et al. [16] and Campiche et al. [27] that did not find any evidence of crude oil price shocks influence on edible oil prices, including palm oil. Yu et al. [16] suggested that if the prices of crude oil continue rising and edible oils become a growing source of biodiesel, the influence of crude oil price on edible oil prices would be raised.

6 Conclusion

This paper attempts to expose the determinants of crude palm oil price in Thailand from 1996 to 2015. The key determinants of the Thai CPO price that was focusing in this paper are Malaysian CPO price, world soybean oil price, and world crude oil price. The analysis unveils the significant positive relationship between Malaysian CPO and Thai CPO prices. Moreover, the finding for the dynamic relation reveals that approximately 34.92% of total disequilibrium in the long-run relation between Thai CPO price and its determinants is corrected each month. Further, Granger causality results indicate bidirectional causality between the prices of Malaysian CPO to the Thai CPO.

With this, it can be concluded that Malaysian CPO price is an impacting factor to CPO price in Thailand. A high level of Malaysian CPO price leads to a high level of Thai CPO price. Likewise, any shocks to Malaysian CPO price will affect on Thai CPO price in the same direction. Therefore, to maintain the welfare of domestic consumers and create sustainable industry, stabilization of the Thai CPO price and monitoring changes in global prices are necessity. These could be achived by setting daily price change ceilings and maintaining a balance of the domestic stock as well as providing alternative sources to supply when there is an excess of CPO. Because CPO is an intermediate material of various commodities, it would help to balance the prices of numerous products.

References

1. D. F. Larson, Policy Research Working Paper 1654, Washington DC, (1996).
2. A. Rifin, IAAE, Beijing, (August, 2009).
3. M. Harjanti, *The Impact of Palm Oil Price Swings on Household Welfare: The Case of Indonesia*, (2012).
4. S. Nazlioglu and Ugur Soytas, Oil price, agricultural commodity prices, and the dollar: A panel cointegration and causality analysis, *Energy Economics*, 34.4 (2012)
5. OAE. *Thailand Foreign Agricultural Trade Statistics 2016* (2017).

6. Office of Agricultural Economics Research. *Strategic of Oil Palm Industry 2004-2029* (2004).
7. J. Tracey-White, *Market Infrastructure Planning: A Guide for Decision Makers*, (2015).
8. S. D. Applanaidu, F. M. Arshad, M. N. Shamsudin, Z. Yusop, ICBER, Sarawak, (March, 2010).
9. R. Jiménez-Rodríguez, Macroeconomic structure and oil price shocks at the industrial level, *International Economic Journal*, **25** (1), 173-189 (2011)
10. S. Rafiq, P. Sgro and N. Apergis, Asymmetric oil shocks and external balances of major oil exporting and importing countries. *Energy Economics*, **56**, 42-50 (2016)
11. S. P. A. Brown, M. K. Yücel, QREF, **42**, 193-208, (2002).
12. S. Lardic, and V. Mignon, Oil prices and economic activity: An asymmetric cointegration approach. *Energy Economics*, 30(3), 847-855 (2008)
13. R. Balach, B.T. Matemilola, L. Chin, A. Baloch, EEL, **15**, 15-24, (2016).
14. T.H. Yu, D. A. Bessler, S. Fuller, AAEA Annual Meeting, California, (July, 2006).
15. K. Chuangchid, A. Wiboonpongse, S. Sriboonchitta, C. Chaiboonsri, IJMS, **4**, 54-65, (2012).
16. R.B. Barsky & L. Kilian. Oil and the Macroeconomy since the 1970s. *Journal of Economic Perspectives*, **18** (4), 115-134 (2004)
17. A. A. A. Hameed, F. M. Arshad, IBBC, Sabah, (December , 2008).
18. R. Abdullah, M. B. Wahid, *World palm oil supply, demand, price and prospects: Focus on Malaysian and Indonesian palm oil industry*, (2010).
19. V. Rungreunganun, S. Talabkaw, S. Cherdchoo-ngarm, *A Mechanism to Analyze the Price of Palm Oil*, (2015).
20. R.F. Engle & C.W. Granger, Co-integration and error correction: representation, estimation, and testing. *Econometrica: journal of the Econometric Society*, 251-276 (1987)
21. D.L. Hoffman & R.H. Rasche, Assessing forecast performance in a cointegrated system. *Journal of Applied Econometrics*, 11(5), 495-517 (1996)
22. S. Johansen, Statistical analysis of cointegration vectors, *Journal of economic dynamics and control*, 12(2-3), 231-254 (1988)
23. S. Johansen, K. Juselius, Oxf. Bull. Econ. Stat, **52**, 169-210, (1990).
24. C.J Granger, Developments in the study of cointegrated economic variables, *Oxford Bulletin of economics and statistics*, 48(3), 213-228 (1986)
25. J. L. Campiche, H. L. Bryant, J. W. Richardson, J. L. Outlaw, AAEA Annual Meeting, Portland, (July-August, 2007).

Utilizing local microorganisms with different dosages and duration of fermentation towards digestibility and rumen activity of oil palm frond in vitro

Randi Mulianda, *Ma'ruf* Tafsin, *Nevy Diana* Hanafi*, dan *Yulinas*

Faculty of Agriculture, Universitas Sumatera Utara, Medan20155, Indonesia

Abstract. This research entitles "utilizing local microorganism with different dosages and duration of fermentation process towards digestibility and rumen activity of oil palm frond in vitro" has been done with purpose to know the effect of fermented feed processing technology towards digestion value and fermentability in vitro. The design used in the research was the factorial completely randomized design. The research results showed that the utilization of MOL with can significantly ($P<0.05$) improve digestibility and rumen activity in vitro where be indicated by increasing of DMD, OMD, and VFA total, N ammonia of rumen fluid in vitro, while duration fermentation had no significant effect ($P>0.05$) on OMD, and VFA total, N ammonia of rumen fluid in vitro. Application of MOL at the level 0,6% with duration of fermentation 7 days show the efficient treatment to increase quality of oil palm frond.

3 Introduction

The oil palm fronds are by-products of palm fruit harvesting. When it is seen from its availability, palm fronds are very potential to use as animal feed. According to [1], palm pruning cycles are every 14 days, it is about 3 fronds pruned in every cycle, and 1 frond is about 10 kg.

One ha of field is planted about 148 palm trees so that it produces ±4.440 kg/ha inevery14 days or 8.880 kg/ha in a month. The content of dry matter in palm fronds is about 35% so that the total of dry matter in palm fronds is 3.108 kg/ha in a month. The analysis results of feed and nutrition science laboratory, Livestock Department faculty of agriculture University Sumatera Utara (2003), the palm fronds contain 6.50% crude protein, 32.55% crude fiber, 4.47% crude fat.

The analysis results show that the content of crude protein in palm fronds is low enough, which is 6.5 % and crude fiber is high enough, which is 32.55%. Low quality feeding with high lignin content will cause rumen condition and function not good enough so that a technology is needed to improve it. Fermentation is a technology to improve feeds origin of waste quality because the involvement of microorganism in degrading the crude

* Correspondingauthor: nevydiana@yahoo.co.id

fiber, decreasing lignin content and anti-nutritious compounds will increase digestion value of feeds origin of waste [2]. This research objective is to find out utilizing local microorganism with different dosages and duration of fermentation process towards digestibility and rumen activity of oil palm frond in vitro with purpose to know the effect of fermented feed processing technology towards digestion value and fermentability.

4 Materials and methods

This research was done in feed and nutrition science laboratory, Animal Science Department, Faculty of Agriculture, University of North Sumatera, Medan. The research was done during three months, started on November until January 2018. In this research, data Analysis used was factorial design (*Hanafiah*, 2003) with two factors, which were factor A (P) local microorganism (MOL) dosage (0%, 0.2%, 0.4%, and 0.6%) and factor B (T) duration of fermentation (7,14 and 21 days) and 3 times repetition. Therefore, the total of treatments was4x3x3= 36 data. The observed parameterswereDry Matter Digestibility (DMD), Organic Matter Digestibility (OMD), and VFA total. Then, the obtained data were measured in form of formula, as follow:

1. Calculation formula of digestibility

   $$DMD\ (\%) = \frac{\text{DM sample} - (\text{DM residue - DM residue blanco})}{\text{DM sample}} \times 100\%$$

   [3]

   Note: DMD = digestion ofdry matter

   Calculation formula of OMD

   $$OMD\ (\%) = \frac{\text{OM sample} - (\text{OM residue - OM residue blanco})}{\text{OMsample}} \times 100\%$$

   [3]

   Note: OMD = digestion of organic matter

2. Rumen activity

   Calculation for
   VFA total = (a-b) x N HCl x 1000/5 ml (Steam destilation) (*AOAC* 1991).
   Note: VFA Total = volatile fatty acid total

   Calculation for NH_3 :
   NH_3 = (ml H_2SO_4 titrasi x N H_2SO_4 x 1000) mM (Owens and Goetsch, 1988)

Measurement proceduresofdry matters digestion coefficient(DMD)

Residues of centrifugeprocess result were added 50 ml pepsin HCL solution to each fermentor tubes and put into waterbathin temperature 39°C for 48 hours. Then, the residues were filtered by using paper filter*Whatman* no 41. Every repetition is made Duplo. The filter result was put into oven in temperature 105°C for 12 hours. After that, it was chilled in desiccatorsfor 15 minutesand weighed.

Measurement proceduresoforganic digestion coefficient(OMD)

OMDmeasurement was done by the way of sample in DMDmeasurement was put into furnace for 6 hours in temperature 105°C so that it became ashes. Then, it was chilled in the desiccators for 15 minutes andweighed.

Measurement procedures of VFA Total

- Five milliliterof supernatant (comes from same tube with supernatantfor analyzing NH3) was put into distillation tube.
- Then, it was added with one ml of H2SO4 15%. The tube wall was rinsed by aquadestand covered immediately with rubber cap linked by distillation pipe with ± 0.5 cm in diameters.
- After that, the other tip of the pipe was linked to Leibig condenser.
- Next, distillation tube was put into boiling flask contained boiling water without touching its surface.
- Then, distillation result was held by Erlenmeyer flask 500 ml contained by 5 ml NaOH 0.5 N.
- The distillation process finished when the holding distillate was 300 ml.
- Finally, the holding distillatewas added with phenophtalein (PP) indicator 2-3 drops and titrated with HCl 0.5 N until the color changes from pink until transparent.

5 Results and discussion

3.1 Dry Matters Digestion Coefficient (DMD)

Table1.Average of DMD (%)

Fermentation (days)	Dosage(%)				Average±SD
	0	0.2	0.4	0.6	
7	37.64^{Ba}	40.90^{Ab}	43.65^{Ac}	60.10^{Ad}	45.57 ± 1.002
14	39.29^{Ba}	41.12^{Aa}	44.57^{Ab}	61.69^{Ac}	46.66 ± 1.114
21	33.95^{Aa}	41.83^{Ab}	45.53^{Ac}	64.88^{Bd}	46.54± 0.724
Average	36.96	41.28	44.58	62.55	

Description: Superscripts with capital letter to the row showed the significant differences(P<0.05).
Superscripts with lowercase letter to the column showed the significant differences(P<0.05).

The result of statistical analysis showed that palm fronds fermented with different dosages and durations of fermentation and the interactions had the significant effect ($P < 0.05$) towards dry matters digestion level. The highest average of DMDwas in treatment of palm fronds by giving MOL in dosage of 0.6%MOL, which was 62.55%. The dosage of fermentation of 0.6% and duration of fermentation in7 days had affected digestion level in 60.10%.It means that the digestion value of dry matter above is still normal. It is based on [4], who stated that digestion value of dry matter is normal if it is between 50–60%, while according to [5], digestionvalue of dry matter is about 50.63-56.30%. On the other hand, the lowest percentage of DMD was in treatment of palm frondswas in treatment of palm frond by giving MOL in dosage of 0%, which was36.96%.

The result of Duncan test analysis showeed that the highest average of DMD was in treatment of palm fronds with giving MOL in dosage of dosage 0.6% and duration of fermentation in 21 days was 64.88 %. Meanwhile, the lowest average of DMD was in treatment of palm fronds with giving MOL in dosage of 0% and duration of fermentation in 7 days was 37.64%. Microorganisms, like *Aspergilusniger* and *Sacharomycescerevisae* have degraded palm fronds so that lignin in palm fronds decreases. It is also caused by *Aspergilusniger*'s activities during fermentation produce some enzymes, like cellulase, amylase and catallase that are able to degrade complex compound into the simplest compound so that cells grows. It is in line with [6], who stated that through fermentation, substrate decomposition occurs by some particular enzymes towards indigestive matters, like cellulose and hemicelluloses into simple sugars.

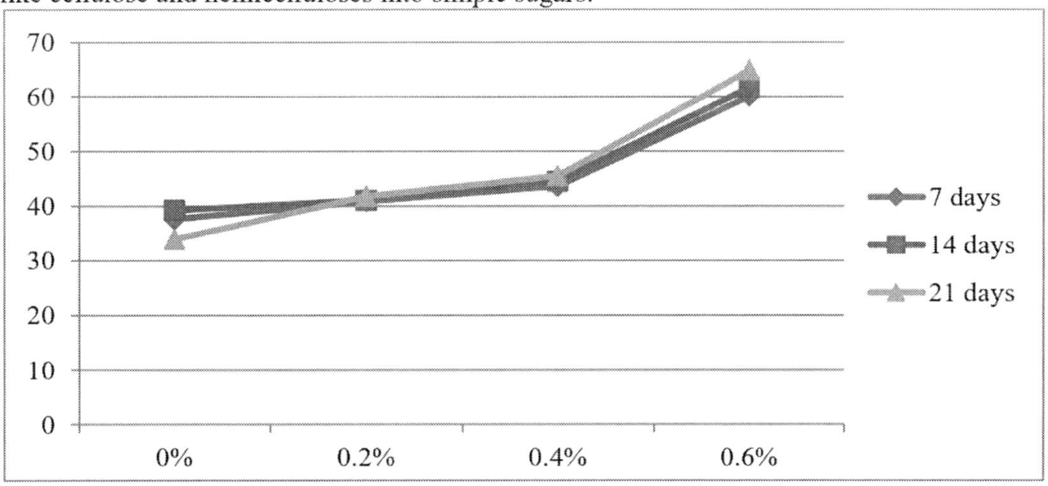

Fig.1. Dry Matters Digestion Coefficient (DMD)

In this research, the fermentation by using MOL is able to increase DMD (Figure 1). It is in line with [7], who stated that something that affects digestibility is feed composition. The feed with complete nutrition composition can improve its digestibility.

[8] reported that palm frond fermentation by using *Aspergilus niger*, *sacharomyces cerevisae* and buffalo rumen isolate can increase the DMD 58.86%. Meanwhile, in this research, the average value of DMD is higher than [8].

3.2. Organic Digestion Coefficient (OMD)

Table 2.Average of Organic Digestion Coefficient OMD(%)

Fermentation (days)	Dosage (%)				Average±SD
	0	0.2	0.4	0.6	
7	36.80^{Ba}	41.59^{Ab}	45.74^{Ac}	59.40^{Ad}	45.90± 2.216
14	38.79^{Ba}	42.05^{Ab}	44.05^{Ab}	62.94^{Bc}	46.95± 1.114
21	33.03^{Ab}	42.26^{Ab}	45.55^{Ac}	65.78^{Cd}	46.43 ± 0.637
Average	36.20	41.96	48.11	62.70	

Description: Superscripts with capital letter to the row showed the significant differences($P<0.05$).

Superscripts with lowercase letter to the column showed the significant differences(P<0.05).

The result of statistical analysis showed that palm fronds fermented with different dosages and interaction with duration of fermentation gave the significant effect (P < 0.05), how ever duration of fermentation didnot give the significant effect (P >0.05) towards OMD. The highest average of OMD was in treatment of palm fronds by giving MOL in dosage of 0.6%, which was 62.70%. According [5], OMD of complete feeds is about 48.32-53.75%. In general, the digestion of organic in the treatment of dosage of 0.6% is still normal, while the lowest percentage of OMD was in treatment of palm frond with giving MOL in dosage of 0%, which was 36.20%.

The result of Duncan test analysis showed that the highest average of OMD was in treatment of palm fronds with giving MOL in dosage of 0.6% and the duration of fermentation in 21 days was 65.78%, in 14 days was 61.69%, and in7 days was 60.10%. Meanwhile, the lowest average of OMD was in treatment of palm fronds by giving MOL in dosage of 0% and the duration of fermentation in 7 days was36.80%.

[8] reported that fermentation of palm fronds by using *Aspergilus niger*, *Sacharomyces cerevisae* and buffalo rumen isolate able to increase OMD 73.88%. In this research, the average value of OMD was 62.43%, which means it was lower than *Tafsin* et al's. The organic digestion value has similar scheme to the dry digestion value. The organic digestion has higher value than the dry digestion. It is caused by dry matters still contain organic in it, while organic does not contain dry matters. [9] stated that organic digestion value is in line with dry matter digestion value. It is caused by organic is a part of dry matters. The height of organic digestion value is also caused by high composition of crude protein. As a result, the growth of microorganisms which digest the feeds also increases. Graphic 2 below illustrates the organic digestion coefficient (OMD).

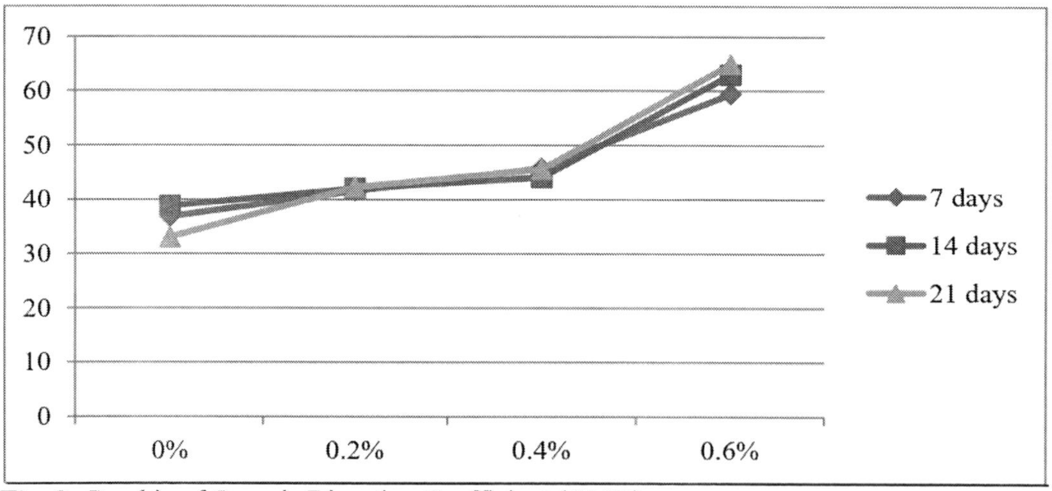

Fig. 2. Graphic of Organic Digestion Coefficient (OMD)

3.3. VFA Total

Table 3. Average of VFA Total (mM) Concentration

Durationof Fermentation	Dosage%				Average ± SD
	0	**0.2**	**0.4**	**0.6**	
7	68.17	80.39	93.48	111.55	88.39 ± 9.273
14	68.93	87.16	103.09	114.61	93.44 ± 13.405
21	71.52	91.10	110.03	122.51	98.79 ± 16.395

Average	69.54^{d}	86.21^{c}	102.20^{b}	116.22^{a}	

Description: Different superscriptsin the same line show significant differences($P< 0.05$)

Volatile Fatty Acid (VFA) is a final product of carbohydrate fermentation and rumen origin primary energy source. Beside the VFA, carbohydrate fermentation in rumen produces CO2 and CH4 [13].

The result of statistical analysis showed that the oil palm fronds fermented with different dosages gave the significant effect ($P < 0.05$), however duration of fermentation and their interaction didnot show significant effect ($P < 0.05$) towards VFA total concentration. The highest average of VFA total was in treatment of palm fronds by giving MOL in dosage of 0.6%, which was 116.22 mM. The result of Duncan test analysis showed that the highest average of VFA was in treatment dosage of 0.6% and duration of fermentation in 21 days (122.51 mM). On the other hand, the lowest average of VFA was in treatment of palm fronds by giving MOL in dosage of 0% and duration of fermentation in 7 days was 68.17mM.

The result of the research showed that there is an increasing of the average of VFA total concentration from 69.54 mM to 116.22 mM with incresing dosage of MOL. According to [10], the normal range of rumen liquid VFA which supports microbes' growth is 80-160 mM. In addition, [11] stated that *Aspergillus niger* has been known to produce enzymes to degrade fiber. It is supported [12], who stated that *Aspergillus niger* can produce extracellular enzymes like cellulase, amylase, pectinase, amyloglucosidase, glucoseoxyidase and catalase. The enzyme which affects to the increasing of VFA total concentration of fermentation using *Aspergillus niger*.

In this research, production of the best VFA total was in range of 111.26 – 122.53 mM.It is still above the range of VFA concentration produced by rumen microbes in normal condition, which is 80 – 160 mM [10]. Besides that, in this research, ammonia concentration was in range of 93.47 – 122.53 mM, which means the concentration still suffices to microbes' needs which is in range of 6.0 – 17.65 mM [13]

3.4. Amonia (NH_3-N)

Table 4.Average of Amonia (NH_3-N) Concentration (mM)

Duration of Fermentation (days)	Dosage %				Average ± SD
	0	0.2	0.4	0.6	
7	5.36	6.11	6.53	6.10	6.425 ± 0.435
14	5.55	6.13	6.92	7.70	6.622 ± 0.515
21	5.99	6.29	7.16	7.89	6.385 ± 0.853
Average	5.63^{c}	6.17^{bc}	6.87^{ab}	7.23^{a}	

Note: Different superscripts in the same row showed significant differences ($P< 0.05$)

The results of the analysis showed that the MOL dosages gave a significant effect ($P <0.05$), while the fermentation time and their interaction did not give a significant effect ($P>0.05$) on the concentration of ammonia of rumen fluid. The lowest ammonia concentration showed by dosage 0%, while the highest ammonia concentration showed by dosage 0,6%. Increasing dosage of MOL increased the ammonia concentration on rumen fluids. The amount of ammonia produced is still within the normal range of ammonia required for growth of the microbial, optimum NH 3 concentration required to support microbial growth is 4-12 mM (average 8 mM).

6 Conclusions

The utilization of MOL with different dosages can improve digestibility and rumen activity in vitro where be indicated by increasing of DMD, OMD, and VFA total, N ammonia of rumen fluid in vitro. Application of MOL at the level 0,6% with duration of fermentation 7 days show the efficient treatment to increase quality of oil palm frond.

Acknowledgements

The author want to say thank to Directorate General of Higher Education for supporting fund through DRPM 2018 for this research.

References

1 E. Wina, Teknologi Pemanfaatan Mikroorganisme dalam Pakan untuk Meningkatkan Produktivitas Ternak Ruminansia di Indonesia. Sebuah review , *Wartazoa* **15** (4): 173-186 (2005)

2 J.M.A. Tilley and R.A. Terry, A two stage technique for in the in vitro digestion of forage crops. *J. Grassland* Soc. **18**:108 (1963)

3 T. Sutardi, A.S. Amirroenas, T.D. Jalaludin, Penggunaan Pod Coklat dan Leguminosa Pohon serta Supplementasi Analog Hidroksi Metionin dan Defaunasi Pada Ruminansia. Dipresentasikan dalam Forum Komunikasi Hasil Penelitian Bidang Peternakan di Yogyakarta, 23 – 25 November 1993 (1993)

4 J. Firsoni, A.S. Sulistyo, T. Suharyono. *Uji Fermentasi In Vitro terhadap Pengaruh Suplemen Pakan dalam Pakan Komplit.* Pusat Aplikasi Teknologi Isotop dan Radiasi BATAN. Fakultas Peternakan Institut Pertanian Bogor.hal :

5 P. Sembiring, Biokonversi Limbah Pabrik Minyak Inti Sawit Dengan Phanerochaete Chrisosporium dan Implikasinya Terhadap Performans Ayam Broiler. Disertasi. UNPAD. Bandung (2006)

6 A.D. Tillman, H. Hartadi, S. Reksohadiprodjo, S. Prawirokusumoand S. Lebdosoekojo. *Ilmu Makanan Ternak Dasar* (Gadjah Mada University Press, Yogyakarta, 1991)

7 M. Tafsin, In vitro digestibility of oil palm frond treated by local mocroorganism (MOL).*IOP Publishing.* **10**.1088 (1755-1315) (2018)

8 J. Andayani, *Evaluasi Kecernaan In Vitro Bahan Kering, Bahan Organik, Protein Kasar Penggunaan Kulit Buah jagung Amoniasi dalam Ransum Ternak Sapi,*(Laporan Penelitian. Universitas Jambi. Jambi, 2010)

9 T.M. Enari, *Microbial Cellulase*, (Applied Science Pub, New York, 1983)

10 F.W. Purwanti, Kualitas Nutrien Onggok yang Difermentasi *Aspergillusniger* dengan Penambahan Level Urea dan Zeolit yang Berbeda. Institut Pertanian Bogor. Bogor (2012)

11 P. Mc.Donald, R.A. Edwards, J.F.D. Greenhalgh and C.A. Morgan, Animal Nutrition Sixth Edition, (Ashford Colour Press, Gosport, 2002)

BOD Pressure in the sustainability of the Citarum River

Resha Ayu Putri Belinawati[1], *Tri Edhi Budhi* Soesilo[1], *Herdis* Herdiansyah[1*], and *Intan Nurul* Aini[1]

[1]School of Environmental Science, Universitas Indonesia, Salemba, Indonesia

Abstract. As one of the 10 most polluted rivers in the world, Citarum river pollution has become the world's spotlight. The pollution that occurred along the Citarum River in West Java has been the concern of the local government. Pollution not only comes from a plant and household, but also from industries. In this study the authors use descriptive quantitative method, where researchers will describe and compare the existing variables to see the possibilities that arise. Variable used is the number of industries that exist and how the impact against BOD. Biochemical Oxygen Demand (BOD) is a measuring instrument that is widely used to see the quality of oxygen levels in the water. In this paper shows that there is a possibility if BOD increases if the number of industries increases.

1 Introduction

Citarum is the longest river in West Java has quite a lot of functions for the people around, especially the people around the watershed. Most of them using Citarum river for daily using, Eg. Drinking, irrigation for farmers. There is 56% of the Upper Citarum Watershed belongs to the administrative district of Bandung, West Java Province [1]. Citarum River is one of Indonesia's strategic rivers, from which the three multi-purpose dam [2].

By the time, the development along the River Ciliwung increasing. The increasing number of residents impacted a large number of settlements and also the industry along the Citarum River. However, after the construction of a large number of factories along Citarum River, Citarum River is no longer a source of water for residents along the Citarum watershed. No less than 1,500 plants located upstream freely discharge waste to creeks or Citarum River, making the Citarum River Basin more polluted [3]. Waste discharged into the Citarum River in liquid or solid form, has exceeded the capacity possessed by the Citarum River. "The concentration of BOD_5 which is a pollution indicator shows that almost all observation stations have BOD concentrations that exceed the class III water quality standard (> 6 mg/L), BOD pollution load from some of the highest observed Citarum River from Ciwideuy River and the smallest is from Cisangri River [4].

There are still many industries that dispose of waste into factors supporting the high level of BOD along the Citarum watershed. The effect of industrial waste water on the water quality in Citarum River is seen from the increase of organic matter content (BOD)

* Corresponding author: herdis@ui.ac.id

and the decrease of dissolved oxygen (DO) from upstream to downstream after passing industrial zone [5]. Near the textile industry area, the value of Biochemical Oxygen Demand (BOD) and Chemical Oxygen Demand (COD) exceeds the standard for all classes of water, at reference point, BOD ranges from 1.7 mg/L, while in downstream river BOD values reach 9.36 mg/L to 523.00 mg/L[6]. Without treatment. The Citarum River is located in western part of Java, and its basin covers 5960 km^2, the total length of the main river is about 315 km and flows from its water sources in mount Wayang to the Java sea in the north [7]. The designation of the Citarum River is divided by region. Upper watershed is still dominated with natural forest, tea, and for the middle water shed is dominated Village, Vegetable Garden, Paddy Field, and the lower watershed is dominated with industries [8].

The Government of Indonesia in 2015-2019 have National Medium Term Development Plan/Rencana Pembangunan Jangka Menengah Nasional called RPJMN. The plan states that the 2015-2019 RPJMN targets to achieve universal access or 100% access coverage for drinking water and sanitation to secure drinking water. Citarum Bestari (Bersih, Sehat, Indah, dan Lestari) is a program created by West Java Provincial Government in solving the Citarum River pollution problem, at: Governor Regulation No. 78 Year 2015 About Citarum Bestari. This program is a priority program for West Java Provincial Government and it is expected that this program can restore the function of river basin Citarum as before.

In this study the authors want to see how the industry can increase the load of BOD, and how the role of government in handling these problems, as the result of this research we will know how far the role of government in the supervision of toxic waste disposal seen from the load of BOD found.

The return of watershed functions is also supported by the agreement of countries that have participated in the signing of SDGs (Sustainable Development Goals). SDGs has 17 Goals, these goals support to meet the overall objectives of SDGs. SDGs has a time period of up to 2030 where each country must reach before the year is reached. The purpose of SDGs No. 6, specifically discusses water and sanitation. The big goals from SDGs 6 Water and sanitation is "ensuring that communities have access to clean water". One of these indicator from the goals is to reduce pollution and reduce the amount of waste disposed into the river. Biochemical Oxygen Demand (BOD) is a measuring instrument that is widely used to see the quality of oxygen levels in the water. BOD parameters are generally used to determine the level of waste water pollution. BOD determination is essential to trace pollution flow from upstream to estuary [9]. BOD checks are needed to determine pollution loads from waste water and to design a biological treatment system [10]. The aerobic bacteria consume and oxidise the organic matter; the BOD test analyzes how much oxygen is being consumed [11].

Biological oxygen demand is defined as the amount of oxygen required by organisms at the breakdown of organic matter, under aerobic conditions, breakdown of organic matter means that this organic material is used by the organism as a foodstuff and its energy is obtained from the oxidation process [12]. Biochemical oxygen demand (BOD) is one frequently used tools for measure the level of pollution in the water. Indonesia has regulations relating to waste water pollution which are regulated in: Government Regulation No. 82 of 2001 on Water Quality Management and Water Pollution Control. The legislation explains that BOD has 4 classes, a range of contaminants are present in the river, from both industrial and domestic sources [13]. Industrial waste is a waste generated by the industry resulting from the results of the industrial activity. Household waste is waste that comes from household activities such as plastic waste, wet garbage, detergent foam and others. both of these wastes can be indicated to increase the BOD load we find along the Citarum River, that can reduce the quality of the Citarum River.

2 Material and method

2.1 Quantitative Descriptive

In this study the authors use descriptive qualitative methods to know how the number of industry can impact on BOD level in the river Citarum. This descriptive approach explains using existing data, and in pour in narrative form. Quantitative descriptive often undertaken with the intention of relating the findings to other work in the field [14]. In this study the authors want to compare the findings that have been there before with the current situation. Quantitative descriptive thus lead logically from an exploratory, hypothesis generating phase to analytical confirmatory hipotesis testing phase [15].

2.2 Sustainability

This principle of sustainability is fundamental to running SDGs as well as development. Sustainability is the most widely used buzzwords of the past two decades, there is nothing, it seems, that cannot be described as 'sustainable': apparently everything can be either hyphenated or paired with it [16]. Sustainability is the ability of earth and human nature to survive, evolve, and adapt to changes in environmental conditions over a long period of time [17]. In general, sustainability is focused on increasing the quality of life with respect to environmental, social and economic considerations, both in the present and for future generations [18]. Sustainability is the ability of a natural, human or mixed system to withstand or adapt to, over an indefinite time scale, endogenous or exogenous changes perceived as threatening [19].

3 Results and discussion

3.1 Result

The simulation is done to predict the BOD load with the amount of industry produced during the production process, The figure illustrates that there is a possibility if the number of industries increases then the number of BOD increases. If the table shows that the number of BOD is not influenced by the number of industries. Industry positions do not have a major impact on the high BOD. It can be said the increase in the number of BOD is different in each industry and can be increased every day. Increasing industrial activity may indicate an increase in the number of BOD available. The higher the number of industries, the greater the BOD load. With the increase in the number of BOD indicates that the improvement of human quality in the principle of sustainability has not been fully implemented. The side of the environment in this case is still in the absence of it is proved still high number of BOD.

Table 1. Debit Simulation and Actual Concentration of Industry [20].

Number of Industries	BOD Load (kg/hr)
20	78.4
20	147
30	325.5
30	430.5
30	491.4
30	540
30	630
30	720
30	810
29	870

3.2 Discussion

This Citarum River problem encompasses many aspects, not only environmental problems affected, but also social and economic problems. Economically fishing communities in particular lost their livelihood, because the fish they get is already getting lost due to waste issued by the factory. Socially the previous River Citarum has a social function, one of which is social interaction. This citarum river problem can be said quite complex, because the impact is quite a lot.

Water is the most dominant element on earth. 71% of the earth's surface is covered by water in the liquid form of the ocean and ocean, and in the solid form of ice in the Arctic (Arctic) and Antarctic (south pole)[21]. In everyday life humans are also not separated from to meet their daily needs, such as eating, drinking, washing and others. Therefore, it is of utmost importance to maintain the sustainability of water resources, so that these resources can be utilized by humans and others, now and also in the future [22]. The world is currently faced with SDGs (Sustainable Development Program). The principle of "no one left behind" [23] is of concern to developed and developed countries. Implementation of SDGs in Indonesia depends on supporting policy. The Indonesian government has a policy that supports the objectives of the SDGs, which are called RPJMN. RPJMN is applied to programs owned by the local government of West Java, to meet the goal no 6 one of them.. The West Java Provincial Government created a program to support that goals, namely Citarum Bestari. Citarum Bestari has a goal that is not much different from the destination SDGs are, to reduce the waste of waste discharged into the river. The implementation of the program, is conducted coordination and consultation between stakeholders, prioritizes community, planning, and implementation [24].

Based on the above results, it can be seen that the relationship between the number of industries with a BOD load of 55%. There is a possible correlation between the numbers of industries with the BOD load. The BOD load can increase as the number of factories increases. The problems that arise between factories and society are quite large in number. For examples: Citarum River passes Karawang area, which was formerly referred to as rice granaries because there are still many rice plants planted there. Over time Karawang is no longer a rice granary but instead has turned into an industrial center. Construction of the factory began to bloom massively. The construction of factories alongside the river has a good objective, one of which is to absorb labor. Yet it is unfortunate that the disposal of industrial waste is not in accordance with the procedure. Factory waste that is directly disposed of the river will have a lot of impact for the people living in the proximity.

Especially for the fishermen and farmers who are still dependent on the Citarum River. The contaminated water will affect the fish population, and the irrigation of rice fields that use Citarum river water, which in turn will affect the quality of the rice.

The impact that comes from waste pollution is quite a lot. Like: Increasing the amount of waste, In 2016 BBWS Citarum mentions that household waste disposal reaches 60%, liquid chemical industry waste disposal reaches 30%, and 10% for agricultural and livestock waste disposal, total waste in the upper reaches of the Citarum River, around Bandung Regency has reached 500.000 m^3/year [25]. In addition, water deficit also threatens Jakarta and Bandung. Drinking water deficit for Bandung Raya area. This happens because the forest's ability to absorb rainwater drops dramatically because of deforestation / deforestation. If this continues to happen then the future of future generations can be said to be worrying.

Citarum Bestari that has been made from 2015 is expected to reduce waste disposal to the Citarum River but to date there is still pollution. The waste found along the river citarum is not only industrial waste that is not visible directly, and also household waste is still found along the river Citarum. The role of government in formulating policy and policy implementation is very important but it needs considerable supervision aspect. In addition, there needs to be social intervention involving the community in river management [25]. For example the community engagement approach undertaken along rivers, especially in urban areas can help maintain river sustainability [26]. Now the issue of community participation in maintaining the environment has been widely applied in various environmental issues.

4 Conclusion

The high number of industries along the Citarum Section, proves that economic development is increasing, evidenced by the shifting of functions that used to be rice fields turned into industrial plants. Numbers of industries along the Citarum River can increase the load of BOD. The results showed there was a correlation or influence between the numbers of industries with the high load BOD on the Citarum watershed, with the correlation coefficient of 0.557. The result is, the load of BOD will increase if the supervision aspect of the industry is still minimal. Basically increasing the BOD load is not influenced by the amount of industry or even influenced.

Supervision aspect is needed, because in fact the increase in the BOD load is not only based on the number of industries but also household waste. Polution along Citarum River, show that the Increasing the load of BOD is influenced from the lack of supervision in the implementation. Supervision and execution are of particular importance in terms of a policy. Cooperation between sectors is a good collaboration in the supervision of a policy because basically to make the Citarum River Bestari, required cooperation between sectors. Good supervision management will improve the quality of the better Citarum river.

Acknowledgements

This research is funded by the Grant of Indexed International Publication for Final Project of Students/*Publikasi Terindeks Internasional Untuk Tugas Akhir Mahasiswa* (PITTA) Universitas Indonesia 2018 with contract number 2579/UN.R3.1/HKP.05.00/2018

Reference

1. A. Mufubi, Y. Setiawan, H. Effendi. Land use/land cover change detection in an urban watershed: a case study of upper Citarum watershed, West Java Province, Indonesia. *Procedia Environmental Sciences* **33,** (2016).
2. Suharyanto, J. Matsushita. Integrated basin-based wastewater management system or water pollution control in an enclosed waterbody of the upper Citarum River Basin, Indonesia: Case study of Saguling Reservoir. *Lakes & Reservoirs, Research and Management* **16, (**2011).
3. A. Prijono. Mastaka Citarum. Jawa Barat: Balai Besar Konservasi Sumber Daya Alam Jawa Barat (Central Conservation of Natural Resources West Java). (2015)
4. BPPKSI. Marine and Fisheries Ministry. Retrieved from http://bp2ksi.litbang.kkp.go.id/index.php/8-halaman-depan/50-penelitian-teknologi-pengendalian-pencemaran-dan-modeling-kualitas-air-di-das-citarum-jawa-barat
5. N.T. Bukit, I.A. Yusuf. Beban pencemaran limbah industri dan status kualitas air Sungai Citarum. *Jurnal Teknologi Lingkungan* **3,** 2 (2002)
6. Sunardi, D. Sumiarsa. Bahan beracun lepas kendali. Greenpeace and Walhi (2012).
7. Nana Terangna Bukit. Water quality conservation for the Citarum River in West Java. *Water Science and Technology* **31**, 9 (1995)
8. Sunardi. Distribution of Fish in the upper Citarum River: an adaptive response to physico-chemical properties. *HAYATI Journal of Biosciences* **19**, 4 (2012)
9. Salmin. Oksigen terlarut (DO) dan kebutuhan oksigen biologi (BOD) sebagai salah satu indikator untuk menentukan kualitas perairan. *Oseana* **XXX**, 3 (2005).
10. G, Alaerts, S.S. Santika. Metoda Penelitian Air (Water Research Methods) (Usaha Nasional, Surabaya, 1987)
11. A. Berlemann. Using a water quality index to determine and compare creek water quality. *American Water Works Association* **105**, (2013)
12. PESCOD, M. D. A.I.T. Bangkok (1973).
13. Green Croos and Black Smith Institute. Citarum River, Indonesia. (2013)
14. P. Huxley. 'Quantitative–Descriptive' Articles in the British Journal of Social Work, Volumes 1–14. *The British Journal of Social Work* **18**, 2 (1988)
15. H.J.B. Birks. Recent methodological developments in quantitative descriptive biogeography. Finnish Zoological and Botanical Publishing Board Stable **24**, (1987)
16. I. Scoones. Sustainability. *Development in Practice* **17**, (2007)
17. G.T. Miller, S. Spoolman. *Environmental Science* (Cengage Learning, Boston, 2014)
18. E.R.D. Marchese. Science of The Total Environnent 613-614 (2017)
19. S.R. Dovers. *Sustainability: demands on policy* (Cambridge University Press, 1996)
20. Director General of Pollution Control and Environmental Degradation. Citarum River Pollution Load Allocation. Jakarta: Director General of Pollution Control and Environmental Degradation) (2014).
21. M.F. Imansyah. Studi umum permasalahan dan solusi das citarum serta analisis kebijakan pemerintah. *Jurnal Sosioteknologi* **11**, 25 (2012)
22. I. Juwana, N. Muttil, B.J.C. Perera. Uncertainty and sensitivity analysis of West Java Water Sustainability Index – A case study on Citarum catchment in Indonesia. *Ecological Indicators* **61,** (2016)
23. D. Satterhwaitte. A new urban agenda? *Environment and Urbanization* **18,** (2016)

24. CitaCitarum. Tentang Citarum (About Citarum). http://citarum.org/tentang-kami/tentang-cita-citarum.html
25. D. Asteria, H. Herdiansyah, I.W.A. Apriana. Women's environmental literacy as social capital in environmental management for environmental security of urban area. *IOP Conference Series: Earth Environmental Science* **30** 012014, (2016)
26. A. Brotosusilo, D. Utari, A.A. Satria. Sustainability of water resources in the upstream watershed-based community engagement and multistakeholder cooperation. *IOP Conference Series: Earth Environmental Science* **30** 012018, (2016)

E3S Web of Conferences 52, 00038 (2018) https://doi.org/10.1051/e3sconf/20185200038
CSSPO 2018

Sustainability: Citarum River, government role on the face of SDGs (water and sanitation)

Resha Ayu Putri Belinawati[1], *Tri Edhi Budhi* Soesilo[1], *Donna* Asteria[2,*], and *Riza* Harmain[1]

[1]School of Environmental Science, Universitas Indonesia, Salemba, Indonesia
[2]Communication Department, Faculty of Social and Political Science, Universitas Indonesia, Depok, Indonesia

Abstract. Citarum River is one of the most polluted river in the word. The pollution is not only derived from plant waste, but also household waste to livestock. West Java Government has a policy to handle Citarum problems, called Citarum Bestari Program. Citarum Bestari expected to solve pollution problems in Citarum River. This paper seeks to attempt the position of the West Java Government policy in the face of SDGs, particularly in this research will using qualitative methods, which is supported by the concept of David Easton political system. The author would like to see any input that may affect the policy-making in the local government with a political ecology approach. Support and demand for local governments of West Java is quite a lot already, not only from the international community but also among local residents and NGOs. This happens because many environmental impacts that occurred lately, such as flooding. Pollution that occurred along the Citarum River should be of particular concern for the Government of West Java, takes a neutral policy and strict supervision. Not only policies that benefit one of the parties (producers) but also to the other party (the environment and society).

1 Introduction

The high population and economic growth in Indonesia, especially in West Java has become the full attention of the Regional Government of West Java. However with the high population growth in West Java has not been balanced with good facilities and infrastructures. It is proven by the position of the Citarum River that became one of the 10 most polluted rivers in the world [1].

Issues about the Citarum River and SDGs is not the first time discussed, there have been many studies about Citarum River and the connection to the SDGs especially for water and sanitation, as well as the local governments awareness and need to execute programs based on the SDGs. This has been highly documented. Some examples about SDGs: "Sustainable Development Goals (SDGs) to be successful, a major shift in conceptual thinking about water management is needed. It must be recognized that there is no single approach for sequencing portfolios of investment to improve water security and that a high degree of flexibility will be needed to embrace the unique economic, social and environmental

* Corresponding author: donna@ui.ac.id

circumstances that will determine a nation's pathway towards water security and prosperity" [2] and the key success of water and sanitation goals is "indicators for this target are defined and monitored will be key to ensuring that resources are most effectively deployed to benefit those currently without access to safe drinking water" [3]. There are also some example about The Citarum River: "Citarum River has been damaged and polluted heavily by sediments and various wastes Therefore it is necessary to formulate an environmental management strategies and technologies to rehabilitate the Citarum River so that its role can be sustained" [4].

The Citarum River is a watershed (river) longest in western Java. The length of coverage Citarum River is also divided into three parts: Indonesia the Citarum River catchment occupies an area of approximately7400 km2, which can be divided into three parts; upper (1771 km^2), middle (4242 km2) and lower (1387 km^2) [5]. Citarum River can be seen also the streams of Citarum. Along Citarum River, there are 3 major dams, there are: Saguling, Cirata and Jatiluhur, storage dams were built on this river and are used for hydropower, irrigation, water supply [6].

The data from the Black Smith Institute and Green Cross Switzerland in 2013 showed that Indonesia referred one of 10 countries that have the most polluted rivers in the world and the Citarum River is one of them. Citarum River pollution is not just at one point but have occurred along the Citarum River from upstream to downstream. Sources of problems that occur in the upper Citarum River are dominated by the public and government low concern about the preservation of nature. Some problems can be formulated as follows [7]:

1. The shift area protected areas (forest and non-forest) into residential areas, agriculture, and industry.
2. The increase in the number of critical land area as a result of planning and poor supervision, 3. pollution of rivers by domestic waste, is waste water originating from settlements, agriculture, and industry
3. The level of groundwater out of control (exploited excessively) causing subsidence and increase the potential of flood-prone areas.

The other data identify the problems of Citarum River, there are:

Table 1. Identification of Citarum River problems [5].

Citarum River Problems	Resolved/Not
Industrial Waste Pollution	Not Resolved
Livestock / Livestock Waste Dissemination	Not Resolved
Domestic waste contamination (household waste)	Not Resolved
Change of Land Use and Critical Land	Not Resolved
Changes in Community Behavior	Not Resolved
Damage / Reduction of Water Sources	Not Resolved
Law Enforcement	Not Resolved

Pollution coming from this household is quite dominant considering the number of human population living along the river Citarum is increasing. One of them is cooking oil, which comes from palm oil. The contribution given by cooking oil in the contamination of the Citarum River is (0.1%) [8].

The impact from polluted Citarum River are; they will lose their livelihood. West Java government should play an active role in addressing the problems Citarum River. This is already become a concern for the Government of West Java. in 2015 West Java Government have a program called "Citarum Bestari". Provincial Government of West Java (West Java Government) issued Governor Regulation No. 78 Year 2015 is Citarum Bestari (Bersih : Clean, Sehat : Healthy, Indah : Beautiful, and Lestari : Lestari). Citarum Bestari is expected to be a solution for the West Java government in facing pollution that occurred along the Citarum River, because the pollution that occurs is quite heavy.

The return of watershed functions is also supported by the agreement of countries that have participated in the signing of SDGs (Sustainable Development Goals). SDGs has 17 Goals, these goals support to meet the overall objectives of SDGs. SDGs has a time period of up to 2030 where each country must reach before the year is reached. The purpose of SDGs No. 6, specifically discusses water and sanitation. The big goals from SDGs 6 Water and sanitation is "ensuring that communities have access to clean water". One of these indicator from the goals is to reduce pollution and reduce the amount of waste disposed into the river. Problem about water and sanitation is become important because, the growing global population requires ever- larger quantities of food, energy, and other resources, which must be met by agricultural, economic, and infrastructure systems already straining planetary boundaries [9]. Water and sanitation is one of the SDGs goals, goals number 6.3 is: Ensure availability and sustainable management of water and sanitation for all [10]. The SDGs for water and sanitation are mentioned indirectly but have considerable coverage.

2 Material and methods

In this study, the authors would to see how the SDGs can influence the policy and how policy can support the Local Government goals for SDGs 2030. In this study the authors would use a qualitative approach. The authors would like to see how the symptoms, events and facts to form a conclusion. Also in this study the researchers also used the approach to environmental science. [11] Approach to environmental science is a science that integrates the natural sciences, social and holistic studies around us.

The authors would looking at how the position of the West Java Government in achieving SDGs. In particular for the water and sanitation (point 6.3) focus on reducing toxic waste into rivers and households. The author would see in terms of policy making, which the policies made by the Government of West Java is in conformity with these achievements without prejudice to sustainability and environmental concepts.

2.1 Sustainability concept

Sustainability is the ability of the earth and the human nature to survive, develop, and adapt to changing environmental conditions in the long term [12]. According to Engger & Smith is an act of sustainable development to meet the needs of the present without compromising the ability of future generations to meet the needs of life on Earth's renewable resources without destroying the ecological processes that support [13]. According to (IUNN/UNEP/WWF) Sustainable development is to improve the quality of life, without damaging nature.

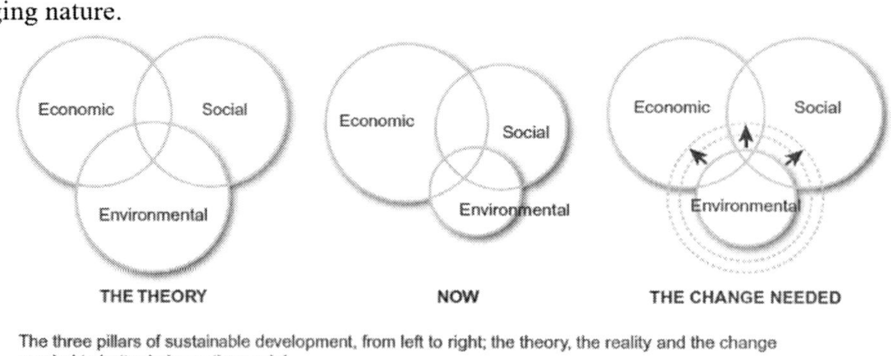

Fig. 1. Sustainable development [14].

From figure 1, the concept of sustainable development is given the IUCN explained that the environmental aspects should be included in the social and economic aspects. This happens because nowadays the environmental aspects tend to be excluded. Indonesia is a developing country, the type of developing countries is to improve the infrastructure, without looking at the environmental aspects. It should not have happened, because aspects of environmental, economic and social have the same parts. Sustainable development is agreed upon as a development that meets the needs of the present without sacrificing the right to meet the needs of future generations. It contains two important ideas: (a) the idea of "necessity" that is essential to sustain human life, and (b) the idea of limitations that stem from technological conditions and social organization's environmental capacity to meet present and future needs [15].

2.2 Political systems theory David Easton

David Easton's systems theory is a simple way to see a political system starting from how the proposed policy as a policy makers and how it will impacts to the community. in Figure 2 shows that the policy has born because there is a support in the environment and the demands of the society. Then they processed in a policy process, and generate a policy, but it does not stop in a policy, we would like to see the impact of the policies to the public, which made by the Government. According to Easton outside and behind the political system there is other systems or the environment, either physical, biological, social, psychological and so that could be the cornerstone of distinguishing between the political systems with the system eat it [16]. Easton views political life as a system consisting of interrelated activities. The activation finds its systemic relationships or links from the fact that they affect how an authoritative decision is formulated and implemented. If political life is viewed as an activity system, then there is a certain consequence of how to conduct an analysis of the operation of a system [17]. In this study the authors will looking at the factors that influence the local government of West Java in creating policies that correspond to the achieve of SDGs. The author would like to see the input factors and the process, and how they impact the local government of West Java.

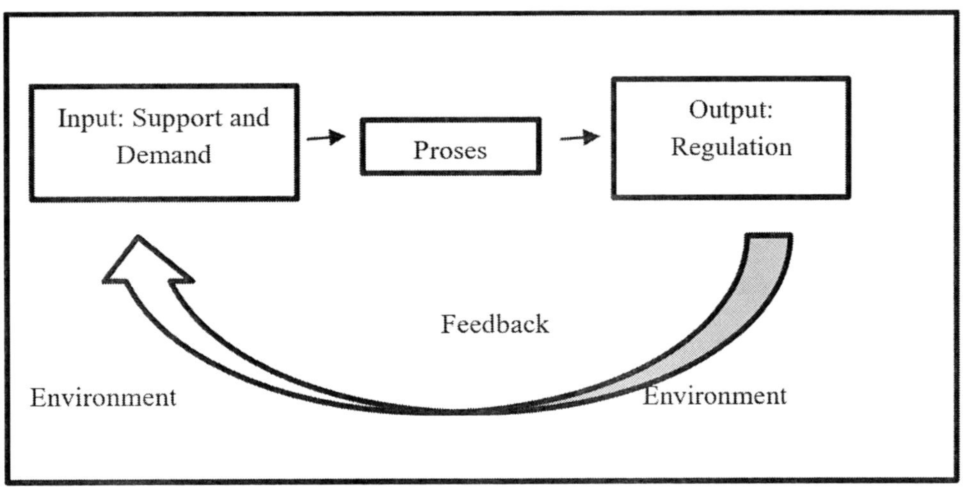

Fig. 2. Political systems theory David Easton

3 Result and discussion

Figure 3 illustrates that the inputs of this demand and support have a large effect on processes and outputs. Currently Indonesia is faced with several demands such as SDGs that have started running and must be achieved in the know 2030, the economic needs and human needs. To fulfil the whole, the Indonesian government must prepare the appropriate policy. This is done so as not to focus on one goal of the economy. The sources of the contaminants come from various kinds: industrial waste, household waste (plastic, cooking oil) contribute to water pollution especially in West Java. To handle it eat though on-stage process and out program that handling pollution problem of Citarum River.

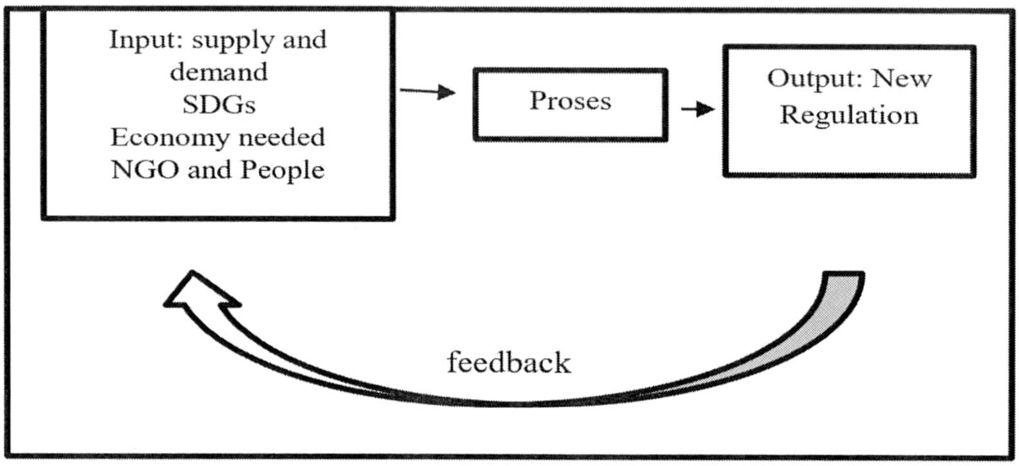

Fig. 3. Implementation of Political Systems Theory David Easton with the case

Sustainable development is a major challenge for Word country, and as the action the make a regulation to solve the issue/problem, called SDGs. Indonesia, is one of the word country need to achieve the goals, in West Java Government (Indonesia) is faced with a major problem that is polluted in Citarum River. This is of particular concern for local governments because current West Java West Java Government are faced with the new goal, there is: SDGs. That is a continuation of the MDGs. When the MDGs era, West Java Government has a program to settle the problems of the Citarum River, called *Citarum Bestari*. However, to achieve the goals of the SDGs, West Java Government need a new strategy to solving those problems. Because the previous program was made for the MDGs, is not change a lot of Citarum River condition.

Policy-making based on supply and demand, which the West Java Government was faced with the demands of SDGs. In SDGs the function of Central Government dis controlling, because the scheme used in SDGs is the contribution of local government to meeting these goals. In the previous, Government of West Java also has a program called Citarum Bestari. Currently the West Java Government must move fastest in resolving this issue. The theory of David Easton's political system help to understand how the the government make a policies/regulation. The demands is not only from the SDGs, but also from communities and NGOs, surrounding communities assisted by NGOs pushing West Java Government to make a policy in addressing the problems of the Citarum River. In the case of the Citarum River catcher the input of experts is needed, not only from environmentalists but from social and economic experts as well. The government needs to hear the advice and input that experts provide, for better Citarum purposes [18].

4 Conclusion

Policies are made by the West Java local government cannot be separated from demands and support. The Demands not only politically issue but also ecological and economic. From the Politics: Government of Indonesia in this case the Government of West Java should meet the objectives of the SDGs, as it will improve the bargaining position of Indonesia in the international world, especially in achieving the SDGs Goals 6.3 (water and sanitation), from an economic standpoint: economic development is still necessary for local residents, pollution is happening currently profitable one, but harmful to the other, such as waste disposal textile, household waste and others, will have an impact there irrigation for farmers along the Citarum River, in terms of ecology, flood already widely occurred in West Java not only that access to clean water is also reduced, which before the residents rely on water from the Citarum River , and now, they didn't use that, because the water conditions are not suitable for consumption.

Acknowledgements

This research is funded by the Grant of Indexed International Publication for Final Project of Students/Publikasi Terindeks Internasional Untuk Tugas Akhir Mahasiswa (PITTA) Universitas Indonesia 2018 with contract number 2579/UN.R3.1/HKP.05.00/2018

References

1. Blacksmith Institute And Green Cross Switzerland. The worlds worst 2013, the top ten toxic threats cleanup, progress, and ongoing challenges. (2013)
2. M. Ait-Kadi. Water for development and development for water: realizing the Sustainable Development Goals (SDGs) Vision. *Aquatic Procedia* **6**, (2016)
3. G.Y. Soetrisno. status and characteristics of pollution in the Citarum Cascade Reservoir. *Journal of Environmental Technology* **2**, 2 (2001)
4. I. Juwana, N. Muttil, B.J.C. Perera. Uncertainty and sensitivity analysis of West Java Water Sustainability Index – A case study on Citarum catchment in Indonesia. *Ecological Indicators* **61,** (2016)
5. Kurniasih, N. Journal of Environmental Technology, **5** (2002).
6. K. Noda, K. Yoshida, H. Shirakawa, U. Surahman, K. Oki. Effect of land use change driven by economic growth on sedimentation in river reach in Southeast Asia. *Journal of Agricultural Meteorology* **73**, (2017)
7. T.F. Sabrina, M. Halimi, & K. Komalasari. Implementasi Gerakan Citarum Bestari (GCB) dalam menumbuhkan civic responsibility terhadap lingkungan. *Jurnal Civicus* **20,** (2016)
8. J.T. Trimmer, R.D. Cusick, J.S. Guest. Amplifying progress toward multiple development goals through resource recovery from sanitation. *Environmental Science and Technology* **51**, (2017)
9. D. Satterhwaitte. A new urban agenda? *Environment and Urbanization* **18,** (2016)
10. P. Thomsona, J. Koehlera. Performance-oriented monitoring for the water sdg – challenges, tensions and opportunities. *Aquatic Procedia* **6**, (2016)
11. W.P. Cunningham, M.A. Cunningham. *Environmental Science: A Global Concern, 12th ed*. (McGraw-Hill, New York, 2012)
12. G.T. Miller, S. Spoolman. *Environmental Science* (Cengage Learning, Boston, 2014)

13. E. Engger, B.F. Smith. *Environmental Science: A Study of Interelationship 12th ed.* (McGraw-Hill, New York, 2010)
14. The IUCN Programme 2005_2008: Many voices, one Earth, adopted at the World conservation Congress, Bangkok, Thailand, 17-25 November 2004
15. Akhmad, F., Alex, O. Journal of Development Economics **15,** (2014)
16. Solong, H. A. Read Journal **8,** (2015).
17. Magriasti, L. *Proceeding the National Symposium on Regional Autonomy* (2011).
18. A. Brotosusilo, D. Utarai, A.A. Satria. Sustainability of water resources in the upstream watershedbased community engagement and multistakeholder cooperation. IOP Conference Series: Earth and Environmental Science **30** 012018, (2016)

Juridical review on recruitment of foreign workers

Rizkan Zulyadi* and *Muhammad Yusrizal Adi* Syaputra

Faculty of law, University of Medan Area, Medan, Indonesia

Abstract. Manpower development is an integral part of national development based on Pancasila and the 1945 Constitution, which is implemented through the development of a reliable Indonesian workers for the realization of a prosperous Indonesian society. Manpower development must be fairly regulated so as to provide comprehensive protection for Indonesian workers. On March 26, 2018 President Jokowi signed the Presidential Regulation No. 20 of 2018 concerning Use of Foreign Workers, this regulation is considered to harm Indonesian workers. This research use a normative method and juridical approach with Theory of Popular Sovereignty by John Locke. Normative legal research is a research that discusses the legal aspect, by conducting library research in terms of comparative law or legal history oriented to the applicable laws and regulations. Based on Theory of Popular Sovereignty, Presidential Regulation 20 of 2018 contradictory to the Manpower Act especially about the Plan of Use of Foreign Workers. Even the regulation does not pay attention to the interests of the people to get a job and only prioritize the entry of investment to Indonesia.

1 Introduction

Indonesia is the largest archipelagic country in the world with abundant natural wealth. This natural wealth is the basic capital given by Allah the Almighty God to the nation of Indonesia as a managed resource to improve the quality of life and welfare of the Indonesian nation. As a legal state based on Pancasila and the 1945 Constitution, the management of natural resources must be implemented based on the basic ideas and concepts of the state contained in Pancasila and the 1945 Constitution.

The meaning of the state is not merely a territorial nation that has full sovereignty because the territory is only a small part of the country. The largest and most substantial part of the state is the unity of politics. The state is the basic institution of political power. Therefore, the state is a law, a territory, a constitution, a society, and a purpose.

The state is not something concrete, but an abstract concept of political attitudes and power contained in it.[1] According to Mirza Nasution†,_the state is an organization among a group or several groups of people collectively inhabiting a certain region that recognizes the

* Corresponding Author: rizkanzulyadi.uma@gmail.com

existence of a government that takes care of their order and salvation. In a region, the state organization is not the only organization, but there are other organizations (religious, party, community that have their own personalities apart from state problems). Among these organizations, the state is the primary organization in a region because it has an authorized government and is able to intervene in the field of other organizations. An understanding of the existence of the state and the existence of human beings, similar to the view that the social community means the diverse unity of individuals and the actions of a number of individuals in the background of the need for state norms. This is what affirms that the state is not merely a legal reality, but a social reality. The reality of law requires the existence of a social reality recognized by law, while the reality of law as a reality to be obeyed by social reality.[2]

The Indonesian nation has poured the goals and ideals of Indonesian independence into the preamble on the fourth paragraph of the 1945 Constitution which reads "....... to form a government of the state of Indonesia which shall protect all the people of Indonesia and all the independence and the land that has been struggled for, and to improve public welfare, to educate the life of the people and to participate toward the establishment of a world order based on freedom, perpetual peace and social justice, therefore the independence of Indonesia shall be formulated into a constitution of the Republic of Indonesia which shall be built into a sovereign state based on a belief in the One and Only God, just and civilised humanity, the unity of Indonesia, and democratic life led by wisdom of thoughts in deliberation amongst representatives of the people, and achieving social justice for all the people of Indonesia ... .. ", and more firmly stipulated in Article 27 Paragraph (2) of the 1945 Constitution which states" every citizen shall have the right to work and to earn humane livelihood ". Therefore it is the duty and obligation of state (government) to provide employment and prepare the development of competent Indonesian workers and able to compete in the global era, in order to advance the general welfare of the Indonesian nation.

Manpower developmenti s an integral part of national development based on Pancasila and the 1945 Constitution, which is implemented through the development of a reliable Indonesian workers for the realization of a prosperous Indonesian society. Manpower development must befairly regulated so as to provide comprehensive protection for Indonesian workers. Manpower development aims to develop the potential of human resources, improvement of productivity and competitiveness of Indonesian workers. The government as the state organizeris obliged to prioritize the needs of the Indonesian workers in the manpower policy.

Peter L. Berger says that there is no life of a nation that does not face any problems at all, such as the problems faced by developing countries such as hunger, illness, high mortality, ignorance, backwardness, shelter needs, and lack of the number of job availability.[‡] Development is a set of human efforts to direct social and cultural change in accordance with the objectives of national and state life, which is to achieve the growth of civilization of social life and culture on the basis of established targets. Thus development is a matter for the government.[3]

On March 26, 2018 President Jokowi signed the Presidential Regulation No. 20 of 2018 concerning Use of Foreign Workers, to make it easier for foreignworkers entering Indonesia in the hope that Presidential Regulation will increase investment and national economy. However,the Presidential Regulation is considered potentially violating Act Number 13 of 2003 concerning Manpower (Manpower Act). The policy is also considered to violate Article 27 paragraph (2) of the 1945 Constitution which states that "Every citizen is entitled to decent work and livelihood for humanity".

In the Presidential Regulation Number 20 of 2018 mentioned, every employer of foreign workers who use foreign workers must have a Plan for the Use of ForeignWorkers (PUFW) approved by ministers or appointed officials, but the employers of foreign workers are not required to have PUFW to employ foreign workers if Shareholders who are members of the board of directors or members of the board of commissioners to the employers of foreign workers; Diplomatic and consular officers to representatives of foreign countries; and foreign workers on the type of work the government needs.

Author assume the Presidential Regulation issued by President Jokowi is a political policy that only gives benefits to the businessmen and foreign investors. This political policy had weaken the spirit of Indonesian workers who need jobs.Presidential Regulation which is a political policy of President Jokowi should be made to strengthen the rules of the use of foreign workers in Indonesia and reduce foreign workers entry, not to facilitate and causes many foreign workers comes to Indonesia.

2 Formulation of the Problem

The problem of this research is formulated as: how is the consequences of Presidential Regulation number 20 of 2018 to Indonesian people especially the workers based on Popular Sovereignty Theory?

3 Research Method

This research is a normative legal research. Normative legal research is a research that discusses the legal aspect, by conducting library research in terms of comparative law or legal history oriented to the applicable laws and regulations.[4] This normative legal research use statutory approaches which relevant to this legal research.[5] The data were analyzed using a comparative approach by conducting comparative law study which oriented to legal theory approach. The method used was comparing a certain legal system with another, which has more or less similarities, so that the similarities and differences between the legal systems were found.[6]

The data were collected using a library research method by documenting relevant literature and provisions of legislation related to the problems to be answered in this research.

4 Literature Review

The Popular Sovereignty Theory by John Locke argues that the people are sovereign in a country. It is the people who decide the will of the state, the people decide who becomes the state government is.[§] This theory explains at the beginning, humans live freely, then humans were joined to form a social contract. In this social contract, each individual hand over some personal affairs to the community, and the community then gave his authority to the king. So the king actually got his authority from the people.The king got the authority from the people, thus who has the highest authority is the people, or in other words the sovereign is the people. The position of the king is only as the executor of the will of the nation which has been decided or desired by the people. According to popular sovereignty theory, the people as the unity formed by individuals, these individuals formed unity through an agreement.[**] The unity of society has formed a general opinion called *Vontole Generale*, *Vontole Generale* is considered as a reflection of the will and the general will.[7]

Based on popular sovereignty theory, the purpose of the state is to enforce the law and guarantee the freedom of its citizens. Freedom here is freedom in accordance with applicable laws. The law itself is an incarnation of the people. Thus, the law is the embodiment of the will of the people. So the people are the highest authorities.

The will of the people is usually collected through the implementation of General Election. General Elections can determine who is the ruler and govern the lives of the people, and can also determine who will be the representative of the people in the House of Representatives and determine what the will of the people. Popular sovereignty theory is affirmed in Article 1 paragraph (2) of the 1945 Constitution "Sovereignty is in the hands of the people and implemented according to the Constitution".[††]

5 Result and Discussion

Increasing the number of foreigners who entering Indonesia is not only have positive effect but also a negative effect, it is necessary to have a selective regulation from the government to provide the terms and conditions for the foreigners to entering territory of the Republic of Indonesia. The foreigners entering Indonesia must be in accordance with a selective policy, which includes entry, residence, and exit from the territory of Indonesia, based on this principle only foreigners who benefit Indonesia alone who can be granted entry, permission to be in Indonesia, while for permission out after settle his obligations in Indonesia then he can get out of Indonesia.

Supervision of foreigners includes the entry and the exit of foreigners to and from the territory of Indonesia, and the presence and activities of foreigners in the territory of Indonesia. Oversight of a foreigner as a series of activities has basically been initiated and undertaken by representatives of the Republic of Indonesia abroad when receiving a visa application. Subsequent monitoring is carried out by immigration officials at immigration checkpoints when immigration officials with their autonomous authority decide to deny or grant residency permits in accordance with their visas. Furthermore, the supervision shall move to the immigration office whose working area covers the residence of the foreigner. Of all established immigration procedures, it should be understood that its operations are carried out on the basis of a selective immigration law policy.

Presidential Regulation Number 20 of 2018 concerning the Use of Foreign Workers is expected to facilitate foreign workers entering Indonesia which leads to increased investment and improvement of national economy. In this Presidential Regulation mentioned, every employers of foreign workers who use foreign workers must have a Plan for the Use of Foreign Workers (RPTKA) endorsed by ministers or appointed officials. However, the employers of foreign workers are not required to have RPTKA to employ foreign workers who are: Shareholders who are members of the board of directors or members of the board of commissioners to the employers of foreign workers; Diplomatic and consular officials to representatives of foreign countries; Foreign workers on the type of work the government needs. For emergency and urgent work, according to Presidential Regulation of Use of Foreign Workers, employers of foreign workers may employ foreign workers by applying for RPTKA approval to the appointed minister or official no longer than two working days after the foreign workers work. In this regulation also affirmed, every foreign worker who work in Indonesia is obliged to have a limited residenced visa (vitas) for work petitioned by employer of foreign workers to the minister in charge of government affairs in the field of law and human rights or appointed officials by attaching notification and proof of payment.

Presidential Regulation number 20 of 2018 concerning the Use of Foreign Workers is clearly contradictory to Article 42 to Article 49 of Act Number 13 of 2003 concerning Manpower (Manpower Act). Manpower Act requires the employers of foreign workers to have a RPTKA indiscriminately, but in Article 10 paragraph (1) Presidential Regulation of the Use of Foreign Workers stipulates that foreign workers employers are not required to have RPTKA to hire foreign workers which is; Shareholders who are members of the board of directors or members of the board of commissioners to the employers of foreign workers; Diplomatic and consular officials to representatives of foreign countries; Foreign workers on the type of work the government needs.

The State is directly responsible for the fulfillment of state obligations to citizens. If the rights of a citizen are violated by an act against the law committed by the state organs, then the purpose of the process may be the demand for the cancellation of the unlawful act.[8]Hans Kelsen through the Stufentheori argues that a legal system is a hierarchy of laws in which a certain legal provision derives from other higher legal provisions. As a higher provision it is Grundnorm or a basic norm that is hypothetical. The lower provisions are more concrete than the higher provisions.[9]

One legal norm is always valid, sourced, and based on higher legal norms above it, and higher legal norms also always refer to higher legal norms. And so on to a fundamental norm (Staatsfundamentalnorm) of the Republic of Indonesia, that is Pancasila. Lower regulations should not be contrary to the higher regulation. The system of Indonesian legal norms underlines that Pancasila is the highest legal norm or source of all sources of state law. Levels in Pancasila at the same time occupy the top of the hierarchy of legislation in Indonesia is the 1945 Constitution as the basic rule of the state (Staatsgrundgezetz).[10]

Presidential Regulation Number 20 of 2018 concerning the Use of Foreign Workers who are considered contrary to the 1945 Constitution and the Law, and to open wide-ranging entrances to foreign workers, may result in conflict within the country. The conflict occurred because the people, especially the Indonesian workers who felt threatened by the Presidential Regulation and the existence of foreign workers which resulted in the termination of employment and difficulty to get a job.

Conflict as a result of the social mobility phenomena is characterized by a clash between values and interests. Usually the collision will occur because a society there are some who are not ready to accept the changes caused by social mobility.[‡‡] On the one hand there are individuals or groups actively engaged (directly) in social mobility wanting to renew certain values that are currently in force aimed at legitimizing their positions.[§§] On the other hand, there are some members of society who still hold the old values and can not simply move on to new values. Thus, these symptoms show the existence of differences in interests between individuals or between groups of citizens and communities. This distinction of interest creates clashes between interest groups in which each party involved in the conflict maintains mutual values and the other seeks to frustrate the interests of others.[11]

Conflict is a social phenomena that is present in social life, so that conflict is inherent, meaning conflict will always exist in every space and time, anywhere and anytime. In this view, society is an arena of conflict or an arena of conflict and integration that continues to take place. Conflict and social integration, therefore, is a symptom that always fills every social life. The things that encourage conflict and integration are the similarities and differences of social interests. In every social life there is not a single human having exactly the same, whether of ethnicity, interests, will, purpose, etc. From each conflict there are several that can be resolved, but some can not be resolved resulting in some acts of violence. Violence is a symptom that can not overcome the root of the conflict, causing violence from the smallest model of violence to war.

The nature of one social group overlaps with the nature of other social groups. Thus there is a relationship that affects each other.[12]Social mobility is also the dominance of a social group compared to other social groups. This phenomena appears from the demands of new treatment of a social group of rights and obligations, competition between social groups to seize domination and oppression of another group. It should be remembered that social groups are formed by social differentiation in society. This phenomenon leads to a shaky social stability. Conflict between social groups can be:[13]

1. The reaction of the oppressed group to the ruler;
2. Violent acts committed by a social group caused by victims of fanaticism; and
3. Protest a social group against the authorities to obtain fair treatment.

Presidential Regulation Number 20 of 2018 gives intimidation to Indonesian workers, Indonesian workers feel the sustainability of their livelihood and their work will be threatened and disturbed by the existence of Presidential Regulation which is facilitate foreign workers to work in Indonesia. The fear may extend to hatred of a certain ethnic group, for example, due to the large number of Chinese foreign workers entering Indonesia, this will create an anti-Chinese sentiment that also affects all Indonesian citizens of Chinese descent. It is still fresh to remember how the 1998 tragedy caused considerable sentiments and damage to Indonesian citizens of Chinese descent. In addition to anti-Chinese sentiments, there will also be public distrust of the government, which will arise public perception that the government is too liberal and pro to foreign. Although Presidential Regulation of the Use of Foreign Worker is assessed to increase investment and economic growth, it must be reconsidered the social impact that will occur in the community, the social impact will trigger a conflict that gives greater damage than the positive impact generated by the Presidential Regulation.

6 Conclusions

Based on Popular Sovereignty Theory, Presidential Regulation of the Use of Foreign Workers should pay attention to the Indonesian Workers who interest to get a job, rather than just prioritizing the entry of investment into Indonesia. And based on the Stufentheori Theory if there are lower regulations contrary to higher regulation then the lower regulation can be judicial review to be canceled entirely or partially canceled.

Although Presidential Regulation of the Use of Foreign Workers assessed to increase investment and economic growth, it must be reconsidered the social impact that will occur in the community, the social impact of course will trigger a conflict that could have damaged from the conflict is greater than the possibility of published The Presidential Regulation.

References

1. Zulkarnaen & B.A. Saebani, *Hukum konstitusi*, p. 55-56(Pustaka Setia, Bandung, 2012)
2. Zulkarnaen & B.A. Saebani, *Hukum konstitusi*, p. 55-56 (Pustaka Setia,Bandung, 2012)
3. P. L. Berger, *Piramida pengorbanan manusia atau jawaban di antara sosialisme dan kapitalisme*, p. 17 (Iqra, Bandung, 1983)
4. H. Nawawi, *Metode Penelitian Sosial*, p. 12 (Laksbang Grafika, Surabaya, 2010)
5. S. Soekanto & S. Mamudji, *Penelitian hukum normatif suatu tinjauan singkat*, p. 4 (RajaGrafindo Persada, Jakarta, 2003)
6. Soeroso, *Perbandingan hukum perdata,* p. 6(Sinar Grafika, Jakarta, 2005)
7. M. Pakpahan, *Ilmu negara dan politik,* p. 75 (Bumi Intitama Sejahtera, Jakarta Pusat, 2010)

8. H. Kelsen,*Teori umum tentang hukum dan negara*, p. 287-288 (Nusa Media, Bandung, 2011)
9. H. Bakir, *Kastil teori hukum*, p. 110 (PT Indeks Kelompok Gramedia, Jakarta, 2005)
10. A. Syamsuddin, *Proses dan teknik penyusunan peraturan perundang-undangan edisi revisi*, p. 43 (Anak Negeri Publisindo, Jakarta, 2013)
11. E. M. Setiadi dan U. Kolip, *Pengantar sosiologi:pemahaman fakta dan gejala permasalahan sosial: teori, aplikasi, dan pemecahannya*, p. 517(Kencana, Jakarta, 2011)
12. C. E. Welch, *Political modernization*, p. 162 (Wadsworth Publ, Belmont, 1967)
13. E. M. Setiadi dan U. Kolip, *Pengantar sosiologi: pemahaman fakta dan gejala permasalahan sosial: teori, aplikasi, dan pemecahannya*, p. 517(Kencana, Jakarta, 2011)

Legal aid for person with disabilities in access to justice

*Rosmalinda** *and Arif*

Faculty of Law, Universitas Sumatera Utara, Medan, Indonesia

Abstract. Some criminal cases took place in palm oil plantation and agriculture area as the crime scene. Unfortunately, the cases involved children and women. This paper discusses on the research results of a study on the development of legal aid model for marginal people including people with disabilities. There were two research questions namely the situation of PwD and how they access the justice. The data were collected through literature review and field study. There were 52 persons who participated in in-depth interview and 3 Focus Group Discussion. All informants are PwD and people who supported PwD. This study figured out a case of sexual harassment which located in palm oil plantation. The victim is a girl with hearing impairment. The result of the study shows that PwD could not access the justice due to the distance and lack of information concerning legal aid provider. Furthermore, data of PwD is not available properly either at the government office or Community Social Organization (CSO)

1 Introduction

Every citizen is equal before the law. This sentence should be applied equally without exceptions. In fact, there are some cases which marginalize people who facing problem to access justice. They could not access the justice properly due to their condition, for example people who live in plantation and agriculture area. [1] Distance became their problem since the area is lack of public transportation. Other is persons with disabilities. They found that it is difficult to access the justice due to their conditions. [2] This situation must be eliminated especially since Indonesia has ratified the Convention on the Rights of Persons with Disabilities (CRPD) through Law No. 19/2011 and Law Number 8/2016 concerning Persons with Disabilities. Some Researchers and Authors who are concerned with the issue of the fulfilment of the rights of persons with disabilities have also undertaken similar research with this research and published them. Two of the researchers and writers referred to here are Muhammad Joni Yulianto in his article on the Indonesian Legal Aid Handbook in chapter 9 entitled "Rights of Persons with Disabilities"[3] and Rahayu Repindowaty Harahap and Bustanuddin in their article entitled "Legal Protection of Persons with Disabilities Convention On the Rights of Persons With Disabilities (CRPD)" in an Innovative Journal published in 2015. [4]

* Corresponding author: rosmalindarohan@gmail.com

The Journal shows the situations of injustice experienced by persons with disabilities that occurred before 2016. It was when Indonesia has ratified the Convention on the Rights of Persons with Disabilities (CRPD) through Law No. 19/2011. Indonesia just has a new National Law concerning Persons with Disabilities in Law No 8 of 2016. The absence of this National Law turned out to be the reason for Law Enforcement Officials and the government at all levels cannot implement or run programs that can provide justice for Person with Disabilities (PwDs). The situation is almost the same in Children who conflict with the Law (*Anak yang berhadapan dengan Hukum*/ABH) as the authors findings when doing research in 2014. That year, the author conducted research on the Application of the last resort principle on Children Criminal Actors in Medan City and Deli Serdang. This study looks at the situation of Children facing the Law (ABH) accessing justice. This is important considering that on August 1 in 2014, Law No. 11 Year 2012 on the Criminal Justice System of Children which well knows as *Sistem Peradilan Pidana Anak* (SPPA) applies. The results show that in 2014 there is no Law Enforcement Officers which well know in Indonesia as *Aparat Penegak Hukum* (APH) capable of providing justice to ABH as mandated in Law No. 11 of 2012. The reason of APH is the absence of obligation of APH to implement the last resort principle through diversion and restorative justice because the law is not yet valid or new without its implementing regulations.

Based on ABH research findings above, in the year 2016 the author conducted a study entitled "Model Legal Assistance for Disabled (People with Special Needs) Facing the Law in Medan & Binjai City." This study aims to see model of legal assistance which provided for marginal people especially PwD and to ensure that PwD access the justice.

2 Research question(s)

The study tried to obtain answers on the research question namely the situation of person with disabilities in Medan and Binjai. Furthermore, the questions figure out the way of the person with disabilities in accessing their right before the law.

These two questions were developed further in order to get clear information for the study for example how many persons with disabilities are in Medan and Binjai?, how many of them faced legal cases?, do they access the legal aid to access justice?, how they get justice through the court decision?.

3 Research Methodologies

The study was conducted using literature studies and field studies. [5] In the literature study, it aims to inventory the current positive law regulation in Indonesia about PwD, legal aid and the model of legal aid provided for the victim. Meanwhile, in field research, it conducted to see the situation of people with disabilities in accessing the legal aid. [6]

This study was conducted in the city of Medan and Binjai based on both cities have Legal Aid Organization or also known in Indonesia as *Organisasi Bantuan Hukum* (OBH) that become partners' of ministry of Justice and Human Rights (KemekumHAM) Sumatra province. Furthermore, these research locations have legal cases with agriculture and plantation area as the criminal scene.

In the literature study, researchers used the Law no. 19 of 2011 concerning the ratification of the Convention on the Rights of Persons with Disabilities (CRPD) and the National Law no. 8 of 2016 on Persons with Disabilities. As for legal assistance, Researchers use Law No. 16 of 2011 on Legal Aid. In addition, the researcher also uses court decisions on cases of persons with disabilities in the city of Binjai namely the Decision of PN Binjai Number: 85 / Pid.B / 2013 / PN.BJ. which is reinforced by the

Medan High Court Verdict Number: 427 / PID / 2013 / PT-MDN. Both of these decisions are important to be included based on the situation of victims with disabilities who lived in plantation area and her parents are farmer.

Researchers conducted in-depth interviews and FGDs to informants which can be classified into 3 groups namely:

1. Group of PwDs; they are PwDs with legal or non-disability.
2. Supporting Groups of Disabling Supporters; They are parents, families and teachers of persons with disabilities
3. Legal Service Provider and Support Group; Law Enforcement Officials such as Police, Prosecutors and Judges, Advocates from Legal Aid Organizations who are partners and not partners of *KemenkumHAM* Sumatera Utara Province

In detailed, the field study was conducted through in-depth interviews and involved 17 informants that can be categorized as below.

Table 1 Information of Informants.

No	Group	Number of In-depth Interview Informants		Number of FGD Informants		Total
		Male	Female	Male	Female	
1	Person with Disabilities (PwD)	5	3	13	9	8
2	People who supported PwD	2	3	3	4	5
3	Legal Aid provider	2	2	3	3	4
		9	8	19	16	52

4 Results and discuss

The document study noted that the notion of persons with disabilities is almost the same between Law no. 8 of 2016 on Persons with Disabilities and Law no. 19 of 2011 on the ratification of the Convention on the Rights of Persons with Disabilities (CRPD) see Table 1. Having similarities, persons with disabilities include physical, mental, intellectual or sensory for long periods of experiencing similar problems and must obtain equality as citizens.

Table 2 Definition of Person with Disabilities.

Law No. 19/2011 concerning the Ratification of Convention on The Rights of Person with Disabilities (CRPD).[7]	**UU No. 8/2016 concerning Person with Disabilities.**[8]
Persons with disabilities include those who have long-term physical, mental, intellectual or sensory suffering where interactions with multiple barriers may make it difficult for full and effective participation in society on an equal basis with others (Article 1)	Persons with disabilities are anyone who experiences long-term physical, intellectual, mental, and / or sensory limitations in interacting with the environment may experience barriers and difficulties to participate fully and effectively with other citizens on the basis of equal rights (Article 1 Number 1)

Based on this definition, study found out the answer of research question is Medan and Binjai. Medan is the capital of Sumatera Utara province consisting of 21 (twenty one) sub districts and 151 urban-communities. Based on data from the Central Bureau of Statistics (BPS) in North Sumatera in Figures 2015, stated that the population of Medan amounted to 2,191,140 people (1,081,797 male and 1,109,343 female) while Binjai as the closest city to

Medan which bordered with Deli Serdang Regency. Binjai City consists of 5 sub-districts which then divided into 37 villages and villages with a population of 261.40 people (130,551 male and 130,938female).[9] Some urban communities in Medan and Binjai are bordered with palm oil plantation area. After carrying out research for 1 year in Medan and Binjai. The study obtained results which will be discuss as below.

4.1 Data of PwD and legal cases

There were not clear data on the number of persons with disabilities in Medan and Binjai in the duration year of 2015 to 2016 see Table 1. Until the end of the study, there was only data on the number of persons with disabilities based on institutions where people with disabilities engage in activities. For example in Medan; *Yayasan Pendidikan Tunanetra* (Yapentra) and Social Institution titled Bina Daksa Bahagia Sumut. This institution is a program which provided by ministry of social affair in Jakarta. Meanwhile, in Binjai, data is only provided by public extraordinary school (*Sekolah Luar Biasa*/SLB). There was not fitted official data of Pwd in Medan dan Binjai which provided by government institution such as district social departments.

There was only data of PwD in 2006 while for 2016, there is not any data. The staff of district social department in Binjai just mentioned that there are 850 persons with disabilities in Binjai. However, there is not any further explanation about the data. Similarly, in the Social Service and Manpower Office of Medan, after several bureaucratic problems, the study only obtained information about the number of persons with disabilities who received program assistance without further explanation, see Table 3.

Table 3 Number of students in SLB Negeri Binjai in 2016

No	Type of Disabilities	sex		Total
		Male	Female	
Primary School				
1	Vision	2	9	11
2	Hearing impairment	18	11	29
3	Mentally Disable	42	29	71
4	Physical disability	8	0	8
5	Autism	8	0	8
Junior high school				
1	Hearing impairment	10	18	28
2	Mentally Disable	7	3	10
3	Vision	0	1	1
4	Physical disability	1	0	1
Senior high school				

1	Hearing impairment	4	5	9
	Total	106	81	187

Based on the results of the research mentioned above, it strengthens the results of research and mentoring on people with disabilities that already exist namely the absence of an official data of PwD in Medan. This opinion was submitted by Director of United Evangelical Mission (UEM) in December 2015 in one daily in Medan City. [10] In this study, although the team tried to through the process of bureaucracy to obtain data from the Social Service and the Manpower of Medan City and Binjai City Social Service but there was not any suitable data. The absence of this data according to the Social Service staff of the two cities is based on the large number of SKPD work program target groups. Another cause of the absence of data of PwD is the absence of disaggregated data from institutions that are concerned with PwD either by sex, age or residence.

According to data from informants offices such as Police, legal aid offices, in the period of 4 latest years namely from 2013 to 2016, there were 5 legal cases which involved persons with disabilities in Medan and Binjai. The cases are sexual harassment. All of the cases positioned children as victims.

Table 4 Legal Case Involving Person with disabilities.

No	Year	Type of Cases	Status	Police Area	Legal Aid Organization
Kota Medan					
1	2015	Sexual harassment	Witnesses (Parents Victim)	Polsek Sunggal	PKPA
2	2015	Incest	Perpetrator	Polresta Medan	PKPA
Kota Binjai					
1	2016	Rape	Victim	Polres Binjai	PKPA
2	2014	Sexual Harassment	Perpetrator	Stop at the police desk	-
3	2013	Rape	Victim	PT Medan; Province Court	Lukman & Association

Based on data which stated above, one of the legal case is PN Binjai Number: 85 / Pid.B / 2013 / PN.BJ. This legal case reinforced by appeal court in Medan High Court (PT Medan) Number: 427 / PID / 2013 / PT-MDN. It declared the defendant proven legally and convincingly guilty of committing a crime which located in a palm oil plantation area in Binjai. This fact is supported World Rainforest Movement finding on the situation of palm oil plantation in Indonesia and Malaysia. It states that the number of sexual harassment including rape is increased since the area changed to be a palm oil plantation.[11]

4.2 PwD and accessing the legal aid

Based on in-depth interviews and Focused Discussions it is known that knowledge and experience of accessing legal aid are different among PwD and the supporting people of PwD. This difference was mentioned by informants in FGD which involved SLB teachers, police and parents of PwD. There are 2 main reasons why people with disabilities did not access legal aid although there is a legal aid provider which registered based on Law Number 16/2011 concerning legal aid,[12] namely:

1. Victims and witnesses of PwD or families of victims and perpetrators of PwD are afraid to advocate their right due to the cost is very expensive. They explained further, since they lived in the palm oil plantation area which is far from the city centre, they need to pay more for their transportation.
2. Victims who is PwDs could not communicate properly with APH or legal counsel for two reasons;
   2.1. Both APH and legal aid provider staffs could not speak sign language which use by people with hearing impairment. They need an interpreter. Unfortunately. The victim was never sent to school because as farmers they do not have enough money to send their daughter to the SLB in Binjai. Lack of public transportation from their house which is in the palm oil plantation becomes the main consideration to not send the girl to SLB. As consequences, the girl is lack of education. Furthermore, when the girl accessed legal aid, no interpreter could understand her sign language. She used informal sign language while the interpreter used the formal one.
   2.2. Actually, family such as her uncle understand her informal sign language. However it was 'taboo' in the community for elder people talked directly to the girl about the incident of sexual harassment.

Lacks of education is a factor of poverty. In specific, being illiterate inhibiting women to access their rights.[13] It can be seen in the legal case process of PN Binjai case Number: 85 / Pid.B / 2013 / PN.BJ. and the appealed case Number: 427 / PID / 2013 / PT-MDN. The victim could not access legal aid service by herself. She needed someone in her family, unfortunately due to her condition as PwD and never sent to school, she faced big problem while through the process of justice especially in communication.

Other study result shows 3 factors of people who supported PwD which contributed to the difficulties of PwD in accessing the justice.

1. The distance; both police and legal aid provider mentioned that they do not have special fund to go to the remote area such as Plantation and agriculture area to provide legal services. Police add some information concerning fund, although they have fund to go to remote area in order to provide security for society, they could not do it usually. They had to share the fund to other cases in different area.
2. Limited capacity of human resources; there was not any police who understand the sign language. When they handled the case of the girl with hearing impairment, they need to hire an interpreter from SLB Binjai. Similar problem also appeared among the staff of legal aid provider. No one in the legal aid office understands the sign language.

This study supported the opinion of Muhammad Joni Yulianto which mentions "... In general, the rights of PwDs are violated by investigators and judges including equality before the law and the right to a fair trial. [2] In the PN Binjai case Number: 85 / Pid.B / 2013 / PN.BJ. and the appealed case Number: 427 / PID / 2013 / PT-MDN shows that the rights of PwD are violated by the law enforcers due to their lack of capacity and facility in sign language. As a citizen, the victim should have proper interpreter while the law process starting from the police office to the court.

Another writer, Rahayu Repindowaty Harahap and Bustanuddin in his article entitled "Legal Protection of Persons with Disabilities by the Convention on the Rights of Persons with Disabilities (CRPD)" mentions the same thing as Muhammad Joni Yulianto regarding Legal Services for PwDs in Indonesia who are still poor. Two basic reasons are: First, there are no rules of gender and disability. Secondly, the availability of the budget especially to provide referrals or referrals of expert witness which related systems including translators particularly for women and girls with disabilities of sexual violence. [2]

This second reason is reinforced by the findings of researchers in another case accompanied by the Legal Aid Organization (OBH) in Medan. Based on the results of interviews Researchers with OBH staff found that currently, OBH is assisting Disabled Persons (Witness), parents of victims of sexual abuse by the victim's parents. Perpetrators of PwD, visually impaired persons working in the Disabled Massage Home of the Disabled Persons, parents of 3-year-old child victims. This incident was reported by the victim's parents after previously reported the incident to the Polsek or sub-district police office. But the case stalled because the victim's parents found barriers to mobilization to OBH and Polsek. On the other hand both OBH and Polsek do not have specific human resources and budget to provide intensive assistance to PwDs (witnesses), parents of victims of sexual violence.

4.3 Justice in the court decision

It stated above that PwDs are violated by people who work as law enforcers. It appear not only due to lack of communication while the legal process but also appears in the court decision. According to the case as stated above which involved a hearing impairment girl as victim of sexual harassment located in palm oil plantation? This study shown that the rape case is worse due to the victims of PwDs was still 16 years old and never attended the SLB. Victims were raped by two perpetrators. The perpetrators, 3 other witnesses and the victim were burning corn near the palm oil plantation in Binjai. In specific, the criminal scene was near to the soccer field of the palm oil plantation area. Although, the perpetrator has been found guilty in accordance with the Decision of PN Binjai Number: 85 / Pid.B / 2013 / PN.BJ. Final court, PT Medan, decided a decision Number 427 / PID / 2013 / PT-MDN which imposed the criminal sanction against the defendant with imprisonment for 8 (eight) years and a fine of IDR. 100.000.000, - (one hundred million rupiah). The court provided other substitute punishment that is if the fine is not paid by the defendant then it can be replaced with 2 (two) months prisons.

The study found that the punishment which was given to the offender is not comparable with the impact that must be burdened by the victim. As stated in this writing, today, the victim is caring her child who was born from the defendant's actions. She does not have any help from the defendant. The victim is a single parent when she was 19 years old. Moreover, based on the court decision, the perpetrator got punishment for the crime committed, while the victim also get restitution rights for the acts that he experienced. In fact, it was not happened. As a basis for implementing the right of restitution for the victims there is mentioned in Law no. 31 Year 2014 about Amendment to Law no. 13 of 2006 on the Protection of Witnesses and Victims. For this purpose the government should help the victims to get their restitution right. [14] In fact, the victim does not get the restitution as stated in the court decision.

This case is really different compare to a case of sexual assault of sexual abuse and rape which conducted 6 times and involved a PwD victim in Sukoharjo. The victim is a student who attended Public High School (SMU) SLB and was 22 years old. The perpetrator is a teacher who teaches in the field of art and computer studies. This crime occurred in 2012 and after going through a judicial process that is almost 1 year, finally the trial in Sukoharjo District Court was won by the victim with the enactment of decision number 28 / Pid.B / PN.SKH with the verdict 8 years 6 months. Subsequently the defendant filed an appeal and obtained Decision No 244 / Pid 2013 / P.T.Smg which strengthens PN Sukoharjo.[2]

5 Conclusions

There are three conclusion and suggestion. First, although, there is not any official data concerning PwD in the district level, government should communicate with persons and institution that have concern on PwD for example conducting a regular meeting at the government office. Furthermore, legal aids are needed for marginal society including people with disabilities in palm oil plantation. Both legal aid and legal enforcers such as police, prosecutors and judges need to gain skill in sign language especially for the law enforcers who are pointed as special officer for cases of children and women. Last but not least, justice should belong for everyone including PwDs. When she accessed for justice, it is a must for law enforcers, judges in particular to accommodate her wills. As impact, she will be able to be independent on her war to achieve prosperity.

Acknowledments

We would like to thank the Research and Service Institute of Universitas Sumatera Utara who supported the research program through HIBAH BERSAING Scheme of fiscal year 2016 with based on the Rector letter No. 154 / UN5.2.3.1 / PPM / SP / 2016.

References

1. Agus Purwoko, Ketut Sukiyono, and Basuki Sigit Priyono, *The Incidence Probability of Violence Against Women: Case Study At Agricultural Society In District of North Bengkulu,* JS, **13,** 58-75 (2011)
2. Rosmalinda, Arif, Ainul Mardiah, *Pendampingan Hukum Bagi Penyandang Disabilitas di Kota Medan dan Binjai (Studi Kasus Pelaku Kekerasan Seksual)*, SINDN USU, 209 (2016)
3. Muhammad Joni Yulianto, Hak-Hak Penyandang Disabilitas, in *Panduan Bantuan Hukum Di Indonesia*, 252-282 (2014)
4. Rahayu Repindowaty Harahap and Bustanuddin, *Perlindungan Hukum Terhadap Penyandang Disabilitas Menurut Convention On The Rights Of Persons With Disabilities (CRPD,* JI, **VIII**, 28 (2015)
5. Solly Lubis, *Filsafat Ilmu dan Penelitian,* (2013)
6. Sulistiyowati Irianto, *Praktik Penelitian Hukum: Persfektif Sosiolegal dalam Metode Penelitian Hukum; Konstelasi dan Refleksi*, 297-315 (2011)
7. Undang-Undang No. 19 tahun 2011 tentang Ratifikasi *Convention on The Rights of Person with Disabilities*(CRPD)
8. Undang-Undang No. 8 tahun 2016 tentang Penyandang Disabilitas
9. Badan Pusat Statistik, *Sumatera Utara dalam Angka 2015* (2015)
10. http://www.tribunnews.com/regional/2015/12/03/tak-ada-jumlah-pasti-penyandang-disabilitas-di-medan,
11. Jutta Kill Y Winfridus Overbeek, *13 replies to 13 lies about oil palm monoculture plantation*, 13 (2018)
12. Undang-Undang No 16 tahun 2011 tentang Bantuan Hukum
13. Lies Marcoes, *Achieving Gender Justice In Indonesia's Forest and Land Governance Sector, How civil society organisations can respond to mining and plantation industry impacts,* 4 (2015)

14. http://www.hukumonline.com/berita/baca/lt57504006f0138/ingat-korban-kekerasan-seksual-juga-punya-hak-restitusi,

The implementation strategy of customary law aspect in protecting local environment

Ratih Lestarini[1,*], *Tirtawening*[1], *Riza* Harmain[2], *Succi* Wulandhary[2], and *Dyah* Utari[3]

[1]Faculty of Law, Universitas Indonesia, Depok, Indonesia
[2]School of Environmental Science, Universitas Indonesia, Salemba, Indonesia
[3]Faculty of Health Science, Universitas Pembangunan Nasional Veteran Jakarta, Indonesia

Abstract. Environmental protection efforts have been made by the people, especially those still governed by customary law. Customary law serves to regulate society in all aspects of life, including the relationship between humans and between humans and the environment. The customary law also serves to regulate the use and the management of the environment efficiently for the purpose of preventing environmental damages and maintaining sustainability for the future. The communities which still have strong customary law will tend to preserve the environment and limit usages that cause damage to the environment. Cases of environmental damage that occurred today is very worrying. The formal law created is not sufficient to control it, so customary law is needed as an alternative to complementing the lack of formal law in order to effectively prevent the environmental damage. This research is aimed to determine the best way to protect the environment using customary aspect as the instrument. The method used in this paper is SWOT analysis that determines the strategies of the customary law in environment protection. The selection of the strategy was determined by the score and the weight of the pre-defined components. The results showed that the customary law, which has its own legal logic with aspects of local knowledge developed in the community for a long time as it relates to the values inherent culture and spiritualism, has a chance to be implemented in protecting environment. Customary law can be instrumental in preventing and protecting the environmental damage because the goal is to create harmony with nature.

1 Introduction

Policies or regulations are designed to protect people and society from the adverse effects arising from environmental problems [1]. The neighborhood is a place and role of human beings between living beings and other living components of the environment while environment is the unity of space and all objects, power, situation, and living creatures, including humans and their behavior, that affect the sustainability of our life. Environmental problems happening today are influenced by the growth of population and society, the high demand of natural resources that are exploitative in its extractions, and the

*Corresponding author: ratihlestarini@yahoo.com

diverse development of technology and cultures between countries. In Indonesia,the protection of environment and the addressing of environmental problems is regulated in Law No. 23 of 2007 on Environmental Management. However, the implementations of the Law have not been able to maximally prevent environmental damage. Customary law serves to regulate the use and management of natural resources. This law is intended for the people to maintain and preserve their natural resources [2].

Almost all regions in Indonesia have customary law. The customary law regulates diverse aspects including on environmental management such as regulation on deforestation, wildlife hunting, forest-produce extraction, and forest protection. Customary law communities will manage the forest that serves as the source of life, using all its wisdom [3].

Customary law serves to regulate the use and management of natural resources. This law is intended for the people to maintain and preserve their natural resources [4]. One example of the use of customary law in addressing deadlock environmental cases is Customary Oath in Dayak Community. A customary oath is done by praying to their ancestors and asked them to punish anyone who changed their borders and territory [2]. Another example is the Sasi Law in Maluku Community. The Sasi Lawprohibits Sasi community to extract natural resources in a set period of time. Violation to this regulation will be punish by their ancestral spirits in the form of illness or it can be a physical punishment such as caning, fines, forced labor, and excluded from society. Sasi Law is still effective in preserving and protecting the environment. With the types of sanctions given, the community will not violate the law and will continue to keep the environment [5].

Environmental protection is an action to protect the environment from all forms of exploitation and in doing so there should be attention from the governments, academia, and individuals [6]. Environmental protection is a major issue in the discussion of environment sustainability. In the process of achieving sustainable development, environmental protection is an important integral development process and cannot be separated with sustainable development [7]. Environmental protection needs to make the laws of nature into consideration, in addition to environmental protection also need to be integrated with state laws and policies [8]. Laws of nature are considered regarding environmental protection to become the basic rule of human life, which later became the norms that develop within the public sphere. The norms evolving locally in the community will become customary law and be obeyed by the community and will have sanctions for those who violate them [9, 10]. Customarylaw on environmental protection is potential to effectively suppress and reduce environmental damages caused by humans, and is also considered appropriate to support the lives of generations to come.

2 Materials and Methods

This research is aimed to determine the best way to protect the environment using customary aspect as the instrument. This research used SWOT analysis which consist of Strength (S), Weakness (W), Opportunity (O), and Threat (T). This research used SWOT Matrix to determine the conditions of the SWOT factors. The use of SWOT analysis in the study of law is to identify the strategic factors in the process of policy formulation [9]. The paradigm of this research is descriptive analysis and works on inductive logic. Tools to develop such a strategy is the SWOT matrix. SWOT matrix to produce four types of strategy, namely strategies SO (Strength-Opportunity), strategy WO (Weakness-Opportunity), the strategy of ST (Strength-Threats), and WT strategy (Weaknesses-Threats) [11]. By using SWOT analysis will additionally receive four types of strategies through internal and external factors that aggressive strategy (SO), competitive strategy (ST), a conservative strategy (WO), and defense strategy (WT). Determination of the weight is done through the deep interview from many experts, like stakeholders, religious leaders and environment activists, to get their judgment correspond to the reality. Therefore, the weight of the material which based on experts judgement is expected to correspond with the actual conditions and fit to reach the objective of the research.

3 Result and Discussion

In the case of strategy of implementation of customary law in protecting the environment thereats are strengths, weakness, opportunities, and threats to be addressed as follows:

a. Strengths: customary law can control people behavior and the interactions with nature. Customary law has strong connection between traditions and its believers.
b. Weakness: the role of customary law in environmental protection is still low. Customary law is not used as the main regulations for environmental protection instead of formal law.
c. Opportunity: customary law can protecting environment, customary law can strenghten the environmental law. Connection between traditions and believers in customary law expected to protect the environment effectively.
d. Threats: environmental sustainability can be threatened without a strong law. Strong connections in customary law can be used to reach environmental sustainability.

Based on the weighting that has been done, the matrix SWOT strategy of the dominant variable component consists of:

Table 1. SWOT matrix.

	S (Strength) Internal strength factor (Tradition, religion, and faith)	W (Weakness) Internal vulnerability factor (powerless and unclear mechanism)
O (Opportunity) External opportunity Factor (Community participation)	SO strategy: create strategies using control people's behaviour and the interaction with nature to protecting the environment	WO strategy: create strategies that enhance the role of customary law to protecting the environment
T (Threat) External Threats Factor (Wisdom and value)	ST strategy: create strategiesusing control prople's behaviour to prevent the environmental problems	WT strategy: create strategies that enhance the role of customary law to prevent the environmat problems

Table 1 about SWOT matrix describes how to define strategy based on Strength, Weakness, Opportunity, and Threats factors. Figure 1 describes the quadrant in the SWOT analysis. Quadrant 1 supports offensive strategy based on SO factor, quadrant 2 supports rational strategy based on WO factor, quadrant 3 supports diversification strategy based on ST factor, quadrant 4 supports defensive strategy based on WT strategy.

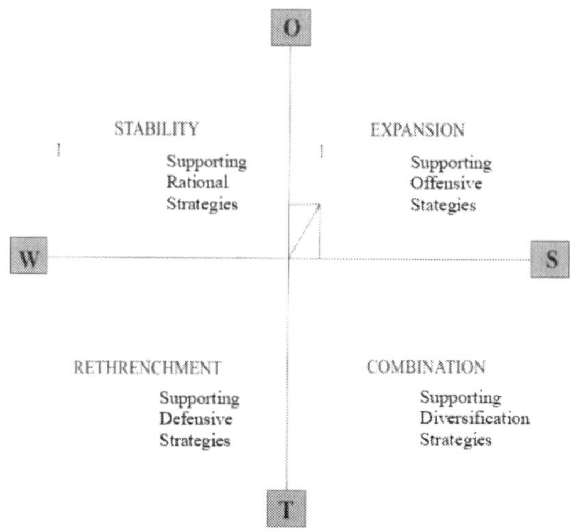

Fig. 1. SWOT matrix.

The result based on the SWOT analysis are the role of customary law in environmental protection is still weak, due to the assumption that customary law does not have strong connection, the assumption that customary law is left behind by formal law, and the absence of a clear mechanism which has effective systems and roles in the formal law towards the customary law which is a particular law. The weak role of customary law causes it can not control human behaviour and their interaction with the environment. Strategies are needed to reduce the weakness of customary law in environmental protection process by opening opportunity for the community to take part in environmental management, especially where the formal law is uncertain and unable to resolve environment problems. The expansion position can be done because the community's local wisdom can be used to create new approach in doing sustainable environmental management.

Other aspects of customary law that can be used in the protection and management of environment are tradition and religious aspect embedded in customary law. The autonomy regime also gives opportunity for customary law to be used in environmental protection. Customary law that is born and evolve from the community is significant to regulate and control the behavior of the community and their interactions with nature. This leads to a harmonious and balanced life. [12].

A minimum role of customary law in environmental protecting causes the meaning of customary law been ignored in environmental law, but the evidence that the existence of customary law is very important[13]. Customary law is a unique normative value that is always there and carried by the communityy, and always remain, but the customary law as a product that is not static and historic from the interaction between the actors and the environment. In another sense, the customary law has a power that is difficult to change because it has been present for generations and is formed through the relationship between human and nature. Customary law to question the way the interpretation, application, and determination of the results from the available resources, but the interpretation of customary law is more complex than the interpretation of state laws and other formal laws. [12]. Interpretation of customary law is more complex because it departs from myths or things that are not real, but has a binding power, while formal law departs from rational thinking by individuals and groups[14]. The customary law regulates many issues related to indigenous peoples property regulated by the state, for example, indigenous forests [16].

In addition to organizing indigenous forests, the customary law also gave the norms and customs of the sacred values of preservation and protection of life enjoyed by indigenous peoples. The customary law becomes an important element in addressing disputes and environmental issues. Environmental protection can be very effective when using the customary law. People living in a community with strong customary law will avoid to damage the environment. They will carry out their daily activities in the traditional way and in accordance with the customary law. This is considered as a form of public adherence to customary law in protecting their environment to remain sustainable.

Customary law is potential to be used because it has been present in the community for generations. The transfer of customary law knowledge from the previous generation to the next is done with full conscience and preservation. The values and norms in customary law is not in contrary with state laws therefore customary law is very feasible to be used in the management and protection of the environment]. Management and environmental protection that are based on customary laws are more flexible and made by consensus. It became the umpteenth factors which reinforce the position of customary law to be used in the protection and management of the environment [17]. However, of course, the customary law should receive extra attention by multi-stakeholder cooperation in environmental protection including environmental activists is one way to support the sustainability of the environment and the parties involved in it [18]. If these factors are considered properly, then the customary law can be used maximally and is potential as a role model for environmental protection laws.

4 Conclusion

Increasingly rampant cases of environmental destruction, such as illegal logging, illegal mining, illegal fishery etc, across the country alerts many people. The formal law available for environmental protection and management is not sufficient to reduce the destruction. However, there is an alternative to the lack of law regulating environment protection and management that is the use of local wisdom. Local wisdom is able to integrate both internal value and external value, to protect local value from outside interference and to control the development of the local community. This local wisdom needs to be studied further. The role of customary law that is based on local wisdom can fill the gaps in the formal law in preventing damage to the environment and this can be effectively enforced. The weak enforcement of formal law, such as blunt upward and sharp downward, gives an opportunity for customary law to be used in environmental protection. Through these conditions, gradual environmental sustainability can be manifested.

Acknowledgements

This research is funded by Ministries ff Research, Technology and Higher Education/Kemristekdikti 2018 with ProgramPenelitian Dasar Ungguan Perguruan Tinggi (PDUPT) 2018, contract number 275/UN.2.R3.1/HKP05.00/2018

References

1. C. Davis, Ext. Ind. Soc. **4**, 63-68 (2017)
2. Z. Judge, M. Nurizka. Peranan hukum adat sasi laut dalam melindungi kelestarian lingkungan di Desa Eti Kecamatan Seram Barat Kabupaten Seram Bagian Barat. Lex Jurnalica **6**, 1 (2008)
3. M. Sirait, C. Fay, A. Kusworo. Bagaimana hak-hak masyarakat hukum adat dalam mengelola sumber daya alam diatur. Southeast Asia Policy Research. Working Paper **24**, (2000)

4. Philip Jacobson. Inside's Indonesia highest-profile land conflict. Retrieved April 4, 2018 from https://news.mongabay.com/2015/12/an-award-from-the-un-in-paris-but-fighting-for-survival-in-borneo/
5. J. M. Ubink, *In the land of the chiefs: customary law, land conflicts, and the role of the state in peri-urban Ghana* (Leiden University Press, 2011)
6. A. Perreau-Saussine, J.B. Murphy, *The nature of customary law* (Cambridge University Press, 2007)
7. F.P. Maganga, Incorporating customary laws in implementation of IWRM: some insights from Rufiji River Basin, Tanzania. Physics and Chemistry of the Earth **28**, (2003)
8. H.Y.S.H. Nugroho, A. Van Der Veen, A. Skidmore, Y.A. Hussin. Theoretical framework for spatial planning and forest management in Indonesia: securing the basic rights for adat people. *Journal of Forestry Research* **29**, 2 (2018)
9. P. Hill. *Environmental protection: What everyone needs to know* (Oxford University Press, 2017)
10. D.K. Anton, D. Shelton, *Environmental protection and human rights* (Cambridge University Press, 2011)
11. L. Lavrysen. Environmental Law Syllabus. Ghent University
12. F.R. David, *Strategic Management Concepts and Cases* (Prentice-Hall Publishing, Francis Marion University, Florance, South Carolina, 2011).
13. M. Samejima, Y. Shimizu, M. Akiyoshi, N. Komoda. *SWOT Analysis support tool for verification of bussiness strategy* (Graduate School of Information Science and Technology, Osaka Univesity, Japan, 2006)
14. S.M Hatefi. trategic planning of urban transportation system based on sustainable development dimensions using an integrated SWOT and fuzzy COPRAS approach. *Global Journal of Environmental Science and Management* **4**, 1 (2018)
15. M.I.F. Rahayu. Aspek hukum peran serta masyarakat adat dalam pengelolaan lingkungan hidup. *Ethos* **1**, 1 (2003)
16. I.N. Sirtha, *The Participation of Customary Law in The Sustainability of Regional Environment* (Universitas Udayana, Bali, Indonesia, 2015)
17. N.L.K. Ni'mah, H. Herdiansyah, T.E.B.Soesilo, E.F. Mutia. Strategy for increasing the participation of masyarakat peduli api in forest fire control. *IOP Conference Series: Earth and Environmental Science* **126** 012148, (2018)
18. D. Asteria, Budidarmono, H. Herdiansyah, N.L. Ni'mah. Gender sensitive education in watershed management to support environmental friendly city. *IOP Conference Series: Earth and Environmental Science* **126** 012146, (2018)

The Estimation Loss and Gain of Mammal Species Diversity due to Oil Palm Plantations: A Case Study of BPME Estate, Riau, Indonesia

Satrio Suryadi Nugroho*, *Yanto* Santosa

Ecology and Wildlife Management Division, Forest Resource Conservation and Ecotourism Department, Faculty of Forestry, Bogor Agricultural University, Bogor, Indonesia

Abstract. The existence of palm oil plantation development allegedly resulted in a ecological losses, including the sustainability of mammalian diversity. So far the ecological losses to the diversity of wildlife species, especially mammals, have not been widely known. Therefore, in February 2018 a direct observation was conducted at 5 observation paths with Strip Transect Method and simultaneous data retrieval, three repetitions at 8.00-10.00 a.m. and 16.00-18.00 p.m. In addition, there are 10 Rodentia traps and 15 camera traps on each observation path. The analysis results obtained the total number of species and wealth of the highest species on oil palm plantation cover. The highest similarity index was obtained from young-aged oil palm (IS = 0.67) and the lowest commonality index was in old-aged oil palm (IS = 0.50). On the other hand, the old-aged oil palm has the highest value of mammal diversity and the lowest loss value, while the young-aged oil palm has the lowest mammal diversity value. Overall, land cover changes to oil palm plantations have a positive impact due to the addition of two species of mammals and the value of species diversity did not change significantly.

1 Introduction

Indonesia is known as a country with a wealth of very high biodiversity, both in ecosystem scale, species and genetics to attract the attention and admiration of people around the world [1];[2];[3]. However, Indonesia is also noted as a country that has a remarkable vulnerability to the decline/loss of potential biodiversity loss. Among the many factors causing biodiversity loss, the development of oil palm plantations (which makes Indonesia the world's largest palm producer) is considered to be one of the determinants [4];[5];[6]. The Indonesian state also has over 80% of the remaining primary forests in Southeast Asia, where many endemic species are threatened with extinction by some of the increasing levels of global deforestation [7];[8];[9];[10]

Biodiversity has been recognized globally, as one of the determinants of the sustainability of an ecosystem [11]. When the expansion of oil palm plantations is accused of threatening wildlife such as Orangutans, Sumatran Tigers, and Rhinoceros (biodiversity loss), a more in-

* Correspondent Author: satriosuryadinugroho@gmail.com

depth study of the issue of biodiversity loss is needed, as research results also show that oil palm plantations provide habitat for another biodiversity (biodiversity gains). Changes in land cover (forests and shrubs) due to human activities and development to oil palm plantations will also impact on the changing conditions and potential for biodiversity. The change can take the form of loss or even loss of potential biodiversity (biodiversity loss) that was present or easy to find. However, it may be the opposite, that is, due to changes in land cover that initially in the form of land (open land cover) which is considered not enough to have the potential of biodiversity (plant and wildlife) into oil palm plantation, it is believed to be beneficial for the development of some wildlife especially predatory animals such as owls and python [12].

Thus, the potential impacts of oil palm expansion on tropical forests and biodiversity are the major conservation concerns of the world [13];[14]. It also raises concerns over oil palm plantations due to rapid and widespread agricultural expansion, posing a serious threat to natural ecosystems around the world [15] including mammalian diversity. Mammals play a role in pest control, seed dispersers, predators, and pollinators for plants [16];[17];[18]. The existence of mammals plays an important role in the balance of natural ecosystems. In addition, mammals also have several ecological functions, namely a soil fertilizer, flower pollinator, seed dispenser, an organic fertilizer producer, biological pest control, helping the process of recycling minerals from plants, indicating the destruction of forest habitat and indicators of air pollution by heavy metals [19]. Therefore, it is necessary to conduct research to find out the diversity of mammal species in various land cover in oil palm plantation, as it is still relatively little research on this matter.

2 Experimental details

2.1 Time and location

The research was conducted at BPME Estate, Riau Province in February-March 2018.

2.2 Data collection

Materials and equipment used in this research include stationery, tally sheet, drone DJI Mavic, rodentia trap, binocular, regional map, GPS, digital camera, camera trap, timer and mammal observation manual. The object of the study was the type of mammals and the number of individuals found in each habitat. The data taken are primary and secondary data. Primary data include mammal species in each habitat, number of individuals, time, the location of direct encounter and indirect encounter data (sound, footprint, and feces). The study used a 1-km strip transect method with right and left widths of 50m each. Considering the impact is the difference between the condition of species diversity before being turned into oil palm plantation with the condition of oil palm plantation, this research is conducted on five land cover. Based on the results of literature searches, interviews of BPME Estate managers and communities around the palm plantation, as well as the analysis results of landscaping interpretation 2 years before the palm oil plantation, was built, the type of land cover in the area is dominated by shrubs and coconut plantation. Therefore, observation on the area before the oil palm plantation (accumulation between shrubs and coconut plantation), while the observation path as a comparison of palm oil plantation, observation is done in the area of young-aged oil palm, middle-aged palm plantation and old-aged palm plantation. In addition, there are also 10 rodentia traps and 15 camera traps on each observation path. Trap bait used is a mixture of shrimp paste and peanut butter. Observations were made in the morning (06.00-08.00 a.m.) and evening (04.00-06.00 p.m.) as three repetitions on each path in each land cover.

2.3 Data analysis

2.3.1 To estimate the impact of species diversity as a result of oil palm plantations, it is used to calculate the diversity index and species richness as follows:

2.3.1.1 Diversity Index (H')

The species diversity in each land cover type uses the Shannon-Wiener Diversity Index (Magurran 1988), by the formula:

$$H' = -\sum Pi \,.\, ln(Pi) = -\sum \left(\frac{ni}{N}\right) . \, ln\left(\frac{ni}{N}\right)$$

Information :
H' : Index of diversity
Ni : Number of individual types i
N : Number of individuals of all types

2.3.1.2 Species Richness Index (Margalef Index)

The species richness index serves to determine the species richness of each species within each community encountered. The following formula is used [20]:

$$D_{mg} = {(S-1)}/{\ln(N)}$$

Information:
Dmg : Index of type riches
S : Number of types
N : Total number of individuals across species

2.3.2 To estimate the impact of species composition as a result of oil palm plantations, the Sorensen Type Similarity Index [20] is used:

$$IS = \frac{2c}{a+b}$$

Information:
a: Number of species found in community A
b: Number of species found in community B
c: Number of species found in communities A and B

3 Result and discussion

3.1 The Impact of oil palm plantations on the diversity of mammals

Based on observations on the five land cover types in BPME Estate, it was found 30 individuals from 6 species consisting of five families: Sciuridae (1), Muridae (2), Suidae (1), Viverridae (1), and Felidae (1). Based on observations on five predetermined paths, the data of species number, diversity index and index of species richness on each land cover (**Table 1**). The data show that the type of land cover that has the highest number of species is found in land cover prior to oil palm plantations (accumulation between shrubs with coconut plantation) and old-aged oil palm as many as four (4) types. A high number of species in both locations also caused the two sites to have the lowest grade index value compared to another land cover. Meanwhile, the encounter of the lowest mammal species (S = 2), is in the cover of young-aged oil palm plantation followed by the value of the species diversity index (H ') which decreased to 0.69.

Table 1. The types and distribution of mammals encountered in oil palm plantations BPME Estate

No.	**Species Name**	**Scientific Name**	**Family**	**Observation Areas**			
				Before Oil Palm Plantation	**Oil Palm Plantation (OPP)**		
					YOP	**MOP**	**OOP**
1	Plantain Squirrel	*Callosciurus notatus*	Sciuridae	4	0	2	10
2	Malayan Field Rat	*Rattus tiomanicus*	Muridae	0	0	1	1
3	Giant Sunda Rat	*Sundamys muelleri*	Muridae	1	0	0	0
4	Eurasian Wild Pig	*Sus scrofa*	Suidae	5	1	1	0
5	Common Palm Civet	*Paradoxurus hermaphroditus*	Viverridae	0	0	0	1
6	Leopard Cat	*Prionailurus bengalensis*	Felidae	1	1	0	1
S				**4**	**2**	**3**	**4**
H'				**1.16**	**0.69**	**1.04**	**0.79**
Dmg				**1.25**	**1.44**	**1.44**	**1.17**

Information: YOP (Young-aged Oil Palm); MOP (Middle-aged Oil Palm); OOP (Old-aged Oil Palm)

The high number of species on the land cover before oil palm and old-aged oil palm has a positive impact because both sites have the highest number of species (S=4), so if it is assumed that shrub and coconut plantation changes into oil palm plantations do not provide significant change, where the value can be seen in **Table 2.**

Table 2. Impact of oil palm plantations on the diversity of mammals in each land cover

	Young-aged Oil Palm	**Middle-aged Oil Palm**	**Old-aged Oil Palm**
Impact on Number of Types	-2	-1	0
Impact on H'	-0.47	-0.12	-0.37
Impact on Dmg	+0.19	+0.19	-0.80

Information: (-) decreased, (+) increased

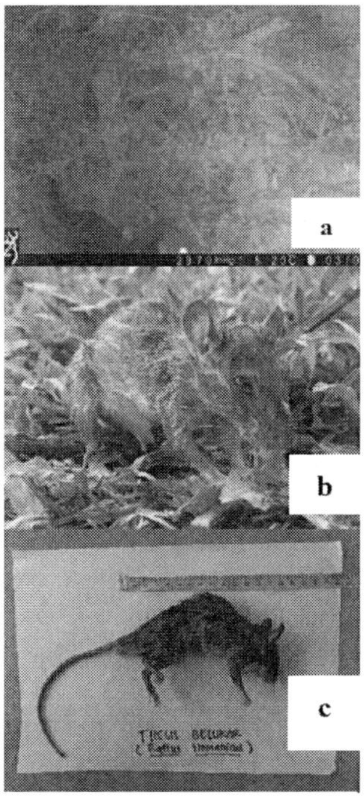

Fig. 1. a *Paradoxurus herma-phroditus*
b *Sundamys muelleri*
c *Rattus tiomanicus*

Oil palm plantations impacted a decline of species in young-aged oil palm and middle-aged oil palm, where the largest differences is in the young-aged oil palm and the smallest difference in the middle-aged oil palm. Both sites show less number of species compared to land cover prior to oil palm. In contrast, the difference between the index of species richness in both sites increased by 0.19. On the difference of index of species diversity showed different things again, in both locations also have decreased value of the index of species diversity and also proportional to the decrease in the number of types. Meanwhile, in the old-aged oil palm also decreased the value of the diversity index, but not directly proportional to the number of species due to the different number of species and the number of individuals at each location. This is supported by [21], which states that the value of species diversity is not only determined by the number of species but also determined by the number of individuals of each species. When examined again, young-aged oil palm and middle-aged oil palm land cover have been decreasing the number of species due to several factors. One of the factors that can affect species diversity in a habitat is the level of disturbance from humans [22]. In palm oil plantation areas, the level of human activity is quite high as there are palm fruit harvesting and spraying activities to remove pests and weeds. Based on interviews to the manager, the average spraying is done once every three months and the harvesting of the oil palm fruit is done every two weeks, so that with many human activities also affect the diversity of species in a habitat.

Overall, the comparative impacts of oil palm plantation cover (accumulation of young-aged oil palm, medium-aged oil palm, and old-aged oil palm) with shrubs and coconut plantation do not make a significant difference, but instead provide new species such as Common Palm Civet (*Paradoxurus hermaphrodites*) and Malayan Field Rat (*Rattus tiomanicus*), but has also eliminated the Giant Sunda Rat (*Sundamys muelleri*) species that are not found throughout the palm oil plantation observation path, as can be seen in Figures 1 and 2.

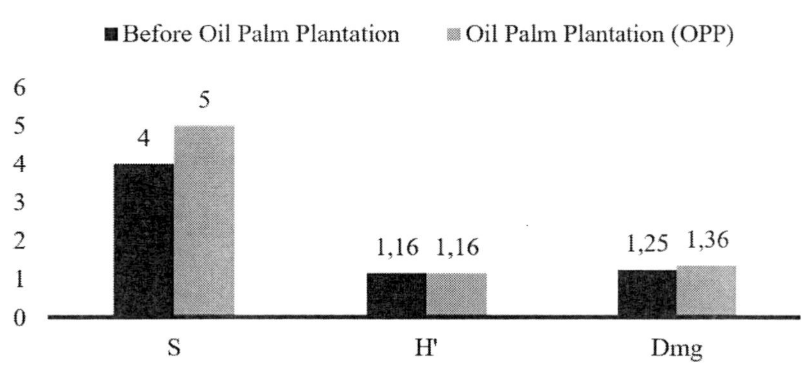

Fig. 2. Impact of oil palm plantations on the diversity of mammals

After merging all the paths on the oil palm plantation cover, there was an increase in the diversity of mammals. Described in Figure 2, the land cover after palm has a higher number of species and higher grade index values. Furthermore, there is no change in the index of species diversity, so in that case, it shows that the change of shrubs and coconut plantation into oil palm plantation area did not give significant change. The difference in value in each location and the high number of species is also influenced by the size of the habitat [23], where this research on oil palm plantation cover has a larger area than the area before being made oil palm plantation. Furthermore, the high number of species, the index of species diversity and the index of species richness is influenced by the variation of the level of vegetation diversity. In land cover prior to oil palm plantations (shrubs) generally vegetation only in the form of seedlings and saplings that resulted in the absence of mammals due to a limited source of feed. The results of this study are consistent [24] hypothesis which suggests that an increase in the number of different habitats may lead to an increase in species diversity. In addition, the high index of mammal diversity is related to the high diversity of vegetation types in oil palm plantations [25].

3.2 The Impact of Oil Palm Plantations on Mammal Number of Species

Comparison of the Sorensen Similarity Index from shrubs and coconut plantation with all types of land cover lines in oil palm plantations (accumulation of young-aged oil palm, medium-aged oil palm, and old-aged oil palm) shows that the value at each location ranges from 0.50 to 0.67 indicating the similarity between cover palm oil plantations that tend to be different (Figure 3).

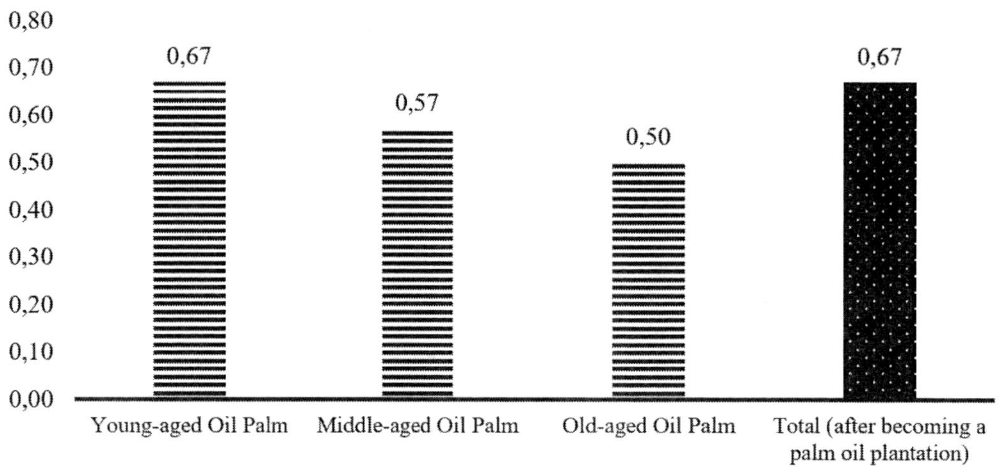

Fig. 3. Comparison of the Sorensen type similarity index between land cover before being used as an oil palm plantation area after being converted to oil palm plantation

The species similarity index between shrubs and coconut plantation with the young-aged oil palm has the highest value (IS = 0.67), the high value is due to the same number of species between the two locations and the distance between sites is close together, causing high IS value. The species similarity index between shrubs and coconut plantation with middle-aged oil palm has a smaller value than the previous value (IS = 0.57), the value is obtained because the same number of species between two locations there are two types, as well as the distance between locations, is quite far apart. Furthermore, the similarity type index between shrubs and coconut plantation with the old-aged oil palm has the lowest value (IS = 0.50), the value is due to the same number of species between the two locations and the distance between the locations is far from each other, thus causing low IS value. Meanwhile, if the value of the similarity index of the species is calculated as a whole, land cover before oil palm plantation with oil palm plantation area shows a high enough value is also equal to 0.67 with the same mammal species in both locations as much as three (3). The change of shrub and coconut plantation into oil palm plantations have affected the similarity index of species, but based on research there is still the same type between the two locations. This indicates that there is still a common characteristic of composition between shrubs and coconut plantation with oil palm plantations. The composition of the two communities said to have a similar degree of composition if it has a species equality index value> 0.50 [26]. Two similar communities are generally no more than 2/3 of the same species can be found. The IS scores that range from 0.25 to 0.50 already show the great species similarity [27]. A value above 0.50 indicates that both locations are the same community, while a value less than 0.25 indicates that the community is different.

There is an increase in the type of mammal species of Malayan Field Rat (*Rattus tiomanicus*) and Common Palm Civet (*Paradoxurus hermaphroditus*) due to changes to palm oil plantations. However, in detail, the loss or addition of mammals to each type of land cover can be seen in Figure 4.

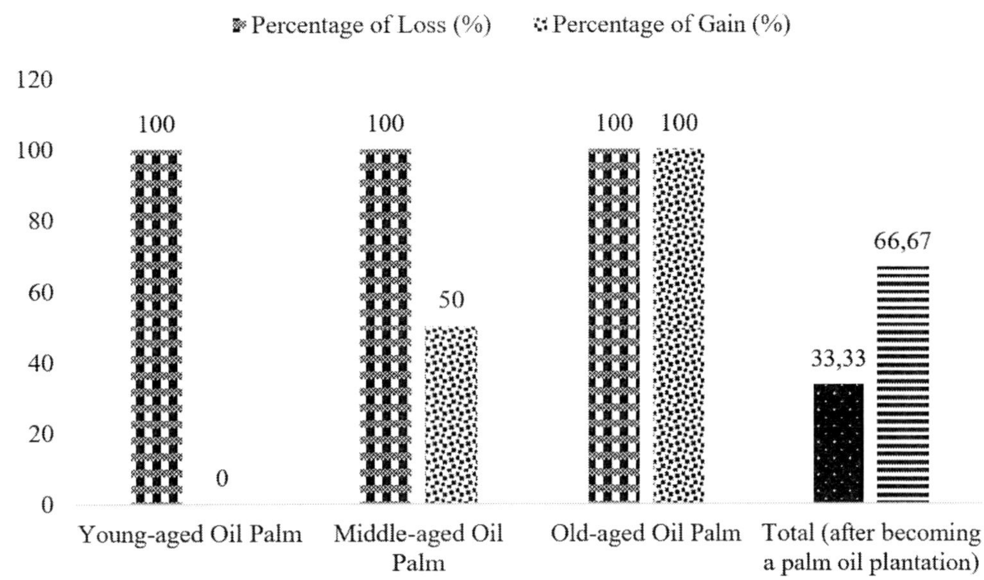

Fig. 4. Comparison of percentage loss and acquisition of mammal species on each land cover

Figure 4 explains that shrub and coconut plantation changes into oil palm plantations (accumulation of young-aged oil palm, medium-aged oil palm, and old-aged oil palm) have eliminated certain types of mammals. The number of mammals which lost on land cover change before oil palm plantations become a young-aged oil palm, resulted in decreasing the index of species diversity to H '= 0.69 and increasing the species wealth index by Dmg = 1.44. The occurrence of changes in the index value has resulted in the loss of species amounting to two (2) types or as many as 100% of mammals such as Plantain Squirrel, Malayan Field Rat, and Giant sunda Rat are not found in the young palm area (Figure 4). The absence of the Rodentia order is due to the limited source of feed in the young-aged oil palm, such as the unproductive fruit of oil palm due to the young-aged oil palm (Table 3). The order of Rodentia like the type of squirrel is commonly found eating the raw palm fruit in oil palm plantations [28]. As a result of the loss of Rodentia, order animals can increase the activity of herbivores such as insects. The type of order Rodentia possesses as a detrivore, insectivore and granivore [29].

Changes in shrubs and coconut plantation to middle-aged oil palm are causing declining index of species diversity to H '= 1.04 and increasing the species richness index by Dmg =

1.44. As a result of the change in the index value, it has resulted in the loss of species of Leopard Cat and Giant Sunda Rat with 100% percentage, but from the change also on the new species change that is Malayan Field Rat with 50% (Figure 4). Loss of Leopard Cat may affect the balance of ecosystems as they can increase Rodentia order animals such as rats and squirrels that often feed on palm fruits and are considered pests (Table 3). Carnivorous species (Javan Mangoose, The Javan ferret-badger, The Sunda stink badger, Common Palm Civet and Leopard Cat) have a role as a balancing ecosystem mainly as a predator of small animals such as rats and squirrels [30]. This small carnivore can be relied upon as a key species capable of preventing the explosion of rat populations and other small vertebrate animals. The presence of a new species (Malayan Field Rat) has provided the benefits of lowering herbivore [29]

Furthermore, changes in shrubs and coconut plantation to old-aged oil palm have resulted in decreasing of species diversity index to H '= 0.79 and decreasing index of species richness by Dmg = 1.17. As a result of the change in the index value, causing the loss of two species of Giant Sunda Rat and Eurasian Wild Pig with percentage 100%. However, from the change also the change of new species namely Common Palm Civet and Malayan Field Rat with the same percentage that is equal to 100% (Figure 3). The absence of Eurasian Wild Pig in the old palm, Eurasian Wild Pig and Leopard Cat only found in oil palm plantations and not found in the forest [31]. The situation is presumably due to the location of adjacent oil palm directly with the coconut plantation, so that Eurasian Wild Pig can move easily from oil palm plantation to coconut plantation and vice versa. The presence of Common Palm Civet in the old-aged oil palm plantations due to abundant feed sources, it is supported by the discovery of Rodentia animals (rats and squirrel) as many as 11 individuals. The number of individuals Rodentia animals is the largest number of individuals compared with all land cover, so it becomes one of the emergences of Common Palm Civet. Species Common Palm Civet has a role as a predator of small animals such as rats and squirrels [30]. In addition, with the presence of Civet Mongoons can give advantages to suppress pests such as squirrels and rats as a fruit-eating palm.

The missing mammal species, as well as new mammal species due to changes in land from shrubs and coconut plantation into oil palm plantations in detail, can be seen in Table 4 below.

Table 3. Types of mammal loss and gain due to shrub and coconut plantation changes into palm oil plantations of BPME Estate

No.	Type of Change	Loss	Gain
1	Before Oil Palm Plantation – Young-aged Oil Palm	Plantain Squirrel, Malayan Field Rat and Giant Sunda Rat	0
2	Before Oil Palm Plantation – Middle-aged Oil Palm	Leopard Cat and Giant Sunda Rat	Malayan Field Rat
3	Before Oil Palm Plantation – Old-aged Oil Palm	Eurasian Wild Pig and Giant Sunda Rat	Common Palm Civet and Malayan Field Rat

Overall, changes in shrub cover and coconut plantation into oil palm plantations in BPME Estate have removed one (1) mammal species, Giant Sunda Rat and the change of one (1) new mammal species namely Common Palm Civet. Furthermore, when viewed from the number of individuals in each species found, an increase in the number of individuals occurs in 3 species of mammals namely Plantain Squirrel, Malayan Filed Rat and Leopard Cat due to the change. However, the number of individuals in the wild boar species experienced a 60% decrease in the encounter.

Based on the data in Figure 3, shrubs and coconut plantation into oil palm plantations (accumulation of young-aged oil palm, medium-aged oil palm, and old-aged oil palm) have taken certain mammals. The types of missing mammals occur in young-aged oil palm areas, where there are two species of mammals: Plantain Squirrel and Giant Sunda Rat are not found in the young-aged oil palm area. However, shrubs and coconut plantation into old-aged oil palm have bought a new type of Common Palm Civet dan Malayan Field Rat. Furthermore, from the results of mammalian observations showed that found one type of protected mammals based on Government Regulation No. 7 of 1999 namely Leopard Cat (Table 4).

Table 4. Status of Species Protection and Scarcity

No.	Species Name	Scientific Name	Family	Conservation Status		
				IUCN	CITES	PP 7/1999
1	Leopard Cat	*Prionailurus bengalensis* (Kerr 1792)	Felidae	LC	AP I	P
2	Malayan Field Rat	*Rattus tiomanicus* (Miller 1900)	Muridae	LC	TT	NP
3	Giant Sunda Rat	*Sundamys muelleri* (Müller 1838)	Muridae	LC	TT	NP
4	Plantain Squirrel	*Callosciurus notatus* (Boddaert 1785)	Sciuridae	LC	TT	NP
5	Eurasian Wild Pig	*Sus scrofa* (Müller 1838)	Suidae	VU	TT	NP
6	Common Palm Civet	*Paradoxurus hermaphroditus* (Pallas 1777)	Viveridae	LC	AP III	NP

Information: PP RI No7/1999; P (Protected); NP (Not Protected); AP (Appendix CITES); VU (Vulnerable); LC (Least Concern); NI (Not Included)

Based on CITES category, Leopard Cat are categorized as Appendix II and Common Palm Civet, including Appendix III category. In the IUCN category, the mammal species found in the oil palm plantation belong to the Vulnerable (VU) category for the wild boar species and the remainder is categorized as Least Concern (LC) (Table 3).

4 Summary

1. Overall, the change from shrubs and coconut plantation into oil palm plantations of BPME Estate have increased the number of mammals and no significant change impacts on species diversity.
2. The impact of oil palm plantations leads to differences in the composition of mammal species. The highest species composition difference is found in old oil cover with the lowest species equality of 0.50, while the lowest composition difference is shown by young-aged oil palm with the highest species equality of 0.67. Overall, the compilation of land cover prior to oil palm plantations (shrubs and coconut plantation) with oil palm plantations (accumulation of young-aged oil palm, medium-aged oil palm, and old-aged oil palm) has a similar species value of 0.67.
3. The change from shrubs and coconut plantation into oil palm plantations has eliminated one type of the Giant Sunda Rat with a percentage of 33.33%. However, these changes also acquire new types such as Malayan Field Rat and Common Palm Civet with a percentage of 66.67%.

References

1. Bappenas *Strategy and Action Plan for Indonesian Biodiversity 2003-2020* [National Document]. (National Development Planning Agency, Jakarta, 2003)

2. I. Maryanto, M. Noerdjito, T. Partomihardjo, *Mount Slamet Ecology: Geology, Climatology, Biodiversity and Social Dynamics.* (LIPI Press, Jakarta, 2012)
3. Suhartini, *Study Local Wisdom Society in Natural Resource Management and Environment.* Proceedings of the National Seminar on Research, Education and Application of Mathematics and Natural Sciences. (Department of Biology Education FMIPA Yogyakarta State University, Yogyakarta, 2009)
4. H. Purba, Menhut Vs Kapolri, Why is unsustainable palm oil expansion a major environmental problem?. Agro Observer. Downloaded from *http://palmoilandtheenvironment.blogspot.com*, (2007)
5. Y. Syaukat, *Creating Economic Competitiveness and Environment of Indonesian Palm Oil Industry.* (Agrimedia MB-IPB, Bogor, 2008)
6. T.M. Li, Social impacts of oil palm in Indonesia: A gendered perspective from West Kalimantan. Occasional Paper 124. (CIFOR, Bogor, Indonesia, 2015)
7. N.S. Sodhi, Southeast Asian biodiversity: an impending disaster. *Trends Ecol Evol.* 19: 654-660 (2004).
8. FAO, Global Forest Resources Assessment 2005. (FAO, Wahington, 2006).
9. N.S. Sodhi, B.W. Brook, *Southeast Asian Biodiversity in Crisis.* (Cambridge University Press, Cambridge, 2006).
10. W.F. Laurance Forest destruction in tropical Asia. *Curr Sci.* 93:1544–1550. (2007).
11. M.L. Pant, A. Lehikoinen , L. Uusitato, R. Venesjarvi R, How to value biodiveristy in environmental management? *Ecological Indicators* 55: 1-11 (2015).
12. A. Muin, *Plantation of Oil Palm Plantations with Conservation Insight.* [tesis]. (Bogor Agricultural University, Bogor, 2013).
13. L.P. Koh and D.S. Wilcove, Cashing in palm oil for conservation. *Natur.* 448: 993-994 (2007).
14. J.P.W. Scharlemann and W.F. Laurance, How green are biofuels?. *Science.* 319: 43–44 (2008).
15. D. Tilman, J. Fargione, B. Wolff, Forecasting agriculturally driven global environmental change. *Science.* 292: 281–284 (2001).
16. N. Novriyanti, Take control of forest pests [Internet]. [downloaded 2017 Jan 23]. Available on: https://www.researhgate.net/publication/307966325_kendalikan_hama_hutan. (2016).
17. I.W.S. Anggraeni, Long-tailed monkey population and habitat (*Macaca fascicularis*) in mangrove ecosystems of Wonorejo and surrounding areas, Surabaya [essay]. Bogor (ID): Bogor Agricultural University (2013).
18. H. Mutiarani, Planning and testing of traps, testing of rodenticidal type in controlling *Rattus tiomanicuss* mill, *Rattus rattus* diardii linn, and *Rattus argentiventer* rob & klo [essay]. Bogor (ID): Bogor Agricultural University (2009).
19. A. Suyanto, *Mammals in Halimun Mountain National Park West Java.* Bogor (ID): BCP-JICA (2002).
20. A.E. Magurran, Why diversity?. In Ecological diversity and its measurement (pp. 1-5). *Springer*, (1988).
21. C.J. Krebs, A review of the Chitty hypothesis of population regulation. *Canadian journal of zoology*, **56**(12), 2463-2480, (1978).
22. J.C. Welty and L. Baptista, The life of birds, (1988).

23. H.S. Alikodra, Dasar-dasar pembinaan margasatwa. Fakultas Kehutanan Institut Pertanian Bogor, (1979).
24. R.H. MacArthur and J.W. MacArthur, On bird species diversity. *Ecology* **42**: 594-598 (1961)
25. A.P. Kartono, The diversity and abundance of mammals in PT Sukses Tani Nusasubur Kalimantan Timur plantation. *Media Konservasi*. **20** (2): 85-92 (2011).
26. Mawazin, A. Subiakto, Diversity and composition of natural regeneration of logged peat swamp forests in Riau. *J Forest Rehabilitation*. **1**(1):59–73 (2013).
27. D.I. Junaedi, The diversity of plant communities in Gunung Ciremai National Park. *Buletin Kebun Raya Indonesia*. **11**(2): 25-32 (2008).
28. G. Davies, J. Payne, *A Faunal Survey of Sabah*. (WWF-Malaysia, Kuala Lumpur, 1982).
29. Y. Santosa, A. Perdana, The role of high conservation value areas in the conservation of mammal species diversity in oil palm plantations: case studies of Riau Province. *Pros Sem Nas Masy Biodiv Indon*. **3**(1): 81-87 (2016).
30. I. Maryanto, M. Noerdjito, T. Partomihardjo, *Mount Slamet Ecology: Geology, Climatology, Biodiversity and Social Dynamics* (LIPI Press, Jakarta, 2012).
31. S. Yue, Effect of oil palm plantation mammal communities in northeastern Borneo[Thesis]. (2014)

Uang tutup babah as a customary dispute resolution in gayonese community

Shira Thani* and *Alvi* Syahrin

Doctoral Program in Law, Graduate School, Universitas Sumatera Utara, Medan, Indonesia

Abstract. There are several customary sanctions which are often imposed in customary dispute resolutions in Gayonese community. One of them is *uang tutup babah* (hush money). It is a payment to resolve customary dispute between parties. However, it is not yet regulated in laws and regulations related to customary law. This paper will discuss the legal strength of decision of *uang tutup babah* in a customary dispute resolution and its impact. Theory of sociological jurisprudence, theory of harmonization and theory of criminalization were used in this study. In general, *uang tutup babah* means hush money. It is a customary dispute resolution in some customary communities in Gayo. Although the sanction, *uang tutup babah*, it is not yet regulated by law on customary institutions, it is often used by parties involved in the dispute to resolve their problem because it has a positive impact on communities. It is also believed that it gives a deterrent effect to the perpetrator. Since this has been done repeatedly and community accept it as a sanction, it should be included in the law to ensure that it gives legal certainty and has legal strength.

1 Introduction

According to Ter Haar, customary law community is a group of community which has a regular community entity, lives in a certain area, has authorities as well as tangible and intangible wealth in which members of each entity has a normal life in community in accordance with law of nature and none of them has a will or a tendency to escape from the bond or leave it forever. [1]

According to Hazairin, customary law community is a community entity which has equipments to stand on its own feet i.e. having legal entity, authority entity and environmental entity based on a collective right to land and water for all of its members. [2]

The local customary law regulates human behavior to ensure that one's behavior will not harm others. Sadly, although the customary law is so good, it is still very often violated by community which, of course, is harmful to others. Therefore, they try to solve their problems by establishing a customary law institution which is subject to local customary law and has authority to regulate and resolve dispute. It also contains sanctions that can be imposed on the law breaker.

* Corresponding Author: shirathani21@yahoo.com

Penal mediation is one of forms of alternative dispute resolutions outside the court or appropriate dispute resolution. In general, alternative dispute resolution is used to resolve civil cases and not criminal cases. According to the applicable laws in Indonesia (positive laws), in principle, a criminal case cannot be resolved outside the court, although in certain cases, it is possible to resolve it outside the court. [3]

Some people in Central Aceh District prefer to resolve their criminal case through a customary institution in which the case is settled by a customary institution called *"Sarak opat*". Under Point f, Article 1 of Qanun of Central Aceh District Number 10 Year 2002 on Gayonese Customary Law, *sarak opat* is a Gayonese customary institution consisting of *Reje, Imem, Petue* and *Rayat Genap Mufakat*. Although they know that there is a formal court which can settle the case, some community members prefer to resolve criminal case through *sarak opat*.

Crimes in Gayonese customary law can be divided into four categories, namely: *Maas, Rujok, Died* and *Bela*. [4] The imposition of the sanction, among others, aims to recover balancewhich has been affected by the committed offense and create peace and harmony in society, not to decide win or lose. There are several customary sanctions which are often imposed in customary dispute resolutionsin customary court. One of them is *uang tutup babah* (hush money). It is a payment to resolve customary dispute between parties. Although the sanction is not regulated in Qanun on Gayonese Customary Law, it is often imposed.

Based on situation above, the problems are: How is the legal strength of decision of *uang tutup babah* in a customary dispute resolution? and what is the impact of the decision of uang tutup babah in a customary dispute resolution on community?

2 Research Method

This dissertation research is a juridical empirical research. A sociological-legal approach was applied in the research. Primary and secondary data was collected in this research. The research was conducted in six sub-districts in Central Aceh District. The primary data was collected through an in-depth interview with the research respondents and informants. Meanwhile, the secondary data was collected througha documentary study. Then, the collected data was analyzed by using a qualitative method.

3 Discussion

3.1 Concept and Basis of Customary Law

Ter Haar gives the definition of customary law based on his theory,*Beslissingen Leer*. It is the overall regulations which transform into the decisions of law funtionaries (in a broad meaning) that have authority (*macht*) and influence (*invloed*) and in its implementation, it is spontaneous and is obeyed sincerely.[5]

According to Bushar Muhammad, customary law is a law which regulates public order and tranquility in achieving peace and calm which does not harm other party. [6]

According to Sobhi Mahmassani, a customary law will only be recognized as a customary law if it meets the following requirements: 1) It must be accepted by good character and logic. 2) It must happen repeatedly and is spread widely. 3) It is still applicable in *muamalat,* such as in shopping by using a currency in a state where it is still valid. 4) There is no difference in requirements resulting from implicit requirement, and 5) It may not against provisions in *nash* or *fiqh*. [7]

Customary law has the following characteristics: 1) It has a sense of togetherness or a strong communal bond, meaning that human, according to customary law, is a creature which has a strong communal bond; the sense of togetherness embraces all fields of customary law. 2) It has a religious or magic character which is related to way of life in Indonesia. 3) It is encompassed by concrete mind, meaning that it the many and repeatation of concrete life relation, and 4) it is visual, meaning that legal relation will only be established by a visible bond. [8]

According to our law scholars, the sources ofcustomary laws in Indonesia are habits and customs which are based on people's tradition which is handed down from generation to generation and is obeyed and followed by community as a good habit and is useful to manage communal life. These sources can be seen from their habits such as 1) implicit and explicit customary sayings, 2) customary jurisprudence (judge's decisions in resolving customary disputes, 3) documents containing customary provisions such as certificates and regulations, 4) law books or documents issued by the king, and 5) research reports on customary laws. [9]

There are some legal sources of customary institutions in Aceh such as Paragraph (2), Article 18B of the 1945 Constitution, Law Number 44 Year 1999 on the Implementation of Speciality of Special Province of Aceh, Paragraph (1) Article 98 of Law Number 11 Year 2006 on Aceh Governance, Aceh Qanun Number 9 Year 2008 on Guidance for Indigeneous Life and Customs, Aceh Qanun Number 10 Year 2008 on Customary Institution and Qanun of Central Aceh District Number 10 Year 2002 on Gayonese Customary Law.

3.2 Criminal Sanction in Gayonese Customary Law

Chapter VI, Article 13, Paragraph (1) of Aceh Qanun Number 9 Year 2008 on the Development of Indigenous Life and Customs clearly lists forms of disputes that can be resolved by a customary institution such as 1) domestic dispute, 2) family conflict related to *faraidh*, 3) communal dispute, 4) *khalwat meusum*, 5) property right dispute, 6) theft in family (petty theft), 7) *sehareukat* property dispute, 8) petty theft, 9) cattle theft, 10) violations of customs on cattle, agriculture and forest, 11) conflicts in sea, 12) conflicts in market, 13) non-serious physical abuse, 14) forest fire (in a small scale which harms customary community), 15) Sexual harrassment, libel, incitement and slander, 16) environmental damage (a small scale), 17) threat (depending on the threat) and 18) other disputes related to norms and customs.

Crimes in Gayonese customary law can be categorized into four categories, namely: *maas, rujok, died* and *bela*. *Maas* is a form of sanction for a non-serious crime which can be settled only by sitting together and forgiving each other. Apolozing is perceived as a sanction because generally it is hard for someone toadmit that he or she is guilty. This is particularly true when it must be done in front of customary leaders. *Rujok* means to reunite or return. This sanction is more severe than *maas* because the crime cannot be settled only by sitting and chewing betel nuts together. The perpetrator must provide foods for *peusyejuk* or *kenduri*. *Died* is a fine imposed on someone or some people who are found guiltyof physically abusingother people. If the victim gets a wound from waist below, the perpetrator will be finedsome chicken. If the victim gets a wound from waist to neck, the perpetrator will be fined a goat. However, if the victim gets a wound from neck to head, the perpetrator will be fined a buffalo or a cow. Bela is a punishment which is equal to the result of the criminal act. According to customary law, "*mate berbela, rusak berpeniri*" or "*rusak bersalin, rayohberpeniri*". It means that if someone kills other people, the punishment is *bela* or *qishash* (beheading) and if the victim gets a wound, he or she will be given a pair of clothes. [4]

3.3 Legal Strength of Uang Tutup Babah as a Customary Dispute Resolution and Its Impact on Customary Community

Any crime which occursin a village is usually reported to village government officials. How it will be resolved greatly depends on its seriousness. Usually, it is only a non-serious crime which can be resolved at village level. The seriousness of the crime also greatly depends on the situation, not on indicators set in the criminal code.

Sarak opat, as an institution which is responsible for the implementation or enforcement of customary law, also functions as a controlling body for public order and security. It deals with social issues and serves as a mediatorin dispute resolution within community. [10] Any problem within Gayonese community will be solved by village government officials known as *sarak opat*. [11]

Since community has developed, they cannot be bound by village governanceanymore. This, of course, has affected the improvement of public services in rural areasbecause a village cannot be separated from another. Therefore, Central Aceh District Government has established customary institution, *sarak opat*, at district, sub-district and village/kelurahan levels. [12]

Currently, physical or corporal punishment is rarely or never given in the implementation of customary law. The customary sanction usually only takes the form of an isolation. However, if the perpetrator still behaves badly although he or she has been given advice, community will report him or her to police. However, community usually prefer to handle their case outside the court. Village government officials usually do not give community an opportunity to directly report the case to police because if they directly reports it to policebefore reporting it to their village head, this will make them feel ashamed.

Gayonese community often use customary sanction known as *uang tutup babah* to solve their problems. Research conducted in some villages found that there were two meanings of *uang tutup babah*in community: general and specific meanings.Generally,*uang tutup babah* means hush money. [13] It is given to someone unofficially to ensure that information about his or her problem will not spread widely to public. The problem is solved through the provision of an amount of money or a number of goods. However, if village government officials, in fact, know the problem, they will not investigate it because the problem has been solved and the case is closed. [14]Specifically, *uang tutup babah* meanscustomary court in which the provision of moneyis considered as a fine imposed on the perpetrator. The payment is done through customary court. According to a government official from Simpang Tiga Uning Village, *uang tutup babah* is given by someone who has made a mistake in his or her village. The punishment is given to the perpetrator through a customary court. After that, the perpetrator also has to apologize in front of public. [15]

Customary court decision is a result of discussion to resolve a dispute between two parties. It may take the form of a customary sanction, starting from a light one to a severe one. The two parties must voluntarily agree the customary decision and the punishment.

There are some forms of customary punishments or sanctions in customary community such as advise, warning, apology in front of public, compensation, eviction from the village, customary title deprivation, boycott and isolation from community. [16]

Actually, *uang tutup babah* is not a customary sanction within Gayonese community. It is only one of ways for dispute resolutions. It has a same deterrent effect like customary court or formal court. So, it is hoped that after the imposition of the punishment on someone for his or her unlawful act, others will not do the same act although some still do that.It depends on themselves whether or not they will do the same act.

According to a government official from Mendale Village, dispute resolution through the provision of *uang tutup babah* usually only involves parties. Village government officials usually only act as a mediator and they are not involved in customary court. The problem usually can be solved based on the agreed solution. So, although *Reje Kampung* knows the problem, he or she will not interfere it because he or she thinks that the problem has been solved by the parties and there is no need to discuss it again. [17]

It is hoped that through the imposition of the sanction or the punishment, the perpetrator will not reoffend and others will not do the same criminal act.*Uang tutup babah* also gives a deterrent effect to the perpetrator although the effect varies from one perpetrator to another. According to a government official from Bewang Village, *uang tutup babah* actually can give a deterrent effect because it may cause shame or disgrace on the perpetrator. They will inform the crime to the perpetrator's parents or family. If necessary, they can inform it to his or her Reje Kampung. This is a form of social saction. This, of course, has a positive impact on community. Unfortunately, the imposition of the sanction,*uang tutup babah*, does not have any legal strength because it is not regulated in Qanun on Gayonese Customary Law. Parties can file a lawsuit against this anytime they want although up to now there is no party which has an objection to this saction, *uang tutup babah*. [18]

Therefore, *uang tutup babah* can be included as a sanction in customary law because an act can be recognized as a customary law when it can be accepted by community, is done repeatedly and is spread widely. [7]

4 Conclusions

In general, *uang tutup babah*means hush money. It is a customary dispute resolution in some customary communities in Gayo.Although the sanction, *uang tutup babah*, it is not yet regulated by Qanun on Gayonese Customary Law, it is often used by parties involved in the dispute to resolve their problem because it has a positive impact on communities. It is also believed that it gives a deterrent effect to the perpetrator because he or shesometimes has to spend much money to pay customary leaders or those who have been involved in solving the problem. Since it has been done repeatedly and community accept it as a sanction, it should be included in the law to ensure that it gives legal certainty and has legal strength.

References

1. B. Muhammad, *Asas-Asas Hukum Adat* (Pradnya Pramita, Jakarta, 1975)
2. Hazairin, *Tujuh Serangkai Tentang Hukum (*Jakarta, 1974*)*
3. B. N. Arief, *Mediasi Penal, Penyelesaian Perkara diluar Pengadilan* (Pustaka Magister, Semarang, 2010)
4. M. Din, *Stimulasi Pembangunan Hukum Pidana Nasional Dari Aceh Untuk Indonesia* (UNPAD Press, Bandung, 2009*)*
5. T. Haar, *Asas dan Susunan Hukum Adat* (Pradya Paramita, Jakarta, 1974.)
6. B. Muhammad, *Asas-Asas Hukum Adat* (Pradnya Pramita, Jakarta,1975)
7. S. Mahmassani, *Filsafat Hukum dalam Islam* (Al-Ma'ruf, Bandung, 1997)
8. S. Wignjodipuro, *Pengantar dan Azas-Azas Hukum Adat* (Penerbit Alumni, Bandung, 1973)
9. Bisri, *Sistem Hukum Indonesia: Prinsip-Prinsip dan Implementasi Hukum di Indonesia* (Rajawali Perss, Jakarta, 2010)
10. Syukri, *Sarakopat* (Hijri Pustaka Utama, Jakarta, 2006)

11. M. Ibrahim and H. A. Pinan, *Syariat dan Adat Istiadat Jilid 1* (Yayasan Maqamam Mahmuda, Takengon, 2002)
12. M. Ibrahim and H. A. Pinan, *Syariat dan Adat Istiadat Jilid 2* (Yayasan Maqamam Mahmuda, Takengon)
13. Sabirin, Simpang Empat Village Head, Kebayakan Sub-District, Central Aceh District, *Interview*, on 10 December 2017
14. Ahmad, Pinangan Village Head, Kebayakan Sub-District, Central Aceh District, *Interview*, on 10 December 2017
15. Sabirin, Simpang Empat Village Head, Kebayakan Sub-District, Central Aceh District, *Interview*, on 10 December 2017
16. Muhammad, Simpang Tiga UningVillage Head, Linge Sub-District, Central Aceh District, *Interview*, on 8 December 2017.
17. Supriadi, Tensaren Village Secretary, Bebesen Sub-District, Central Aceh District,*Interview* on 11 December 2017
18. Afriandi, Bewang Village Secretary, Bintang Sub-District, Central Aceh District. *Interview* on 12 December 2017

A contrast among farmers' ethnic groups: is this a social polarization tendency?

Siti Aida Adha Taridal[1,1], *Nur Isiyana* Wianti[2], *Muhammad* Arsyad[3], *and Arifiana Shima* Ekaputri[4]

[1]Department of Agribusiness, Faculty of Agriculture, University of Halu Oleo, Kendari 93232, Indonesia
[2]Department of Agricultural Extension, Faculty of Agriculture, University of Halu Oleo, Kendari 93232, Indonesia
[3]Department of Agricultural Socio-economics, Faculty of Agriculture, Hasanuddin University, Makassar 90245, Indonesia
[4]Marketing Department, University of Birmingham, United Kingdom

Abstract. The aim of the study was to capture the differences between Tolaki, Javanese, and Bugis ethnic in relation to their livelihoods in paddy upland farming activity. The study uses the post-positivism paradigm. In our research setting in Sulawesi drylands ecology, Tolaki local farmer, inhabit most areas of dry land, tend to be sub-subsistence and subsistence-oriented, their planting system is shifting cultivation, or the swidden farming, which was the legacy of their ancestors. Javanese tend to be subsistence and supra-subsistence oriented, while the Bugis farmer is expansive and supra-subsistence oriented. Another important message showed the symptoms of social polarization between the Tolaki and the Bugis as ethnic immigrants. The climax conditions will lead to social conflicts between Tolaki local farmer, Javanese, and Bugis migrants as a result of the emergence of social polarization.

1 Introduction

Agriculture serves as a valuable source of income, contributing to poverty reduction through global food security [1]. For Indonesian, rice is one of valuable income, besides two seasons of secondary food crops (*palawija*). Rice production in many parts of Indonesia is still dominated by wet rice (*sawah*), while paddy on dryland and cultivation in which perennial export crops (i.e. cashew, cocoa, and pepper) play only a secondary role.

About 791.000 acres of the area in Sulawesi is drylands ecology [2]. This drylands area is around 3.15 percent of the total drylands area in Indonesia. The Drylands rank among the geographically largest, biologically least productive and demographically fastest growing biomes on earth [3]. Livelihoods in these semi-arid and arid regions have evolved under variable, unpredictable and extreme environmental conditions [4]. Not only the ecology vulnerability but also social economic vulnerability. Roden *et al.* [5] argued that dry land areas of open access have led to heightened insecurity and ongoing conflicts over land-use rights, with livelihoods tending towards an unsustainable use of the environment. It

[1] Corresponding author: aidataridala@yahoo.com

signifies that the dry land ecology is like two sides of a coin, on one hand, it is open, and on the other hand it also creates risks and challenges.

While progress has also been made on the poverty front, 1.4 billion people still live in extreme poverty. Seventy-five percent live in rural areas of developing countries, especially sub-Saharan Africa and southern Asia [6]. Altieri [7] said that throughout the developing world, resource-poor farmers (about 1.4 billion people) located in risk-prone, marginal environments, remain untouched by modern agricultural technology. Globally, there are approximately 2.5 billion people involved in full- or part-time smallholder agriculture, managing an estimated 500 million small farms. Studies by Rigg *et al.*,[8] showed that poverty rates are higher in rural areas –generally between two and four times higher- and, moreover, that this gap has usually widened rather than narrowed.

Later, Rigg *et al.* [9] said that the World Bank's view noted above about the need to modernize farming begins to gain traction. Increases in farm productivity have been critical in reducing absolute poverty across East Asia over the last four to five decades, because such growth is powerfully pro-poor and spatially and socially inclusive for the very reason that the smallholder has been the dominant farm type. In Asian countries like Cambodia, the Lao PDR and Myanmar, and in regions of Indonesia, there is considerable remaining scope to pursue agriculture-based strategies for development and poverty alleviation.

Peasant conditions even more difficult because many government policies do not support the efforts in reducing vulnerability. Pokorny *et al.* [10] showed that discourses regarding the development of the Amazon region highlight the importance of the local cultures, local knowledge and participation of smallholders, such as indigenous people, traditional communities and small-scale colonists. Current policies, however, still pursue a development model that is oriented towards global commodity markets and the capacity of well-qualified entrepreneurs with the capital required for large-scale investments, despite a growing consensus on its ecological incompatibilities, social limitations and economic risks. Later, Jacobsen [11] found that the implementation of the land tenure policy of the Vietnamese government has affected the agricultural system, livelihood strategies and food self-sufficiency of Thai farmers in a remote upland village.

In Sulawesi's dry land paddy, the farmer was separated by the ethnic dimension, reflect their types of smallholder, and asserted their attention for persistence and subsistence life as a peasant, or make deeper engagement with the market relation as a farmer. This article focuses on smallholders in upland of South-Konawe regency with their differences. The aim of this study was to capture how the differences between Tolaki, Javanese, and Bugis are, in relation to their livelihoods in paddy upland farming activity.

2 Method

This study used the post-positivism paradigm. Methodologically, two different approaches are incorporated into the unity of mutual support. A first approach is a quantitative approach, using the survey as a method of data collection in the field. A second approach is a qualitative approach using non-survey approach oriented to study the cases, in the realm of qualitative research. Thus, we take a case study as an instrument to get the social fact with a view to capturing much more precisely on the different ethnicities. Like in Baito, farmers are divided into Tolaki and Javanese. While in Laeya, farmers are divided into Tolaki and Bugis.

The unit of analysis is the household. The study population was all upland rice farming households in four sub-districts: District of Wolasi, District of Baito, District of Kolono, and District of Laeya, South of Konawe Regency, Indonesia (Fig. 1). Respondents were chosen randomly (simple random sampling). The total number of respondents were 395 households (33,3% of total population). The number of respondents in the District of

Wolasi were 70 households (5,9% of total population), in the District of Baito were 156 households (13,2%), in District of Kolono were 67 households (5,7%), and in the District of Laeya were 116 households (8, 61%). The research was conducted on May-July 2016.

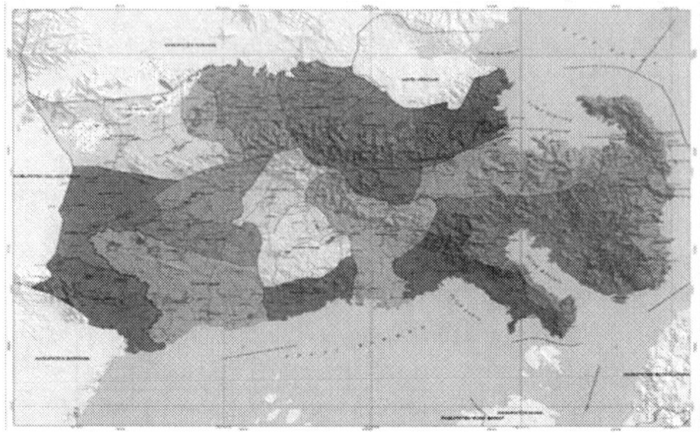

Fig. 1. The study site [12].

3 Results and discussion

3.1 Planting behaviour by ethnic

Ethnicity is a social category that shows the differences in cultural value orientation and then reflected in social behaviour. Stratify peasant farmer into age, class, ethnicity and gender while often describing poverty as an absence of entitlements, no land rights, no ability to employ assets or simply as a condition of weak capabilities [13]. In this paper, we will clarify the tendency of the ethnic dimension of farmers which will be associated with farming behaviour applied by the farmer households in each ethnicity.

We found that shifting cultivation system with *huma* system performed by almost all Tolaki peasant, while the cropping pattern system only used by a small part of Javanese peasant. Javanese farmers with large land used *huma* system. However, for Javanese smallholders who have small plots of land, cultivation system is farming permanently. The term of land used by the peasant of Tolaki are two years, and then they would move on to a new land. While the Javanese smallholders with no land to cultivate did shifting cultivation without a certain period. If the crop harvesting is not a satisfactory, Javanese farmers will soon move to other land in the next planting season. The soonest move is made every year, while the latest is three years.

The results also show a different trend in the behaviour of upland rice crop cultivation among Tolaki and Javanese farmers. For example, Tolaki farmers began planting in December to February, while the Javanese farmers start planting in October-November. Nowadays, Lampung upland rice varieties innovation has been largely accepted by 62 % of farmers in the district Baito. Farmers tend to quickly adopt the Lampung varieties. Almost 92 % of the Javanese farmers have now been using upland rice varieties types of Lampung on their own lands. Meanwhile, some Tolaki farmers (approximately 25 %) have begun to change this type of rice paddy fields to the type of Lampung. Symptoms of farming behavioural changes could not be separated from the imitative behaviours Javanese peasant for Tolaki people managed to increase their production of rice fields far above the amount of production produced by Tolaki farmers. However, 75 % of Tolaki farmers in the district of Baito are using local varieties in the area Baito.

In addition to the differences in the use of paddy fields seed, Javanese and Tolaki farmers also differ in drylands paddy fertilization maintenance behaviour. Almost all (93

%) Java farmers performed plant maintenance twice a month using toxic chemicals (farmers called it the "Rambo", "Noxon", "DMA" and "TMA"). Javanese farmers also tend to use chemical fertilizers, NPK, namely to maintain soil nutrients for rice fields. Some (32 %) Tolaki farmers also use toxic chemicals to kill weeds twice a month, but most of them (68 %) still maintain manually, i.e. mowing weeds that interfere with the growth of paddy field twice in a month and the ways of mystical, i.e. using a "prayer "and water that has been blessed with prayer then sprinkled on the plants that have been grown. It is believed by some Tolaki farmers, especially the elderly, to be able to protect the rice fields from pests. Some (36%) Tolaki farmers has also switched to NPK, but the rest of themin Baito used manure, which is organic waste from livestock (i.e. cows and goats).

Tolaki and Javanese cropping patterns are relatively the same, i.e. intercropping system and stocking seed planting system. The plot of land owned by Tolaki and Javanese are three to five types of plants. Besides paddy,the land fields are planted with corn, soybeans, sweet potatoes, manioc or cassava, vegetables, plants and fruits (i.e. pineapple and papaya). Chickens also roam freely around the plants, since the farmers believe that their excreta provide essential nutrients.

However, there are differences in the types of hedge planted by Tolaki and Javanese. Tolaki peasant tends to plant teak, not only as a hedge but also as the "symbolic" plants which serves as markers of land that had been prepared for the planting in the next season. In addition, teak considered as savings or future income. Meanwhile, the types of crops grown by Javanese farmers tends to be a commercial crop such as coconut, cocoa, and cashew nuts. Most of Javanese small peasant plant teak as a hedge.

Thus, another difference between Tolaki and Javanese smallholding is in cropping activities such as harvesting and post-harvesting. As a result of the use of different varieties among Tolaki and Javanese, resulting in the difference on the mechanism of harvest and post-harvest inclination and orientation of production for the households. As discussed previously, Javanese smallholders use Lampung upland rice varieties that are superior in terms of results and post-harvest, as well as the innovation that was introduced by the government through the Department of Agriculture South Konawe.

Despite that, some farmers (i.e. Tolaki peasant) still survive using local varieties. The Tolaki respondents told us that the indigenous variety has become a habit and does not need money to buy the seed. The Lampung variety is easily cropped in harvest time. This makes the smallholders use the crescent (farmers call it the "sickle") at the harvesting time, which perceived to be more effective compared with the cutting stems using "*ani-ani*". Javanese smallholders who use the sickle-like systems will tend to blow the system when the harvest arrives. In this system, Javanese farmers open opportunities for certain parties sharing system for harvesting by using a sickle. Yields then divided into four sections. Three parts to owners of agricultural land, while one part goes to the helpers that help with the harvest.

Meanwhile, most of the Tolaki who uses *ani-ani* using "*Bawon*" system like traditional paddy harvest in rural Java before the green revolution. Bawon system gives the opportunity to all members of the community to participate and harvest the rice by using *ani-ani* and receive a certain part of the results:3 parts go to the land owner and 2 parts to the helpers. For the Tolaki peasant with no land ownership (known as "*tunakisma*"), this system would help the household in terms of food security.

In addition to that, after harvesting, Javanese smallholder then drying and threshing the grains by beating them or using grain threshing machine, and storing the grains for a month in the house. The Javanese said that the advantage of Lampung or Batutegi variety that has been developed by Javanese farmers is it is easy to process to become rice that ready to eat. The crops were then sold to the middlemen, or directly to the consumers. Most of the rice stock is stored at home for personal consumption.

For the Tolaki, after the paddy were harvested, it is then dried relying on sunlight. After it is dry, it is then tied up and stored in the barn. If they want to consume it, they would stomp and pound the paddy traditionally using "*alu-alu*" to separate the relatively thicker skin than Lampung variety. This collective process had an impact on the division of the results of rice processing. The entire crop after divided by five, three parts for the owner are stored in the granary. Storage is intended for their own use (subsistence), and preparation of the seeds for the next planting season.

The Bugis and the Tolaki are similar. Bugis Farmer and Tolaki peasant plant wetland paddy seed in their fields and half of the Tolaki peasant plant their land with local seed. The majority of both ethnics plant their field with Lampung as superior varieties which was introduced by the local government which have better productivity levels than local varieties. However, more Bugis than Tolaki farmers use this variety. This behaviour occurs from experiments in Bugis farmers to plant species of paddy fields belonging to their recognized Bugis farmers of higher productivity levels than the paddy fields. This behaviour was also due to the limitations and unavailability of irrigation channels with an adequate water supply for paddy rice farming activities. For upland, paddy planting activity carried out once a year, while for paddy wetland is grown twice a year. Both Bugis and Tolaki use dryland paddy only for subsistence, while the wetland paddy is used more for commercial purposes (to be sold to earn money).

Another important and crucial finding shows that shifting cultivation (*huma*) was also carried out by Tolaki farmers in District Laeya. In each plot of land cultivated up to two planting seasons or two years, then Tolaki farmers will leave the farm and seek new farm land. Land that has been cultivated is then planted two years of plantation crops such as cocoa, pepper and cashew nuts. Agricultural land is then burned and left for two weeks, before they start cultivating the land. Most Tolaki peasant (in Laeya, and south Konawe as a general) never use the chemical fertilization, their reliance on the remains of litter which works as organic fertilizer.

Bugis farmers are performing the settled cultivation system. This trend occurs not only because of the width of the land plantation area but also because of their work ethic. They utilise the big area by dividing it and planted it by similar plants without *tumpangsari* cropping (monoculture). Bugis farmers are focusing on the establishment of the benefits of farming systems in drylands. Grains derived from the plot of commercial cash crops such as pepper, cocoa, and cashew nuts as well as a land plot of rice crops grown on a dry land. While paddy fields with plots that tend to be more narrowly directed at meeting the subsistence household needs for food.

The next differences between Tolaki and Bugis farmers are related to harvest and post-harvest. On dry land rice harvest activities, all farmers who plant Bugis Lampung (*batutegi*) variety harvested by using "*sabit*" (a scythe) or "*arit*" (sickle). Meanwhile, Tolaki farmers who grow local ricevarieties harvested using *ani-ani* and this activity is predominantly done by women. Once they done harvesting, the Bugis Farmers will put the grains (which falls easily) into a sack to dry, and then store it in the house. To make it into rice, Bugis farmers would bring the grains to rice mill. Before storage, the harvest should be given to those who helped with the blow harvesting system. Bugis people do not easily allow all parties to help harvesting like "*bawon*" systems which are widely adopted by Tolaki farmers. They believe if they apply "*bawon*" system, it will reduce the profit earned from the harvest of the particular type of paddy, or will reduce the consumption of subsistence rations for these types of paddy fields. Bugis farmers use the used paint bucket (size 25 litres) as measurement in sharing the yield to those who helped. According to key informants, the harvest results usually obtained approximately 10 paint buckets. Eight buckets are for the owner of the land, and two buckets for those who helped to harvest.

Most of the peasant of Tolaki do the same with Bugis farmers during harvest and post-harvest, especially the use of superior varieties Lampung (*batutegi*), but the ones who still use local varieties will behave differently. Tolaki farmers who survive using local varieties will use the *ani-ani* to harvest. This labour-intensive system uses the revenue sharing system "*bawon*", which is letting anyone come to help and enjoy the harvest obtained. Once the harvest is obtained, grains are then tied up (1 bunch = ± 5 litres of rice). If there are five ties were obtained, two of them go to the helpers, and remaining three go to the land owners. After the harvest, the owner will split the grains into two purposes: the one to be soon consumed, and the other to be put in a good barn for food stock in the following months. Both are to be seeds in the next planting season. The grain for immediate consumption should be trampled upon beforehand to separate the skin with the grains. The next stage is to pulverize them by using "*alu-alu*" (a mortar-pestle). As for the grains stored in the barn, they are dried for a few weeks before kept in the granary. Almost the entire crop of paddy fields used by Tolaki people to meet household subsistence.

3.2. Agrarian structure and social polarization by ethnic

In Baito district, there is a tendency related to differences in ownership of agricultural land between Tolaki and Javanese farmers of rice fields. In general, the ownership of agricultural land by Tolaki farmers comes from inheritance. On the other hand, the Javanese smallholders acquire their lands by buying from Tolaki farmers. Tolaki farmers tend to lose their land, in contrast to agricultural land expansion as that of the Javanese farmers. The tendency shows that when Tolaki farmers in Baito are facing a pressing needs, for example to finance their child's marriage, they would prefer to release their agricultural land to buyers.

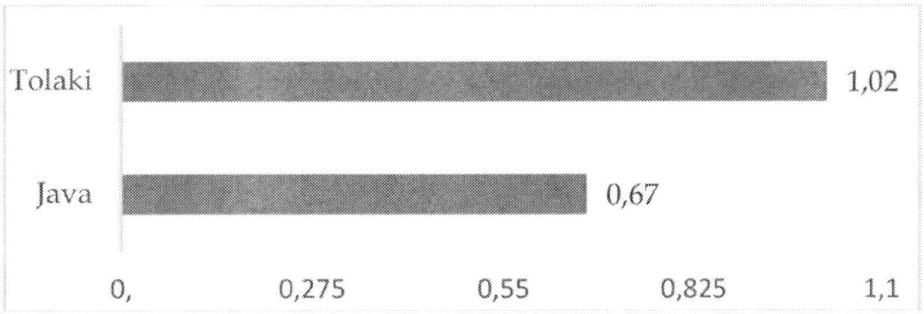

Fig. 2. The average of plantation land that owned and controlled by the Tolaki and the Javanese in Baito District.

Figure 2 showed that the average Tolaki farmers farmland area of paddy fields are larger than the Javanese farmers'. The average area of agricultural land of rice fields owned by farmers of rice fields in the district of Baito i.e. about 0.98 ha. Javanese farmers or homesteader immigrant communities have upland rice farms with an average area of about 0.67 hectares owned or below the overall average of respondents in District Baito. Meanwhile, Tolaki farmers have agricultural land above the overall average of respondents in District Baito with the average value of the agricultural land owned area of approximately 1.02 hectares. Based on this fact, the projection of the shrinkage of agricultural land owned by Tolaki farmers must be done to show the burgeoning of Tolaki farmers with no land ownership ("*tunakisma*"), which is in contrast to the Javanese farmers with their ever-expanding agricultural land. This might be caused by the increasingly expensive cost of farming. If this keeps happening, the polarization or class differentiation among Tolaki and Java farmers could arise.

We analysed using the gap between the farmer by Morton [14], which he called a continuum of smallholders and subsistence farmers and design three fluid groupings mainly pertaining to food security: sub-subsistence; subsistence; and supra-subsistence. It is a salient feature of the 'three types' of peasant farmers that 'sub-subsistence' refers to peasant farmers who are chronically food insecure and rely on irregular cash income from various off-farm sources while 'subsistence' refers to those who are periodically food insecure relying on irregular cash income from diversification into a range of sources. Actors in this two categories end to be risk evaders. Finally, 'supra-subsistence' refers to peasant farmers who have adequate resources but are constrained in times of a severe drought or ill-health. Some of those may operate as opportunity seekers [15]. A strong assumption occurred that every smallholder could be rational in both cultural values or customs and economic reason.

Results of research give us the important message that most Tolaki farmers are as peasants who orientated for sub-subsistence and subsistence. It is characterized by the Tolaki farmers who only has little and adequate land, no capital, few physicals assets, using collective action to reduce risk, using members of the family as labour, and agricultural activities by *huma* system. The Tolaki elders in their upland do not engage much in commercial activities and market interaction. We found that crops and land meanings for Tolaki are embedded with believe, culture, core as local people identity, and privilege from Olawi as sharing food for others in the village. Thompson and Scoones [16] said that even though rural populations at large are relatively poorer than urban populations, they are not equally poor because rural livelihood worlds are diverse. On the other hand, the Javanese smallholders tend to be actors who are at subsistence orientation and supra-subsistence. For example, Javanese farmers are already using farming equipment, using seeds of better productivity than the local seeds (i.e. Loiyo); all of these leads to a surplus of land-based agriculture. A major implication of this finding showed that the Javanese farmers are more progressive because [17] argued that in Java, the land availability is getting very scarce, and its role as a transferable asset is changing.

There is a social phenomenon which is similar to the sub-district Baito but tend to show the phenomenon of social polarization between the two groups of peasant and farmer with different ethnic (i.e. Tolaki and Bugis) that occurred in the Laeya District. This figure shows the social polarization occurred in the Laeya District.

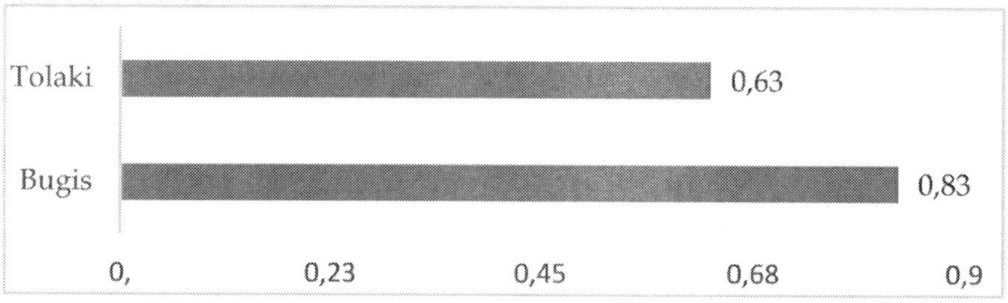

Fig. 3. The difference on the average area of agricultural land owned by Bugis and Tolaki in Laeya District, 2016.

Based on figure 3, the research found that the average area of agricultural land for paddy farming activity owned by Bugis farmers is wider than those of Tolaki farmers. Bugis farmers have an average area of 0.83 hectares of agricultural land, while Tolaki peasants controlled an average area of 0.63 hectares. Most of Bugis have large farms, while Tolaki farmers' land tends to be sinking in terms of land status: from the owner of extensive land becoming landless farmers. In fact, in some cases, some of the peasants are landless or called "*tunakisma*". They cultivated the land belongs to *PT. Kapas* called *Hutan Tanaman Industri* or HTI (industrial timber estates), which those smallholders called it as "*pinjam-pakai*". The respondent said that there is no obligation for the peasant for the government on the loan use or "*pinjam-pakai*".

Longacre and Hermes [18] argued that inequalities in rice landholdings and yields were ameliorated through the household exchange of pottery. Households with the highest rice productivities (rice yield divided by field area) received pots from households with lower productivities. There is a clear inverse relationship between household investment in rice farming and ceramic exchange to other community households. By tracing out the exchange networks, we find that village division have an influence on who exchanges with whom.

The fact that Bugis farmers as immigrant communities tends to have larger farms today than the Tolaki, could not be separated from the indigenous Tolaki farmers in Laeya District who easily release his own agricultural land to fulfil their urgent economic needs. The research found that Bugis farmers obtain agricultural land by buying from Tolaki farmers as well as from inheritance.

Not only the symptoms of social polarization between Tolaki and Bugis farmers, differences were also shown in the system of planting, maintenance, harvesting, and post-harvesting. The results showed that Bugis farmers tend to be very expansive in managing assets in the form of land, more expansive than Tolaki farmers. Bugis farmers manage agricultural land other than paddy rice paddy fields: crops and commercial plantation crops such as cocoa, clove, and cashew nuts. Bugis Farmers, unlike Tolaki farmers who grow by cropping system, they focus on the plots of land for growing similar plants. We found that Bugis farmer was an entrepreneurial farmer who are profit-oriented, business-knowledgeable, and skillful in business expansion.

Acciaioli [19] said that not all cases of Bugis migration exhibit such an orderly sequence of differentially organized migration stages. Later, he found in many ways that the Bugis movement tothe Lindu plain in Central Sulawesi displays a less orderly, simultaneous invocation of different bases of recruitment. Although the coasts, valleys and upland plains of Central Sulawesi have long constituted a frontier for Bugis expansion, the privations imposed by a prolonged state of warfare between guerillas battling for the creation of an Islamic state in eastern Indonesia and the largely Japanese national army from the early 1950s to the mid-1960s [20], an era labeled locally the *gerombolan* ('gangs'), transformed this continual dribble into a veritable stream.

Based on the results, we believed that the sum of this situation would cause Tolaki farmers to slowly lose their soil due to land selling culture for fulfilling their daily and social needs (i.e. for the cost of their children's marriage), while Javanese and Bugis farmers keep expanding their agricultural land. If the indigenous Tolaki farmers are eliminated from local land ownership, there would be food insecurity, which could trigger chronic poverty and the burgeoning of crime as its result. The climax could be the social conflicts among Tolaki local farmers, Javanese and Bugis migrants as a result of the emergence of social polarization. This indicates that the gap truly affects social polarization.

4 Conclusion

In our research setting in Sulawesi drylands ecology, Tolaki local farmers who inhabit most areas of the dry land, tend to be sub-subsistence and subsistence-oriented. Their planting system is shifting cultivation or the swidden farming, which they learned from their ancestors. In Baito district, Tolaki farmers and Java smallholders are dominating. Javanese farmers tend to be subsistence and supra-subsistence oriented. While in Laeya, the Bugis farmers is expansive and supra-subsistence oriented. Furthermore, the results showed the symptoms of social polarization between Tolaki as locals and Bugis as ethnic immigrants in

Laeya. This study recommends the local governments to assert land tenure system and provide regulations to protect the locals (especially Tolaki) with no land ownership or "*tunakisma*" in shifting cultivation on their property of "*ulayat*" rights. The government should also set strict regulations in order to protect the land as the core of Tolaki's culture from social changes which emerged from the new land tenure and new business system from outsiders.

References

1. IFAD. *Smallholders, Food Security, and the environment.* IFAD and UNEP. (2013)
2. Sukarman, I.G. Subiksa, S. Ritung. Identification of potential dry land for food plant development in prospects for dryland agriculture for supporting food security. Agricultural research and development agency Republic of Indonesia. (2012)
3. Millennium Ecosystem Assessment (MEA). Ecosystems and Human Wellbeing: Synthesis (Island Press, Washington DC, 2005)
4. E. Huber-Sannwald, M.R. Palacios, J.T.A. Moreno, M. Braasch, R.M.M. Pena, J.G. de A. Verduzco, K.M. Santos, Philos. Trans. R. Soc. Lond. Ser. BBiol. Sci. **367**, 1606, 3158-3177 (2012)
5. P. Roden, C. Bergmann, A. Ulrich, M. Nusser, Journal of Arid Environments. **124**, 239-248 (2016)
6. IFAD. *Smallholders, Food Security, and the environment.* IFAD and UNEP. (2013)
7. M.A. Altieri. *Agroecology: the science of natural resource management for poor farmers in marginal environments.* *Agriculture, Ecosystems & Environment* **93**, 1-3 (2002)
8. J. Rigg, A. Salamanca, E.C. Thompson. The puzzle of East and Southeast Asia's persistent smallholder. *Journal of Rural Studies* **41,** (2016)
9. J. Rigg, A. Salamanca, E.C. Thompson. The puzzle of East and Southeast Asia's persistent smallholder. *Journal of Rural Studies* **41,** (2016)
10. B. Pokorny, W. de Jong, J. Godar, P. Pacheco, J. Johnson. From large to small: Reorienting rural development policies in response to climate change, food security and poverty. Forest Policy and Economics **36**, (2013)
11. J. Jacobsen, K. Rasmussen, S. Leisz, R. Folving, N. V. Quang. The effects of land tenure policy on rural livelihoods and food sufficiency in the upland village of Que, North Central Vietnam. *Agricultural Systems* **94**, 2 (2007)
12. Central Statistic Biro (CBS). Konawe Selatan dalam Angka (CBS, Kendari, 2015)
13. A. Sen. *Development as Freedom* (Oxford University Press, Oxford, UK, 1999)
14. J.F. Morton. The impact of climate change on smallholder and subsistence agriculture. *Proceedings of the National Academy of Science* **104**, (2007)
15. A. Jerneck, L. Olsson. More than trees! Understanding the agroforestry adoption gap in subsistence agriculture: Insights from narrative walks in Kenya. *Journal of Rural Studies* **32,** (2013)
16. J. Thompson, I. Scoones. Addressing the dynamics of agri-food systems: an emerging agenda for social science research. *Environmental Science & Policy* ***12***, 4 (2009)
17. T. Bottema, M. Sirega, H. Madiadipura. Family Life History as a Tool in The Study of Long-term Dynamics of Poverty: an Exploration. Proceeding of National Seminar Land and Household Economy 1970-2005 "Changing Road for Poverty Reduction". Indonesian Center for Agricultural Socio-Economic and Policy Studies (ICASEPS) and ESCAP. (2009)
18. W.A. Longacre, T.R. Hermes. Rice farming and pottery production among the Kalinga: New ethnoarchaeological data from the Philippines. *Journal Anthropological Archaeology* **38**, (2015)

19. G. Acciaioli. Kinship and debt; the social organization of Bugis migration and fish marketing at Lake Lindu, Central Sulawesi. In: Bijdragen tot de Taal-, Land-en Volkenkunde, Authority and enterprise among the peoples of South Sulawesi 156, no: 3, Leiden, (2000)
20. G. Acciaioli. Kinship and debt; the social organization of Bugis migration and fish marketing at Lake Lindu, Central Sulawesi. In: Bijdragen tot de Taal-, Land-en Volkenkunde, Authority and enterprise among the peoples of South Sulawesi 156, no: 3, Leiden, (2000)

Technical efficiency of pesticide application on tomato, chinese cabbage and cauliflower in simpang empat sub-district Karo Regency

Sri Fajar Ayu[1*] and *Destanul* Aulia[2]

[1]Faculty of Agriculture, Universitas Sumatera Utara, Medan, Indonesia
[2]Public Health Faculty, Universitas Sumatera Utara, Medan, Indonesia

Abstract. Prior research by authors on the use of pesticides and their impact on 75 vegetable farmers in Simpang Empat subdistrict, Karo district, North Sumatra showed that 60% of respondents applied pesticides in excess of the packaging instructions. Respondents were also found not to use a complete protective device. Almost 70% of respondents have mild toxicity complaints, and based on blood cholinesterase levels are also found most respondents have mild toxicity. Pesticide residue test results also found 0.728 residues of chlorpyrifos and 0.321 profenofos in vegetables produced in this area. This led to continued review by checking the technical efficiency in the use of pesticides. The method used is a quantitative method with the Data Envelope Analysis (DEA) approach to analyze technical efficiency of pesticide on each commodity. The population is the whole vegetable farmer who grew tomatoes, Chinese cabbage and cauliflower at the time of the study. The entire population is used to be a sample of 35 tomato and cauliflower farmers and 20 Chinese cabbage farmers. The results of the study indicate that the uses of pesticide as production inputs on tomato commodities, Chinese cabbage and cauliflower are not technically efficient. It is suggested to farmers to use pesticide input according to the rules, proper dosage, on target, on time, the right way so that pesticide used is technically efficient.

1 Introduction

The current pattern of pesticide use by farmers is a problem. There are many farmers, who use pesticides are incompatible with the recommended dosage and procedure, resulting innegative impacts on the economy and the environment. According to Lakitan (1995), the pesticide application can cause disturbance to the stability and balance of ecosystem. Secondary pest explosion is one of a side effect of pesticide use. The diversity of species is also sometimes reduced as a negative impact of pesticide use. Applications of pesticides can contaminate the physical environment, including soil and water. In addition, the unwise use of pesticides also directly affects farmers' income. Where the current prices of pesticides are quite expensive, and this will certainly affect the cost of production of

* Correspondent Author : srifajar.ayu@gmail.com

farmers. The fact that the rising cost of production is not always offsetted by the increased income of farmers. And this leads to inefficiency in vegetable farming in Karo District.

One of the areas in Karo Regency which became the largest horticulture producer is Simpang Empat Sub-district. In 2016, Simpang Empat Sub-district produced 10,090 tons of cabbage flowers, 8,778 tons of petsai and 9,870 tons of tomatoes [1]. Efforts to increase horticulture production continue to be done in line with the increasing demand due to population growth. These efforts are intensification and extensification programs recommended by the government. On a farmer scale, increased technical production is more inclined towards superior seed selection, planting techniques and pesticides management techniques. Scientific studies have shown that the use of pesticides may increase farm production. Increased production will have an impact on the increase in income received by farmers in every harvesting activity [2, 3, 4, 5, 6].

Increased pest attacks or plant pest organisms become one of the reasons farmers adopt pesticide application techniques that are incompatible with established standards doses on pesticide packaging. The use of excessive doses results in the presence of pesticide residues attached to horticultural products, soil and water. All pesticide chemicals in general inhibit the important metabolic processes of an organism [7]. Therefore, pesticides are considered to be toxic compounds. This statement should be a concern, especially for vegetable consumers to do right action before consuming vegetables and fruits.

Pesticide residue test conducted on samples of tomato plants, Chinese cabbage and cauliflower obtained from the study area showed varying results. The test results can be seen in the following table:

Table 1. Results of Pesticide Residue Test Using Gas Chromatography

Plant Names	Weight	Pesticide	Working System	Active Materials	Large of Land	Spot/ point	Residue Detected (mg/kg)	BMR* (mg/kg)
Cauli flower	2 Kg	Dursban	Direct Contact	Klorpirifos	125 m^2	6	Metidation (0)	0
Chinese cabbage	2,5 Kg	Antracol Dursban	Direct Contact	Propineb Klorpirifos	175 m^2	7	Klorpirifos (0,728)	1
Tomato	3,6 Kg	Dursban Curacron	Direct Contact	Klorpirifos Profenofos	175 m^2	10	Profenofos (0,321)	10

*Laboratory of BPTPH Sumatera Utara 2017

Testing of pesticide residues in chinese cabbage showed different results with the cauliflower. Chinese cabbage showed positive results of pesticide residues. The residue detected was Klorpirifos of 0.728 mg/kg, with the maximum limit of residual (BMR) of 1 mg/kg (as determined by the Agriculture Service of Sumatra Utara), it was seen that the detected residue had almost reached BMR. However, the chinese cabbage which become the sample of research still save to be consumed. The pesticide residue present in the vegetable can be reduced by giving the treatment of doused with clean running water and washed with detergent or vegetable soap. The next vegetable is tomato (Solanum lycopersicum) which also positively contains pesticide residues. The material detected through the test is Profenofos as much as 0.321 mg/kg, with the pesticide residue limits (BMR) of 10 mg/kg then the tomato being sampled in this study is still classified in the category of save for consumption.

In addition to residual issues, excessive use of pesticides would require higher costs, resulting in inefficient management of production inputs. Efforts to increase production can also be done using technical efficiency calculations. Increased production through technical efficiency is now an important alternative, because it can add to the potential production that is available to vegetable farmers. This study was aimed to analyze the technical efficiency of pesticide use in tomato, Chinese cabbage and cauliflower.

2 Method

This study was conducted with quantitative methods, namely technical efficiency analysis of pesticide on tomato, Chinese cabbage and cauliflower. Data analyzed using the Data Envelopment Analysis (DEA) approach. Primary data will then be processed with the help of Microsoft Excel and DEAP. DEA method is a job evaluation tool that requires one or more kinds of inputs and produces one or more output types. In a simple way, the measurement is expressed by the ratio of output to the input which is the unit of measurement of efficiency or productivity, which can be expressed partially (Cooper *et al.*, 2006). The DEA approach uses fixed weighting across all inputs and outputs of each evaluated DMU. The DEA Variable Return to Scale model approach is more appropriate to be used in this study because the farming of tomatoes, Chinese cabbage and cauliflower at the research sites barely reaches the optimal scale. The assumption of this model is that the ratio between input and output additions is different (variable return to scale). That is, the change in input and output of a DMU takes place in a linear manner, allowing increased (increasing returns to scale) and decreasing (decreasing returns to scale) value of pesticide input efficiency in each commodity. The population is the whole vegetable farmer who grew tomatoes, Chinese cabbage and cauliflower at the time of the study. The entire population is used to be a sample of 35 tomato and cauliflower farmers and 20 Chinese cabbage farmers.

3 Result and Discussion

Technical efficiency is a combination of capacity and ability of an economic unit to produce up to the maximum output level of input quantities and technologies. The results of calculations on the technical efficiency of pesticide input use in tomatoes are as follows:

Table2. Technically Efficient of Tomato

No.	CRS	VRS	RTS
1	0,001	0,001	Drs
2	0,000	0,001	Drs
3	0,000	1,000	Drs
4	0,001	0,002	Drs
5	0,001	0,001	Irs
6	0,000	0,001	Irs
7	0,000	0,001	Irs
8	0,000	0,001	Drs
9	1,000	0,001	Drs
10	0,000	1,000	-
11	0,000	0,001	Irs
12	0,000	0,002	Drs
13	0,000	0,001	Drs
14	0,000	0,000	Drs
15	0,000	0,001	Drs
16	0,000	0,000	Drs
17	0,000	0,000	Drs

No.	CRS	VRS	RTS
18	0,000	0,001	Drs
19	0,000	0,001	-
20	0,000	0,001	Drs
21	0,000	0,001	Drs
22	0,000	0,001	Drs
23	0,000	0,001	Drs
24	0,000	0,001	Drs
25	0,000	0,001	-
26	0,000	0,000	Drs
27	0,000	0,001	Drs
28	0,000	0,001	Drs
29	0,000	0,001	Drs
30	0,000	0,001	-
31	0,000	0,001	Drs
32	0,000	0,001	Drs
33	0,000	0,001	Drs
34	0,000	0,001	Drs
35	0,000	0,001	Drs
Mean	0,029	0,058	

Table 2 shows that the efficiency of pesticide use in tomatoes with Constant Return to Scale (CRS) assumption is 0,029. The technical efficiency value shows that the average sample can reach 0,29%. The average efficiency of pesticide the assumption of CRS is $0.029 \leq 1$ which means that the use of pesticide input in tomato plants is not efficient from a technical point of view. The analysis showed that 34 farmers (97, 14%) were technically inefficient in return for Scale (CRS) and only 1 farmer (2.86%) were technically efficient in the use of pesticide in tomatoes. From the results obtained only DMU 9 has an efficient value of 1,000. While using the assumption of Variable Return to Scale (VRS) obtained the average value of technical efficiency of 0.058, technical efficiency value indicates that the average sample can reach 0.58%. The average efficiency of pesticide use using VRS assumption is $0.058 \leq 1$ which means that the use of pesticide input in tomato plants is not efficient from technical point of view. Calculation of analysis with Variable Return to Scale (VRS) resulted in 33 inefficient farmers (94.28%) and only 2 farmers (5.72%) technically efficient in the use of pesticides in tomatoes. So, it can be concluded for tomatoes, only one farmer (2, 86%) found technically efficient by CRS and two farmers (5, 72%) efficient by VRS. This implies that a large number of respondents face serious problems in terms of technical efficiency.

Table 3. Technically Efficient of Chinese Cabbage

No	CRS	VRS	RTS
1	1,000	1,000	-
2	0,735	1,000	Irs
3	0,478	0,482	Drs
4	0,359	0,445	Irs
5	0,381	1,000	Irs
6	0,184	0,667	Irs
7	0,570	0,597	Drs
8	0,560	0,676	Irs
9	0,959	0,969	Drs
10	0,804	1,000	Irs

No	CRS	VRS	RTS
11	0,523	0,537	Drs
12	0,161	0,556	Drs
13	0,459	0,800	Drs
14	0,708	0,720	Drs
15	0,666	1,000	Drs
16	0,654	0,804	Irs
17	0,611	0,619	Drs
18	0,924	1,000	Drs
19	0,851	0,869	Drs
20	0,116	0,381	Drs
Mean	0,585	0,756	

Table 3 shows that at the average efficiency of pesticide used in Chinese cabbage using the assumption of Constant Return to Scale (CRS) is 0,585. The technical efficiency value indicates that the average of sample can reach 0, 585%. The average efficiency of pesticide use using the assumption of CRS is 0,585 ≤ 1 which means that the use of pesticide input is inefficient from technical point of view. The analysis showed Constant Return to Scale (CRS) as much as 19 farmers (95%) are technically inefficient and only 1 farmer (5%) is technically efficient. While using the assumption of Variable Return to Scale (VRS), the mean value of efficiency of 0.756 indicates that the average sample can reach 7.56%. The average efficiency of pesticide use VRS assumption is 0.756 ≤ 1, which means that the use of pesticide input on mustard is inefficient from technical point of view. Calculation of analysis with Variable Return to Scale (70%) found 6 farmers (30%) were technically efficient in the use of pesticides, which is the farmer or DMU samples 1,2,5,10,15 and 18 with value of 1,000. So, it can be concluded that only 1 farmer (5%) efficient by CRS and 6 farmers (530%) efficient in VRS. This implies that a large number of respondent's farmers face serious problems in terms of technical efficiency. The technical efficiency value of pesticide use on the cauliflower is presented in Table 4.

Table 4. Technically Efficient of Cauliflower

No	CRS	VRS	RTS
1	0,002	0,173	Drs
2	0,001	0,007	Drs
3	0,047	0,083	Irs
4	0,020	0,294	Drs
5	0,001	0,005	Drs
6	0,251	0,444	Irs
7	0,014	0,048	Irs
8	0,576	1,000	Irs
9	0,041	0,071	Irs
10	0,795	0,924	Irs
11	1,000	1,000	-
12	0,002	0,002	-
13	0,002	0,392	Drs
14	0,002	0,002	-
15	0,182	1,000	Drs
16	0,061	0,079	Irs
17	0,662	0,770	Irs
18	0,041	0,071	Irs
19	0,041	0,143	Irs
20	0,003	0,983	Drs
21	0,041	0,143	Irs

No	CRS	VRS	RTS
22	0,003	0,281	Drs
23	0,002	0,002	-
24	0,003	0,923	Drs
25	0,003	1,000	Drs
26	0,002	0,048	Drs
27	0,002	0,470	Drs
28	0,025	0,043	Irs
29	0,029	1,000	Drs
30	0,190	0,667	Irs
31	0,002	0,456	Drs
32	0,003	0,005	Drs
33	0,002	0,223	Drs
34	0,002	0,493	Drs
35	0,003	0,006	Drs
Mean	**0,115**	**0,377**	

Table 4 shows that the efficiency of pesticide usage using the assumption of Constant Return to Scale (CRS) is 0.115. The technical efficiency value shows that the average sample can reach 11.5% and the value ≤ 1, which means that the use of pesticide input is inefficient from technical point of view. The analysis shows that 34 farmers (97.15%) are technically inefficient in return for Scale (CRS) and only 1 farmer (2.85%) is technically efficient in the use of pesticides in cauliflower. While using the assumption of Variable Return to Scale (VRS) obtained the average value of technical efficiency of 0.377 that the average sample can reach 3.77%. The average efficiency of pesticide use using VRS assumption is 0.377 ≤ 1, which means that the use of pesticide input on the flower crop is not efficient from a technical point of view. The calculation of analysis with Variable Return to Scale (VRS) resulted in 30 farmers (85.27%) found inefficient and only 5 farmers (14.28%) were technically efficient. From the results obtained samples (flower farmers) that are efficient use of pesticides technically are respondent or DMU 8,11,15,25 and 29 with value of 1,000. So, it can be concluded that only one farmer (2, 85%) efficient by CRS and five farmers (14, 28%) efficient by VRS. This implies that a large number of respondent farmers face serious problems in terms of technical efficiency.

4 Conclusion and Recommendation

The uses of pesticide inputs by farmers on tomato, Chinese cabbage and cauliflower are not technically efficient. With the assumption of Constant Return to Scale (CRS) is only one farmer found technically efficient use of pesticide while with the Variable Return to Scale (VRS) assumption, there are two farmers in the use of pesticides on tomatoes, six farmers on Chinese cabbage and five farmers in cauliflower. It is advised for farmers to use pesticides in accordance with the instructions given, appropriate doses, on target and on time.

References

1. [BPS] Badan Pusat Statistik, *Kecamatan Simpang Empat dalam Angka*, (CV E'Karya: Karo, 2017)
2. L. Darmayanti, Factors affecting production, income and employment in wetland paddy farm in irrigation area Parigi Moutong, *SEPA* Vol. 9 No. 2, Februari, (2013)

3. Suprayitno, Faktor-faktor yang mempengaruhi pendapatan usahatani cabai merah, *E-Journal Unigoro*, (2015)
4. S.H.S. Kalauw, N.R. Timisela and M.T.F. Tuhumury, Analisis faktor-faktor yang mempengaruhi produksi sayuran buncis (*Phaseolus Vulgaris L*) di Dusun Telaga Kodok Kabupaten Maluku Tengah, *Jurnal Agrilan* **3 (2),** (2015)
5. W. Sujana, Analisis Pendapatan dan faktor produksi yang mempengaruhi usahatani tomat di Desa Lebak Mundung, Bandung, IPB: Bogor, Agribisnis FEM IPB, (2010)
6. A. Hartati, and S. Kabul, The Efficiency Level of Production Factors on Potato Farming in Sub District Karangrejo Purbalingga, Central Java, *Jurnal Agrin Volume 16 No. 1 April, ISSN: 1410-0029*, (2012)
7. B.Y.H. Nugroho, S.Y. Wulandari and A. Ridlo, Analisis residu pestisida organofosfat di perairan Mlonggo Kabupaten Jepara, *Jurnal Oseanografi* **4**, 3 (2015).

Assessment of value added and development opportunity of agroindustry activity based on marine resources in Sub District of Panarukan, District of Situbondo

Syamsul Hadi*, *M. Iwan* Wahyudi, *M.* Hazmi, and *Henik* Prayuginingsih

Agriculture Faculty of University of Muhammadiyah Jember, Jember 68121 East Java, Indonesia

Abstract. Sub-District Panarukan, Distric of Situbondo has many food processing and non-food processing centers based on marine resources, all of which reaches 197 business units which include: Craft, Fish Cracker, Fish Chips, Shrimp paste and rengginang fish. The socio-economic phenomena of coastal communities should provide added value, absorb a lot of unemployment, but the people are still not prosperous in economically. The purposes of this study are to: Analyze the value added of resource-based on agroindustry, and to know the opportunity of marine resource development based agroindustry. The type of research used is descriptive with survey technique. Furthermore, the excavated data consists of qualitative and quantitative data which collected by using depth interview technique and the determination of respondents is done by simple random sampling technique. In order to answer the first, and second, the Hayami value-added analysis, and cost efficiency (R/C-Ratio) are used. The result of the research revealed that the ratio of added value of agribusiness activity of marine basis is 37,5% (value added ratio 15% - 40%), then the added value is classified. While the business opportunity of agribusiness activity in the research area is very big which indicated the value of R / C reach 1.95. If marine resource-based agroindustry activities in the research area are intervened with a touch of modern science-based management, then the value-added ratio potentially reaches more than 40% (> 40% high added value categories).

1 Introduction

The number of industrial companies in District of Situbondo in 2014 can be explained as follows: small industries and household handicrafts 12.726 units (up 3.00% from 2013) with a workforce of 44,725 people (up 5.00% from 2013). Of the 12.726 units of small scale industries and household handicrafts, 4,436 businesses or 34.86% employed 19.989 workers, Textile / Sewing 681 businesses or 5.35%, tile 435 businesses or 3, 42% and bamboo woven 253 businesses or 1.99% and other industries under 1%. Furthermore, in District of

* Correspondent Author : syamsul.hadi@unmuhjember.ac.id

Situbondo, the number of Scallop Craft is 367 business units with 1,307 workers and investment value of IDR 43.7830.000, -, the production value is IDR 4,405,824,000, - and the raw material value reaches IDR 1.938.499.000 (BPS Situbondo, 2015). Meanwhile, the number of shrimp raw materials processing industry reached 18 business units absorbing labor 139 people with investment value IDR 75.483.000, -, production value IDR 663.705.000, - and raw material value IDR 349.209.000, -. Based on the development of Industrial Enterprises by as much as of Business Unit, Manpower, Investment, Value of Production and Raw Material of agroindustry based on marine resources are respectively reached Agro 4.452 business unit, 20.057 people, IDR 7.999.202.000, - IDR 51.108.545.000, - and IDR 22.771.979.000, -.

Panarukan is one of the sub-district of Situbondo, located adjacent to the city of Situbondo with the height of the village from sea level is between 3 m to 15 m. The width of Sub-district of Panarukan is 54,55 Km2 or 5.455 H consisting of 8 villages having many food processing and non-food processing centers based on marine resources. Because four of the eight villages in Sub-district of Panarukan, Kilensari Village, Peleyan Village, Duwet Village and Gelung Village have coastal areas. The number of business units, absorbed labor, and the value of production of food and non-food processing industries based on agricultural commodities and marine resources in Subdistrict of Panarukan, District of Situbondo are presented in table 1 below.

Tabel 1 Distribution of Food and Non-Food Processing Industry Based on Agricultural Commodities and Marine Resources in Subdistrict of Panarukan, District of Situbondo Year 2014

No	Type of Agroindustry	Quantity of Business unit	Quantity of Labor (Person)	Production Value (IDR)
1	Craft shells	96	470	1.258.573.000
2	Fish crackers	46	310	182.732.000
3	Coconut shell bag	32	70	110.666.000
4	Fish chips	6	50	124.023.000
5	Cracknel	17	95	338.164.000
Total		**197**	**995**	**2.014.158.000**

Source: BPS District of Situbondo,Year 2015

Agroindustry activities in Soekartawi's opinion (2008) can increase the income of agribusiness actors, able to absorb labor, increase foreign exchange earnings and most importantly able to encourage the emergence of other industries. The indication of the opportunity for the development of agroindustry in the coastal area is supported by the marketing system built up so far, the products are sold directly to the nearest traditional market and the final consumer and the direct collectors come to the craftsman's house. The marketing opportunities of fishery-based agro-industry products in coastal areas are also increasingly wide open. This is because Pathek Beach in Village of Duwet and Gelung and fish harbor in Village of Kilensari is reconstructed into a national-type harbor and Peleyan Village has Fish auction to support the marketing of catch fish for Fish auction in Kilensari which was first established.

Marketing Efficiency and Prospect of Agroindustry Development of Fish Cracker in Village of Gelung, concluded that marketing channel of fish cracker agroindustry consist of three channel that is zero level channel, first level channel and second level channel but the most efficient is one-tier marketing channel of 0.2%. The prospect of fish cracker agroindustry is based on the results of SWOT Analysis in the position of White Area with emphasis on S-O strategy (Stength-Opportunities). However, the added value problem and the prospect of marine resource based agroindustry activities are not yet known. So there is no information on the matter that can be conveyed to the policy makers as a material consideration in the development effort in the field of food processing industry and handicrafts. Therefore, it is deemed necessary to study more deeply about the study of added value and prospect of marine resource based agroindustry considering Sub District of Panarukan has been known for centuries as the biggest fish producing area in District of Situbondo and as a supplier area for other districts in East Java. Processed food products such as fish 'rengginang', fish crackers, fish chips, and souvenir products such as craft shells is a potential development of agroindustry that can provide added value on export quality marine resources.

The hope is to expand employment opportunities and increase the income of coastal communities living on the line and / or below the poverty line of ± 21.15%. Although the potential of marine resources in the region contains tremendous expectations in the growth and movement of the economy, there are still many obstacles encountered, among others (Soejono, 2008): (1) marketing conditions that do not encourage the development of consumption, distribution of fishery products; (2) quality standards or quality of fish catch and sanitation; (3) relatively low equipment condition and application of technology in sea fishing activities; (4) the ratio of the amount of capital and the level of capital cost required is still very low; (5) Environmental issues are a lot of complaints and are used as an excuse to reject or restrain the marketing of fish; and (6) lack of development of the processing industry of fishery products in the coastal areas. Starting from the various problems faced by the fishermen community, one of the alternative options for problem solving that can be done is the development of agroindustry of the catches processing which the organization and its management are rationally designed to get maximum commercial value added by producing the product according to market demand.

The potential of agro-industry development through the development of marine resource-based household industry Sub-district of Panarukan, District of Situbondo is faced with various problems that start from the main problem that is the level of knowledge and skill is still limited, the ownership of business capital is relatively limited, scattered production activities in non-group, business capital loans, technology and markets are relatively difficult. Such conditions lead to productivity due to work motivation to be weak, which in turn implies the movement of downstream become slow. Craftsmen engaging in agro-industry activities are driven only by routine orientation due to traditional factors without considering performance improvement. Based on the above phenomenon, then the aims of the study are: 1) to analyze the value added of agroindustry based marine resources in Subdistrict of Panarukan, District of Situbondo; and 2) to know the opportunity of marine resource development based agroindustry in Subdistrict of Panarukan, District of Situbondo.

2 Research Methods

2.1. Research model

The study employed survey method to see how far the condition of agroindustry activity and its development opportunities in research area objectively and factually, potential and existing problem [1]

2.2. Location, Time and Research Population

The location of this research is the Subdistrict of Panarukan, district of Situbondo precisely the location of the village which has the center of marine resource based agroindustry which includes Village of Gelung, Kilensari, Paowan, Duwet, Peleyan and

Wringinanom. The termination of research location is determined by purposive sampling on the consideration that in that location is a center of food processing and non-food processing industry based on marine resources. This research has been done in July till September 2017. The study population is craftsmen agroindustry based marine and non-food resources (non-fishermen) are determined intentionally through techniques of non-probability sampling. The research sample is determined proportionally stratifield random sampling based on the type of agroindustry they do.

2.3. Resources and Data Collection Techniques

Based on the source that the data collected in this study includes primary data and secondary data. Primary data were obtained from the craftsmen collected by combining several complementary data collection techniques, namely: FGD and Indepth Interview. This data collection technique is guided by a list of questions prepared beforehand for the purpose of collecting data information can be directed and systematic and can answer the issues under review. The secondary data were collected from related institutions with this research.

2.4. Data analysis

Answer the First Goal

The amount of added value that can be known by Hayami method that can also be used to find information about the amount of income for direct labor and profits regardless of fixed costs. This research uses value-added calculation method for processing with [2]. There are three indicators of value added ratio [3] that are:

1) If the value added ratio <15%, then the added value is low;
2) If the value added ratio is 15% - 40%, then the added value is medium; and
3) If the value added ratio> 40%, then the value added is high.

Answer the Second Objective

The R / C Ratio analysis shows the amount of business opportunity where R / C stands for Revenue Cost Ratio, which is the ratio between acceptance and cost as the following formulation [4] R / C =Total Revenue (TR) / Total Cost (TC), Where:

R = TR = Acceptance = P.Q
TC = TFC + TVC
FC = Fixed Cost
VC = Variable Cost
The criteria used in the analysis of R / C ratio as follows:
- If the R / C ratio of> 1 business is said to be viable and profitable,
- If the R / C ratio of <1 business is said to be unfit to continue,
- If the value of R / C ratio = 1 business is said to be break even (no profit and no loss).

3 Results And Discussion

3.1 Value Added Study Agroindustry Based on Marine Resources

Based on the potential in the research area, some commodities and agroindustry products that can be developed include sea fish and non fish. In the opinion of that value added analysis is a value addition to a product after further processing. Value-added analysis will provide information on factors of the production process that produce or increase added value and vice versa. The following will discuss the results of value-added analysis of agroindustry based on marine resources as presented in Table 2 below.

3.2 Output, Input, and Price

Respondents of marine resource based agroindustry are referred to in this research is divided into five groups, the respondents group agroindustry craft souvenirs shells, rengginang shrimp paste, shrimp paste, dried fish and fish crackers. Actually there are still other craftsmen who are not includes in this research respondents, namely seaweed craftsmen and sea fish abon because of time, cost and energy. Most of the respondents (85.25%) obtained their raw materials in agro-industry activities by buying from fishermen or other parties, and the rest (14,75% of respondents) came from sea fish catch themselves.

The result of value added analysis using Hayami and Ruttan model, 1989 as presented in Table 1 reveals that from marine resource based agroindustry activities through five types of products produced value added as much as IDR 14.222,25 per kg per production process . It is known that the production process in the past business activity takes between 1 - 21 days, in which the making of shell souvenir takes 7 days, rengginang need 4 - 5 days, shrimp paste takes 21 days, dried fish and fish crackers each takes 5 days and ± 9 days. The added value ratio of marine resource-based product processing process reaches 34.95% and is classified as moderate (15% - 40%) according to [3] and higher than jackfruit chips agroindustry from in Lumajang and in Cikeas, Bogor Regency with value added ratio of 21,79% and 26,52%, respectively.

When compared with the value added value of marine fish based agroindustry activities in Dusun Payangan Sumberejo Village Ambulu District Jember Regency by [5] which is high (40%), then the value added ratio in this research area is still lower. This is because the agroindustry activity in this study area average raw material price (IDR 15.528,69 / kg) is higher than the result of research in Payangan Hamlet (IDR 10.952,38 / kg), while the price of handicraft product in this research area (IDR 37.606,56 / kg) is lower than in the area of Dusun Payangan (61.023,81kg). It can be explained that at the time of this research, the craftsmen respondents in Subdistrict of Panarukan enter the famine season, so they have to buy raw materials from outside the region with high price, except craftsmen rengginang where raw material glutinous rice which is relatively stable even though they have to buy relief ingredients shrimp paste and fish is also relatively high.

Tabel 2 Results of Value Added Analysis of Marine Resource Based Agroindustry Subdistrict of Panarukan, District of Situbondo Year 2017

	Output, Input, Price	**Formula Value**	**Formula**
1	Amount of production (kg /pp)	49.81	A
2	Raw material (kg / pp)	46.03	B
3	Labor (MWD)	4.77	C
4	Conversion factor (1 / 2)	1.08	A / B = M
5	Koefisien of Labor (3 / 2)	0.10	C / B = N
6	Product Price (IDR / Kg)	37,606.56	D
7	Average wages (IDR / MWD)	30,336.07	E
	Income and Profits *		
8	Raw material Price (IDR / kg)	15,528.69	F
9	Donations of other inputs (IDR / kg)	10,942.62	G
10	Revenue (4x6) (IDR / kg)	40,693.56	M x D = K
11	a. Added Value (10-9-8) (IDR / kg)	14,222.25	K – F – G = L

Output, Input, Price		Formula Value	Formula
	b. Ratio Added Value (11.a / 10) (%)	34.95	(L / K) % = H%
12	a. Employee benefits (5x7) (IDR / kg)	3,143.80	N x E = P
	b. Parts of labor (12.a. / 11.a.) (%)	22.10	(P / L) % = Q%
13	a. Profits (11a – 12.a) (IDR/kg)	11,078.45	L - P = R
	b. Profit rate (13.a / 11a) (%)	77.90	(R / L) % = O %
Balas Jasa Untuk Faktor Produksi**			
14	Margin (IDR / kg)	25,164.87	K – F = S
15	Direct labor income (12a / 14) x 100%	12.49	(P / S) x 100% = T
16	Donations of other inputs (9 / 14) x 100%	43.48	(G / S) x 100% = U
17	Corporate profits (13a / 14) x 100%	44.02	(R / S) x 100% = V

Table 2 above illustrates that the average labor allocation of 4,77 MWD on the use of raw material input is 46,13 kg per production process with an average wage of labor of IDR 30.336,07 / Men's Working Days (MWD) per production process. It can be explained that one MWD is 7,5 working hours or one adult work day in a day with 8 hours applicable in the research area. The amount of labor coefficient value shows the amount of labor donation needed to process one kilogram of raw materials of sea fish and shellfish. Meanwhile, the conversion factor value is the ratio of output to input, where in this study the average conversion value of 1,08 and the average value of labor coefficient of 0,10. On the basis of the average conversion value and labor coefficient can be interpreted that to process 100 kilograms of raw materials to be 108 kilograms of processed products required direct labor as much as 100 MWD. From the value of the coefficient of labor can be seen whether the entrepreneur has been efficient production or not. The smaller the value of the coefficient of labor the more efficient the producer produces. In this context the use of labor as much as 4.77 MWD to process raw materials of sea fish as much as 46.13 kg to produce processed production of 49,67 kg is not efficient in one production process. Similarly, the activities of agroindustry in this research area are relatively the same level of efficiency with the result of [6] research in Payangan Hamlet Sumberejo Subdistrict of Ambulu Sub-district of Jember Regency.

The share of labor received per kg of processed raw material only gained a share of 22,10% per kg, where the percentage is higher than the result of research in Lumajang Regency about value added of nangka chips agroindustry which percentage reaches 17.40% . This is due to the power conversion factor (0,10) in this study is lower than the results of his research which reached the value of 0,18. These conditions further indicate that the use or allocation of labor in agroindustry activities in this research area is relatively more efficient, so it will have implications on the level of profit that can reach 77,90%. If the profit level is compared with the result of Hadi and Farhiyah research (2016) in Dusun Payangan Desa Sumberejo Subdistrict of Ambulu of Jember Regency which reaches 91,30%, the profit level in this research area is still lower.

Revenue and Profit

The result of calculation of value added analysis obtained from marine resource based agroindustry in research area shows that the average value added is IDR 14.222,25 per kilogram of raw material. Value added calculation aims to know the value added value in one kilogram of marine fish after processed into pindang fish, smoke and shrimp paste. The amount of added value depends on the cost incurred including the average cost of raw material purchases amounting to IDR 15,528.69 per kilogram ranging from IDR 12.000, - s.d Rp 55.000, - per kilogram and other input contribution of IDR 10.942.62 per unit. Other input contributions consist of all variable costs except raw material costs and labor costs. Other inputs in it include the cost of purchasing auxiliary materials (herbs, salts, and additives), fuels, glue (Autosol), dyes, rubber tires, strings, straps, polishes, and packaging.

Table 2 above also reveals that the average profit earned from agro-industry activities is IDR 11.078,45 per kilogram for return on capital and management with an average profit rate of 77.90% of the added value of the product which means every kilogram of processed products. The average profitability and profitability in the research area is much lower than that of [5] in the above Payangan Hamlet, each of which reached IDR 39.870,74 in return for capital and management and 91,30% kg of raw materials. This condition is caused by the price of raw materials in this research area is much higher, and the price of handicraft product is much lower although the labor usage is relatively the same. Similarly, when compared with research results in Lumajang on value added analysis of jackfruit agroindustry, the average profit rate is 82,60% or higher than the average profit rate in this study area. This condition is caused by the conversion factor is 1,8 and the labor coefficient of 0,13, so it will affect the amount of added value.

Reply Services for Production Factors

Production factor services consist of remuneration for labor production factors, other inputs, and rate of return. Margin is the difference in price or value of the product with the input value of the raw materials used. Margin will be distributed in return for labor, other input contributions, and profits of the company or craftsman. The average margin of processed production is IDR 25.164,87 per kilogram of raw material consisting of 12,49% direct labor income, 43,48% contribution of other inputs, and 44.02% of craftsman profits. The company's profit margin is the largest and largest margin compared to the labor income margin. It shows that marine resource based agroindustry effort in research area is labor intensive business. In other words, the agro-industry activities can absorb a lot of manpower and become the main alternative in increasing population income and reduction of unemployment in coastal area.

3.3 Development Opportunity of Agroindustry Based on Marine Resources

Revenue Cost Ratio Analysis (R / C Ratio) is an analysis tool to see the relative profit of a business in one year to the cost used in the activity or better known as cost efficiency analysis to know the chances of a productive economic enterprise. The result of financial analysis of marine resources based agrooindustry business is shown in Table 3.

Tabel 3. Result of Finansial Analysis of Business Opportunity of Agroindustry Activity Based on Marine Resources at Subdistrict of Panarukan of District of Situbondo Year 2017 (in one process producer)

No	Descriptions	Unit	Quantity	Price/Unit	Value/Account (IDR)
A	**Operational Cost:**				**958,932**
	1. Depreciation Tools	Ls	1	15,000	15,000
	2. Raw Materials	Kg	46.03	15,529	714,829
	3. Helper Material	Ls	2	10,943	21,885
	4. Fuel	Ls	2.5	17,000	42,500
	5. Labors	MWD	4.77	30,336	144,718
	6. Letters (Packing)	Unit	2	10,000	20,000
B	**Amount of production**	Kg	**49.81**		
C	**Price of Agroindustry Product**	Kg		**37.607**	
D	**Total Revenue (TR)**				**1.873,238**

No	Descriptions	Unit	Quantity	Price/Unit	Value/Account (IDR)
E	Opportunity (R/C)				1.95

Table 3 above shows that the marine resource based agroindustry business opportunity in the research area is quite large, as evidenced by the R/C ratio of 1,95 in one production process. This means that each fisherman's expense per IDR 1, - will generate income of IDR 1,95, -. When compared with [6] in Puger where the value of R/C ratio on agroindustry activity of shrimp paste, fish cracker, dried fish, fogging (fumigation) fish and fish meal are 1.82, 1.32, 1.31, and 1.17, respectively. Even in 2012 Pudjanarso's research on value-added study and the break-even point on fish stocks in the same area, the R/C ratio is only 1.21. Along with the increasingly skilled sea fish craftsmen in the research area, the level of processing cost efficiency is higher in addition to the price of processed products is also higher. Therefore, the value of agroindustry development opportunities in the research area is increasingly high.

Unlike the results of research [5] in Dusun Payangan Sumberejo Ambulu Jember regency, where it can be disclosed that the group of respondents who act as fishermen as well as crafters have average R / C ratio on craft activities reached 3.84, while for the group of respondents only acting as a craftsman only average R / C ratio reaches 5,31. This condition caused in the first group of respondents intensification of agroindustrial business is not running maximally because the management concentration is split on fishing activity. While in the second group of respondents, more concentration on the management of one activity only with no fear of lack of raw materials because it has established strong cooperation with some fishermen.

The condition of socio-cultural factors of coastal communities in each region has different traditions and economic endeavors. This is because the environment is very strong influence on the characteristics of the way of management of a business one's economy besides influenced by geostrategic factors, geopolitics, and local government regulations. Following in detail will be discussed about financial analysis to reveal how big business opportunities based agroindustry marine resources by type of craft business in the study area.

4 Conclusion

1. The result of value added analysis using Hayami model in this research reveals agroindustry activity based on marine resources through five types of products produced gives added value as much as IDR 14.222,25 per kg per production process. The average price of raw materials is IDR 15.528,69 / kg, the price of handicraft products is IDR 37.606,56 / kg with the average labor allocation of 4,77 MWD on the use of raw material input of 46,13 kg per production process with average wage the average workforce of IDR 30.336,07/MWD per production process and other input contributions amounted to IDR 10.942,62 per unit.
2. The average fishing business opportunity in the research area is quite large, it is proved by the R/C ratio of 1,95 in one production process. Detailed business opportunity of Agroindustry Souvenir of Leather Shell reach 3.83, Rengginang agroindustry activity financially have big enough business opportunity with R/C Ratio 2.02, Shrimp paste reach 1.56, Dry fish reach 1.66, and Fish Cracker of 2,76. The profit level obtained by the five types of agroindustry is enough with an average of IDR 914,306, - each production process with a range from IDR 106.195, - to IDR 1.737.770, -.

Acknowledgements

Thanks to the authors say to the Institute of Research and Community Service UM Jember who has supported research funds through Internal Research Scheme Fiscal Year 2017. In addition, we would also like to thank the Director of Intellectual Property Management of Directorate General for Research and Development Reinforcement Ministry of Research, Technology and Higher Education through the Overseas Seminar Aid Program, so that we can participate in CSSPO International Conference 2018 in Sarawak, 09 - 11 July 2018. This conference offers a great opportunity for academics and researchers to exchange ideas and also to build a network.

References

1. M. Singarimbun and S. Effendi, *Survey Research Methods*, (LP3ES, LP3ES, 2005)
2. Hayami and Ruttan, *Agricultural Development An International Prespective*, (The John Hopkins Press. Baltimore, London, 1989)
3. M. Hubeis, Toward a Small Professional Industry in the Era of Globalization through Empowerment of Industrial Management. Scientific Oration Professor of Permanent Management Science Industry. Faculty of Agricultural Technology. IPB. Bogor. (1997)
4. F. Hernanto, *Agricultural Science*, (Self-Help Publishers, Jakarta, 1996)
5. S. Hadi and N.F. Fauzi, Development Opportunity of Sea-Based Agroindustry in Payangan Village Sumberjo Subdistrict of Ambulu Jember Regency, *Journal of Marine Fisheries* **7**, (2) (2016).
6. D. Soejono, Pattern of Agroindustry-Based Marine Fisheries Development in Puger, District of Jember, *In J-SEP* **2**, (1) (2008)

Empowerment of small and medium (SMEs) enterprises through the provision of credit with the guarantee of movable objects

Syarifah Lisa Andriati* and *Tan* Kamello

Doctoral Program in Law, Graduate School, Universitas Sumatera Utara, Medan, Indonesia

Abstract. Small and medium enterprises have an extremly important position in national economy and social life. Especially when it is associated with the economic empowerment of the community and the amount of labor that the small and medium enterprises can absorb. In addition to having a strategic meaning for development, small and medium enterprises also serve as a means to state the results of development that has been achieved. Empowerment of Small and Medium Enterprises (SMEs) is usually accompanied by capital needs. One of the obstacles faced by SMEs in developing their business related to the weakness of capital structure and lack of access to strengthen the capital structure. There is a need for a legislation that regulates related to the granting collateral of movable objects to facilitate SMEs regarding capital loans.

1 Introduction

The national development aims to create a just and prosperous society which is evenly and spiritually based on the Pancasila of the 1945 Constitution, especially in the economic field can be seen by increasing national production, opening employment opportunities, maintaining financial stability, abroad in order not to deficit, to seek an increase in domestic income and equal distribution of income. [1]

Pancasila and the 1945 Constitution of the State of the Republic of Indonesia is the primary philosophy underlying the activities of the state and nation especially in the economic development of Indonesia. In Article 33 of the 1945 Constitution of the State of the Republic of Indonesia, the foundation for the implementation of the national economy states that the economy is compiled and developed as a sustainable joint effort of all people based on the principles of justice, efficiency and economic democracy to bring prosperity, prosperity and social justice for all Indonesian people.

Small and medium business sector plays a significant role in building the economy of society. This is especially when it is associated with the economic empowerment of the community and the amount of labor that the small and medium enterprises can absorb. In addition to having a strategic meaning for development, small and medium enterprises also

* Correspondent Author: Syarifah_lisa_andriati@usu.ac.id

serve as a means to state the results of development that has been achieved. Central Bureau of Statistics (BPS) defines SMEs based on labor quantity.

From BPS data and Ministry of Cooperatives, from all business class shows that small scale business in Indonesia occupies about 99% portion, almost means all market in Indonesia is a small business, only 1% only medium and big business. The growth and growth of SMEs are quite good from year to year. Almost every government emphasizes the empowerment of SMEs. [2]

Table 1. SMEs classification based on Law no. 20/2008

Value Of Bussiness	**Assets**	**Turn Over**
Micro Bussiness	Minimal 50 Millions	Maximal 300 Millions
Small Bussiness	> 50 Millions – 500 Millions	Maximal 3 Billions
Average Bussiness	> 500 Millions– 10 Billions	> 2,5 – 50 Millions

Sources : UU No.20 years 2008

One of the obstacles faced by SMEs in developing their business related to the weakness of a capital structure and lack of access to strengthen the capital structure. There are still many perpetrators of SMEs who have difficulty obtaining credit access from banks, for example, do not have enough collateral. [3]

Business actors of SMEs have not realized their role optimally in the national economy. Therefore, to achieve the function and development of SMEs should be supported by financing, especially banking credit. If there is no role of government in regulating the regulation related to the funding especially banking credit, it is feared SMEs will have difficulties in capital so that cannot develop its business.

2 Problem

The problem in this study: how to empower SMEs through the provision of credit with a guarantee?

3 Research methods

This research applied a normative legal approach. A normative legal method is a legal research that treats law as a norm system. The data was secondary data consisting of: a) primary legal materials in the form of legislation; b) secondary legal materials in the form of writings, both in the form of books and articles containing comments or analysis related to the subject matter; c) tertiary legal materials in the form of dictionaries. The data was collected using a literature review and document analysis. The data was analyzed using a qualitative method based on deductive reasoning.

4 Discussions

4.1 The economic system in Indonesia

Indonesia is a legal country. Therefore, the financial system must also refer to Pancasila and the 1945 Constitution which became Pancasila Economic System namely the economic system that is imbued by Pancasila and based on the 1945 Constitution. [4] In the financial order of Pancasila industrial relations among economic actors based on ethical or moral Pancasila with the ultimate goal of realizing social justice for all Indonesian people. [5]

Ningrum Natasya Sirait argues that at present there is not a single country free of the issues concerning politics, economics and democratization efforts, although the level of the problem is different from one another. Of the various economic systems that exist, then each country will implement a system that is considered appropriate and in accordance with the national interests of the country. [6]

Therefore Zweigert and Kortz interpret ideology as a legal system as the meaning of political or economic doctrine or religious belief. Fukuyama in his book says in most regions of the world, now there is no more ideology with pretense to the universality that is in a position of challenging liberal democracy and no longer the principle of universal legitimation except the sovereignty of the people. [7]

The patterns of global migration, both old and new, and the exchange between different countries, economies, societies and legal systems, varying degrees and through various methods over time have created a cross-national legal environment that is naturally plural, multi-ethnic and multicultural, increasingly emerging in today's reality. [8]

In economic system with the fundamental principle of economic democracy trying to avoid free liberal financial system, command economy system and unhealthy competition. [9]. Pancasila Economic System or also called the populist financial system is generally interpreted as a system that combines constitutional ideology (Pancasila and the 1945) of the Indonesian nation with a mixed economic system (managed market economic system) embodied through democracy as well as described in the financial steps that are aligned and empowerment of all layers of a just and prosperous society. [10]

In the regulation Small and Medium Enterprises also adheres to the principles of economic democracy. In the Elucidation of Article 2, it states that the law of economic freedom is the empowerment of micro, small and medium enterprises held as a unity of national economic development to realize the prosperity of the people. And Article 3 of Law No.20 Year 2008 on Micro, Small and Medium Enterprises which states that micro, small and medium enterprises aims to grow and develop business in order to build a national economy based on a just economic democracy.

4.2 Funding Credit SMEs

Indonesia is still at the level of developing countries, is actively building the economy through micro, small and medium enterprises (SMEs). Various policies made by the government to continue to encourage the growth and development of SMEs one of them in the field of capital.

In Indonesia, the definition of SMEs is regulated under the Law of the Republic of Indonesia Number 20 the Year 2008 on Micro, Small and Medium Enterprises. Definitions according to Law no. 20 of 2008 are: (1) Micro Business is productive business owned by individual and individual business entity fulfilling the criteria of Micro Business as regulated in this Law. (2) Small-scale industry shall be a stand-alone productive economic enterprise, carried out by an individual or business entity which is not a subsidiary or not a branch of a company owned, controlled, or a part, directly or indirectly, of a medium-sized or large-fulfill the criteria of Small Business as referred to in the Law; (3) Medium-sized Enterprises shall be stand-alone productive economic enterprises, carried out by an individual or business entity which is not a subsidiary or branch of a company owned, controlled, or partakers either directly or indirectly with a small or large-scale business by the amount net worth or annual sale proceeds as stipulated in the Act.

In the Salmek Saragih Research, Tan Kamello et al. Argues that the procedure of credit granting cooperatives and small businesses still consider it quite complicated; it is indicated by the many stages that must be passed. Micro, Small and Medium Enterprises generally rely on their capital in running their business, and sometimes they are caught in the lenders' attachment given the lack of accessibility to formal financing sources. [11]

In his research entitled "Development and Strategy of SMEs Financing," Lestari mentioned some obstacles experienced by SMEs in fulfilling their capital needs. One of the barriers is the complexity of procedures that must be undertaken so that SMEs increasingly difficult to get credit from the banks. It is necessary to improve the system of credit procedures from the financial institution so that SMEs can feel credit facilities.

Loans given by the bank are based on trust, thus crediting is a gift to the customer. Ability to execute a grant or enter into a loan with a pledge whose payments will be made within an agreed timeframe. According to Law Number 10 the Year 1998 concerning Banking, the Bank is a business entity that collects funds from the public in the form of savings and distributes to the community in the way of credit and other types to improve the standard of living of the people. One form of channeling of funds is to provide credit to the community. Bank as a guarantee in lending based on confidence, besides doing penicillin to debtor based on 5C that is: character, capacity, capital, collateral, condition. [12]

Credit also provides protection to economically weak groups in the development of their business. The credit sector is one means of capital generation for the business community. For banks, every gift distributed to entrepreneurs always carries risks. Therefore, the need for security elements in return. In terms of convincing the bank that the customer is truly trustworthy, then before the credit is given in advance the bank conducts credit analysis. Credit analysis includes the background of the customer or the company, its business prospects, guarantees and other factors. The purpose of the study is for the bank to be confident that the credit provided is entirely secure. [13]

The security element is one of the primary principles in lending credit other than the aspect of suitability and profitability. Banks also always ask for collateral or collateral. The form of securing credit in banking practice is done by binding of the guarantee. Guarantees have essential positions and benefits in supporting economic development. Because the existence of this institution can provide benefits to creditors and debtors. The creditor's interest is the realization of security against closed trading transactions as well as providing legal certainty for creditors. [14]

Rachmadi Usman argues that general guarantees are addressed to all creditors and to all debtor properties. Each creditor has the same right to obtain debt repayment from the proceeds of sale of all material possessed by the debtor. [15] About the guarantee as a grant of security of funds or credit then in broad outline there are two kinds of warranties, namely individual guarantees and real guarantees. The most preferred guarantee by the bank and the other party as the creditor is the guarantee of materiality.

In the development of the Indonesian economy, credit guarantee is one of the solutions needed by micro and small enterprises to obtain banking credit facilities. In essence, credit guarantee is required as a fulfillment of technical requirements for micro and small enterprises that have an excellent prospective business, but not enough assurance so that technically does not fulfill credit requirement from a bank. In other words, credit guarantee is a bridge for those who are feasible, but not yet bankable. [16]

The debtor is not able to return the principal of credit and interest, the bank or the owner of the capital may execute against the guarantee object. The value of the corresponding object is usually at the time of making the estimated cost higher when compared to the principal of interest in arrears. However, for SMEs material security in the form of land is very burdensome for SMEs who do not have property or certificate. Therefore, regarding providing convenience in providing collateral then by merely moving goods owned by SMEs such as cars, inventory, machinery and so forth.

5 Conclusions

Capital needs usually accompany the development of Small and Medium Enterprises (SMEs). In the case of capital loans from banks always apply the various requirements of credit security guarantees channeled which is often requested by the Bank in the form of material security with the guarantee of the land. This case is very burdensome SMEs. There is a need for a legislation that regulates related to the granting collateral of movable objects to facilitate SMEs regarding capital loans for example Bank Indonesia regulation.

References

1 Hafni R. Tujuan Pembangunan Nasional dan Pertumbuhan Ekonomi. Diakses dari http://www.materisma.com/2014/09/tujuan-pembanguan-nasional-dan-ekonomi.html, (2014).

2 BPS, Retrieved December, 26th 2017, from http://www.bps.go.id

3 BI Retrieved March, 3rd 2018 from http://www.bi.go.id/umkm/kelayakan/Contents/Default.aspx

4 M. Siregar, *Hukum Ekonomi Indonesia dan Pembangunan Nasional*, dalam Butir-Butir Pemikiran Guru Besar dari Masa ke Masa, Pidato Pengukuhan Jabatan Guru Besar Fakultas Hukum USU 1979-2001, penyunting Tan Kamello Pustaka Bangsa Press, Medan, (2003).

5 Mubyarto, *Moral Ekonomi Pancasila*, (Yayasan Idayu, Jakarta, 1982).

6 N.N. Sirait, *Indonesia dalam Menghadapi Persaingan Internasional*, Pidato Pengukuhan Guru Besar Tetap dalam Bidang Ilmu Hukum Internasional, Universitas Sumatera Utara, Medan, (2006)

7 P. de Cruz, *Perbandingan Sistem Hukum Common Law, Civil Law dan Socialist Law*, (Nusamedia, Jakarta, 2017)

8 W. Menski, *Perbandingan Hukum Dalam Konteks Global, Sistem Eropa, Asia dan Afrika*, (Nusamedia, Bandung, 2016).

9 A. Madjid dan S. E. Wasono, *Wawasan Ekonomi Pancasila*, (Universitas Indonesia Press, Jakarta, 1988)

10 J. Assiddiqie, *Konstitusi Ekonomi,* (Kompas, Jakarta, 2010).

11 S. Saragih, Tan Kamello dkk. "Melakukan kajian pemberian kredit modal bergulir untuk koperasi dan usaha kecil dalam rangka pemberdayaan ekonomi rakyat di Sumatera Utara", *Mercatoria* **1**, (1) (2008).

12 M. Djumhana, *Hukum Perbankan Indonesia,* (Citra Aditya Bakti, Bandung, 1997).

13 Kasmir, *Dasar-dasar Perbankan*, (Raja Grafindo Persada, Jakarta, 2000)

14 T. Kamello,*Hukum Jaminan Fidusia: Suatu Kebutuhan yang didambakan*, Alumni, Bandung , (2004).

15 R. Usman, *Hukum Jaminan Keperdataan*, (Sinar Grafika, Jakarta, 2008).

16 E. Mulyati, *Kredit Perbankan: Aspek Hukum dalam Pengembangan Usaha Mikro Kecil dalam Pembangunan Perekonomian Indonesia*,(Refika Aditama, Bandung, 2016).

Legal protection for women environmental activists in urban areas

Tien Handayani Nafi[1], *Ratih* Lestarini[1,*], *Inayati*[2], *Tirtawening*[1], *Succi* Wulandhary[3], *Intan Nurul* Aini[3], and *Dyah* Utari[4]

[1]Faculty of Law, Universitas Indonesia, Depok, Indonesia
[2]Faculty of Administrative Science, Universitas Indonesia, Depok, Indonesia
[3]School of Environmental Science, Universitas Indonesia, Salemba, Indonesia
[4]Faculty of Health Science, Universitas Pembangunan Nasional Veteran, Jakarta, Indonesia

Abstract Men and women have equal rights, duties, and degrees in protecting and fighting for the rights of environment. However, in fact, it is men who dominate most conservation and environmental management activities. Even though, women's roles in environmental conservation activities are also no less great than men's given their high sense of empathy and care for the environment. This makes women taking more actions to protect the environment itself, but they often get so many threats, physical harass, even murder threats from those who feel threatened by their actions; and it already happened for a long time, and it still does until now. This should be overcome as soon as possible to prevent more women be the next victim. This research used SWOT analysis to determine the strategies for protecting women environmental activists in urban areas. The SWOT calculation results indicate that there is a need for progressive rules in guaranteeing women's rights so as to be able to engage in the process of environmental conservation in their respective regions.

1 Introduction

Issues about women, the last few decades to the present, many fill the discourse in the middle of the life of nation and state in addition to political and economic discourse. This issue is becoming increasingly attractive as awareness of injustice among men and women is increasingly high among our society. In addition, one of the seventeen objectives of the SDGs is gender equality, making it more interesting to continue discussions on this issue of gender equality. Especially at this time the number of women greater than the number of men. On the other hand, women have not yet filled out and occupy influential public sectors in determining the decisions and key state policies [1].

Gender equality is one of the most important aspects of sustainable development. Globally, the phenomenon of gender equality has been long-standing and this can be seen in the achievement of gender equality in some countries of the world from different continents. Until 2015, countries with the highest gender equality are dominated by countries from Continental Europe. However, there are two Asian countries that are able to achieve a very high gender equality, one of which is the ASEAN country, namely

* Corresponding author: ratihlestarini@yahoo.com

Singapore [2]. In the pursuit of gender equality, gender justice is the main foundation. Gender equity is a fair process for women and men, to ensure that the process is fair for women and men to take action to stop things that socially and historically discourage women and men from participating in and enjoying the results and the role it plays [3].

Given the value of the UNDP Gender Inequality Index (GII) with the scope of ASEAN, by 2015, gender inequality can be said to have begun to decline. For example in Singapore, gender inequality is ranked 11th in the global scope and the lowest in ASEAN. There is also a Malaysian country whose value of gender inequality lies at the bottom 2 of Singapore. Unfortunately, the value of gender inequality index in Indonesia is ranked second after Cambodia. This means that there are still gaps and gender imbalances in the country of Indonesia in various sectors of life, one of them in the environmental sector. Gender inequalities in the environment sector in Indonesia can be seen in many activities in the name of 'environmental conservation' and the management and sustainable use of the male-dominated environment [4] [5].

The strong encouragement and motivation of women to fight for environmental sustainability considering their high empathy towards the environment has to be paid with life. According to Global Witness records, by the end of 2016, there were about 200 environmental activists, including female activist, from various countries who died of assassination. Even according to Global Witness, the murder of environmental activists is more 'widespread' than 'developing', which means they must be more vigilant because for the next few years, such killing events will occur in more regions in other countries. Even the number of killings in the year 2016 alone has expanded to 24 countries, more than in 2015 which occurred in only 16 countries [8].

Surely things like this should be addressed as soon as possible so that the number of victims stops growing. Many ways can be done to protect the right of life of environmental activists, one of them by establishing a regulation of protection against them. These regulations may apply in regional and international environments. The hope, with the enactment of this regulation, is to minimize the number of victims as well as the possibility of similar events happening in the years to come so that environmental activists can still fight for rights and obligations to the environment without the need to feel threatened. Based on what had been explained before, the research questions are how is the existing condition of urban women as the environmental activists so far and how do women feel about the enforcement of legal protection which it protects women environmental activists from torture and assasination threats nowadays. The goals of this research are to explore the existing condition of urban women as the environmental activists and also to find out how women feel about the enforcement of legal protection which it protects women environmental activists from torture and assasination threats nowadays.

The term gender is used to describe the differences in roles, functions, status, and responsibilities between men and women formed from socio-cultural constructions. The concept of gender equality is a time when women and men have the same status and conditions to fully realize human rights and potential in development in all areas of life. While gender equality is a fair condition that removes barriers to play a role for women and men through cultural and policy processes [6]. Women's organizations in the field of environment and development have been active in the last fifteen years and have grown considerably over the last decade. The lesson to be learned is the importance of working in an established organization with committed and experienced women with different backgrounds [6].

Results of research by Asteria et. al (2013) indicates the important role of women in overcoming the environmental conflict. Female activists prioritize communication to create harmonious relationships with communication styles that tend to be cooperative considering women have an orientation to maintain relationships in the future and for the welfare of society as a whole. Female activists tend to choose mediation and negotiation efforts in environmental conflicts. Female activists in Indonesia will prioritize deliberations to reach agreement on resolving environmental conflicts. It relates to the concept of a self-oriented gender femininity to sustain society sustainably in their community [9].

Not only making some negotiations and mediations to solve the environmental conflicts, women also promoting healthy lifestyle in such many ways and forms as their efforts of the environmental activists. For example, women in Europe has launched a web that promotes healthy lifestyle by giving some tips on maintaining healthy environment for children and baby. The tips are build a baby room with no toxin materials and buy some toxin-free baby and children stuffs to protect children and baby from toxin exposures. Women in developing countries, such as Burkina Faso, also maintaining and purchasing sustainable and healthy lifestyle by making environmentally friendly products with no plastics and no metals [10]. The movement of feminism is a movement of social conflict aimed at breaking down the old values protected by functional structural tradition. Ecofeminism is a movement that serves to restore human consciousness to care about environmental damage due to the disappearance of quality nurturing and maintenance (feminine quality) [9].

2 Material and method

SWOT analysis is a method of determining strategic factors in four areas: Strength (S), Weakness (W), Opportunity (O), and Threat (T) [10]. The SWOT Matrix is a tool to help develop four types of strategies: SO (Strength-Opportunity) strategies, WO (Weakness-Opportunity) strategies, ST strategies (Strength-Threats), and WT (Weakness-Threats) strategies [11]. In the SWOT analysis, the SO (Strength-Opportunity) strategy is the best and for achieving that strategy, and hence relevant factors [12] are required. SO strategies use internal strength to take advantage of external opportunities. The WO strategy aims to improve internal weakness by taking external advantage. The ST strategy uses the company's power to avoid or mitigate the impact of external threats. WT strategy serves to reduce internal weakness and avoid external threats. By using SWOT analysis, four types of strategy will be obtained through internal and external factors, namely aggressive strategy (SO), competitive strategy (ST), conservative strategy (WO), and defense strategy (WT) . The deep interview from many experts to get the judgment to the reality was conducted to support SWOT analysis.

3 Results and discussion

In the case of legal protection for women's environmental activists, there are stronger strengths, weaknesses, opportunities, and threats to be addressed as follows.

1. Strength : The magnitude of the role, caring attitude, and empathy attitude of women to conservation and environmental management activities.
2. Weakness : Limited opportunities for women to be actively involved in conservation and environmental management activities.
3. Opportunity: Women become more active in voicing their rights as citizens who are also entitled to participate in maintaining and managing the environment even seem to threaten the existence of private projects that tend to damage the environment.
4. Threat : The right to life of women environmental activists is threatened, ranging from just a threat to actually happening cases of torture even to murder.

The following table presents the SWOT on legal protection cases for women environmental activists.

Table 1. SWOT Matrix

	S (Strength) Internal power factors	W (Weakness) Internal weakness factors
O (Opportunity) External opportunity factors	SO strategy: creating a strategy for women's involvement in conservation activities	WO Strategy: creating strategies to increase women's active involvement and role in conservation activities
T (Threat) External threat factors	Strategy ST: create a strategy using the power of life to deal with threats	WT Strategy: Creating a strategy of flexibility of involvement and active participation of women in conservation activities through the granting of a right to life

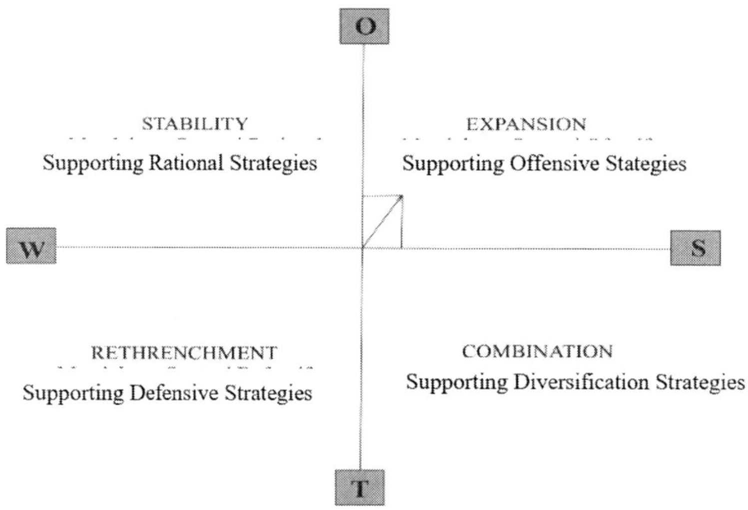

Fig. 1. Cartesian diagram of SWOT analysis

The results showed that there are still limitnesses of women's rules in environmental protections and conservations, whereas their roles and perceptions about environment are also as big as men's. Those limitnesses make women to more utter their rights as the environmental activists equally as men are. Unfortunately, many of women activitists still feel unsafe about their whereabouts because of their actions. They often get some threats from someone who feel threatened of their actions against environmental destructions. It's getting worse considering the regulation to protect women from those threats are still strongly unbonding and unprotecting. From here, it is very clear that there has to be some strategies to be made to decrease the limitness towards women's rules in environmental conservations and protections.

In accordance with the results of SWOT analysis based on each strength, weakness, opportunity, and threat, then for this case SO strategy is used. In dealing with and resolving this case, it is necessary to create a strategy for women's involvement in conservation activities. The involvement of women in conservation activities has great potential to realize the main objectives in this activity. The role of women in conservation activities is essentially the same as the conservation actions undertaken by men. However, women have the nature of caring, thorough and meticulous. [13] This is what will eventually complement each other with the male roles in environmental conservation activities and actions. Certainly it should also be supported by rights and legal protection for women to participate in conservation activities.

The need for a legal protection strategy for women's environmental activists is also important. Through this, there will be no act of women who feel their lives are threatened their due to their involvement in voicing rights and aspirations to the conservation and damage of the environment. Men's gender dominance in the environmental management sector does not necessarily indicate that men have a great empathy and concern for environmental management because, in terms of empathy, it is women who have greater empathy in the environment and the biotic components. Because of its great empathy, women tend to preserve the environment by not exploiting the existing natural resources. This is somewhat at odds with the behavior of men who tend to have a sense of 'power' or 'superiority' to their environment so that the possibility of exploitation of nature and its biotic components by men is greater [14].

However, many women still feel that their opportunities to participate actively in environmental activities are still limited or that they can participate as members of environmental groups only [15], whereas in Law Number 32 Year 2009 on Environmental Protection and Management Article 2 has stated that gender differences do not affect management activities or environmental protection because both the rights, obligations, and degrees of men and women are the same in the eyes of the law [16]. Considering the existence of deviations between the policy and its implementation in the field, the emergence of female figures who act as environmental activists aims to achieve gender equality in terms of environmental conservation as stated in the relevant legislation.

In the future, if women are given full rights to be able to contribute to environmental conservation, certainly greater and wider potential of environmental sustainability can be realized. The more elements of society that contribute to environmental conservation activities, both oversight and enforcement will certainly ensure the sustainability of the environment. There should be no more parties that restrict the rights of women to get involved in the life of nation and state. Citizens all have inherent rights and it is not justified for those to be restricted. Therefore, gender equality in the above case is very important to be realized. Legal protection for women environmental activists is also the case. Legal safeguards will provide wider space for conservation activities (both coercion and supervision) and contribute to reducing the burden on women themselves. In this context the burden in question is an impact of injustice on the law imposed on women's space. The freedom of space owned by women's environmental conservation activists has become a cornerstone of human rights and the potential for massive environmental conservation. Anticipative action is also required, for example in the aspect of community empowerment so that all agencies together to keep environmental activists. So that the future of women can help maintain and improve the welfare of society and increasing environmental awareness in all aspects of life.

4 Conclusion

Legal protection for women engaged in the effort to fight for their rights as citizens entitled to legally manage and protect the environment by channeling their aspirations as environmental activists requires clear instruments. As long as they become environmental activists, the state must be able to protect the existence and rights of women. SWOT calculation results show the importance of women's involvement in conservation of the limb. To achieve this, there needs to be a progressive regulation in ensuring women's rights so as to be able to engage in the process of environmental conservation in their respective regions. This research hopes that woman activist are treated equally and more protected by law in conducting conservation and environmental management movements.

Indonesia actually has its own legal protection already to protect women activist's right and it is well said on Paragraph 66 of Law No. 32 of 2009 on the Protection and Management of the Environment. It says that no environmental activists can be punished by any legal laws. Unfortunately, the implementation of the law is terrible considering there are terrors, physical harass, and even murder threats that still happen to environmental activists, including women, until now; and also the law itself has no explanation on how to protect those activists technically, so most citizens don't know how to protect people they know, who happens to be women environmental activists. That's why, to overcome this problem, there's no need to making any new protection regulations anymore because the problem solving can be done starting with strengthen the laws and its implementation that already exist. First things first, the state should revised the law they already made with putting some technical moves on how to protect activists from threats and physical harass. The state then could increase the public awareness of how important the women presence in any 'environmental caring, protecting, and keeping' activities can be by making any public discussions or public demonstrations and citizen, publicly, can participate; hoping that from these strategies, public will aware on how dangerous the risk of being environmental activists could happen and together, they will help to protect the activists from those dangerous risks and the activists themselves are free from those harms that addressed for them and they still can fight for environment without feeling any worries about their safety anymore.

Acknowledgements

This research is funded by Ministries of Research, Technology and Higher Education/ Kemristekdikti 2018 with Program Penelitian Dasar Ungguan Perguruan Tinggi (PDUPT) 2018 with contract number /UN2.R3.1/HKP05.00/2018

References

1. Marzuki. Studi tentang kesetaraan gender dalam berbagai aspek. (UNY, Jogjakarta, 2008)
2. United Nations Development Programme. 2015. Human Development Reports: Table 5. Gender Inequalty Index. Retrieved April 3, 2018 from http://hdr.undp.org/en/composite/GII
3. N.M.D. Widayani, S. Hartati. Kesetaraan dan keadilan gender dalam pandangan perempuan Bali: studi fenomenologis terhadap penulis perempuan Bali. *Jurnal Psikologi Undip* **13**, 2 (2014)
4. I.D.A. Nurhaeni, R. Sugiarti, S. Marwanti, R.D. Pratiwi. Disparitas gender dalam pembangunan pariwisata ramah lingkungan (Gender Disparities in Ecologically Friendly-Tourism Development). PALASTREN Jurnal Studi Gender **10**, 1 (2017)
5. Global Witness. 2016. Defenders of the Earth Report: Global Killings of Land and Environmental Defenders in 2016. Retrieved April 3, 2018 from https://www.globalwitness.org/documents/19122/Defenders_of_the_earth_report.pdf.
6. H. Puspitawati. Konsep, teori dan analisis gender. Departemen Ilmu Keluarga dan Konsumen-Institut Pertanian Bogor. (2013)
7. I. Dankelman. Climate Change: Learning from gender analysis and womens's experiences of organising for sustainable development. Gender and Development **10**, (2002)

8. D. Asteria, E. Suyanti, D. Utari, D. Wisnu. Model of environmental communication with gender perspective in resolving environmental conflict in urban area (study on the role of women's activist in sustainable environmental conflict management). *Procedia Environmental Sciences* ***20***, (2014)
9. M. Samejima, Y. Shimizu, M. Akiyoshi, N. Komoda. *SWOT Analysis support tool for verification of business strategy* (Graduate School of Information Science and Technology-Osaka Univesity, Japan, 2006)
10. F.R. David. Strategic Management Concepts and Cases (Prentice-Hall Publishing, Francis Marion University, Florance, South Carolina, 2011)
11. S.M. Hatefi. Strategic planning of urban transportation system based on sustainable development dimensions using an integrated SWOT and fuzzy COPRAS approach. *Global Journal of Environmental Science and Management* **4**, 1 (2018)
12. T. Sofiani. Eksistensi Perempuan Pekerja Rumahan dalam Konstelasi Relasi Gender. *Muwazah* **2**, 1 (2010)
13. J. Graca, M.M. Calheiros, A. Oliveira, T.L. Milfont. Why Are Women Less Likely to Support Animal Exploitation than Men? The Mediating Roles of Social Dominance Orientation and Empathy. *Personality and Individual Differences* **129**, (2018)
14. H. Ankesa, S. Amanah, P.S. Asngari. Partisipasi kelompok perempuan peduli lingkungan dalam penanganan sampah di Sub DAS Cikapundung, Jawa Barat. *Jurnal Penyuluhan* **12**, 2 (2016)
15. Undang-Undang Republik Indonesia Nomor 32 Tahun 2009 Tentang Perlindungan dan Pengelolaan Lingkungan Hidup. Retrieved April 3, 2018 from http://jdih.menlh.go.id/pdf/ind/IND-PUU-1-2009-UU%20No.%2032%20Th%202009_Combine.pdf
16. M. Saleh. Partisipasi perempuan dalam pengelolaan lingkungan hidup. *Musawa* **6**, 2 (2014)

Econometric analysis of technical efficiency of pearl millet farmers in Kano State, Nigeria

Umar Mukhtar[1&2*], *Zainalabidin* Mohamed[1], *Mad Nasir* Shamsuddin[1], *Juwaidah* Sharifuddin[1], and *Muhammad* Bala[1&3]

[1]Department of Agribusiness and Bioresource Economics, Faculty of Agriculture, Universiti Putra Malaysia, 43400 UPM Serdang, Selangor Darul Ehsan, Malaysia
[2]Department of Agricultural Economics and Extension, Faculty of Agriculture, Federal University Dutse, P.M.B. 7156, off- Ibrahim Aliyu Bye-Pass, Jigawa State-Nigeria
[3]Department of Agricultural Economics and Extension, Faculty of Agriculture, Taraba State University, Jalingo-Nigeria.

Abstract. In Nigeria, particularly the northern parts, pearl millet is a traditional crop that is very important to nation's food security. Empirical evidence indicates that the main problem in pearl millet production in Nigeria is associated with low productivity and inefficiency in allocation of resources. This study therefore sought to analyse the Technical Efficiency (TE) of pearl millet farmers in Kano State of Nigeria, being the state with highest concentration of pearl millet farmers. SFA approach is used on a data collected from 256 randomly sampled pearl millet farmers during 2013/2014 cropping season. The study discovers that all the production inputs/variables included in the Cobb-Douglas Stochastic Frontier function with exception of agrochemicals were positively and significantly associated with TE. The TE range of 21% to 94% and mean TE of 73% were obtained from the analysis. The study also reveals that access to credit, education, extension and household size were the major contributors to TE. This therefore calls for formulation of appropriate policy that would extend these key determinants of TE which would go a long way in improving farmers' efficiency level and economic benefits. This study contributes the first SFA approach in estimating TE of pearl millet farmers in Kano State, Nigeria, which can effectively help in addressing food insecurity problem in the country.

1 Introduction

Pearl millet is grass in the family of Panacea that originated from western tropical Africa about 3,000 years ago. It spreads throughout eastern and southern Africa, also some parts of America and Australia. According to [1], pearl millet, the sixth most important cereal crop in the world, providing sustenance to about 500 million masses in the arid and semi-arid, predominantly in Asia and Africa.

Nigeria was once the second world largest millet producer after India, where it is widely grown for its grain to make flour and other materials. In Nigeria, especially in the northern

*Correspondence: umarmukhtar79@gmail.com

parts pearl millet is one of the most important staple crops both in terms of production and c onsumption, and it is considered as staple food crop for over 40% of the populace. However , performance of pearl millet among smallholder farmers has either stagnated or progressed at a very slow pace, placing the average yield for this vital crop at 1-1.5 t/ha as against the potential yields of 2.5-4t/ha [2,3]. Although a lot of efforts were made by both governments and non-governmental organizations to improve the productivity of pearl millet in the country through introduction of improved varieties and associated technologies, but the yields realize by resident farmers is still far below the average potential yield. This positions the country's yield at 30th in the world ranking [3].

This low productivity can be ascribed to inefficiency among farmers, which usually caused by poor understanding of efficient agricultural practices and inappropriate inputs combination. It is asserted that gap between actual yield realized on farmers' fields and the potential farm yield is largely attributed to management practices, which depend on socioeconomic characteristics of farmers [4]. This means that agricultural performance is not only a function of the productivity of technical inputs, but increased yield is also directly associated with socioeconomic factors of the farmers. As productivity and output are positively related to efficiency, it is imperative to estimate the TE levels and to also determine the socioeconomic factors responsible for these TE levels among pearl millet farmers in the study area which this study is set to achieve.

2 Materials and Methods

The sampling frame of the study is pearl millet farmers in Kano State of northern Nigeria. Structured questionnaire was used as a research instrument to collect the data from 256 randomly selected pearl millet farmers in the study area. Stochastic Frontier Analysis (SFA) was used during 2013/2014 cropping season.

3 Empirical Model

The Stochastic Frontier Analysis (SFA) was used to analyze the technical efficiency of pearl millet farmers. The stochastic frontier production function to measure the technical efficiency of production was independently developed by [5, 6]. The Stochastic Frontier Production Function is more appropriate for measuring technical efficiency because it overcomes the inadequate characteristics of the assumed error term in traditional (conventional) production functions which have limitations on statistical inference of the parameters and the resulting efficiency estimates.

The production technology of the pearl millet farmers in Kano State, Nigeria was assumed to be specified by the Cobb-Douglas frontier production function. This according to [7, 8] has been used by many empirical studies, particularly those relating to agriculture in developing countries. The Cobb-Douglass frontier production function of the model in mathematical form can be specified as follows:

$$lnY_i = \beta_0 + \beta_1 \ln X_{1i} + \beta_2 \ln X_{2i} + \beta_3 X \ln_{3i} + \beta_4 \ln X_4 i + \beta_5 \ln X_{5i} + \beta_6 \ln X_{6i} + v_i - u_i$$

Where:
ln = the natural logarithm
Y_i = Quantity of Pearl millet output (kg)
X_1 = Farm Size (hectares)
X_2 = Fertilizer (Kg)
X_3 = Manure (ox-cart)
X_4 = Labour (man-days)

X_5 = Seeds (kg)
X_6 = Agrochemicals (litres)
β_0 = Intercept
β_1 - β_6 = Parameters to be estimated
i = number of farm
V_i = Stochastic component of error term beyond the control of farmers
U_i=non-negative random variables assumed to account for technical inefficiency in production process.

The farm specific inefficiency variables

The technical inefficiency model was used to identify factors that influence the efficiency among pearl millet farmers in the study area, and based on [9]. the model was specified as follows:

$$U_i = \propto_0 + \propto_1 Z_{1i} + \propto_2 Z_{2i} + \propto_3 Z_{3i} + \propto_4 Z_{4i} + \propto_5 Z_{5i} + \propto_6 Z_{6i} + \propto_7 Z_{7i} + E_i \quad (2)$$

Where:

U_i=technical inefficiency
Z_1 = Age (years)
Z_2 = Level of Education (years)
Z_3 = Farming experience (years)
Z_4= Cooperative membership (1=member, 0=otherwise)
Z_5 = Access to credit (1=access, 0=otherwise)
Z_6 =Household size (number)
Z_7= Extension contact (1=access to extension, 0=otherwise)
α_1-α_7 =scalar parameters to be estimated
α_o = constant or intercept
E_i= stands for the random variables defined by the truncation of the normal distribution with mean 0 and variance, σ_u^2, such that the point of truncation is $-Z_i \propto$, i.e. $E_i \geq -Z_i \propto$. These assumptions are consistent with U_i being a non-negative truncation of the $(Z_i \propto, \sigma_u^2)$ distribution [9].

4 Results and discussion

The frequency distribution of the stochastic production frontier model is shown in Table 1.

Table 1 Technical Efficiency Distribution of Pearl Millet Farmers for Kano State

Technical Efficiency	Number of Farmers	Percentage
0.21-0.30	9	3.52
0.31-0.40	12	4.69
0.41-0.50	17	6.64
0.51-0.60	16	6.25
0.61-0.70	40	15.63
0.71-0.80	57	22.26
0.81-0.90	53	20.70

0.91-0.99	52	20.31
Total	**256**	**100.00**
Minimum TE	**0.21**	
Mean TE	**0.73**	
Maximum TE	**0.94**	

Results of the analysis reveal that TE of pearl millet production in the study area ranged from a minimum of 21% to a maximum of 94%, with an average of 73%.

The results further indicate that for the average farmer in the sample to attain the TE of his/her most efficient peers, he/she could achieve about 22.34% {1-(73/94) x100} cost savings. Similarly, the least technically efficient farmer will have 77.66% {1– (21/94) x 100} cost savings to become the most efficient farmer.

Maximum likelihood estimates (MLE) of the parameters of the stochastic frontier function

The result of the maximum likelihood estimates of the Stochastic Frontier of pearl millet production function is presented in Table 2. The sigma-square was 0.155 ($p<0.01$), confirming the good fit of the model and the correctness of the distributional form assumed for the composite error term. The value of gamma (γ) was 0.818 and was statistically significant at 1%, implying that 82% of the variation in pearl millet output of the farmers was due to technical inefficiency term (u_i) rather than random error (v_i) associated with uncontrollable factors which affect production. This explains why ordinary least squares (OLS) or an average production function was not a suitable specification for sampled pearl millet farmers.

Table 2 Maximum likelihood estimates of the parameters of the stochastic frontier function

Variables	Coefficient	std-error	t-ratio
Constant	1.879	0.852	2.205**
LnFarmsize	0.232	0.051	4.549***
LnFertilizer	0.391	0.166	2.355**
LnManure	0.184	0.092	2.000**
LnLabour	0.526	0.105	5.010***
LnSeed	0.243	0.054	4.500***
LnAgrochemicals	-0.428	1.262	-0.339^{NS}
Inefficiency Effects			
Age	0.7746	0.2634	2.294**
Level of education	-0.1449	0.0268	-5.408***
Farming experience	0.2556	0.5632	0.454^{NS}
Coopt. Membership	0.8428	0.3936	2.141**
Access to credit	-0.1485	0.0387	-3.834***
Household size	-0.8606	0.4552	-1.89*
Extension Contact	-0.2372	0.4289	-5.531***
Diagnostic Statistics			
sigma-squared(δ^2)	0.155	0.042	3.690***

Variables	**Coefficient**	**std-error**	**t-ratio**
gamma (γ)	0.818	0.124	6.597***
Log-likelihood Function	7.848		

Note: NS=Not Significant. ***, ** and *=significant at 1%, 5% and 10% levels, respectively.

The estimated coefficient for farm size (0.232) was positive and statistically significant at 1%. This implies that a 1% increase in farm size will *ceteris paribus* lead to an increase of 0.232 % in output. This increase is inelastic because a 1% increase in the input leads to a less than 1% increases in the output. This result is not at variance with those obtained by [10] and [11].

The production elasticity of output with respect to a quantity of fertilizer (0.391) was positive and statistically significant at 1% level. This implies that a 1% increase in the amount of fertilizer applied on the farm will increase output by 0.391 % *ceteris paribus*. Similar results were found by [12] and [13].

The estimated coefficient for manure was positive and significant at 1% which implies increased output, because as the quantity of manure increase by 1%, the yield will also increase by 0.184 %, *ceteris paribus*. This result is in tandem with the one obtained by [14] who found animal manure to have a positive and significant effect on TE among Banana farmers in Uganda. It is also in conformity with [13] and [15].

The estimated coefficient for labour (persons in man-days) was positive and statistically significant at 1%. This implies that increasing labour by 1% will increase output by about 0.526 % *ceteris paribus*. The findings correspond to those obtained by [10] and [16] but are not in accordance with findings of [17] and [18]

The estimate for seed (0.243) is positive and significant at 5%. The significance of seed quantity is due to the fact that seed determines to a large extent the output obtained. Output tends to be low even when other inputs are in abundance if quality seeds and accurate seed rates are not used. . The result was again indistinguishable to that of [19] and [20].

Factors Explaining Inefficiency

The parameters of the explanatory variables in the inefficiency model were simultaneously estimated in a single stage using a computer program, FRONTIER 4.1. Since the dependent variable of the function stands for inefficiency, the sign of the coefficients of these variables has important policy implications as the positive sign indicates a negative effect on technical efficiency whereas negative sign signifies a positive effect on efficiency. The results of the inefficiency effects model are also shown in Table 2.

The positive sign of farmers' age show that TE decreases with age of pearl millet farmers and is significant at 5% level of probability. This implies that aged farmers were less technically efficient in pearl millet production Kano State. This may be due to the fact that younger pearl millet farmers adaptive to modern and improved technologies of farm production and management which result in increased farm efficiency. In contrary, aged farmers are often reluctant to try new farm technologies and mostly stick to traditional and conservative farming techniques which are usually associated with inefficiency.

The coefficient of cooperative membership indicates unexpected sign, it has significant positive effect on the technical inefficiency of pearl millet farmers. This implies that farmers who did not belong to any cooperative society are more technically efficient. It is agreed that the key impetus of farmers in cooperative societies is to access vital information and farm inputs at a subsidized rate for improved productivity. This can be attributed to the fact that pearl millet farmers in the study area never actively participated in the activities of cooperative societies which contributed to the inability to improve their knowledge and skills in crop production management.

The coefficient of education is negative and significant at 1%, suggesting that investment in human capital is a strong device to improve technical efficiency in pearl millet production in the study area. The result further signifies that smallholder pearl millet farmers that have acquired education at a certain level would be more technically efficient than those that have not. The reason might be that such farmers would more likely follow and apply the recommended principles underlying the use of improved technology at their farms and result in efficient resource use.

The coefficient of extension contact was found to be significant and negative, whose implication is that increased contacts with extension officers decrease technical inefficiency. This implies that easy access to extension contact would also expose the pearl millet farmers in the study area to some modern farming techniques that assure higher yield and productivity.

The coefficient of access to credit is significant and negatively associated with technical inefficiency. That is, increased access to credit increases technical efficiency level of the pearl millet farmers. An assertion by [21] also explains that farm credit boosts diversification of agricultural systems that stabilize and possibly improve farm productivity, if it is appropriately extended, managed and utilized.

The household size variable is negatively related to technical inefficiency, but weakly significant only at 10%. This means farmers with large family size are less technically inefficient. This is probably because farmers that have large household size tend to endeavour to obtain higher output in order to meet their subsistence necessities. Furthermore, large household size has labour endowment required to implement farm management decisions.

Comparison of the Efficient, Average and Least Efficient Farmers in Terms of Inputs used Per Hectare

Optimum use for each input per hectare at farm level and its optimum score are reported in Table 3. The quantities of inputs used by the most technically efficient farmers in the sample to produce an average output of 1372.85kg/ha were fertilizer (82.78kg /ha), manure (15.76 ox-cart/ha), labour (34 Man-days/ha), seed (28. 00 kg/ha) and agrochemicals (1.35 litre/ha). In contrast, an average efficient farmer in the sample used fertilizer (83.37 Kg/ha), manure (13.60 ox cart/ha), labour (33 man days/ha), seed (31. 40 kg/ha) and agrochemicals (1.09 litre/ha) to produce 1123.12 kg/ha of output. Similarly, the least technically efficient farmer in the sample made use of fertilizer (74.23 Kg/ha), manure (18.85 ox cart/ha), labour (33 man days/ha), seed (20. 77 kg/ha) and agrochemicals (1.54 litre/ha) to cultivate an average pearl millet output of 1120.67 kg/ha. These results imply that fertilizer, seed, and agrochemicals were over-utilized, while labour and manure were over-utilized by the average technically efficient farmer in the sample. Therefore, for the average technically efficient farmer to attain optimality, they should reduce the use of fertilizer, seed and agrochemicals by 0.59 kg/ha (0.67%), 3.40 kg/ha (10.83%) and 0.26 litre (23.85) from the current levels, respectively. However, since manure and labour were underused, their use should be increased by 2.16 ox-cart/ha (15. 88%), 1 man-days (3.03%) to ensure efficient pearl millet production.

In the same way, when compared to the most technically efficient farmer, the least technically efficient farmer in the group under-utilized the application of fertilizer, labour, and seed. So, since the utilization of these inputs has not reached the optimum level as used by the most technically efficient farmers for optimal resource allocation in pearl millet production, it is necessary to increase their consumption by 8.55 kg/ha (11.52%), 1 man-days/ha (3.97%) and 7.23kg/ha (34.81%) from the levels currently used by farmers, respectively. However, the least efficient farmers should reduce the use of manure and agrochemicals by 3.09 ox-cart/ha (16.39%) and 0. 19litre (12.34), respectively, since the use of these particular inputs were found to be greater than their optimum levels compared to the most efficient farmer. The findings are somehow similar to results reported by Coelli *et al.,* (2002) [22] where fertilizer and labour were found to be highly over-utilized by rice farmers in Bangladesh.

https://doi.org/10.1051/e3sconf/20185200049

Table 3 Comparison of average inputs used by the most efficient, average and least efficient farmers per hectare per season

	Most efficient farmer	**Average efficient farmer**			**Least efficient farmer**		
Variable	**Quantity of inputs used**	**Quantity of inputs used**	**Deviation from optimality**	**% Deviation from optimality**	**Quantity of inputs used**	**% Deviation from Optimality**	**% Deviation from optimality**
Farm size	1Ha	1Ha	1Ha	1Ha	1Ha	1Ha	1Ha
Fertilizer	82.78Kg	83.37Kg	-0.59 Kg	-0.67	74.23Kg	8.55 Kg	11.52
Manure	15.76 Ox-cart	13.60 Ox-cart	2.16 Ox-cart	15.88	18.85 Ox-cart	-3.09 Ox-cart	16.39
Labour	34 Man-days	33 Man-days	1.00Man-days	3.03	33 Man-days	1 Man-day	3.97
Seed	28.00Kg	31.40Kg	-3.40 Kg	10.83	20.77 Kg	7.23 Kg	34.81
A/chem	1.35 Litre	1.09 Litre	-0.26 Litre	-23.85	1.54 Litre	-0.19 Litre	-12.34
Output	1372.85Kg	1123.12 Kg	249.73Kg	22.24	1120.67	252.18 Kg	22.50

5 Conclusions Policy Implications

The study revealed that the farmers were, on average, 73% efficient, which implies the existence of potentials improve TE in pearl millet production in the area. The above figure also implies that cultivating at about 27% further down the frontier level. This is an evidence that farmers in the study area do not have much score (little scope) to expand the productivity of pearl millet through efficiency under the current conditions of resource use and technology at the farm level. The study further discovered that educational level, extension contact, credit access and household size were found to be the main factors directly associated with TE of pearl millet farmers, while the age of farmers and cooperative membership were inversely related to TE. The study concludes that pearl millet farmers have abundant potential to enhance their production through improving available resources at their disposal. There were also both underutilization and overutilization of resources. The findings are also of paramount importance as they provide detailed and vital information to policy makers on nature of production techniques employed in pearl millet in the area. Hence, since the sources of inefficiency are identified, policy formulations to improve farmers' performance can be effectively done.

References

1. A.G. Gaya, S.A. Adebitan, and A.U. Gurama, Reaction of Pearl Millet Infected with Downy Mildew (Sclerospora Graminicola (Sacc.) Schroet) Intercropped Cowpea with On Days to 50% Heading and Grain Yield in The Savanna Zone of Northern Nigeria. Bay. *J. Pure & Appl. Sci.* **5**, 1– 4 (2012).
2. B.G. Soulé, D. Balami, and R. Blein, *Framework for involvement in rainfed food cro p supply chains development in West and Central Africa: Nigeria's Cereal Economy*, AFD, CIRAD, IFAD, July (2010).
3. FAO, http://faostat3.fao.org/home/index retrivied : December 20, 2015)
4. E.O. Onyuka, J. Kibbet, and C.O. Gor, Socio-Economic Determinants of Groundnut Production in Ndhiwa Sub-County, Kenya. Int. *J. Agric. & F. Res.* **6**, 1-14 (2017).
5. D. Aigner, C.K. Lovell, and P. Schmidt, Formulation and estimation of stochastic frontier production function models. *J. Econ.* .**6**, 21-37 (1977).
6. W. Meeusen, and J. van Den Broeck, Efficiency estimation from Cobb-Douglas production functions with composed error. *Int. econ. rev.* **18**, 435-444 (1977).
7. P.T. Tsue, W.L. Lawal, and V.O. Ayuba, Productivity and Technical Efficiency of Catfish Farmers in Benue State, Nigeria. Adv. *J. Agric. Res.* **1**, 20-25 (2013).
8. K. Ogundari, S.O. Ojo and B. Brummer, Productivity potential and technical efficiency of aquaculture production in alleviating poverty: Empirical Evidence from Nigeria. *J, Fish. Int.* **1**, 21-26 (2006).
9. G.E. Battese, and T.J. Coelli, A model for technical inefficiency effects in a stochastic frontier production function for panel data. Emp. *econ.* **20**, 325-332 (1995).
10. B.E. Bravo–Ureta, and A.E. Pinheiro, Technical, Economic and Allocative Efficiency in Peasant Farming; Evidence from Dominican Republic. *The Dev. Econ..* **35**, 48–67 (1997).
11. S.B. Wassie, Technical efficiency of major crops in Ethiopia: stochastic frontier model. Acad. *J. Agric. Res.* **2**, 147-153 (2014).
12. A. Ali, and A.U. Jan, Analysis of technical efficiency of sugarcane crop in Khyber Pakhtunkhwa: A stochastic frontier approach. Sarhad *J. Agric.* **33**, 69-79 (2017).

13. W. Alwarritzi, T. Nanseki, and Y. Chomei, Analysis of the factors influencing the technical efficiency among oil palm smallholder farmers in Indonesia. *Proce. Env. Sci.* **28**, 630-638 (2015).
14. F. Bagamba, F.R. Ruben, and M. Rufino, *Determinants of banana productivity and technical efficiency in Uganda. An economic assessment of banana _genetic improvement and innovation in the Lake Victoria Region of Uganda and Tanzania: Vol.* ***155***, 109 (2007).
15. K.M.M. Rahman,M.I.A. Mia and M.K.J.A. Bhuiyan, stochastic frontier approach to model technical efficiency of rice farmers in Bangladesh: An empirical analysis. *The Agriculturists*, **10**, 9-19 (2012).
16. Ng'ombe, and T. Kalinda, A Stochastic Frontier Analysis of Technical Efficiency of Maize Production Under Minimum Tillage in Zambia. *Sust. Agric. Res.* **4**, 31-46 (2015).
17. G.M. Kitila, and B.A. Alemu, Analysis of Technical Efficiency of Small Holder Maize Growing Farmers of Horo Guduru Wollega Zone, Ethiopia: A Stochastic Frontier Approach. *Sci. Tech. & Arts Res. J.* **3**, 204-212 (2014).
18. D.K. Mugambi, M. Mwangi, S.K. Wambugu, and A.M. Gitunu, Assessment of performance of smallholder dairy farms in Kenya: an econometric approach. *J. Appl. Biosci.* **85,** 7891-7899 (2015).
19. D. Awunyo-Vitor, J. Bakang, and S. Cofie, Estimation of farm level technical efficiency of small-scale cowpea production in Ghana. Amer-Eur. *J. of Agric. & Env. Sci.* **13,** 1080-1087. (2013).
20. M.H. Ahmed, Z. Lemma and G. Endrias Measuring Technical, Economic and Allocative Efficiency of Maize Production in Subsistence Farming: Evidence from The Central Rift Valley of Ethiopia. *Appl. Std. in Agrib. Comr.* **9,** 63-74 (2015).
21. B.M. Desai, and J.W. Mellor, Institutional Finance for Agricultural Development: An Analytical Survey of Critical Issues, *Food. Pol. Rev* .**1**, (1993)
22. T. Coelli, S. Rahman and C. Thirtle, Technical, Allocative, Cost and Scale Efficiencies in Bangladesh Rice Cultivation: A Non-parametric Approach. *J. Agric. Econ.* **53**, No. 3: p607-626 (2002).

Criminal policy on cyberbullying toward children

Wenggedes Frensh* and *Mahmud* Mulyadi

Doctoral Program in Law, Graduate School, Universitas Sumatera Utara, Medan, Indonesia

Abstract. Currently, the information and communication technology is experiencing a rapid development. Technological advances also trigger a variety of crimes (cybercrime). Cybercrime is a term that refers to criminal activities using computer and its network. One of the developing cybercrimes today is cyberbullying. *The National Conference of State Legislatures* describes this crime as an act of harassing and threatening others intentionally and repeatedly. In developed countries, serious efforts have been made to combat cyberbullying. While the awareness of cyberbullying threat as a crime is still low in Indonesia, there is no clear legal rule regarding this crime. This study used the theory of criminal policy and theory of criminal law reform. There are two impacts caused by the absence of the law. First, the difficulty of running a criminal policy. Second, cyberbullying has a psychological impact on victims such as a frustration, a depression, and even a suicidal intention. The result of the research showed that in terms of a penal policy, Indonesia applies the Electronic Information and Transactions Law (UU ITE) No.19 of 2016 in dealing with cyberbullying with cyberharassment type in article 27 paragraph 3 and cyberstalking in article 27 paragraph 4 after comparing with Arkansas code 2012 § 5-71-217. In terms of a non-penal policy, moral approach, technological approach, and media role are used.

1 Introduction

Globalization creates a limitless relationship between countries around the world. The interconnectedness of all countries in the world leads to the development of information and communication technology. The sophistication of information and communication technology also has an impact on the cybercrime.

Cybercrime is a term used for criminal activities using a computer and its network. [1] The types of a cybercrime include online auction fraud, check fraud, credit card fraud, identity fraud, child pornography, online prostitution and others. [2] One of the developing cybercrimes is cyberbullying.

The National Conference of State Legislatures (NCSLs) describes cyberbullying as an intentional repeated use of cell phones, computers and other electronic communication devices to harass and threaten others. [3]

* Corresponding Author: law_weng@yahoo.com

Based on the statistical data of Indonesian Association of Internet Service Providers, the number of internet users in Indonesia in 2017 was 143.26 million or about 54.68% of the total population of Indonesia (262 million people). [4] The internet users in Indonesia will increase through a demographic bonus. Indonesia will receive a demographic bonuses in 2020-2030 where the number of people aged 15-65 will be 70%, and 30% of the population are non-productive people (under 15 years and above 65 years of age). [5]

The data obtained from a research on the impact of cyberbullying on children in Indonesia showed that out of 353 teenager respondents, consisting of 157 men and 196 women with the age range of 12-15 years, the number of cyberbullying victims among teenagers reached almost half of the respondents, that is 172 respondents (49%). This means that almost one in two respondents had been a victim of cyberbullying. And, the most-widely-used media by cyberbullying perpetrators were Facebook, SMS, and Instagram. The kinds of cyberbullying experienced by the victims were in the form of mockery, slander, threat, and gossip objects. [6] Cyberbullying has a harmful effect on its victims such as a frustration, a depression, and even a suicidal intention. [7] Cyberbullying is a social problem and it needs social science and law to overcome it. [8]

Therefore, a comparison is done with one of the countries that has complete rules and definition of cyberbullying action. Arkansas which is one of the states in America has a cyberbullying rule in Arkansas Code 5-71-217. In addition, technological and communication developments exist in America, so of course America has a policy and cyberbullying crime that uses technology as a means of crime.

In relation to legislation, Indonesia still has no specific rules on cyberbullying. The rules on this matter are still generally accommodated in Law No. 19 of 2016 on Electronic Information and Transactions.

There is a need to compare how other countries apply such legislation to be a reference. Indonesia needs to learn the best practice of other countries in coping with cyberbullying and criminal policies to protect the children.

2 Formulation of the Problem

The problem of this research was: how should Indonesia adopt Penal and Non-Penal Policy on Cyberbullying toward children?

3 Research Method

This research was a normative legal research. Normative legal research is the study of positive legal norms, principles, principles and legal doctrines. In addition, it also used a comparative juridical study by conducting comparative studies of criminal law regulations in other countries in governing cyberbullying. The data of this study was the secondary data consisting of (a) main legal materials in the form of a legislation, and (b) secondary legal materials including books and journals. The data were collected using literature reviews and document analysis relating to penal and non-penal policies. A criminal policy theory was applied in considering that the prevention of cyberbullying uses both penal and non-penal efforts.

4 Discussion

4.1 Penal Policy on Cyberbullying

Wiener (1961)in his book "*Cybernetics or Control and Communication in the Animal and the Machine*" introduced the term cyber in *cyberspace, cybercrime*, and *cyberlaw*. [9] This term evolved from the use of *cybernetic* terminology by the word *cyber*, which stands for *cyberspace* and derived from the word *cybernetics* and *space*. [10] The term *cyberspace* first appeared in William Gibson's novel Neuromancer. [11] In 1990, John Barlow applied the term *cyberspace* to refer to the online world connected to the internet. [12]

Ken Rigby defines *Bullying* as a desire to harm and is shown through actions to cause a person to suffer. The actions are performed directly by a stronger, irresponsible person or group, usually repetitive and with a feeling of happiness. [13] These bullying actions are divided into four types: physical, verbal, relational, and cyber actions. [14]

Cyberbullyingin Arrigo (2014) book *"Encyclopedia of Criminal Justice Ethics"* provides an explanation of an action that includes the use of technology (including messaging, social media, and short message delivery) aimed at hurting or embarrassing other individuals. Although age factor is not important, cyberbullying is usually associated with early age (children). [15] Chadwick (2014) his book "*Impacts of Cyberbullying, Building Social and Emotional Resilience*" describes the following types of cyberbullying: [16]

- *Harassment* or disturbance by sending rough and annoying messages continuously without interruption.
- *Denigration* or defamation lies in spreading misleading information about others in web pages, emails, and short messages or it may come in digitally modified photographs.
- *Flaming* or extreme messaging is an online battle or an extreme exchange of opinions by using electronic messages in chat forums or by sending emails containing vulgar words.
- *Impersonation* or imitation is an entry into an email or social network which uses a person's online identity to transmit rude or embarrassing things about others.
- *Masquerading* or pretense is to pretend to be someone else by creating fake emails or fake short messages.
- *Pseudonyms* or shadow/disguise is using a pseudonym to cover the original identity. The cyberspace can recognize the perpetrator from his pseudonym that may be dangerous or insulting.
- *Outing and Trickery* or deception and dissemination of a secret is the act of publicly displaying or copying and resubmitting text messages, emails or instant messages relating to a person's secrets in order to embarrass him.
- *Cyberstalking* or cyber prowling is a form of disruption that uses the virtual world to threaten the safety of others. Usually, messages are sent via emails, text messages as well as personal communication tools.

4.1. 1 Law No. 19 of 2016 on Electronic Information and Transaction (EIT)

Based on the provisions of the articles of Chapter XI concerning the criminal provisions of the EIT Law, some prohibited acts (elements of criminal offense) that are closely related to cyberbullying in each article can be identified as follows:

Article 27 Paragraph 3 stipulates the prohibition to distribute, transmit and make accessible any form of Electronic Information or Electronic Document which has contents of an insult or defamation. The word insult or defamation categorized in cyberspace comes under the type of *harassment* and *denigration cyberbullying*.

The elements of crime contained in the Laware as follows:

- *Deliberately.* Deliberately means doing something with a consciousness and knowing the consequences of the deed. In harassment and denigration cyberbullying, an act is done by sending messages insultingly to attack somebody's honor. The act is done continuously without interruption. The act explains that the perpetrator performs his actions intentionally and knows the consequences of his actions. In this case, the element of *deliberately* is met.
- *Without rights* is an act against the law that violates the provisions of the legislation. In cyberharassment, without rights means that the offender has no legal permission to do so.
- *Distribute and/or transmit and/or make accessible any form of Electronic Information and/or Electronic Documents.* Distributing means sharing or submitting something. Distributing here means to send an electronic information or electronic document, either to someone or some parties by utilizing the information and communication technology. The other element, transmitting, include sending or forwarding an electronic information or electronic document to somebody. Distributing shares something to many people, while transmitting to one person only.
- *Having contents of an insult and defamation in the EIT Law* refers to the provisions of an insult or defamation set forth in the Criminal Code (Penal Code). The essence of an insult or defamation in the EIT and the Penal Code is an act of attacking the honor or reputation of others in order to be known by the public.

Article 27 Paragraph 4 stipulates the prohibition to distribute, transmit and make accessible any form of Electronic Information or Electronic Document that has blackmailing or threating contents. The word blackmail or threat used in the cyberspace is categorized into cyberstalking.

The elements of a crime contained in the Law are as follows:

Article 27 Paragraph 4 of the Law on EIT refers to the blackmailing and threating acts regulated in Chapter XXIII Book II of the Penal Code. Blackmailing and threating in cyberspace are not done physically but by forcefully threatening somebody to defame or disclose his secrets by using information technology as a tool. This type of cyberbullying is referred to as cyberstalking.

4.1. 2 Arkansas Code 2012 § 5-71-217 on Cyber bullying

Zweigert and Kotz explain the comparative law of an intellectual activity, where a law acts as an object and comparison as a process. [17] Comparative law provides an understanding of the nature of a law and its development. [18] Cyberbullying legislation is regulated in one of the American States called Arkansas. The definition of cyberbullying and what kind of actions categorized into cyberbullying can be found in Arkansas Code 2012 § 5-71-217. So, it is necessary to know the rules of cyberbullying in Arkansas as part of a comparative law study, given that the information and knowledge have connected people around the world. [19]

Arkansas Code 5-71-217 describes that the cyberbullying occurs when someone transmits, sends, or posts a communication by electronic means in order to frighten, force, intimidate, threaten, abuse, humiliate, and warn others. Transmitting here means continuing to engage in severe, repetitive or hostile behavior toward others.

Posting communication means the communication of a person's information without any change in the form or content of the information sent and received. Whereas electronic means refers to textual, visual, written, or oral communication through the use of online computer services, internet services, telephone, or other means of communication, which also covers bulletin board services, internet chat rooms, electronic mails, social networking sites and online messaging services.

Transmitting, sending or posting communication by electronic means has been regulated in each criminal provision of the EIT Law. If the act is intended to frighten, force, intimidate and threaten, then the act comes under Article 27 Paragraph 4 on blackmailing and threating. As to insulting act, it has been regulated in Article 27 Paragraph 3 on insult and defamation.

Performing cyberbullying repeatedly is categorized into the element of deliberately. A deliberate act means doing something well or repeatedly and knowing the consequences arising from the act. The use of electronic means by making textual, visual, written or oral communications in the EIT Law comes under the electronic information. When electronic information enters the virtual world by using online computer services, internet services, telephones and other means of communication, it becomes an electronic document.

4.2 Non-Penal Policy on Cyberbullying

4.2. 1 Moral Approach

Efforts to implement moral education can be done by families through education. Borba(2008) in her book "Building Moral Intelligence" explains that families can give 7 moral teachings to children. Moral education is very important considering that cyberbullying is done by insulting and slandering someone in the cyberspace. It is important to teach children to know empathy; to have a conscience, self-control, mutual respect; to nurture goodness; to maintain tolerance; and to be fair. [20] If a moral approach is done, then a sense of a mutual respect and creating a welcoming environment will avoid cyberbullying.

4.2. 2 Technological Approach

Parental control applications and filters are used to protect children's security on the internet. [21] Parental control applications can be installed such as Qustodio, K9 Web Protection, Kakatu and DNS Nawala. [22] The above software is used to know the activities of the children in a cyberspace when connected to the internet such as what sites they often log in. This software gives a warning if the sites visited have a malicious content.

4.2. 3 The Role of Media

4.2.3.1 Television

Appearances of the contain that contain harsh, disrespectful, sexual and other negative views will affect television viewers. [23] Children as spectators will tend to imitate everything seen from the television media. [24] If a child imitates negative things such as abusive, disrespectful, sexual and other negative things, they will trigger actions related to cyberbullying. The reason why many television shows today are less educational is because the censorship institutions, television and Indonesian Broadcasting Commission (KPI) are paying less special attention to this matter. [25] So, television shows are often loaded with a lot of harsh word contents.

4.2.3.2 Online Media

Here are some media on the internet which are often used and visited by children that require parental supervision in order to prevent the occurrence of cyberbullying:

4.2.3.2.1 Social Networking

Cyberbullying cases are dominated by social networking and many suicidal cases occurred due to the introduction of social networking. [26] Any person who uses the network can be a victim of cyberbullying. [27] Most internet users among teenagers use social networks like Facebook, Instagram, and Twitter. [28] The regulation of ethics when using social networking is very important. [29] It is intended that every social network users, especially children can feel safe. Ethical arrangements in everyday life which have been done with a moral approach must also be supported with ethics in the cyberspace.

4.2.3.2.2 Video Web (Video Hosting Service)

Video Web is a video-sharing site provided to share both private and public documentaries. One of the famous video websites is *Youtube*. [30] In a video web like *Youtube*, there is a great number of uploaded videos with contents of insulting and abusive words. Videos with contents of insulting words will have a negative impact on the audience, which are mostly children who often imitate the things seen. [31]

4.2.3.2.3 Online Game

Online games, which are facilitated by communication buttons, open up opportunities for *cyberbullying* if players use harsh words and intimidate other players. One of the most frequent online games which is subject to cyberbullying is a game that comes from the New Zealand called *smallworlds*. [32] This game comes under a friendship category which connects many players from various countries around the world. Providing education to children related to the use of good and safe online games is important to protect them from cyberbullying.

5 Conclusions

5.1 Summary

The findings show that the existing criminal policies in Indonesia on cyberbullying, both in terms of penal and non-penal policies, can be used in overcoming cyberbullying problems. In terms of a penal policy, Indonesia applies the EIT Law No. 19 of 2016 in dealing with cyberbullying with cyberharassment type in article 27 paragraph 3 and cyberstalking in article 27 paragraph 4 after comparing with Arkansas code 2012 § 5-71-217. In terms of a non-penal policy, moral approach, technological approach, and media role are used.

5.2 Suggestions

The penal policy using the EIT Law No. 19 of 2016 should be implemented firmly as a repressive effort in combating cyberbullying. In addition, non-penal policies should take precedence as preventive measures in tackling cyberbullying, meaning that non-penal policies should be improved by using the moral approach, the technological approach, and the role of the media.

References

1. R. Moore, *Cybercrime: Investigating High-Technology Computer Crime*, p. 15 (Anderson Publishing, Mississippi, 2011)
2. D. Juju and F. Sulianta, *Hitam danPutih Facebook,* p. 73 (PT Elex Media Komputindo, Jakarta, 2010)
3. C. D. Marcum, *Cybercrime*, p. 100 (Wolters Kluwer Law & Business, New York, 2015)
4. K. Ayuwuragil, Pengguna Internet IndonesiaCapai 143 Juta. www.cnnindonesia.com/teknologi/pengguna-internet-indonesia-capai 143-juta *appdev*.2018.04.007
5. A. Rosari, Bonus Demografi dan Dampak Terhadap Indonesia. *www.kompasiana.com*/andhinirosari/bonus-demografi-dan-dampak-terhadap-indonesia *appdev*.2018.04.007
6. Sartana & N. Afriyeni, Perilaku Perundungan Maya Cyberbullying Pada Remaja Awal, *Jurnal Psikologi Insight,* Vol.**1**,25-41 (2017)
7. R. Luppicini, *Ethical Impact of Technological Advancements And Applications in Society*, p. 121 (Information Science Reference, United State, 2012)
8. M. Andrew, *Information Technology Law: The Law and Society*, p. 157 (Oxford University Press, New York, 2013)
9. N. Wiener, *Cybernetics or Control and Communication in the Animal and the Machine*, p. 11(MIT Press, Massachusetts, 1961)
10. J. Sitompul, *Cyberspace, Cybercrimes, Cyberlaw: Tinjauan Aspek Hukum Pidana*, p. 4 (Tatanusa, Jakarta, 2012)
11. R.C. Lee and S.C Wong, *Asian America.Net: Ethnicity, Nationalism, and Cyberspace*, p. 3 (Routledge, New York, 2003)
12. J.M. Norman, *From Gutenberg to the Internet: A Sourcebook on the History of Information Technology*, p. 6 (Historyofscience, California, 2005)
13. P.R. Astuti, *Meredam Bullying: 3 Cara Efektif Mengatasi Kekerasan Pada Anak*, p. 3 (Grasindo, Jakarta, 2008)
14. O. Saracho, *Contemporary Perspectives on Research on Bullying and Victimization in Early Childhood Education*, p. 188 (Information Age Publishing, Charlotte, 2016)
15. B. Arrigo, *Encyclopedia of Criminal Justice Ethics*, p. 220-221 (SAGE Publication, California, 2014)
16. S. Chadwick, *Impacts of Cyberbullying, Building Social and Emotional Resilience*, p. 4-5 (Springer, North Ryde Australia, 2014)
17. P. De Cruz, *Perbandingan Sistem Hukum: Common Law, Civil Law, and Socialist Law*, Trans.N. Yusron, p. 4 (Nusa Media, Jakarta, 2016)
18. A. Watson, *Legal Transplants: An Approach to Comparative Law*, p. 16 (University of Georgia Press, Georgia, 1993)
19. W. Menski, *Perbandingan Hukum Dalam Konteks Global Sistem Eropa, Asia, dan Afrika*, p. 10 (Nusa Media, Bandung, 2016)
20. M. Borba, *Membangun Kecerdasan Moral (Tujuh Kebajikan Utama Agar Anak Bermoral Tinggi)*, p. 7-8 (Gramedia Pustaka Utama, Jakarta, 2008)
21. N. A. Hassan, *Digital Privacy and Security Using Windows: A Practical Guide*, p. 84 (Apress, New York, 2017)

22. D.W. Ningrum, Alat Bantu Untuk Pantau Anak Internetan, *https//inet.detik.com/konsultasi-cyberlife/d-3202659/alat-bantu-untuk-pantau-anak-internetan appdev*.2018.04.008
23. M. Chatib, *Orangtuanya Manusia: Melejitkan Potensi dan Kecerdasan Dengan Menghargai Fitrah Setiap Anak*, p. 190 (PT Mizan Pustaka, Bandung, 2012)
24. I. Kartila, Anak-anak Cenderung Meniru Adegan Di Televisi, *https://tekno.kompas.com/2012/05/28/22490165 appdev*.2018.04.009
25. E. Kristian, KPI dan Kemen Kominfo Diminta Tegas agar Stasiun TV Jera, *https://nasional.kompas.com/read/2016/01/28/00034711/KPI.dan.Kemen.Kominfo.Diminta.Tegas.agar.Stasiun.TV.jera appdev*.2018.04.009
26. L. Victor and F. Ana, *Towards Suicide Prevention : Early Detections of Depression on Social Media in Social Media*, p.428-436 (2017)
27. M. Horowitz and D. M. Bollinger, *Cyberbullying in Social Media Within Educational Institutions: Featuring Student, Employee, and Information*, p. 1 (Rowman & Littlefield, London, 2014)
28. K. E. Dill-Shackleford, *How Fantasy Becomes Reality: Information and Entertainment Media in Everday Life*, p. 82 (Oxford University Press, United States,2016)
29. M. Khosrowpour, *Cyber Behavior: Concepts, Methodologies, Tools, and Applications*, p. 934 (Information Science Reference, United States, 2014)
30. J. A. Decesare, *Streaming Video Resources for Teaching, Learning, and Research*, p. 12 (Library Technology Reports, Chicago, 2014)
31. S. Feinstein, *From the Brain to the Classroom: the Encyclopedia of Learning*, p. 515 (Greenwood, California, 2014)
32. E. Reed, Cyberbullying on Smallwords, *appdev*.2018.04.012

The variation of bird diversity in different land cover at oil palm plantation: case study at Asm Oil Palm Estate in Central Kalimantan, Indonesia

Yanto Santosa[1*], *Windi* Sugiharti[2], and *Erniwati* [3]

[1]Ecology and Wildlife Management Division, Department of Forest Resource Conservation and Ecotourism, Faculty of Forestry, Bogor Agricultural University. Jl. Raya Dramaga, Bogor 16680, West Java, Indonesia.
[2]Department of Forest Resource Conservation and Ecotourism, Faculty of Forestry, Bogor Agricultural University. Jl. Raya Dramaga, Bogor 16680, West Java, Indonesia.
[3]Department of Forestry, Faculty of Agriculture, Universitas Bengkulu. Jl. W.R. Supratman, Kota Bengkulu 38371A, Bengkulu, Indonesia. Tel.: +62-73621170.

Abstract. The expansion of oil palm plantation has been claimed as a major driver of biodiversity loss in the tropics, including birds. To date, only a few studies have been conducted to document the variation of bird diversity in different land cover within oil palm plantation. Therefore, it is necessary to conduct the research to investigate the variation of bird diversity. The objective of this study to identify the diversity of bird in different land cover within oil palm plantation. The observation of bird diversity was done in young stand of oil palm, old stand of oil palm, smallholder oil palm and in secondary forest as high stock carbon as natural reference before converting in to oil palm. The survey was done in August 2017 in Central Kalimantan, Indonesia. Data of bird species were recorded by direct observation using transect method. To calculate the diversity value of birds at different land cover, we used Margalef Index (Dmg) and Comunity similarity (IS). The result showed that the number of species and species diversity was highest at high carbon stock area which the number of species was 25 species and Margalef Index was 5.99. The young stand of oil palm have the highest evenness index (0.93), while the highest similarity index was observed in old stand of oil palm and in smallholder oil palm. Our study concludes that the existence of high carbon stock areas within oil palm plantation had a positive influence in increasing bird diversity. From a conservation perspective, retaining of natural habitat such as high carbon stock within oil palm plantation is one strategy to conserve biodiversity.

1 Background

*Corresponding author : yantohaurjaya@yahoo.co.id

Oil palm plantations cover large areas and dominate plantations in Indonesia, especially in Kalimantan and Sumatra. Oil palm is one of the plantation crops that have an important role for the national economy. In 2017, palm oil became Indonesia's largest foreign exchange contributor (US$ 17.8 billion) and it became a source of income for approximately 2.2 million farm households (1).

The development of oil palm plantations has caused a number of problems. Conversion of forest areas into agricultural land has resulted in decreased soil fertility, increased levels of soil erosion, frequent floods, prolonged droughts, local extinction and declining diversity of flora and fauna species, including birds (2).

Birds are wildlife that has the ability to live in almost all habitat types. Birds can occupy a variety of habitat types, both forest habitat and non-forest habitat (3). Birds have high mobility and adaptability to a wide variety of habitat types. According Darmawan (4), birds are one type of wild animal that has many functions and is often used by humans. Birds have important role in the ecosystems for pollination, seed dispersal, and the control of crop pest (5). Hence, the need for research on the diversity of bird species in oil palm plantations. This study aims to identify the diversity of bird species before and after oil palm plantation is established by calculating the similarity of bird communities.

2 Material and Method

2.1. Study Area

This research was conducted in August 2017 in six different land cover namely; young palm oil (planted in 2016), old palm oil (planted in 2013), shrubs, High Carbon Stock area (HCS), and two smallholding oil palm plantations (planted in 2013) at Andalan Sukses Makmur (ASM) Plantation estate in Central Kalimantan. The tools used in this research are binoculars, digital cameras, timers, stationery, tally sheets, Global Positioning System (GPS) and a MacKinnon and Philips bird field guide in Sumatra, Java, Bali and Kalimantan (6). The observed objects include birds located in various land cover and habitat conditions in observation location in PT Andalan Sukses Makmur (ASM) palm oil plantation area.

2.2 Procedure

Primary data were collected through direct observation of bird species diversity and habitat conditions in oil palm plantations. Data were observed using transect method, each land cover had one transect line with 1 km long with 50 m to the left and the right side of the transect line. Observation were conducted in the morning at 06.00 - 09.00 and afternoon at 16.00 - 18.00. The data collection was repeated five times to maximise the total bird recorded. The observation included bird species, number of bird of species, time of encounter and bird activity. To describe the habitat condition, the characteristic of each land cover was recorded during the observation.

2.3 Data analysis

To estimate species richness, we used the Margalef Index (Dmg) which has the ability to respond to differences in species and has high sensitivity (7). The formula used is:

$$\mathrm{Dmg} = \frac{S-1}{\ln(N)} \quad (1)$$

Dmg : Species richness index
S : The number of species
ln : Logarithm
N : The total number of individual of species

To determine the evenness of species richness between site types, we used an Evenness Index:

$$\mathrm{E} = \frac{H'}{\ln S} \quad (2)$$

E : Evenness index
S : The number of species
ln: Logarithm
H': Species diversity index

The similarity of bird communities among different sites was estimated using the Sørensen Index, using the formula :

$$\mathrm{IS} = \frac{2c}{a+b} \quad (3)$$

IS : The similarity Sorensen Index
a : The number of species found in location 1
b : The number of species found in location 2
c : The number of species found in both location

Moreover, we also categorized the conservation status of bird. The conservation status of the birds found in this research is referred to based on the Indonesia regulation about the preservation on wildlife, IUCN criteria and Appendix CITES.

3 Result and Discussion

3.1 Species Bird Composition

Our result showed that there were 39 species of birds from 26 families in all land cover. The highest number of bird species was found at the HCS area with 25 species. A total of 8 species were found at the young stand of oil palm, 13 species at both the smallholder oil palm area and shrub, and 10 species found in the old stand of oil palm and another smallholder oil palm area. Some bird species can be found in all types of land cover (Table 1).

Table 1. List of bird species found in different land cover in the study site

No	Scientific Name (Common name)	Family	SM	ST	KR1	KR2	HCS	SB
1	*Centropus bengalensis* (Lesser coucal)	Cuculidae	√	√	√	√	-	-
2	*Rhipidura javanica* (Sunda pied fantail)	Rhipiduridae	√	-	√	√	√	√
3	*Pycnonotus plumosus* (Olive-winged bulbul)	Pycnonotidae	√	-	-	√	√	√
4	*Pycnonotus goiavier* (Yellow-vented bulbul)	Pycnonotidae	√	√	√	√	√	√
5	*Streptopelia chinensis* (Spotted dove)	Columbidae	√	√	√	√	-	√

Noted: SM= young stand of oil palm, ST= old stand of oil palm, KR=Smallholder oil palm, HCS=High Carbon Stock area, SB=Shrub area

Pycnonotus plumosus and *Pycnonotus goiavier* are birds of the Pycnonotidae family. According to Partasasmita (8), bird species of the Pycnonotidae family have high adaptability in various types of vegetation, and are highly tolerant of vegetation changes. These bird species can be found in secondary forest habitat types, shrubs, farmland and even rural and urban environments. *Centropus bengalensis* are almost found in all land cover types. Sopiyandi (9) and Permana (10) stated that *Centropus bengalensis* is a type of bird that prefer to be in the open area habitat such as oil palm plantations.

Rhipidura javanica is a type of bird that is active in secondary forest, yard and mangrove forests. Therefore, it is common to find this bird species in almost all land cover in oil palm plantations. The bird species found mostly were insectivorous. The large number of insectivorous birds might be because the of the abundant ground cover as source of food (insects) for the birds. The shrub areas had higher insect diversity compared to other types of land cover (11). *Centropus bengalensis* are insectivorous bird species. Therefore, this bird plays a role as a natural enemy of insects in oil palm plantations.

The comparison of Number of species, species richness and the similarity of bird species

Our result showed that High Conservation Stock area had the highest number of species (25 species) and species richness (Dmg;5.99) compared to other land covers, while the lowest number of species and species richness was at young stand of oil palm with 8 species and the index of species richness was recorded at 2.52. (Figure 1). The highest similarity index (0.93) were recorded in young stand of oil palm.

 https://doi.org/10.1051/e3sconf/20185200051

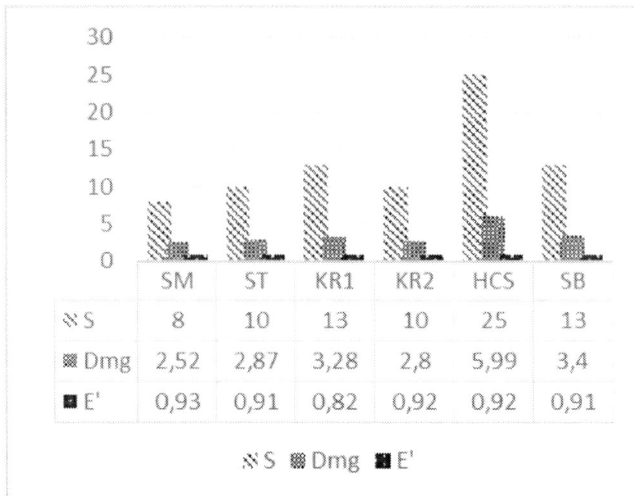

Noted : SM= young stand of oil palm, ST= old stand of oil palm,
KR=Smallholder oil palm, HCS=high carbon stock, SB=Shrubs area

Fig. 1. Number of species (S), species richness (Dmg) and Evenness Index (E') in each land cover.

High conservation stock area had the highest number of bird species and species richness due to its complex vegetation or habitat structure compared to oil palm plantations. The habitat diversity affects bird species diversity. The more diverse the habitat structure the greater the diversity of species (4). These results are similar to studies conducted by Erniwati *et al* in Riau Province (12) which showed that secondary forests had higher number of bird species compared to oil palm plantations. The number of species and species richness found at shrubs and smallholder oil palm mostly were mainly open habitat species. The existence of those species was influenced by condition of the area and the proximity to the to HCS areas. Yoza (13) stated that forests have more complex stratification, thus providing strata variation for bird habitats, as well as relatively heterogeneous composition of vegetation creating varied ecological niches. The types of birds in the young palm are open bird species (*C. bengalensis*, *P. goiavier*, *R. javanica* and insect-feeders) which have adapted to oil palm areas having overgrown shrubs as a source of food.

The Evenness Index in each site ranged from 0.82-0.93 (Figure 1). According to the concept of evenness, if the index of evenness is close to 1, the species are distributed evenly. The data in Figure 1 indicates that the bird species were distributed almost evenly among sites. The highest Evenness Index was found in young age of oil palm (E'=0.93), whereas the lowest Evenness Index was found insmallholder oil palm (E'=0.82). According to Ludwig and Reynolds (14), the value of evenness is influenced by the number of individuals. Santosa (15) stated that if each species has the same number of individuals, it indicates that the evenness of the community has a maximum value, but if the number of individuals in each species is much different, it shows that the evenness of the species have a minimum value.

The similarity of bird community in different site

The index of bird community similarity assesses the similarity in species composition between habitats. The results show that the highest similarity found between old stand of oil palm (ST) and smallholder (KR) oil palm about 0.61. This could be influenced by the presence of old stand of oil palm adjacent to smallholder oil palm and the age of oil palm is the same from both location. Keindeigh (16) said that the factors causing the similarity of species between the two habitats are the distance of adjacent habitat, the similar vegetation composition and other environmental factors.

While the lowest level of similarity found between old stand of oil palm and HCS area is about 0.23, it might be due to differences in vegetation structure, therefore the types of birds found are very different. Ganevan (17) stated that the similarity of bird communities is caused by the same characteristic of vegetation composition and altitudes. (Table 2)

Table 2. The similarity index of bird community in the study site

Land cover	SM	ST	KR1	KR2	HCS	SB
SM	1	0.44	0.48	0.56	0.24	0.57
ST		1	0.61	0.3	0.23	0.43
KR1			1	0.43	0.42	0.38
KR2				1	0.34	0.35
HCS					1	0.32
SB						1

Noted : SM= young stand of oil palm, ST= old stand of oil palm, KR1&2=Smallholder oil palm, HCS=High Carbon Stock, SB=Shrub area

Using the IUCN criteria, there was one species that is classified as Endangered (EN), one species classified as Vulnerable (VU) and two species categorized as Near Threatened (NT). There were a total of ten different bird species that are protected by Indonesian law and five different species listed in Appendix II of CITES. The protected bird species such as hawk eagle (*Elanus caeruleus*) and Buccane eagle (*Spilornis cheela*) from the family of Accipitridae play a role in the ecosystem as bird of prey. According to Sozer *et al.* (5) several species of bird of prey has a function to maintain the balance of ecosystems of pest populations of rats and populations of snakes. Furthermore, the bird species of the Alcedinidae family are protected because they belong to fish-eating birds.The Alcedinidae family itself is used as an indicator of a habitat (4). The bird species along with their conservation status and their distribution by habitat are presented in Table 3.

Table 3. Conservation status of birds in the study area

No	Scientific Name	Indonesian Law	IUCN	CITES
1	*Microhierax fringillarius*	**D**	LC	-
2	*Ciconia stormi*	**D**	**EN**	-
3	*Leptoptilos javanicus*	**D**	**VU**	-
4	*Psittacula longicauda*	TD	**NT**	**II**
5	*Halcyon smyrnensis*	**D**	LC	-
6	*Elanus caeruleus*	**D**	LC	**II**
7	*Spilornis cheela*	**D**	LC	**II**
8	*Rhipidura javanica*	**D**	LC	-
9	*Arachnothera longirostra*	**D**	LC	-
10	*Pericrocotus igneus*	TD	**NT**	-
11	*Loriculus galgulus*	TD	LC	**II**
12	*Gracula religiosa*	**D**	LC	**II**
13	*Anthracoceros albirostris*	**D**	LC	-

Note: Indonesian law number 7 year 1999 about preserving planta and animal species. IUCN = International Union for Conservation of Nature, CITES = The Convention on International Trade in Endangered Species of Wild Fauna dan Flora. D = Protected, TD = Not Protected , LC = Least Concern, EN = Endangered, VU = Vulnerable, NT = Near Threatened

4 Conclusion

The number of spesies and species diversity was highest at HCS area with 25 species recorded with Margalef Index 5.99. The young stand of oil palm have the highest Evenness Index (0.93), while the highest similarity index is in old stand of oil palm and in smallholder oil palm. Our study concluded that oil palm establishment have reduced the bird species diversity. However, the existence of HCS area within oil palm plantations have increased the bird diversity. From a conservation perspective, retaining natural habitats such as high carbon stock areas within oil palm plantation is one strategy to conserve biodiversity. Moreover, the plantation estate should understand the importance of bird species diversity for ecosystem in plantations, especially bird species which are useful as natural pesticides, thereby reducing costs incurred to eradicate oil palm pests.

References

1. Directorate generale of plantations. Tree Crop Estate Statistic of Indonesia 2015-2017. Central Bureau of Statistic, Jakarta.(2016).
2. E. Ruswenti, W. Novarino, Rizaldi. Jenis-jenis burung di perkebunan kelapa sawit PT. Andalas Wahana Berjaya (AWB), Kabupaten Dharmasraya, Sumatera Barat. *Jurnal Biologi Universitas Andalas*. Vol **3**(3) : 219-226 (2014)
3. H.S. Alikodra, *Pengelolaan Satwaliar*. Forestry Faculty Bogor Agricultural University (2002).
4. M.P. Darmawan, Keanekaragaman Jenis Burung Pada Beberapa Tipe Habitat Di Hutan Lindung Gunung Lumut Kalimantan Timur. Forestry Faculty Bogor Agricultural University (2006)
5. R. Sozer, Y. Saaroni, P.F. Nurwatha, *Jenis-jenis burung dilindungi yang sering diperdagangkan* (Yayasan Pribumi Alam Lestari, Bandung, 1999)
6. J. MacKinnon, K. Phillipps, B. van Balen, *Burung-burung di Sumatra, Jawa, Bali dan Kalimantan*. Bogor (ID): Burung Indonesia. (2010).
7. A. Magurran, *Ecological Diversity and Its Measurement*, (Croom Helmed Limited, London, 1988)
8. R. Partasasmita, Komunitas Burung Pemakan Buah Di Panaruban, Subang : Ekologi Makan dan Penyebaran Biji Tumbuhan Semak, Bogor Agricultural University (2009).
9. S. Sopiyandi, Keanekaragaman Burung Pada Berbagai Tipe Tutupan Lahan Di Perkebunan Kelapa Sawit PT Mitra Unggul Pusaka Kabupaten Pelawanan, Riau. Bogor Agricultural University (2016).
10. M.F. Permana, Keanekaragaman jenis burung pada beberapa tipe habitat di hutan Lambusango, pulau Buton, Sulawesi Tenggara, Bogor Agricultural University (2016).
11. J. Setiawan, Keanekaragaman Jenis Burung di Areal PT Indocement Tunggal Prakarsa Tbk. Pabrik Palimanan. Bogor Agricultural University (2013).

12. Erniwati, E.A.M. Zuhud, Y. Santosa, I. Anas, The value of secondary forest patches for bird conservation in palm oil landscapes of Riau, Sumatra, *Biodiversitas* (**17**). (2016)

13. D. Yoza, Keanekaragaman jenis burung di berbagai tipe daerah tepi *(edge)* Taman Hutan Raya Sultan Syarif Hasyim Propinsi Riau. [Tesis]. Bogor Agricultural University (2006).

14. J.A. Ludwig, J.F. Reynolds, *Statistical Ecology: A Primer on Methods and Computing*, (John Wily and Sons, New York, 1988).

15. Y. Santosa, *Teknik Pengukuran Keanekaragaman Satwaliar*. Jurusan Konservasi Sumberdaya Hutan Fakultas Kehutanan-Institut Pertanian Bogor, Bogor. (Unpublished). (1995)

16. S.C. Kendeigh, *Ecology with Special Reference to Animal and Men,* (Prentice Hall of India Private Limited, New Delhi, 1980).

17. G. Ganevan, Keanekaragaman jenis burung pada lima ketinggian tempat di Resort Gunung Putri Taman Nasional Gunung Gede Pangrango Bogor Agricultural University (2016).

Certification of nagari assets in west sumatra

Yontri Faisal* and *Muhammad Yamin* Lubis

Law Program, University of Sumatera Utara, Jl. Abdul Hakim No. 4, Medan, 20155, North Sumatera, Indonesia

Abstract. Nagari in West Sumatra is the same as "customary village" (Article 7 of Act Number 6 of 2014 concerning the Village) which is the spearhead of the lowest administrative region. Nagari as the lowest government also has an assets, both in the form of land and others form. The legal certainty of the Nagari asset in the form of land is held through land certification. In general, Nagari assets are communal lands. with the empowerment of Nagari assets in West Sumatera, through the Regional Government Budget and Expenditure, will provide wider space for the Nagari in utilizing its assets, in an effort to improve the Nagari economy, or at the same time to provide legal certainty regarding land assets. This research use theory of legal certainty. The conclusion of this research, concerning the certainty of certification of Nagari assets in the form of communal land has not provided legal certainty, because communal land does not include land certification object.

Keywords: Legal Certainty, Certification of Communal Land, Nagari Asset.

1 Introduction

Based on the West Sumatera Regional Regulation Number 9 of 2000 concerning the Basic Provisions of the Nagari Government that have been replaced by Regional Regulation Number 2 of 2007 concerning the Principles of Nagari Government. This regulation as an application of Act Number. 22 of 1999 concerning Regional Government which has also been replaced by Act Number 32 of 2004, then Act Number 23 of 2014, then changed with Government Regulation in Liew of Legislation Act Number 2 of 2014 , and last changed with Act Number 9 of 2015 concerning Regional Government.

Fresh Winds concerning the Government of Nagari, since the issuance of Act Number 6 of 2014 concerning the Village which then followed up with some provisions of implementers such as Government Regulation Number 43 of 2014 concerning the Implementation of Act Number 6 of 2014 concerning the Village, which then changed with Government Regulation Number 47 of 2015 concerning the amendment of Government Regulation Number 43 of 2014 concerning the implementation of Act Number 6 of 2014 and Regulation of the Minister of Home Affairs of the Republic of Indonesia Number 83 of 2015 concerning Appointment and Dismissal of Village Devices. then Regulation of the

*Corresponding Author:yontrifaisal3@gmail.com

Minister of Home Affairs of the Republic of Indonesia Number 1 of 2016 concerning Utilization of Village Assets.

The total area of West Sumatra Province is 42,200 km^2 or equal to 2.21% of the total area of Indonesia consisting of 19 second level regions, 648 nagari. Nagari administration in West Sumatra has been known for a long time, with the governance system of Koto Piliang and Bodi Caniago. The formation of Nagari was once determined to be at least four tribes (Nagari Baampek Tribe), with the territory determined first. Thus the existence of nagari and communal system is inherent in Minangkabau culture.[1]

There are 4 Functions of Nagari Government that need to be integrated into Regional Regulation Number 2 of 2007 that is:

1. Nagari as the lowest government unit of Indonesia is equivalent to the Village
2. Nagari as a unit of unitary custom and socio-cultural unity in order to embody the cultural philosophy*Adat Basandi Syarak and Syarak Basandi Kitabullah* (Custom Based on Sharia, Sharia Based on Kitabullah).
3. Nagari as a unit of communal land as a Nagari asset and as a unit of economic based economic entity.Nagari is a legal entity. Nagari can also try in the form of Nagari-owned enterprises or cooperative with the principles of sharia, in addition to the efforts of each villagers
4. Nagari as a unit of security and security units that autonomously can safeguard and protect the people and the wealth of Nagari. The existence of Custom Density of Nagari in the Regional Regulation Number 9 of 2000, replaced by Custom Density and Sharia with the principle of *Adat Basandi Syarak and Syarak Basandi Kitabullah* (Custom Based on Sharia, Sharia Based on Kitabullah). The members of custom density an sharia consist of elements of TTS (Tungku nan Tigo Sajarangan, Tali nan Tigo Sapilin) that is ninik-mamak, Ulama and the scholars plus elements of women and youth.

Nagari assets in the form of land, generally are communal land (tanah ulayat), which consists of three levels of ulayat nagari, tribe and kaum. The ulayat nagari can not be transferred to its ownership. In relation to the ulayat nagari land that has been transferred to the Right of Exploitation from any party needs to be returned to Nagari. Because the Right of Exploitation also can not come from communal rights (Article 28 Agrarian Act).

In Article 1 Paragraph (1) of the Village Act, it is affirmed that the village and customary village or other so-called names, hereinafter referred to as the Village, is a legal community unit having the territorial boundaries to regulate and administer government affairs, the interest of the local community based on the initiative community, origin rights, and/or traditional rights recognized and respected within the system of government of the Unitary State of the Republic of Indonesia. In the paragraph (2) it is mentioned that the Village Government is the administration of the affairs of the government and the interests of the local people in the system of government of the Unitary State of the Republic of Indonesia or Integral Village with the government of the Republic of Indonesia.

Nagari asset management is a series of activities ranging from planning, procurement, utilization, security, maintenance, deletion, transfer, handling, reporting, assessment, guidance, supervision and control of village/Nagari assets.[2] Management of Nagari's government assets have the same provisions with the management of government assets, both in Central Government, Provincial Government, Regency Government and City Government.

In the case of assets in the form of land must be proven by the certificate to ensure legal certainty.[3] The land in question is the surface of the earth, with some types of rights on it (Article 16 a Agrarian Act)[4] which in its use includes also some of the earth's body beneath it and part of the space above it.[5] The implementation of Nagari Government in relation to the land assets (Article 76 paragraph (1) of the Agrarian Act) states that the assets of the Village can be a Village cash, a communal land, ... etc, managed by the Village under the leadership of the Village Head, assisted by the Village Secretary and the asset management.[6] Village assets/Nagari that are strategic one of them is communal land.[7] Other village assets can also be derived from the property owned by the Government and the village-level Regional Government in the village may be granted ownership to the village (Article 76 paragraph (3), in Article 76 paragraph (4), (5) and (6) village's asset the form of village property which form of land must be certified on behalf of the Village Government. The Law concept of communal land certification in the provision of legal certainty concerning Nagari's assets, in the management and development of nagari is very necessary.

2 Formulation of the Problem

The problem of this research is formulated as: how Certification of Nagari asset interacts with customary land rights?

3 Research Method

This research is empirical juridical research. Research based on field data,[8] besides that also use normative juridical method using secondary data from library.[9] Library data is collected from books, regulations and other documentation provided in the library, or private collection.[10] Legal material collected through libraries includes primary legal materials, secondary legal materials, and tertiary legal materials, such as dictionaries, encyclopedias, legal magazines, etc. This study is descriptive, with data collection techniques through interviews with respondents who were determined purposively. The determination of the location of the research is also purposive, with the criteria of Minangkabau origin region, namely Luhak Nan Tou that is Tanah Datar, Luhak Tangah that is Agam Regency, and Luhak Nan Bungsu ie Lima Puluh Kota Regency. Then the Rantau area of Padang City. The nagari's identification is also done by purposive, Nagari Parahiyangan, Nagari Kapau in Luhak Agam, and Nagari Tarantang and Harau in Luhak Nan Bungsu, and the study uses a population that is a set of objects with the same characteristics,[11] and in research is certification of nagari assets.[12] The sample is a smaller part of a population.

4 Literature Review

Legal certainty as one of the objectives of law can be said as part of the effort to realize justice. The real form of legal certainty is the enforcement or enforcement of a law regardless of who does. With the legal certainty everyone can predict what will be experienced if a certain legal action. Certainty is necessary to realize the principle of equality before the law without discrimination.[13]

Certainty is a feature that can not be separated from the law, especially for written legal norms. The law without a certainty value will lose meaning because it can no longer be used as a behavioral guide for everyone. Certainty itself is referred to as one of the goals of the law. The word "certainty" is closely related to the principle of truth, which is something that can be strictly legalized. Through deductive logic, the rules of positive law are placed as major premises, whereas concrete events become minor premises. Through a closed logic system will immediately be obtained conclusion. The conclusion must be something that can be predicted, so everyone is obliged to hold on to it. It is with this grip that society becomes orderly. Therefore, certainty will lead people to the order.[14]

5 Result and Discussion

5.1 Land Registration

Objects of land registration are Right of Ownership, Right of Exploitation, Right of Building, Right of Use, Right of Management, Rights of Waqf, exclusively for state land not issued certificates, but communal land excluding land registration objects.This registration includes:

1. Measurement of land mapping and accounting;
2. Registration of land rights and transfers of those rights;
3. The provision of letters of proof of rights, which applies as a powerful evidentiary instrument.[15]

The certificate is a land title certificate. Registration is an appropriate tool to provide a description and identification of the soil which is the continous recording.[16]The data in the Land Registration Office (LRO) are grouped into 2 (two) parts:[17]

1. Juridical Data Group, which collects data on the name of the right to the land, who the holder, the transfer and charging if any, all of these are collected in the Land Book.
2. Group of Physical Data that is collecting data about the location of the land where, the length/width of the soil and the boundaries of the land are all collected in the Measure Letter.

The implementation of land registration is done by the National Land Agency. Land registration is also done by adat principle principle that is done with clear, real and light.[18] In the registration provisions, the registration object does not include communal land.

Regional assets/property is an important element in the framework of government administration and service to the community.[19] For that, regional governments are required to manage regional wealth professionally, transparently, accountably, efficiently and effectively starting from the planning, distribution and utilization and supervision.[20] For the management of fixed goods in the form of land there must be proof in the form of land certificate in accordance with the type of rights assigned to it.Government assets in the form of land must be certified on behalf of the Regional Government.

One of the difficulties in managing regional government assets is due to the status of the land, its use, the element of interest to the land. The land inventory of an administrated regional government comes from the Dutch government whose land can be bought, forcibly taken, and / or rented. Besides, there are also assets that are not easily traced evidence of ownership.[21]

Article 2 paragraph (4) of Agrarian Act affirms that the right to control the state can be authorized to regions and the communal peoples. Later in Government Regulation Number 40 of 1996, it is stipulated that for the government the right to use and the right of management shall be granted, with the right of management rights being the right of control as stipulated in Article 2 paragraph (4) and be given proof of their right in the form of a certificate.

Land procurement for the government may be used solely for its duties and functions but may also be granted to third parties if its rights are the right of management.Customary rights, although recognized in the Agrarian Act but do not have legal certainty, because it can not be certified with the existing provisions. Use rights and management rights for regional governments can not be granted on customary land, with respect to which communal land certification as a regional government asset requires in depth assessment.

5.2Relation Between Certification of Nagari Asset and the Communal Land Law

Result of recapitulation at West Sumatera, it can be seen that the field of land that has been registered until 2015 only 672983 fields, with an area of 1055692.37 ha. The details of the types of rights are property rights, tenure, building use rights, use rights, management rights, property rights of apartment units, and Waqf land.

Of this amount is seen for private and public use rights of 5901 areas with an area of 292707.94 ha, and the right to manage a total of 55 areas with an area of 775.87 ha. This amount has not separated the regional government land that is licensed to use Rights of Use and Rights of Management. West Sumatra area of 42,297,30 km², newly registered area of 1055692.37 ha, unknown how much communal land, and other land.

Land Procurement For the Land Needs of Local Goverment from communal land, most do not have any rights base, only once given by the community either with or without certificate, while national land agency also did not do the mapping of the certification. Over time, grandchildren appear materilinial and land use situation is quite significant, then certification of regional assets become constrained. There are also those who have been given a kind of compensation, but since there is no written proof, the child's nephew declared that the act never happened.

The procurement of land derived from communal lands is considered to be the same as other rights, often resulting in cospic contradictions, values and norms of the right to land for regional governments. The regional government's asset protection from communal land if not reused does not return to communal land, state land. Land certification of regional government asset, especially the nagari's asset which derived from communal land has not been well mapped in West Sumatera. Recapitulation of land registration in West Sumatra, but not visible communal land, but in fact most of the land is communal. From the list of land in West Sumatra seen a number of 672983 with an area of 1055,692.37 ha, no detailed of communal land.

In the Agrarian Act, expressly recognized customary rights (ulayat rights) (Article 3 Agrarian Act), whereas public legal entities are only possible acquisition of their rights from state lands, not from communal Land. With the Village Act, the nagari administration can propose the certification land to be assigned to its duties and functions with a request for rights. Certification of nagari assets in the form of ulayat right with Government Regulation Number 24 of 1997, destroyed the concept of customary rights themselves. If the implementation of the nagari government lies on customary land of society, the nagari may be facilitated by liberation and/or by grant. Likewise, the nagari administration needs land which has been certificated with individual rights, it can be done by the acquisition of land (Act Number 2 of 2012).[22] Nagari's assets/property is one of the important elements in the administration of government and services to the people in the nagari, therefore Nagari government is required to manage the Nagari wealth professionally, transparently, accountably, efficiently and effectively starting from planning, distribution and utilization and its supervision.[23]

With regard to legal certainty Nagari's assets in the form of land must be certified according to Article 19 of the Agrarian Act. through Government Regulation number 24 of 1997 concerning land registration. The certificate is a proof of legal certainty of Nagari land rights. According to Government Regulation Number 24 of 1997, the types of land rights that are required to be registered are listed, but excluding communal land under Article 3 of the Agrarian Act. Registration is an appropriate tool that provides the description and identification of the land as well as continous recording (continuous recording rather than land rights).[24]In Article 19 of the Agrarian Act it is stated that all land rights must be registered, of course, with the assumption including ulayat land (Article 3 Agrarian Act), but in Article 9 of Government Regulation number 24 of 1997, no ulayat land is seen as the object of land registration. This shows that ulayat land is not a land object that must be guaranteed legal certainty

If communal land is not included in the land registration object, will communal land be registered? Which legal basis will be used? Is it assumed that the ulayat right is the same or equivalent of State land or indivual land, while its existence is expressly regulated in a separate Article in the Agrarian Act. The concept of communal land in Agrarian Act that does not develop in organic rules, then today often lead to conflict. The emergence of the provisions on the Village and to provide legal certainty quiet assets in the form of land becomes become blurred. In other words, nagari's land assets are ulayat increasingly leading to legal uncertainty, if the legal certainty is only proven by the certificate of land rights.

The reality in West Sumatra, Nagari Parahiyangan is awarded as one of the best villages in the world, but its nagari land assets are not registered. Likewise, Kapau Nagari, even with its own uniqueness, no one one citizen allowed to certifying their land in the National Land Agency, but it recorded in its nagari's asset. in Nagari Tarantang and Nagari Harau, formerly a Nagari because it is located in the area of Lima Puluh Kota then the certificate is still on behalf of the Regional Government of Lima Puluh City, while Nagari Harau has not been certified yet. Besides the asset of Village in Batang Kabung Ganting Village, Padang City even though already in the city area but still not certified.

Concerning the asset nagari in Nagari Parahiyangan, Nagari Kapau, and Nagari Harau, declared nagari asset in the form of land is communal land, but none of them are certified. However, in the report of the nagari asset, it is shown that the land is used for the wali nagari's office. In this case, although the legal certainty of asset in the form of certificate is not owned, but there is no legal certainty concerning the nagari asset, because the certainty is based on the sincere and voluntary submission of the owner of the customary right, as long as it is used for nagari governance.

Based on West Sumatera Regional Regulation Number 2 of 2007, nagari is a legal entity, which is integral with the government of Indonesia, the land assets are managed similarly to the management of the land asset of the City Government, the Regency Government, the Provincial Government.

In the Village Act it was suggested that one of the origin of the village/Nagari land asset is communal land. It is seen that the Village Act is also carelessly incorporating village assets with communal lands. Concerning the Nagari asset in West Sumatra does not need a certification of communal land asset, but still guarantees Nagari governance regarding its uncertified assets.

Implementation of the communal land certification of Nagari's assets in the provision of legal certainty. Although the concept of land registration of Nagari's assets derived from communal land, its registration. In West Sumatra Regional Regulation Number 6 of 2008, the registration of communal land is carried out under Government Regulation Number 24 of 1997, which does not mention ulayat land as the object of land registration.

As a result of West Sumatera Regional Regulation of the Ulayat Right of West Sumatra which impose the customary rights of the village is registered with Government Regulation Number 24 of1997, so that it can be granted with the concept of communal right in Article 3 of Agrarian Act.

6 Conclusions

Registration is an appropriate tool that provides the description and identification of the land as well as continous recording (continuous recording rather than land rights).[24]In Article 19 of the Agrarian Act it is stated that all land rights must be registered, of course, with the assumption including ulayat land (Article 3 Agrarian Act), but in Article 9 of Government Regulation number 24 of 1997, no ulayat land is seen as the object of land registration. This shows that ulayat land is not a land object that must be guaranteed legal certainty

The concept of communal land in Agrarian Act that does not develop in organic rules, then today often lead to conflict. The emergence of the provisions on the Village and to provide legal certainty quiet assets in the form of land becomes become blurred. In other words, nagari's land assets are ulayat increasingly leading to legal uncertainty, if the legal certainty is only proven by the certificate of land rights.

In the land certificate of regional government assets derived from communal land, tribes and nagari through land procurement such as do not encounter significant obstacles, but juridically contradict with the concept of value and norm of Communal Right which recognized by Agrarian Act. Juridically, the legal certainty concerning Nagari's assets in the form of communal land must be registered with the provision of land registration (Government Regulation Number 24 of 1997) and proven by the certificate of land rights. But nagari's assets derived from ulayat land can not be registered or certified.

References

1. Y. Mirwati, *Makalah dalam Pendaftaran Tanah Aset Pemerintah di Kabupaten Solok*, p. 10(Sumbar, Koto Baru Solok 2016a)
2. Article 4 Paragraph (6) IndonesianRegulation of the Minister of Home Affairs Number 1 Year 2016
3. Article 4 paragraph (1) Indonesian Act Number 5 of 1960 concerning Basic Regulation on agrarian principles
4. Article 16 paragraph (1) Indonesian Act Number 5 of 1960 concerning Basic Regulation on agrarian principles
5. B. Harsono, *Hukum Agraria Indonesia-Sejarah Pembentukan UUPA*, p.265(Djabatan, Jakarta,1999a)
6. Article 4 Paragraph (5) IndonesianRegulation of the Minister of Home Affairs Number 1 Year 2016
7. Article 4 Paragraph (3) IndonesianRegulation of the Minister of Home Affairs Number 1 Year 2016
8. H. Sutrisnio, *Metodelogi Research jilid 1*,p. 70(Andi Offset, Yogyakarta,1989a)
9. Suratman, *Metode Penelitian Hukum*, p. 52 (Alfabeta, Bandung, 2012)
10.]H. Hadikusuma, *Metode Pembuatan Kertas Kerja atau Skripsi Ilmu*, p, 44 (1995)
11. H. Sutrisnio, *Metodelogi Research jilid 1*,p. 70(Andi Offset, Yogyakarta,1989b)
12. H. Sutrisnio, *Metodelogi Research jilid 1*,p. 40(Andi Offset, Yogyakarta,1989c)
13. S Rahardjo, *Ilmu Hukum,* p. 20 (Citra Aditya Bakti, Bandung 1991a)
14. S Rahardjo, *Ilmu Hukum,* p. 20 (Citra Aditya Bakti, Bandung 1991b)

15. B. Effendi, *Pendaftaran Tanah di Indonesia*, p.14(Raja Wali, Jakarta, 1993)
16. A.P. Parlindungan, *Pendaftaran Tanah Di Indonesia*. P.18 (Mandar Maju, Bandung, 1999a)
17. B. Harsono, *Hukum Agraria Indonesia-Sejarah Pembentukan UUPA*, p.155(Djambatan, Jakarta,1999b)
18. Y. Mirwati, *Makalah dalam Pendaftaran Tanah Aset Pemerintah di Kabupaten Solok*, p. 10(Sumbar, Koto Baru Solok 2016b)
19. M.Yusuf,*Delapan langkah Pengelolaan Aset Daerah Menuju Pengelolaan Keuangan Daerah Terbaik*, p. 11(Salemba Empat, Jakarta, 2010)
20. C. Soleh, & H. Rochmansjah,*Pengelolaan Keuangan dan Aset Daerah; Sebuah Pendekatan Struktural Menuju Tata Kelola Pemerintahan Yang Baik*,Cetakan Edisi Kedua. p. 167(Fokusmedia, Bandung, 2010a)
21. [l]C. Soleh, & H. Rochmansjah,*Pengelolaan Keuangan dan Aset Daerah; Sebuah Pendekatan Struktural Menuju Tata Kelola Pemerintahan Yang Baik*,Cetakan Edisi Kedua. p. 167(Fokusmedia, Bandung, 2010b)
22. M.Yusuf,*Delapan langkah Pengelolaan Aset Daerah Menuju Pengelolaan Keuangan Daerah Terbaik*, p. 11(Salemba Empat, Jakarta, 2010b)
23. C. Soleh, & H. Rochmansjah,*Pengelolaan Keuangan dan Aset Daerah; Sebuah Pendekatan Struktural Menuju Tata Kelola Pemerintahan Yang Baik*,Cetakan Edisi Kedua. p. 167(Fokusmedia, Bandung, 2010c)
24. A.P. Parlindungan, *Pendaftaran Tanah Di Indonesia*. P.18 (Mandar Maju, Bandung, 1999b)
25. A.P. Parlindungan, *Pendaftaran Tanah Di Indonesia*. P.18 (Mandar Maju, Bandung, 1999b)

Land conversion to the poor people in the perspective of utilitarianism theory by bentham

Yusrizal[*] and *Muhammad Yamin* Lubis
Law Program, University of Sumatera Utara, Indonesia

Abstract. The politics of Indonesia's national land law is based on the principle of fulfilling the constitutional rights of every citizen of the land. The land according to Indonesian positive law is used for the general welfare of Indonesian People. In 2001, with the issuance of MPR RI Decree No.IX / MPR-RI / 2001 concerning Agrarian Reform and Natural Resource Management is the momentum of the agrarian reform program. Agrarian reform is implemented gradually by allocating land for the poor. The problem is: the distribution of the land is not right on target, the existence of a land mafia, bribery of the National Land Agency and the annexation of community land by the investor. Karo Regency became one of the areas which received land allocation from conversion forest.This research method is empirical legal research. This research uses primary data. This research use legal benefit theory (utilitarian) by Jeremy Bentham. Based on legal benefit theory, that agrarian reform by giving land conversion to the poor is an alternative effort of government to give decent living and fulfillment of constitutional rights of Indonesian citizen and eradicate poverty.

1 Introduction

Land is a very basic necessity of human life. Humans live and perform activities on the land, so that every time humans are always associated with the land, can be said almost all human life activities either directly or indirectly always need the land. Even after death humans still need the land as a burial place. Once the importance of the land for human life, then everyone will always try to own and control the land.

Due to the increasing need of people to own the land, efforts are needed to optimize the use or utilization of the land in accordance with its ability and attention to the environment. It is necessary to have a land use planning, land use arrangement, land rights management, provision of data or land registration maps with mapping and land registration measurement activities.[1]

Land has a great role in the dynamics of development, then in the 1945 Constitution Article 33 paragraph (3) mentioned that the earth, water and natural resources contained therein are controlled by the state and used for the greatest prosperity of the people. The provisions on the land can also be seen in Act Number 5 of 1960 concerning the Basic Regulations of Agrarian Principles or what we usually call the Agrarian Act.

* Corresponding author: yusrizal_legal@hotmail.com

Talking about the definition of the land we will revolve around Act Number 5 of 1960, but in the provisions of the legislation is not mentioned in detail about the definition of the land. Act Number 5 of 1960 on Article 1 only mentioned about earth, water and space is an inseparable unity which is also known as a concept of understanding of the definition of the Archipelagic Insight. According to A.P. Parlindungan, land means "Surface of the Earth".[2]

The surface of the earth includes water, land and space as well as what is contained therein. This is described because the land as an object given the right of ownership or ownership over it, is not limited only to the definition of the land in the true sense, but also includes visible water (eg ponds above the ground) and water in the soil (such as water obtained from within the soil through a well) is owned by a person holding the rights to the land above the surface of the earth.

The definition of "control" can be used in a physical sense, also in a juridical sense. Juridical definition are based on rights protected by law and generally authorize the right-holders to physically occupy the land in custody. The definition of control is used in the civil sense In the 1945 Constitution and Agrarian Act the definition of "controlled" is used in a public sense, as defined in Article 2 of the Agrarian Act. the definition consist of land rights such as property rights (Article 20), the right to use and the right to collateral for land called mortgages (Articles 23, 33, 39 and 51). While the right to control the state that the public law is the right to control the state that covers all the land without any exceptions.

The legal relationship between the State and the land gives birth to the right to control the land rights by the State. The relationship between indigenous and tribal peoples with their communal lands gave birth to customary rights and a combination of individuals and lands gave birth an individual rights to the land. Ideally, the relationship of the three rights (the right to control the land by the State, customary rights, and individual rights to the land) is harmoniously and equitably aligned, meaning that the three rights are equal and strength and not mutually harmful. However, legislation in Indonesia gives great power and unclear limits to the State to control all the land in the territory of Indonesia.

One of the fundamental perspectives of land management is that all land rights have social functions (Article 6 Agrarian Act), the management of land is in principle a Government affair. Therefore, the social function of land rights can be set forth in the form of policy, regulation, control and supervision, and service. In carrying out these social missions the government considers the availability of land, to be able to meet the needs of stakeholders, justice for all the people, legal certainty and protection for holders of land rights, and sustainability in services, availability and ecosystems.[3]

With the existence of the National Land Law is expected to create legal certainty in Indonesia. To achieve this objective, the government implemented it with the provision of written legal instruments in the form of other regulations in the field of national land law which support legal certainty and implementing the law enforcement through the existing regulatory instruments in the form of effective land registration.

According to Article 9 of Government Regulation Number 24 of 1997 concerning Land Registry, which can be the object of land registration are:

1. Land parcels having the status of hak milik (right of ownership), hak guna usaha (right to cultivate), hak guna bangunan (right of use building), and hak pakai (right of use);
2. Land having the status of hak pengelolaan (right of management);
3. Wakaf land (edified land/land donated for religious purposes);
4. An apartment ownership right;
5. Security title;
6. State land.

In reality there is still the right of eigendom, opstal rights, erfpacht rights and indigenous rights subject to customary law which have no written evidence, which local residents have often called customary lands such as communal land, customary land, Yasan land, gogolan Land, etc.Under the terms of Article 9 above, it is clear that land derived from western rights can not be registered. If these lands can not be registered, this policy will disadvantage the landowners, because they will certainly lose their rights. It is therefore necessary to have a way for this soil to be registered, then the way it can be done is to convert the land from the western right. With the conversion of land from the western rights, it is hoped that the people will not lose their rights because after being converted the right will be able to be registered.[4]

Conversion of land rights is one of the instruments to fulfill the principle of legal unification through Act Number 5 of 1960. The Regulation of the Minister of Land and Agrarian Affairs Number 2 of 1962 regulates the provisions on affirmation of conversion and registration of ex-rights of Indonesia over land normatively. The conversion rule is the implementation of transitional provisions of Act Number 5 of 1960. The purpose of registration of land conversion to provide legal certainty, legal protection to holders of land rights or produce a Certificate of Right which is valid as a powerful evidentiary.[5]

2 Formulation of the Problem

This study examines the implementation of land conversion based on the perspective of Utiliarism to create a welfare state in Indonesia.

3 Research Method

According to Hutchinson, legal research is placed in the Applied Research category, and differentiates legal research into four types:[6]

a) Doctrinal research; research which provides a systematic esposition of the rules governing a particular legal category, analyses the relationship between rules, explain areas of dificulty and, perhaps, predits future development.
b) Reform-Oriented Research: Research which intensively evaluates the adequacy of exiting rules and which recommends changes to any rules found wanting
c) Theretical Research: research which fosters a more complete understanding of the conceptual bases of legal principles and of combined effects of a range of rules and procedures that touch on a particular area of activity.
d) Fundamental research: Research designed to secure a deeper understanding of law as a social phenemonen, including research on the historical, philosophical, linguistic, economic, social or political implication.

This research uses normative legal research with the regulation approach and field research. The analytical method used is prescriptive by placing secondary data sources as sources of legal material comprising primary legal materials, secondary legal materials and tertiary legal materials.

4 Literature Review

Nature has placed mankind under the governance of twosovereign masters, pain and pleasure. They alone point out what we ought to do and determine what we shall do; the standard of right and wrong, and the chain of causes andeffects, are both fastened to their throne. They govern us in all we do, all we say, all we think; every effort we can make to throw off our subjection ·to pain and pleasure· will only serve to demonstrate and confirm it. A man may claim to reject their rule but in reality he will remain subject to it. The principle of utility:[7]

Recognises this subjection, and makes it the basis of a system that aims to have the edifice of happiness built by the hands of reason and of law. Systems that try to question it deal in sounds instead of sense, in caprice instead of reason, in darkness instead of light. But enough of metaphor and declamation! It is not by such means that moral science is to be improved.

The principle of utility is the foundation of the present work, so I should start by giving an explicit and determinatethe principle that approves or disapproves of every action according to the tendency it appears to have to increase or lessen i.e. to promote or oppose the happiness of the person or group whose interest is in question. I say 'of every action', not only of private individuals but also of governments.

By 'utility' is meant the property of something whereby it tends to produce benefit, advantage, pleasure, good, or happiness (all equivalent in the present case) or (this being the same thing) to prevent the happening of mischief, pain, evil, or unhappiness to the party whose interest is considered. If that party is the community in general, then the happiness of the community; if it's a particular individual, then the happiness of that individual.

'The interest of the community' is one of the most general expressions in the terminology of morals; no wonder its meaning is often lost! When it has a meaning, it is this. The community is a fictitious body composed of the individuals who are thought of as being as it were its members. Then what is the interest of the community? It is the sum of the interests of the members who compose it.

It is pointless to talk of the interest of the community without understanding what the interest of the individual is. A thing is said to 'promote the interest' (or be 'for the interest') of an individual when it tends to increase the sum total of his pleasures or (the same thing) to lessen the sum total of his pains.

An action then may be said to conform to the principle of utility. . . .when its tendency to increase the happiness of the community is greater than any tendency it has to lessenit. And the same holds for measures of government, which are merely one kind of action performed by one or more particular persons.

When someone thinks that an action (especially a measure of government) conforms to the principle of utility, he may find it convenient for purposes of discourse to imagine a kind of law or dictate of utility and to speak of the action in question as conforming to such a law or dictate.

A man may be said to be a 'partisan' of the principle of utility when his approval or disapproval of any action (or governmental measure) is fixed by and proportional to the tendency he thinks it has to increase or to lessen the community's happiness.

Of an action that conforms to the principle of utility one may always say that it ought to be done, or at least that it is not something that ought not to be done.

One may say also that it is right that it should be done; it is a right action; or at least that it is not wrong that it should be done; it is not a wrong action. When thus interpreted, the words 'ought' and 'right' and 'wrong' and others of that sort have a meaning; otherwise they have none.

Has the rightness of this principle ever been formally contested?

The utility of human actions, Bentham said In the 1776 preface to *A Fragment on Government*, is their tendency to promote the common end of all such actions – happiness; and from utility in this sense, he went on, we can 'denominate a *principle* . . . a principle that is recognized by all men'. This, he wrote four years later in the opening chapter of *An Introduction to the Principles of Morals and Legislation*, is the principle that is to serve as 'the foundation of that system, the object of which is to rear the fabric of felicity by the hands of reason and of law'. And he then attempted to formulate the principle itself more precisely:[8]

> *By the principle of utility is meant that principle which approves or disapproves of every action whatsoever, according to the tendency which it appears to have to augment or diminish the happiness of the party whose interest is in question*

Happiness, moreover, is to be understood in a way already adumbrated in 1776 but now greatly elaborated in terms of the balance of pleasure over pain. However we are to interpret the celebrated opening words of the *Introduction to the Principles* and Bentham himself implicitly warned against attaching too much weight to 'rhetoric and declamation' we cannot ignore the fact that this is how he chose to open what he plainly envisaged as a major statement of his position

Nature has placed mankind under the governance of two sovereign masters *pain* and *pleasure*. It is for them alone to point out what we ought to do, as well as to determine what we shall do. On the one hand the standard of right and wrong, on the other the chain of causes and effects, are fastened to their throne. They govern us in all we do, in all we say, in all we think: every effort we can make to throw off our subjection, will serve but to demonstrate and confirm it . . . The *principle of utility* recognizes this subjection.

In his discussion of the principle of utility in An Introduction to the Principles of Morals and Legislation, Bentham did not explicitly refer to paraphrase is or his other methods of exposition, yet they form the foundation upon which the whole account is constructed. In the famous opening passage he announced that the perceptions of pain and pleasure constituted the 'real source' both of the principle of utility and of human motivation.[9]

Nature has placed mankind under the governance of two sovereign masters, pain and pleasure. It is for them alone to point out what we ought to do, as well as to determine what we shall do. On the one hand the standard of right and wrong, on the other the chain of causes and effects, are fastened to their throne. They govern us in all we do, in all we say, in all we think every effort we can make to throw off our subjection, will serve but to demonstrate and confirm it. In words a man may pretend to abjure their empire but in reality he will remain subject to it all the while. The principle of utility recognize this subjection, and assumes it for the foundation of that system, the object of which is to rear the fabric of felicity by the hands of reason and of law. Systems which attempt to question it, deal in sounds instead of sense, in caprice instead of reason, in darkness instead of light.

5 Result and Discussion

There are 531 large scale forest concessions provided on an area of 35.8 million hectares, while on the other hand there are 60 HKm, HD, and HTR licenses owned by 257 486 families (1,287,431 inhabitants) on an area of just 646,476 hectares. The forest management in the form partnership (one model of forest management by the holder concessions by way of partnering with local communities) only reaching 11,500 hectares, while *Pengelolaan Hutan Bersama Masyarakat* (PHBM) *Perhutani* which include 5.293 *Kelompok Tani Hutan* (KTH), 1,200 primary cooperatives, and 5,394 villages covering only 2.1 million hectares, but not yet can provide welfare to farm households.

According to BPN (2016) records, the total area of HGU plantations is about 15 million hectares and HGB area of 29,584 hectares. Meanwhile, the 2013 Agricultural Census by the Central Bureau of Statistics (BPS) shows that there are 2,452 large-scale agricultural enterprises in agriculture, 56% ladders only have less than 0.5 hectares of land. The number of farmers in 2013 is 31.7 million people (food crop farmers as much as 20.4 million people) who work on an area of 13.7 million hectares.

The Agrarian Reform or legally formally referred to as Agrarian Reform is the process of restructuring (ownership rearrangement) of ownership, control and use of agrarian resources (especially land). In Article 2 of the Decree of the People's Consultative Assembly of the Republic of Indonesia Number IX / MPR / 2001 it is explained that "The agrarian reform includes a continuous process with respect to the rearrangement of land ownership, use and utilization of agrarian resources, implemented in the framework of achieving legal certainty and protection and justice and prosperity for all people of Indonesia

Land Reform Plus Implementation Process, namely the arrangement of land assets for the community and structuring of public access to economic and political resources that enable the community to utilize the land properly. In the implementation of Land Reform Plus, there are two important things, namely Reform Asset and Access Reform. The Agrarian Reform aims to create an agrarian-based resources for the welfare of the people in order to organize the life of a more just society to improve the sustainability of the Indonesian national and statehood system and to promote community harmony.

The Agrarian Reform aims to create an agrarian-based resources for the welfare of the people in order to organize the life of a more just society to improve the sustainability of the Indonesian national and statehood system and to promote community harmony.

The objectives of the Agrarian Reform are to reduce poverty, create jobs, improve people's access to economic resources especially land, rearrange the inequality of ownership, use and utilization of land and agrarian resources, reduce disputes and conflicts over land and agriculture, improve and maintain quality of the environment, increasing food security and community energy.

The objective of land reform in Indonesia is to enhance the income and living standards of the tiller peasants, as a basis or prerequisite for organizing economic development towards a just and prosperous society based on Pancasila. To achieve the intended objectives is done by holding a fair share of the peasants' livelihoods in the form of land and fair share of results as well, implementing land tenure for peasants, ending the landlord system, and protecting the weak economy. Landreform programs include:[10]

a. Limitation of maximum landholding area;
b. The prohibition of land ownership in so-called "absentee" or "clasp";
c. Redistribution of the remaining lands from the maximum, the lands affected by the "absentee" ban, the former swapraja lands and state lands;
d. Regulation of the return and redemption of pawned agricultural lands;
e. The re-arrangement of land-sharing agreements, and;
f. Determination of minimum area of agricultural land ownership, accompanied by a prohibition to engage in acts that result in the split of the ownership of agricultural lands into too small parts

Current phenomena indicates the still occurrence of land cultivation by certain parties, whereas Article 7 of the Agrarian Act regulates the prohibition of controlling the land beyond certain limits, because it is detrimental to the public interest, due to the limited supply of agricultural land, especially in densely populated areas. The scarcity of land causes the land to have a very high economic value.

Inequality in land ownership and control is increasingly apparent as a result of the uneven distribution of land, this can be seen from the symptoms, the land accumulates in the hands of certain persons or bodies. As stated by Hery Haryadi, a lecturer at the Faculty of Law

Unika that there is a depreciation of the area of productive land. Including land tenure that only a handful of people. The biggest concern is the soaring price of the land. To minimize the aksees of the inequality, he expressed the need for awareness of the rules. Law enforcement concerning the land issue is strongly needed so that the land will not be controlled by certain people or entities. If it continues then there will be a lot of wasted land, because in housing environment not all the land is used. Whereas on the other hand a small group of people find it difficult to find land to grow crops. Presumably the ideal of Agrarian Act to realize the national land system which is devoted to the interests of the people is still far from reality.

In the seminar of Re thinking Land reform in Indonesia held by the National Land Agency with the Land Law Initiative and Rural Development Institute in Jakarta (May 2002), Deputy of Land Administration Agency of the National Land Agency, Heru Wijono stated in its development the implementation of land reform in Indonesia also stagnated, and unfinished, where the main obstacle to the implementation of land reform is the weak political will of the New Order government which is pursuing higher economic growth. This policy is not enough to give support to the weak economic community, including peasants who really need the land. Heru suggested the land policy in the framework of land reform needs to be reviewed.

In its development, the new order of agrarian politics and the reform order no longer use the concept of redistribution, but rather refers to the green revolution, transmigration and modernization. In addition, various laws of legal products of the new order which are of a legal nature no longer associate them with Agrarian Act or even contradictory to the Agrarian Act, where the pattern of control of natural resources is now shifted to state and private ownership (including foreign private).

In the future orientation of this reformation period, land reform is still needed in order to carry out development. The Government of Indonesia can follow the example of other countries that have succeeded in implementing land reform programs in their countries such as Japan, Taiwan and South Korea.

The consequences that we can still perceive today as a result of the non-implementation of land reform are:[11]

a. Land reform creates market or purchasing power. Without the equity of land, there is no power of purchasing power, meaning there is no market power, consequently production will not develop.
b. Peasants without land assets are synonymous with poor peasants who will not be able to create savings. Whereas agricultural savings are needed by every government, to fund agricultural development and other sectors development.
c. Without increasing the farmer's economy, the agricultural tax will remain minimal.
d. Without land reform, there would be no widespread differentiation of the division of labor in the countryside that grew out of the rural needs itself.
e. Without land reform, there will be no investment in agriculture by the peasants themselves. Instead disinvestment occurred because over time many poor peasants lost land and widespread poverty.
f. The land eventually became the object of speculation, because it could not be used productively by its peasants but was looted by the classes in the city for the benefit of speculation and non-productive investment.

As mentioned above by looking at the consequences of the non-implementation of the land reform program in the form of land conversion to economically disadvantaged communities, it is recognized that the land reform program is still needed in order to carry out development in particular to improve the livelihood of peasants. The main obstacle to the implementation of land conversion is the pattern of the current government's political policies that are pursuing economic growth and there is a view among bureaucrats that land conversion is a "taboo" program to be discussed, moreover to be implemented because it is considered a communist product, the concept of land for peasants is a communist concept.

Land reform through land conversion to the poor is a program of land politics policy that needs to be completed in its implementation, in order to anticipate the tendency of emergence of inequality in land ownership. What must be done is to make adjustments to the provisions of land reform with present situation. Because the number and needs of the population and the development of technology, transportation and social economy is different from the situation in the 1960s. The maximum extent of non-agricultural land tenure as intended by Article 17 and Law Number 56 of 1960 stipulated by the government shall be timed to receive clear attention and regulation. It is done is to implement the sovereignty of the people where people have sovereignty to get equal treatment and rights in the field of land for the welfare of his life.

6 Conclusions

The Agrarian Reform aims to create an agrarian-based resources for the welfare of the people in order to organize the life of a more just society to improve the sustainability of the Indonesian national and statehood system and to promote community harmony. The objectives of the Agrarian Reform are to reduce poverty, create jobs, improve people's access to economic resources especially land, rearrange the inequality of ownership, use and utilization of land and agrarian resources, reduce disputes and conflicts over land and agriculture, improve and maintain quality of the environment, increasing food security and community energy.The objective of land reform in Indonesia is to enhance the income and living standards of the tiller peasants, as a basis or prerequisite for organizing economic development towards a just and prosperous society based on Pancasila. To achieve the intended objectives is done by holding a fair share of the peasants' livelihoods in the form of land and fair share of results as well, implementing land tenure for peasants, ending the landlord system, and protecting the weak economy community.

The main obstacle to the implementation of land conversion is the pattern of the current government's political policies that are pursuing economic growth and there is a view among bureaucrats that land conversion is a "taboo" program to be discussed, moreover to be implemented because it is considered a communist product, the concept of land for peasants is a communist concept.

References

1. B. Harsono, *Hukum agraria indonesia sejarah pembentukan UUPA, isidan pelaksanaannya*, p. 47 (Penerbit Djambatan, Jakarta, 2007)
2. A.P. Parlindungan, *Komentar atas undang-undang pokok agraria*, p. 68(Alumni, Bandung, 1998)
3. A. Surono, *Fungsi sosial tanah*, p. 2(Universitas Al-Azhar Indonesia, Jakarta, 2013)
4. U. Hasanah, Status kepemilikan tanah hasil konversi hak barat berdasarkan UU no. 5 tahun 1960 tentang peraturan dasar pokok-pokok Agraria dihubungkan dengan PP No. 24 tahun 1997 tentang pendaftaran tanah, *Jurnal Ilmu Hukum*, **3 (1)**, (2013)
5. A. Raharjo, *Pendaftaran konversi tanah hak milik adat oleh ahli waris*, Tesis, p. 14 (Universitas Diponegoro, Semarang, 2010)

6. P. Mahmud Marzuki, *Penelitian hukum,* p. 52 (Kencana Prenada Media Group, Jakarta, 2009)
7. J. Bentham, *An introduction to the principles of morals and legislation,* p. 6-7 (Batoche Books, Kitchener, 2000)
8. H. Burns, Happiness and utility: Jeremy Bentham's equation, *Journal Utilitas*, **17 (1)**,p. 43(2005)
9. P. Schofield, *Utility and democracy: the Political thought of Jeremy Bentham*, p. 28-29 (Oxford University Press Inc, New York, 2006)
10. Mudjiono, *Hukum agraria*, p. 42-43 (Liberty, Yogyakarta, 1992)
11. Mudjiono, *Hukum agraria*, p. 50 (Liberty, Yogyakarta, 1992)

Model availability of landfill as a conflict mediation process in Central Java

Yulia Pasha[1], *Herdis* Herdiansyah[1,*], *Lugina* Setyawati[2], and *Elok Faiqotul* Mutia[1]

[1]School of Environmental Science, Universitas Indonesia, Salemba, Indonesia
[2]Sociology Department, Faculty of Social and Political Science, Universitas Indonesia, Depok, Indonesia

Abstract. Waste is still a problem in major cities in Indonesia. The high production of waste is not proportional to the availability of the landfill (TPA). The land use plans for waste landfill by local governments have conflict potential between government and community when such infrastructure facilities are not desired by the community. Environmental conflicts/disputes can be resolved through mediation mechanisms. This research uses linear regression modeling system to see the correlation between the amount of waste volume generated with the availability of landfill (TPA) in Central Java Province. The results showed the greater the number of residents the greater the volume of waste generated. In contrast the number of available landfill (TPA) is not affected by the volume of waste because the regression coefficient is only 0.0256.

1 Introduction

Indonesia has the fourth largest population in the world after China, India, and the United States [1]. The total population of Indonesia in 2025 is projected to rise to 284.83 million from 238.52 million in 2010. This number will continue to grow to 305.65 million by 2035. BPS is also projected by 2030 more than 60% of the population Indonesia lives in urban areas. Population growth is followed by increasing demand for land, so the correlation entails the larger the number of people, the less amount of land available.

Human activities and lifestyles always produce unused waste and pollute the environment [2]. Increasing the amount of household waste generated is positively correlated with the number of population, i.e. the larger the number of people in a region, the more the amount of household waste produced [1]. Provinces in Indonesia with the largest population in 2016 are sequentially occupied by West Java, East Java, and Central Java [1].

Garbage and its management are still a problem in major cities in Indonesia [1,3]. Data from the Ministry of Environment mentioned that in 1995 the average person in urban areas in Indonesia produced 0.8 kilograms of garbage per day and continued to increase up to 1 kilogram per person per day in 2000. Estimated waste pile in 2020 for each person per day is 2.1 kilograms, so with an ever-increasing number of residents it is estimated that the

* Corresponding author: herdis@ui.ac.id

waste produced per day is about 500 million kilograms by 2020 or about 190,000 tons per year [3].

Production of solid waste per day is quite high in Java, namely in Jakarta, Surabaya, and Semarang [1]. The high production of waste is not proportional to the availability of land and the number of landfill (TPA) available. Data from the Ministry of Public Works and Public Housing or Kementrian Pekerjaan Umum dan Perumahan Rakyat (PU-PR) in 2015, said that Central Java Province has only 3 (three) landfill (TPA) with a land area of 15.50 hectares and a capacity of 175 cubic meters. Limited land in big cities become one of the causes of waste accumulation increases. The same thing happened in Klaten District in Central Java Province.

Data from the Public Works and Spatial Planning Office of Klaten Regency in 2016 stated that Klaten Regency has 1 landfill (TPA) which is Candirejo TPA with a land area of 0.9 hectares, and 224 local transfer station (TPS) which are 207 permanent models and 17 containers. The collection and transportation of garbage is done once a day. The amount of waste production generated by the people of Klaten Regency reaches 255.4 cubic meters per day, resulting in waste generation of 896 cubic meters per day, with garbage transported from local transfer station (TPS) to landfill (TPA) of 162 cubic meters per day or 18.08%. The Government of Klaten Regency allocates a total of IDR 5.050.000.000,00 (over 5 billion rupiahs) of garbage into TPAS by 2016.

The need for a landfill (TPA) encourages the District Government of Klaten to build a landfill (TPA) in Troketan Village Pedan District with a planned land area of 9,464 hectares. The plan of the Troketon TPA will start operating in 2018, but this is facing challenges with the rejection of the local community in 3 (three) villages in Pedan sub-district ie Kaligawe Village, Troketon Village, and Kalangan Village. The rejection of the local community resulted in conflict between the Government of Klaten district and the community. Environmental dispute can be resolved through Alternative Dispute Resolution mechanism [4]. The most common form of Alternative Dispute Resolution is mediation [5].

2 Literature review

Waste is produced from materials that are not used in human activities and pollute the environment. Waste or solid waste, divided into two types: industrial solid waste generated by mines, farms, and industries that supply goods and services to people; and municipal solid waste is a mixture of solid waste produced by houses and offices other than factories, for example paper, cardboard, food scraps, cans, bottles, yard waste, plastics, metal, glass, and electronic waste [2,6].

Solid waste can be managed by 2 (two) approaches: waste management and waste reduction [2]. Waste management is trying to reduce the environmental impact of urban solid waste without reducing the amount of waste generated. The second way is to reduce waste, which produces less waste and less pollution, and the waste generated is considered a potential resource that can be reused, recycled or composted.

The need for effective solid waste management becomes an environmental management strategy in Indonesia [7] especially since the issuance of Law No. 18 of 2008 on Waste Management (hereinafter referred to as Law 18/2008) of Law 18/2008. Referring to this law, waste management is a systematic, thorough, and continuous activity that includes waste reduction and handling. It is estimated that only about 2% of the provinces in Indonesia are budgeted for waste management [7].

The definition of landfill according to the 18/2018 Act is a place to process and return the waste to environmental media safely for humans and the environment. The waste and the residue of its processing are returned to the environmentally safe medium preceded by processing to change the characteristics, composition and amount / volume of waste [8]. To date there are 4 (four) types) of landfill (TPA) operated in Indonesia, namely open dumping, controlled, sanitary, and modern sanitary [7]. The type of infrastructure development where the landfill (TPA) is best known is open dumping, which results in environmental pollution. This has led to a decline in the quality of urban environments including ground water [7].

The construction of public infrastructure facilities such as substations, landfills, incinerators, power plants, roads, and hospitals [9], often face challenges from local communities when specific locations are submitted [10]. Land use planning or local land use in the site selection process is undesirable by local communities and potentially conflict [11]. The understanding of society is based on the fact that certain facilities will create negative externalities or disturbances that will affect the surrounding environment in the presence of development, but have no significant impact on people and houses around the site [12].

Conflict is caused by differences of perspective or point of view due to differences in economic, social, cultural, background, and values [13,14]. These differences lead to disagreement so that conflict occurs [13]. This can be between two parties individually, communal, or even involving many parties and States [14]. Conflict over a policy is characterized by limited information that are obtained by the community and uncertainty about the process [15]. Mediation can be used in resolving environmental disputes since it is very effective that involve multiple parties or communities for an environmental destruction [16]. In some developed countries such as Japan, the United States, and Canada, environmental dispute resolution places greater emphasis on the use of mediation [17].

3 Material and method

Linear regression is used to analyze the relationship (correlation) between dependent and independent variables based on data provided from the statistical point of view that is random observation data with some size errors. On the other hand, the relationship between dependent and independent variables can be characterized by an unclear functional relationship for unclear data [18].

This research uses linear regression to see the validity level of the model in predicting the number of landfill needed to accommodate the volume of waste, to see the correlation between the number of residents with the volume of waste, and the correlation between the volume of waste with the amount of landfill availability. The linear regression model [19] uses calculations:

$$y = X\beta + \varepsilon \tag{1}$$

i.e. y shows the observation vector of the dependent variable, X denotes the independent variable and the design matrix of the regression model, β denotes the vector of unknown regression parameters, and ε is the interference vector.

4 Results and discussion

The linear regression model to see the correlation between the number of population and the volume of waste, and the correlation between the volume of waste with the number of available landfill (TPA) is done through the collection of population data, the volume of waste, and the number of available TPA. Population data in Central Java Province taken from 2010 to 2014 can be seen in Table 1.

Table 1. Population in Central Java Province 2010 to 2014 adapted from Purba et al (2017).

Year	Gender (people)		
	Male	**Female**	**Total**
2010	16.091.555	163.52.331	32.443.886
2011	16.231.847	16.493.531	32.725.378
2012	16.367.703	16.630.989	32.998.692
2013	16.499.377	16.764.962	33.264.339
2014	16.627.023	16.895.640	33.522.663

Data on the volume of waste per day and the number of TPA available in Central Java Province is shown in Table 2.

Table 2. Waste volume per day in Central Java Province 2010 to 2014 adapted from Department of Public Works of Regency/City throughout Central Java Province (2014).

Year	Waste Volume	Number of Landfill (TPA)
2010	737.168,68	65
2011	24.116,63	57
2012	29.466,47	123
2013	33.337,51	61
2014	34.278,41	59

The result of simulation of correlation between population and waste volume is shown in Figure 1. The simulation is done to predict the volume of waste generated by the population and see the correlation between the variable X (horizontal axis) that is the population with the variable Y (vertical axis) i.e. the volume of waste. Positive correlation (+) occurs if the larger the population, the greater the volume of waste. Negative correlation (-) if the smaller the population the smaller the volume of waste. The correlation between the variables in Figure 1 shows the model in the mathematical equation (formula) is $Y = -0.5254X + 2\varepsilon + 07$ with the determinant coefficient is $R^2 = 0.5022$ indicating the validation rate. The model prediction on 2010 can be calculated if the variable Y known as 737.168,68 waste volume then variable X which is the number of people can be calculated or vice versa. Validation results show there is correlation or influence between the population with the volume of garbage though not dominant. The bigger the population, the greater the volume of waste generated.

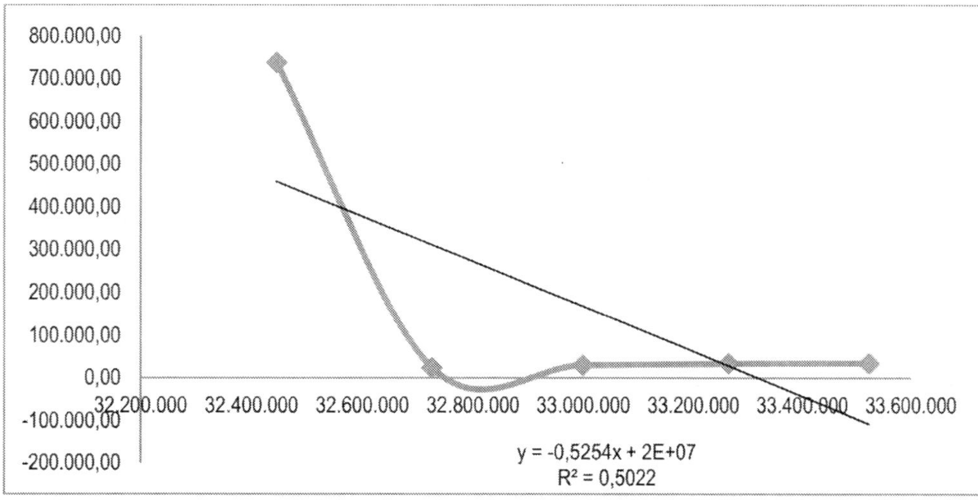

Fig. 1. Linear regression model of population and volume of garbage.

Result of the simulation of correlation between garbage volume with the number of TPA is shown in Figure 2. Simulation is done to predict the amount of TPA needed with the volume of waste generated by the population and see the correlation between variable X (horizontal axis) that is the volume of waste with variable y is the number of TPA. Figure 2 shows the model in the mathematical equation (formula) is y = -1ε-05X + 75,443 with the determinant coefficient is R^2 = 0.0256 indicating the model validation rate. Validation results show no correlation between the volume of waste with the number of landfill available. Greater volume of waste does not affect the number of available TPA.

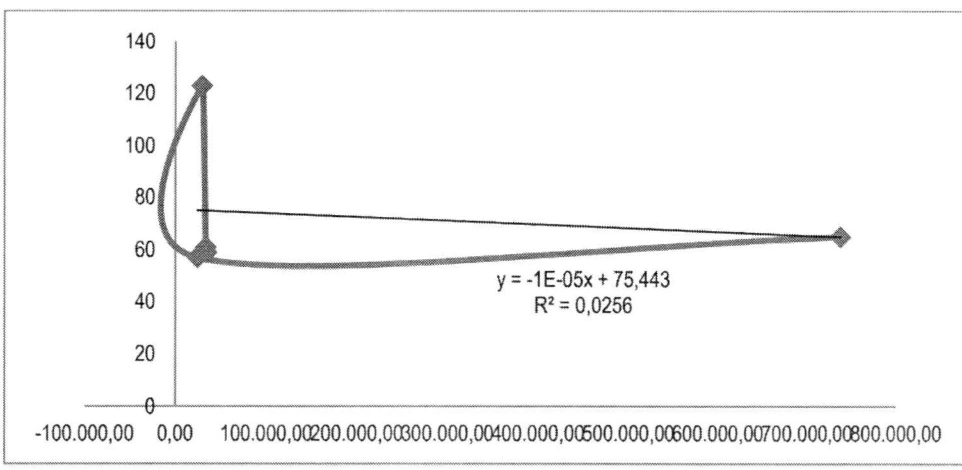

Fig. 2. Linear regression model of waste volume and number of landfill.

Land use planning for waste landfill affects the number of waste landfill availability in Klaten. The location or site of waste landfill has been regulated in local regulation Number 13 year 2012 on Waste Management. The availability policies are often in conflict with others and it can be analyzed using a qualitative approach [20]. Land use for waste landfill has potential conflict as being opposed by local community. The most common form of conflict resolution or environmental dispute is mediation. In some places, the potential of waste can be used as alternative energy [22], as well as urban waste management is now beginning to be done with community engagement approach, hope the empowerment approach will reduce the potential for future conflicts [23], including involving women in the movement of sorting and choosing waste from household [24] [25]

5 Conclusion

The results showed the coefficient is 0.502. It means that the ability of the independent variable explained by the variant of the dependent variable is 50,2% and the variant of the dependent variable explained by other factors is 49,8% (100% - 50,2%). The coefficient denotes that there was a correlation or influence between the population with the volume of garbage though it is not dominant. This shows the greater the population, the greater the volume of waste generated. In contrast the number of available waste landfill is not affected by the volume of waste because the regression coefficient is only 0.0256 or 25,6%. It means that the variant of the dependent variable explained by other factors was bigger than the ability of the independent variable explained by the variant of the dependent variable that is 74,4% (100% - 25,6%).

The land use plans for waste landfill by local governments have conflict potential between government and community when such infrastructure facilities are not desired by the community. The waste landfill conflict can be categorized as an environmental conflict/dispute which can be resolved through mediation mechanisms.

Acknowledgment

This research is funded by the Grant of Indexed International Publication for Final Project of Students/Publikasi Terindeks Internasional Untuk Tugas Akhir Mahasiswa (PITTA) Universitas Indonesia 2018 with contract number 2579/UN.R3.1/HKP.05.00/2018.

Reference

1. Purba, W. Sartika, Safitri, P. Ajeng, R. Andianti. *Indonesia environmental statistics 2017* (Indonesian Central Bureau of Statistics, Jakarta, 2017)
2. G.T. Miller, S.E. Spoolman. *Sustaining the earth* (10th ed) (Belmont, Brooks/Cole., 2012a)
3. M. Suparmoko. *Ekonomi sumber daya alam dan lingkungan: Suatu Pendekatan Teoritis* (Edisi 4 Revisi) (Fakultas Ekonomika dan Bisnis UGM, Yogyakarta, 2016)
4. H. Rhiti. *Panduan lengkap hukum penyelesaian sengketa lingkungan hidup: untuk mahasiswa, umum, dan penegak hukum* (Universitas Atma Jaya, Yogyakarta, 2013)
5. S. Matsumoto. A duration analysis of environmental alternative dispute resolution in Japan. *Journal of Ecological Economics* **70**, (2011)
6. G.T. Miller, S.E. Spoolman. Living in the environment (17th ed) (Belmont: Brooks/Cole, 2012b)
7. Ministry of Energy and Mineral Resources. Waste to energy guidebook (Directorate General of New, Renewable Energy and Energy Conservation Ministry of Energy and Mineral Resources, Jakarta, 2015)
8. Kementerian Pekerjaan Umum dan Perumahan Rakyat. Informasi Statistik Infrastruktur (Pusat Data dan Informasi Kementerian PUPR, Jakarta, 2015)
9. L. Sun, E.H.K. Yung, E.H.W. Chan, D. Zhu. Issues of NIMBY conflict management from the perspective of stakeholders: a case study in Shanghai. *Journal of Habitat International* **53**, (2015)
10. P.J. Davidson, M. Howe. Beyond NIMBYism: understanding community antipathy toward needle distribution services. *Journal of International Journal of Drug Policy*, (2013)

11. I.A. Kaya, N.K. Erol. Conflicts over Locally Unwanted Land Uses (LULUs): reasons and solutions for case studies in Izmir (Turkey). *Journal of Land Use Policy* **58**, (2016)
12. P. Hubbard. NIMBY. *Journal of Elsevier*, (2009)
13. S. Fisher, J. Ludin, S. Williams, D.I. Abdi, R. Smith, S. Williams. Mengelola konflik. translated from working with conflict: skills and strategies for action (Karikasari, S.N., Tapilatu, Meiske D., Maharani, Rita, dan Rini, Dwiati Novita, Penerjemah) (The British Council, Jakarta, Indonesia, 2001)
14. C. Irawan. Aspek hukum dan mekanisme penyelesaian sengketa di luar pengadilan (Alternative Dispute Resolution) di Indonesia. (Mandar Maju, Bandung, 2010)
15. A. Dagnino, N.M. Fraser, K.W. Hipel. Conflict analysis of an environmental dispute. *Journal of Negotiations and Conflict Resolutions*, (1985)
16. Sudiarto. Negosiasi, mediasi, dan arbitrase: penyelesaian sengketa alternatif di Indonesia. (Pustaka Reka Cipta, Bandung, 2015)
17. S. Sunarso. Hukum Pidana Lingkungan Hidup dan Strategi Penyelesaian Sengketa (Rineka Cipta, Jakarta, 2005)
18. P. Guo, H. Tanaka. Dual models for possibilistic regression analysis. *Journal of Computational Statistics & Data Analysis* **51**, (2006)
19. M. Cerny, J. Antoch, M. Hladík, Milan. On the possibilistic approach to linear regression modelsinvolving uncertain, indeterminate or interval data. *Journal of Information Sciences* **244**, (2013)
20. T. Tryfonas, D. Gritzalis, S. Kokolakis. A qualitative approach to information availability. *Journal of Information Security for Global Information Infrastructures*, (2000)
21. Central Java Province of Central Bureau of Statistics. Environmental Data and Information in Central Java 2014. (Central Java Province of Central Bureau of Statistics, Semarang, 2014)
22. Nuryadin, Tb R.R.Nitibaskara, H. Herdiansyah, R. Sari. The model of gas supply capacity simulation in regional energy security framework: policy studies PT. X Cirebon Area. *IOP Conference Series: Earth and Environmental Science* **88 012029**, (2017)
23. A. Brotosusilo, D. Utari, A.A. Satria. Sustainability of water resources in the upstream watershed-based community engagement and multistakeholder Cooperation. *IOP Conference Series: Earth and Environmental Science* **30** 012018, (2016)
24. D. Asteria, H. Herdiansyah, I.W.A. Apriana. Women's Environmental Literacy as Social Capital in Environmental Management for Environmental Security of Urban Area. *IOP Conference Series: Earth and Environmental Science* **30** 012014, (2016)
25. D Asteria, Budidarmono, H. Herdiansyah, N.L. Ni'mah. Gender sensitive education in watershed management to support environmental friendly city. *IOP Conference Series: Earth and Environmental Science* **126** 012146, (2018)

AUTHOR INDEX

AUTHOR INDEX